Physical Asset Management for a Sustainable World

This book describes techniques, tools, and models about engineering maintenance and life cycle models, aimed at the efficient managing of assets. It presents new asset management models, namely the Holistic Diagnostic Model (HDM) and Integrated Asset Management Support Model (IAMSM), including quantitative decision support tools for life cycle management of physical assets. It presents an overview of the life cycle models for replacement through a global model, with emphasis on availability and maintenance costs.

Features:

- Covers important aspects on Physical Asset Management (PAM) and offers a developed perspective on the economic aspects.
- Presents the relationships among some maintenance Key Performance Indicators (KPI), specifically MTTR, MTBF, Availability, and Reserve equipment.
- Analyzes models to calculate the influence of relevant variables on the withdrawal time and in the reserve size of the critical assets.
- Introduces new diagnostic models and Life Cycle decision support models.
- Includes new models for diagnosing organizations including pertinent case studies.

This book is aimed at researchers, professionals, and graduate students in PAM, maintenance management, and industrial engineering.

Physical Asset Management for a Sustainable World

José Manuel Torres Farinha

CRC Press
Taylor & Francis Group
Boca Raton London New York

CRC Press is an imprint of the
Taylor & Francis Group, an **informa** business

First edition published 2025
by CRC Press
2385 NW Executive Center Drive, Suite 320, Boca Raton FL 33431

and by CRC Press
4 Park Square, Milton Park, Abingdon, Oxon, OX14 4RN

CRC Press is an imprint of Taylor & Francis Group, LLC

ISBN: 9781032428352 (hbk)
ISBN: 9781032498379 (pbk)
ISBN: 9781003395690 (ebk)

DOI: 10.1201/9781003395690

Typeset in Times
by codeMantra

Access the Support Material www.routledge.com/9781032428352

Contents

About the Author ix

PART I Life Cycle Models

Chapter 1 Life Cycle Cost of Physical Assets 3

1.1 Introduction 3
1.2 Life Cycle Models 3
1.2.1 Linear Depreciation Method 4
1.2.2 Sum of the Digits Method 4
1.2.3 Exponential Method 5
1.3 Economic Life Cycle 6
1.3.1 UAI Method 6
1.3.2 Minimization of Total Average Cost Method 7
1.3.3 Minimization of Total Average Cost Method Reduced to Present Value 8
1.4 Performance Indicators 8
1.4.1 Availability 9
1.4.2 Mean Time Between Failures 9
1.4.3 Mean Time to Repair 9
1.4.4 Mean Waiting Time 10
1.5 Investment Analysis Methods 10
1.5.1 Net Updated Value 10
1.5.2 Internal Rate of Return 11
1.5.3 Equivalent Uniform Annual Cost 11
1.5.4 Payback Period or Recovery Period 12
1.5.5 Return on Investment 12
1.6 New Economic Replacement Models 13
1.6.1 New Economic Replacement Models – Financial Indicator 13
1.6.2 New Economic Replacement Models – Performance Indicators 14
1.7 Reserve Set 17
1.8 Life Cycle Considering Incomes 18
1.9 Simulation of Methods Application 18
1.9.1 UAI Method 19
1.9.2 Minimization of Total Average Cost Method 25
1.9.3 Minimization of Total Average Cost Method Reduced to Present Value 27
References 43

Chapter 2 Life Cycle Investment of Physical Assets 44

2.1 Introduction 44
2.2 The Model of a Holistic Approach of the Physical Assets Life Cycle 44
2.3 LCI Considering Maintenance and Non-Production 47
2.4 Case Study Simulation 50
2.4.1 Case Study Simulation Considering the Availability 50
2.4.2 Case Study Simulation Considering the Non-Production 53
2.5 Conclusions 58
References 58

Chapter 3 Life Cycle of Physical Assets with Recovery 59

3.1 Introduction 59
3.2 Methodology 59
3.3 Basic Concepts to Support Math Models 59
3.4 Behaviour of the Physical Asset Value along Time 61
3.5 Production Economic Movements 61
3.6 Functioning Economic Movements 63
3.7 Global Movements Analysis 63
3.8 Return on Investment 64
3.9 Models Validation and Discussion 64
3.10 Conclusions 71

PART II Maintenance Methodologies

Chapter 4 Maintenance Types and Strategies 75

4.1 Introduction 75
4.2 Predetermined Maintenance 75
4.2.1 Scheduled 75
4.2.2 Condition-Based Maintenance 76
4.3 Corrective Maintenance 87
4.4 Enterprise Asset Management Systems 87
4.5 Maintenance Performance 88
References 89

Chapter 5 Maintenance Algorithms 90

5.1 Introduction 90
5.2 Stochastic Algorithms 90
5.2.1 Moving Average 92
5.2.2 Exponential Smoothing 93

5.2.3 Autoregressive Moving Average ... 94
5.2.4 Autoregressive Integrated Moving Average ... 95
5.2.5 Seasonal Autoregressive Integrated Moving Average ... 96
5.3 Artificial Intelligence Algorithms ... 96
5.3.1 Some Predictive Algorithms ... 98
References ... 102

PART III Standards

Chapter 6 Maintenance Standards ... 107
6.1 Introduction ... 107
6.2 Characterization of Maintenance Standard NP 4492 ... 107
6.3 Diagnosis of the Maintenance State ... 108
6.3.1 Holistic Diagnosis Model ... 109
6.3.2 Questionnaires ... 111
6.3.3 The Explanation Sheets ... 112
6.3.4 Organization and Analysis of Information Collected ... 115
6.3.5 Elimination Criteria ... 115
6.3.6 Elimination Grid ... 116
6.3.7 Establishment of an Improvement Action Plan ... 116
6.4 Diagnosis to Implement the Standard NP 4492 ... 119
6.5 Economic Implications of Implementing Maintenance Standards ... 121
6.5.1 Life Cycle Cost of Physical Assets ... 121
6.5.2 Life Cycle Investment of Physical Assets ... 160
6.5.3 Life Cycle of Physical Assets with Recovery ... 172
Reference ... 183

Chapter 7 Norms of Physical Assets Management ... 184
7.1 Introduction ... 184
7.2 Main Requisites of Physical Asset Norms ... 184
7.3 Diagnosis to Implement Physical Asset Norms ... 185
7.4 Quantitative Evaluation to Implement Physical Asset Norms ... 185
7.4.1 Life Cycle Cost of Physical Assets ... 187
7.4.2 Life Cycle Investment of Physical Assets ... 224
7.4.3 Life Cycle of Physical Assets with Recovery ... 237

Chapter 8 Risk Norms ... 252
8.1 Introduction ... 252
8.2 Characterization of Risk Norms ... 252

8.3 Diagnosis to Implement Risk Norms ... 254
8.4 Quantitative Evaluation about the Implementation of Risk Norms ... 254
8.4.1 Life Cycle Cost of Physical Assets ... 254
8.4.2 Life Cycle Investment of Physical Assets ... 287
8.4.3 Life Cycle of Physical Assets with Recovery ... 307
Reference ... 321

Chapter 9 Energy Norms ... 322

9.1 Introduction ... 322
9.2 Characterization of Energy Norms ... 322
9.3 Diagnosis to Implement Energy Norms ... 323
9.4 Quantitative Evaluation about the Implementation of Energy Norms ... 323
9.4.1 Life Cycle Cost of Physical Assets ... 323
9.4.2 Life Cycle Investment of Physical Assets ... 355
9.4.3 Life Cycle of Physical Assets with Recovery ... 370

Chapter 10 Environmental Norms ... 386

10.1 Introduction ... 386
10.2 Characterization of Environmental Norms ... 386
10.3 Diagnosis to Implement Environmental Norms ... 387
10.4 Quantitative Evaluation about the Implementation of Environmental Norms ... 387
10.4.1 Life Cycle Cost of Physical Assets ... 387
10.4.2 Life Cycle Investment of Physical Assets ... 427
10.4.3 Life Cycle of Physical Assets with Recovery ... 435
Reference ... 450

Index ... 451

About the Author

José Manuel Torres Farinha has Habilitation in Electrical and Computer Engineering, a PhD in Mechanical Engineering, and a degree in Electrical Engineering, option Energy. He is a National Defence Auditor. He is Full Professor in Coimbra Polytechnic Institute (IPC) and its former President. He is the Coordinator of the Master's Degree in Engineering and Management of Physical Assets in IPC. He has completed several PhD supervisions and dozens of Master's degree orientations. He has published three books, two normal editions and one in an international edition. He has more than 250 publications – articles in indexed and non-indexed international journals, in national magazines and in proceedings of international and national congresses. He has several registered patents (national and international). He is a Specialist Member of Industrial Maintenance Specialization and a Counsellor Member of the Portuguese Association of Engineers.

PART I

Life Cycle Models

1 Life Cycle Cost of Physical Assets

1.1 INTRODUCTION

This chapter presents the traditional models to evaluate the Life Cycle Cost (LCC), to which new approaches, namely maintenance performance indicators and investment analysis methods, have been added. The situation of Physical Assets (PA) is also considered when the evaluation of the reserve of PAs is taken into account.

To make a balance between expenses and incomes and to provide more robust information to support the manager's decision, the profits are also added to the original model.

This chapter is structured in the following main sections:

- Three Life Cycle models;
- Maintenance performance indicators;
- Investment analysis methods;
- New economic replacement models;
- Simulation of scenarios.

1.2 LIFE CYCLE MODELS

The replacement of a PA is a broad concept that ranges from the selection of similar but new assets to replace existing ones, to the valuation of assets that act in completely different ways in the performance of the same function (Raposo *et al.*, 2017, 2018, 2019).

It is important to characterize suitable calculation methods for determining the appropriate time for the replacement of PAs. For this purpose, it is necessary to take into account several variables, such as:

- Initial investment (*II*)
- Assignment value (*AV*)
- Exploration values (*EV*)
  - Cost of maintenance (CM)
  - Cost of functioning (*CO*)
- Inflation rate (i_I)
- Capitalization rate (i_C)

The values of most of the preceding variables are taken from the history, with the exception of the assignment value. In this case, there is a need to be obtained from

DOI: 10.1201/9781003395690-2

the market value for each specific PA, which may prove to be difficult for many PAs. As an alternative, several types of devaluation can be simulated, such as the following:

- Linear depreciation method – The annual decay of the equipment value is constant over time;
- Sum of the digits method – The annual depreciation is not linear but less than that of the exponential method;
- Exponential method – The annual depreciation is exponential over the equipment's life.

The choosing of the appropriate depreciation method depends on the characteristics of the PA, the applicable tax legislation, and the economic policies adopted by the company.

1.2.1 Linear Depreciation Method

The linear depreciation method is one of the simplest and most common methods used to calculate the depreciation of PAs. This method estimates the annual depreciation of a PA over its useful life; it assumes that depreciation occurs at a constant rate over time.

In this model, the annual depreciation is calculated by subtracting the residual value from the original cost of the PA and dividing the result by the number of years of estimated life for the PA. In other words, the annual depreciation has an equal proportion of the PA's original cost that is accounted for as an accounting expense each year.

This method considers that the decay of the value of the PA is constant over the years, being calculated as follows (Equation 1.1):

$$d_l = \frac{II - VC_n}{N} \tag{1.1}$$

where

d_l Annual depreciation quota
II Initial investment
VC_n Residual value of the equipment at the end of N time periods
N Time of life corresponding to VC_n
l $l = 1, 2, 3 \ldots n$
V_n Equipment value in period $n = 1, 2, 3 \ldots n$

The value of the PA, V_n in a period n less than N, is given by Equation 1.2:

$$V_n = II - l * d_l \tag{1.2}$$

1.2.2 Sum of the Digits Method

The sum of the digits method is an accelerated depreciation method that takes into account the fact that most assets tend to depreciate faster at the beginning of their useful lives and more slowly as they age. This method is based on the sum of the years of the asset's useful life, attributing greater weight to the first years.

In this case, the annual devaluation is nonlinear, being calculated as follows (Equation 1.3):

$$d_l = 2 * \frac{N-(l-1)}{N+1} * \frac{II - VC_n}{N} \tag{1.3}$$

where

d_l Annual depreciation quota
II Initial investment
N Time of life corresponding to VC_n
VC_n Residual value of the equipment at the end of N time periods
l $l=1, 2, 3 \ldots n$;
V_n Equipment value in period $n=1, 2, 3 \ldots n$

The value of the PA, V_n in a period n less than N, is given by Equation 1.4:

$$V_n = II - d_l \tag{1.4}$$

1.2.3 Exponential Method

The exponential method incurs an annual declining depreciation charge over the life of the equipment.

The exponential method of depreciation assumes that depreciation occurs more rapidly at the beginning of the asset's useful life and decreases progressively over time. This reflects the fact that many PAs tend to lose value more quickly in the early stages of use.

The exponential method of depreciation of PAs is an approach used to calculate the depreciation of a PA over time, being based on a constant rate applied to the residual value of the PA.

The calculation formula is given by Equation 1.5:

$$d_l = VC_{l-1} * \left(1 - \sqrt[N]{\frac{VC_n}{II}}\right) \tag{1.5}$$

where

d_l Annual depreciation quota
II Initial investment
N Time of life corresponding to VC_n
VC_n Residual value of the equipment at the end of N time periods
l $l=1, 2, 3 \ldots n$;
V_n Equipment value in period $n=1, 2, 3 \ldots n$

The value of the PA, V_n in a period n less than N, is given by Equation 1.6:

$$V_n = II - d_l \tag{1.6}$$

1.3 ECONOMIC LIFE CYCLE

The equipment can be replaced according to various criteria. On the financial side, a common criterion is the "Economic Cycle", which allows determining the optimal period, which minimizes the average total costs of operation, maintenance, and capital immobilization (Farinha, 2018; Raposo *et al.*, 2017).

Another method, usually used, is the "Useful Life", which defines that it ends when its maintenance costs exceed the maintenance costs plus the capital amortization of an equivalent new equipment.

However, even though it is possible, based on the market depreciation values, it is possible to proceed to the analysis of the replacement of equipment, taking into account two other types of variables, mentioned above, which are:

- Inflation rate, designated by i_I;
- Capitalization rate, designated by i_C.

These rates are related according to Fisher's equation (Equation 1.7):

$$i_A = i_I + i_C + i_I \times i_C \tag{1.7}$$

where

i_A - Apparent rate

i_I - Inflation rate

i_C - Capitalization rate

According to the IEC 60300-3-3 (Farinha, 2018) standard, there are several methods for determining the economic cycle of equipment replacement; the most common are as follows:

- Uniform Annual Income Method (UAI);
- Minimization of Total Average Cost Method (MMTAC);
- MMTAC Reduced to Present Value (MMTAC-RPV).

1.3.1 UAI Method

The UAI Method is used to estimate the annual income needed to recover a given investment. It takes various factors into account, such as expenses, inflation rate, and other taxes, to determine the amount of annual income required.

The UAI Method makes use of the following data:

- Equipment acquisition cost;
- Assignment values (calculated in accordance with the methods set out above);
- Maintenance and exploration costs over the years;
- Apparent rate.

The present net value in year *n* (PNV_n) is given by Equation 1.8:

$$PNV_n = II + \sum_{j=0}^{n} \frac{CM_j + CO_j}{(1+i_A)^j} - \frac{V_n}{(1+i_A)^n} \tag{1.8}$$

where

II Initial investment

CM_j Cost of maintenance in year $j = 1, 2, 3 \ldots. n$

CO_j Cost of operation in year $j = 1, 2, 3 \ldots n$

i_A Apparent rate

V_n Value of the equipment over a period $n = 1, 2, 3\ldots N$

The Uniform Annual Income for year n (UAIn) is calculated by Equation 1.9:

$$UAI_n = \frac{i_A(1+i_A)^n}{(1+i_A)^n - 1} * \left(II + \sum_{j=0}^{n} \frac{CM_j + CO_j}{(1+i_A)^j} - \frac{V_n}{(1+i_A)^n} \right) \tag{1.9}$$

Combining Equations 1.8 and 1.9, it results in Equation 1.10:

$$UAI_n = \frac{i_A(1+i_A)^n}{(1+i_A)^n - 1} * PNV_n \tag{1.10}$$

The minor calculated UAI value indicates the respective period (multiples of the year) in which the PA ought to be replaced. This amount is equivalent to a minimum income that the PA would cost annually.

1.3.2 Minimization of Total Average Cost Method

The MMTAC makes it possible to determine the lowest average cost of ownership of the PA and the respective year in which it occurs, which corresponds to the optimal replacement time. Capital costs and inflation rates are not considered. The calculation procedure is given by Equations 1.11–1.13:

$$C'_n = \frac{1}{n} \sum_{j=1}^{n} (CM_j + CO_j) \tag{1.11}$$

$$C''_n = \frac{1}{n}(II - V_n) \tag{1.12}$$

$$C_{n(MMTAC)} = C'_n + C''_n = \min_{n \in \{1, 2, \ldots, N\}} \frac{1}{n} \left(II - V_n + \sum_{j=1}^{n} (CM_j + CO_j) \right) \tag{1.13}$$

where

II Initial investment

CM_j Cost of maintenance in year $j = 1, 2, 3\ldots n$

CO_j	Cost of operation in year $j = 1, 2, 3\ldots n$
V_n	Value of the equipment over a period $n = 1, 2, 3\ldots N$
n	Number of years $n = 1, 2, 3 \ldots n$
$C_{n(MMATC)}$	Average total cost

1.3.3 Minimization of Total Average Cost Method Reduced to Present Value

In the MMTAC Reduced to Present Value method (MMTAC-RPV), the calculation procedure is identical to the previous one, with the exception that capital costs and the inflation rate are considered here. The various maintenance and assignment amounts, over the years, are reduced to their present value, in accordance with the following procedure (Equations 1.14–1.16):

$$C_n^{'} = \frac{1}{n}\sum_{j=1}^{n}\left(\frac{CM_j + CO_j}{(1+i_A)^j}\right) \tag{1.14}$$

$$C_n^{''} = \frac{1}{n}\left(II - \frac{V_n}{(1+i_A)^n}\right) \tag{1.15}$$

$$C_{n(MMTAC-RPV)} = C_n' + C_n'' = \min_{n\,\in\{1,\,2,\,\ldots,\,N\}} \frac{1}{n}\left(II - \frac{V_n}{(1+i_A)^n} + \sum_{j=1}^{n}\left(\frac{CM_j + CO_j}{(1+i_A)^j}\right)\right) \tag{1.16}$$

where

II	Initial investment
CM_j	Cost of maintenance in year $j = 1, 2, 3\ldots n$
CO_j	Cost of operation in year $j = 1, 2, 3 \ldots n$
V_n	Value of the equipment over a period $n = 1, 2, 3 \ldots N$
i_A	Apparent rate
n	Number of years $n = 1, 2, 3\ldots n$
$C_{n(MMTAC\text{-}RPV)}$	Average total cost

1.4 PERFORMANCE INDICATORS

The emphasis of the PA's Life Cycle is placed on its maintenance because it is one of the most important aspects that can be managed aiming to optimize their use. The way to do this is by choosing the most adequate maintenance policy (e.g., condition monitoring maintenance, predictive maintenance, and other).

One pragmatic approach to measure the results of the maintenance policy is by using key performance indicators (KPI), being the most adequate the following:

- Mean Time Between Failures (MTBF);
- Mean Time To Repair (MTTR);
- Availability (A).

1.4.1 Availability

Availability (A) means the probability that a device is operational at time t, knowing that at time $t = 0$ it is operational. For the present case, the Operational Availability given by Equation 1.17:

$$A = \frac{MTBF}{MTBF + MWT + MTTR} \tag{1.17}$$

1.4.2 Mean Time Between Failures

The MTBF is the average time measured between each failure presented by a certain PA (unit or set), within a certain period of time. It represents, in average terms, the expected runtime before the next failure (Equations 1.18 and 1.19).

$$MTBF = \frac{\sum_{i=1}^{n} TBF_i}{n} \tag{1.18}$$

$$MTBF = \frac{MTTR}{\frac{(1-A)}{A}} \tag{1.19}$$

where

TBF Time Between Failures

1.4.3 Mean Time to Repair

The MTTR is the average time to repair a PA (Equations 1.20–1.21):

$$MTTR = \frac{\sum_{i=1}^{n} TTR_i}{n} \tag{1.20}$$

$$MTTR = MTBF * \frac{(1-A)}{A} \tag{1.21}$$

where

TTR Time To Repair

1.4.4 Mean Waiting Time

Sometimes it is necessary to take into account another parameter, that is the Mean Waiting Time (*MWT*), which corresponds to the average waiting time between the identification of a fault and the start of the corrective maintenance intervention (1.22).

$$MWT = \sum_{i=1}^{n} WT \tag{1.22}$$

where

WT Waiting Time

1.5 INVESTMENT ANALYSIS METHODS

In addition to the theoretical replacement models for PAs, presented in the previous section, there are investment analysis methods that also make it possible to determine, depending on the PA's history and its economic life cycle, the most rational time for its replacement.

1.5.1 Net Updated Value

The net updated value (NUV) or present net value (PNV) is a financial technique that compares the inflows and outflows of capital on the initial investment date. Mathematically, the NUV is given by Equation 1.23:

$$PNV_n = II + \sum_{j=0}^{n} \frac{CM_j + CO_j}{(1+i_A)^j} - \frac{V_n}{(1+i_A)^n} \tag{1.23}$$

Or

$$NUV_n = \sum_{j=0}^{n} \frac{CF_j}{(1+i_A)^j} \tag{1.24}$$

where

CF_j Cash flow in year $j = 1, 2, 3\ldots n$

i_A Apparent rate

n Number of years $n = 1, 2, 3\ldots n$

When applied from the perspective of investment analysis, the NUV criterion allows to verify the economic viability of the investment and support decision-making, and the following rejection criteria should be used for this purpose:

- A project with $NUV > 0$ is accepted;
- A project with $NUV < 0$ is rejected;
- Between the two projects, it is chosen the one with the highest NUV.

When applied from the perspective of Life Cycle analysis, the NUV, or PNV, allows the identification of the optimal moment for replacing the PA.

1.5.2 Internal Rate of Return

The Internal Rate of Return (IRR) measures the periodic rate of return on invested capital and corresponds to the moment when the current value of the benefits of an Investment equals the current value of the respective initial investment, that is, the highest interest rate from which the investment is no longer viable. The IRR determines the investment project's internal interest rate, obtaining the PNV expression equal to zero.

Equation 1.25 corresponds to the formula to calculate the IRR:

$$IRR = t_1 + \left[\left(t_2 - t_1\right)\right] * \frac{NUV_1}{NUV_1 - NUV_2} \tag{1.25}$$

where

t_1 is the rate for $NUV_1 > 0$

t_2 is the rate for $NUV_2 < 0$

NUV_1 is the positive net present value

NUV_2 is the negative net present value

The investment is attractive if the IRR is higher than the Apparent Rate.

1.5.3 Equivalent Uniform Annual Cost

For the particular case in which it is necessary to compare investments with different life cycles, the equivalent uniform annual cost (EUAC) can be determined, which is given by Equation 1.26:

$$EUAC = NUV * \frac{i_A\left(1+i_A\right)^n}{\left(1+i_A\right)^n - 1} \tag{1.26}$$

where

NUV Net updated value

i_A Apparent rate

n Number of years $n = 1, 2, 3 \ldots n$

The above equation allows to transform the NUV into an annuity; thus, when determining the EUAC of two different investments, the investment with the lowest EUAC must be chosen. If the EUAC is being determined to decide one possible replacement of a PA, the year in which the EUAC is minimum corresponds to the end of the Economic Life Cycle of the PA.

From the perspective of investment analysis and selection of different investments, the EUAC method allows economically verify which investment is the most attractive and thus which one supports the decision-making. For example, in the context of a PA replacement, this method can help to decide which is the best purchase option between two different PAs. From the perspective of life cycle analysis, the EUAC allows to determine the optimal moment for replacing the PA.

1.5.4 Payback Period or Recovery Period

The payback period or recovery period corresponds to the period of time necessary for the value of the financial benefits generated by the Investment to equal the initial value of the Investment. It does not take into account the time value of money. It is usual to stipulate a maximum payback period and reject investments that exceed this previously established period. This criterion may be useful in prioritizing investments, with priority being given to investments with the shortest payback period. Payback is obtained by dividing the value of the initial investment by the gain in a given period, thus obtaining the estimated time value for the recovery of the investment (Equation 1.27).

$$Payback = \frac{\left|CF_{acum(-)}\right|}{\left|CF_{acum(-)}\right| + \left|CF_{acum(+)}\right|} * (Year_{(+)} Year_{(-)}) + Year_{(-)} \qquad (1.27)$$

where

$CF_{acum.(-)}$ is the last negative value of the accumulated cash flow
$CF_{acum.(+)}$ is the first positive value of the accumulated cash flow
$Year_{(-)}$ is the year corresponding to the last negative value of the accumulated cash flow
$Year_{(+)}$ is the year corresponding to the first positive value of the accumulated cash flow

The investment is accepted when the payback period of the invested capital is less than the number of years of expected economic life.

1.5.5 Return on Investment

To link the maintenance activity to financial management, in this book, the return on investment (*ROI*) is used. This is an indicator to assess the financial performance of the PA. It is used as a "bridge" between maintenance and economics, showing when the PA starts to generate profits or losses for the company. ROI is an indicator that helps to find the present value of the capital investment and the cash flow (*CF*) corrected through the Apparent Rate.

Traditionally, ROI is an index of profitability, effective (real) per unit of invested capital, given by Equation 1.28:

$$ROI = \frac{\sum_{j=1}^{n} \frac{CF_j}{(1+i_A)^j}}{II} \qquad (1.28)$$

where

II Initial investment
CF_j Cash flow

i_A Apparent rate
j $j = 1, 2, 3\ldots N$

Therefore, for each investment unit, the project generates a certain return which, for it to be profitable, will have to be higher than the unit.

ROI measures the amount of return on an investment as a percentage of the value of the Investment. In the present approach, the ROI is used in absolute value, given by Equation 1.29:

$$ROI = \sum_{j=1}^{n} \frac{CF_j}{(1+i_A)^j} - II \tag{1.29}$$

1.6 NEW ECONOMIC REPLACEMENT MODELS

The new economic replacement models present in this section consider and integrate the performance indicators described above and the respective traditional economic replacement models, forming a set of equation systems, which are discussed in the following subsections (Raposo, 2018).

1.6.1 New Economic Replacement Models – Financial Indicator

To this approach, it is necessary to remember the Uniform Annual Income (UAI_n) and the ROI (Equations 1.9 and 1.29):

$$UAI_n = \frac{i_A(1+i_A)^n}{(1+i_A)^n - 1} * \left(II + \sum_{j=0}^{n} \frac{CM_j + CO_j}{(1+i_A)^j} - \frac{V_n}{(1+i_A)^n} \right) \tag{1.9}$$

With ROI

$$ROI = \sum_{j=1}^{n} \frac{CF_j}{(1+i_A)^j} - II \tag{1.29}$$

Equations 1.9 and 1.29 can be placed in the form of a system of equations (Equation 1.30):

$$\begin{cases} UAI_n = \dfrac{i_A(1+i_A)^n}{(1+i_A)^n - 1} * \left(II + \displaystyle\sum_{j=0}^{n} \frac{CM_j + CO_j}{(1+i_A)^j} - \frac{V_n}{(1+i_A)^n} \right) \\ ROI = \displaystyle\sum_{j=1}^{n} \frac{CF_j}{(1+i_A)^j} - II \end{cases} \tag{1.30}$$

Remembering the MMTAC and the ROI (Equations 1.13 and 1.29):

$$C_{n(MMATC)} = C'_n + C''_n = \min_{n \in \{1,2,\ldots,N\}} \frac{1}{n} \left(II - V_n + \sum_{j=1}^{n} (CM_j + CO_j) \right) \tag{1.13}$$

$$ROI = \sum_{j=1}^{n} \frac{CF_j}{(1+i_A)^j} - II \tag{1.29}$$

Equations 1.13 and 1.29 can be placed in the form of a system of equations (Equation 1.31):

$$\begin{cases} C_{n(MMTAC)} = \min_{n \in \{1,\, 2,\, \ldots,\, N\}} \dfrac{1}{n}\left(II - V_n + \sum_{j=1}^{n} (CM_j + CO_j) \right) \\ ROI = \sum_{j=1}^{n} \dfrac{CF_j}{(1+i_A)^j} - II \end{cases} \tag{1.31}$$

Remembering the formulas of MMTAC Reduced to Present Value (MMTAC-RPV) and the ROI (Equations 1.16 and 1.29):

$$C_{n(MMTAC-RPV)} = C_n^{'} + C_n^{''} = \min_{n \in \{1,\, 2,\, \ldots,\, N\}} \frac{1}{n}\left(II - \frac{V_n}{(1+i_A)^n} + \sum_{j=1}^{n} \left(\frac{CM_j + CO_j}{(1+i_A)^j}\right) \right) \tag{1.16}$$

With ROI,

$$ROI = \sum_{j=1}^{n} \frac{CF_j}{(1+i_A)^j} - II \tag{1.29}$$

Equations 1.16 and 1.29 can be placed in the form of a system of equations (Equation 1.32):

$$\begin{cases} C_{n(MMTAC-RPV)} = \min_{n \in \{1,\, 2,\, \ldots,\, N\}} \dfrac{1}{n}\left(II - \dfrac{V_n}{(1+i_A)^n} + \sum_{j=1}^{n} \left(\dfrac{CM_j + CO_j}{(1+i_A)^j} \right) \right) \\ ROI = \sum_{j=1}^{n} \dfrac{CF_j}{(1+i_A)^j} - II \end{cases} \tag{1.32}$$

1.6.2 New Economic Replacement Models – Performance Indicators

It is important to link the replacement models with the maintenance performance indicators, including the MTTR, emphasizing its influence on the economic Life Cycle of PA. This section also includes the integration of ROI with the equipment replacement economic cycle. For this, several systems of equations are presented, starting with the Uniform Annual Income Method (UAI_n) and taking into account the ROI as a function of the MTTR. These systems of equations include several equations discussed in the previous sections (Equations 1.9, 1.20, and 1.29):

$$UAI_n = \frac{i_A(1+i_A)^n}{(1+i_A)^n - 1} * \left(II + \sum_{j=0}^{n} \frac{CM_j + CO_j}{(1+i_A)^j} - \frac{V_n}{(1+i_A)^n} \right) \tag{1.9}$$

$$MTTR = \frac{\sum_{i=1}^{n} TTR_i}{n} \tag{1.20}$$

$$ROI = \sum_{j=1}^{n} \frac{CF_j}{(1+i_A)^j} - II \tag{1.29}$$

Equations 1.9, 1.20, and 1.29 can be placed in the form of a system of equations (Equation 1.33):

$$\begin{cases} UAI_n = \dfrac{i_A(1+i_A)^n}{(1+i_A)^n - 1} * \left(II + \displaystyle\sum_{j=0}^{n} \dfrac{\left(t * MTTR * \dfrac{CM_j}{d} \right) + CO_j}{(1+i_A)^j} - \dfrac{V_n}{(1+i_A)^n} \right) \\ ROI = \displaystyle\sum_{j=1}^{n} \dfrac{CF_j}{(1+i_A)^j} - II \end{cases} \tag{1.33}$$

where

t Number of intervals considered for *MTTR*

d Number of days per year

The MMTAC as a function of the *MTTR* and the *ROI* (Equations 1.13, 1.20, and 1.29):

$$C_{n(MMATC)} = C'_n + C''_n = \min_{n \in \{1,\, 2,\, \ldots,\, N\}} \frac{1}{n} \left(II - V_n + \sum_{j=1}^{n} (CM_j + CO_j) \right) \tag{1.13}$$

$$MTTR = \frac{\sum_{i=1}^{n} TTR_i}{n} \tag{1.20}$$

$$ROI = \sum_{j=1}^{n} \frac{CF_j}{(1+i_A)^j} - II \tag{1.29}$$

Equations 1.13, 1.20, and 1.29 can be placed in the form of a system of equations (Equation 1.34):

$$\begin{cases} C_{n(MMTAC)} = \min_{n \in \{1, 2, \ldots, N\}} \dfrac{1}{n}\left(II - V_n + \displaystyle\sum_{j=1}^{n}\left(\left(t * MTTR * \dfrac{CM_j}{d}\right) + CO_j\right)\right) \\ ROI = \displaystyle\sum_{j=1}^{n} \dfrac{CF_j}{(1+i_A)^j} - II \end{cases} \tag{1.34}$$

Considering the Equations 1.16, 1.20, and 1.29, corresponding to the MATC method with Reduction to Present Value (MMTAC-RVP), MTTR, and ROI:

$$C_{n(MMTAC-RPV)} = C_n' + C_n'' = \min_{n \in \{1, 2, \ldots, N\}} \frac{1}{n}\left(II - \frac{V_n}{(1+i_A)^n} + \sum_{j=1}^{n}\left(\frac{CM_j + CO_j}{(1+i_A)^j}\right)\right) \tag{1.16}$$

$$MTTR = \frac{\sum_{i=1}^{n} TTR_i}{n} \tag{1.20}$$

$$ROI = \sum_{j=1}^{n} \frac{CF_j}{(1+i_A)^j} - II \tag{1.29}$$

Equations 1.16, 1.20, and 1.29 can be placed in the form of a system of equations (Equation 1.35):

$$\begin{cases} C_{n(MMTAC-RPV)} = \min_{n \in \{1, 2, \ldots, N\}} \dfrac{1}{n}\left(II - \dfrac{V_n}{(1+i_A)^n} + \displaystyle\sum_{j=1}^{n}\left(\dfrac{\left(t * MTTR * \dfrac{CM_j}{d}\right) + CO_j}{(1+i_A)^j}\right)\right) \\ ROI = \displaystyle\sum_{j=1}^{n} \dfrac{CF_j}{(1+i_A)^j} - II \end{cases} \tag{1.35}$$

1.7 RESERVE SET

Many organizations have several similar PAs, being to work simultaneously or as redundancy PAs. This redundancy may assume several approaches, being one of the most paradigmatic the reserve fleet in the passenger transport sector or in health area where it is always necessary to have some additional equipment available because not to put the lives of patients at risk.

To help evaluate the Reserve Set of some type of PA, it used the methodology presented by Raposo *et al.* (2018) and Raposo *et al.* (2019) that uses, mainly, the maintenance indicator MTTR to support its dimensioning. However, the Equations referred in the preceding papers were modified to consider the number of functioning days of the PAs along the year.

The Availability depends on the MTBF and MTTR, and these are, namely the last, clearly dependent on the maintenance policies used in the companies and, consequently, influence the dimensioning of the Reserve Set; Equations 1.36–1.38 represent this relationship:

$$RS = \frac{m * MTTR}{365} \times \frac{k}{365} \tag{1.36}$$

where

RS	Reserve Set
m	Number of active Equipment
$MTTR$	Mean Time To Repair
k	Number of working days/year

Using Equation 1.21:

$$MTTR = MTBF * \frac{(1-A)}{A} \tag{1.21}$$

And substituting Equation 1.21 in Equation 1.36, it results in Equation 1.37:

$$RS = \frac{m * MTBF * \frac{(1-A)}{A}}{k} \times \frac{k}{365} \tag{1.37}$$

where

RS	Reserve Set
m	Number of active Equipment
$MTBF$	Mean Time Between Failures
A	Availability
k	Number of days/year

If it is used Equation 1.20:

$$MTTR = \frac{\sum_{i=1}^{n} TTR_i}{n} \tag{1.20}$$

And substituting Equation 1.20 in Equation 1.36, it results in Equation 1.38:

$$RS = \frac{m * \frac{\sum_{i=1}^{n} TTR_i}{n}}{k} \times \frac{k}{365} \tag{1.38}$$

where

RS	Reserve Set
m	Number of active Equipment
TTR	Time To Repair
i	$i=0, 1, 2, 3 \ldots n$
k	Number of days/year

1.8 LIFE CYCLE CONSIDERING INCOMES

The traditional approach of the LCC does not consider the incomes or profits. However, it is important to balance the negative and the positive financial movements aiming to understand the final result that comes out from the PA performance.

With this objective, this section considers the profits or benefits with the reduction to PNV. There are additional variables that may be considered, which will be referred and included in the models later. The main approach about incomes is given by Equation 1.39.

$$GR_n = \sum_{t=0}^{n} \frac{B_t}{(1+i_A)^t} \tag{1.39}$$

where

B_t	Benefit in year i
i_A	Apparent rate
t	$t=1, 2, 3 \ldots N$

To evaluate the final result in the simplest way, considering Benefits and Expenses, it can use Equation 1.40.

$$GR_n = \sum_{t=0}^{n} \frac{B_t}{(1+i_A)^t} - \sum_{t=0}^{n} \frac{E_t}{(1+i_A)^t} \tag{1.40}$$

where

B_t	Benefit in year i
E_t	Expenses (Functioning and Maintenance) in year i
i_A	Apparent rate
t	$t=1, 2, 3 \ldots N$

1.9 SIMULATION OF METHODS APPLICATION

In this section, three simulations of methods application are presented, namely the following:

- UAI Method
- Minimization of Total Average Cost Method
- Minimization of Total Average Cost Method Reduced to Present Value

As will be seen, the time when the Life Cycle of PA reaches the theoretical right time to be withdrawal is different according to the method used. However, an interval of decision of some years may be understood as positive because it permits the manager will have enough time to take the best decision.

1.9.1 UAI Method

To simulate all the methods included in this book, it uses the same type of table adapted to each one. The objective is to simplify the methods' understanding and make the comparison among them.

The first method simulated is the UAI Method, using the data included in Table 1.1. From Table 1.1 some values stand out, like the following ones:

- Initial Investment (*II*) –12000
- Life Time simulated 25
- Residual value of equipment (*RV*) 10
- Functioning (*F*) –4500
- Energy –200
- Maintenance (*M*) –300
- Benefit (*B*) 50000
- Internal Rate of Return (*r*) 0.12
- Risk (*R*) 0.11
- Apparent Rate 0.2432
- Maintenance Increase Rate 0.01
- Availability 0.8929
- *MTBF* 500
- *MTTR* 50
- *MWT* 10

From the simulation presented in Table 1.1, they emphasized some results:

- Figure 1.1 presents the Devaluation curves resulting from the Linear, Sum of Digits, and Exponential Methods.
- Figure 1.2 shows the results from the UAI Method according to the three preceding Devaluation methods; as can be seen, the results converge near year 13.
- Figure 1.3 shows the point of Useful Life; as can be seen, the Useful Life is lower than the UAI.
- Figure 1.4 shows the final balance between expenses and profits.
- Figure 1.5 shows the global result of the PA life cycle.

TABLE 1.1
Data to Simulate the Uniform Annual Income Method

Time	0	1	2	3	4	5
Initial Investment (*II*)	−12000					
Initial Investment (*II*)	−12000					
II Present Value (IIPV) – Return	−12000	−9652.51	−7764.25	−6245.37	−5023.63	−4040.88
Life Time simulated	25					
Residual value of equipment (RV)	10					
Linear Depreciation Method (Vn)	−12000	−11519.60	−11039.20	−10558.80	−10078.40	−9598.00
LDM Present Value (Vn)		−9266.09	−7142.59	−5495.30	−4219.18	−3232.03
Sum of Digits Method (Vn)	−12000	−11076.15	−10189.26	−9339.32	−8526.34	−7750.31
SDM Present Value (Vn)		−8909.39	−6592.66	−4860.63	−3569.43	−2609.84
Exponential Method (Vn)	−12000	−9036.79	−6805.29	−5124.83	−3859.33	−2906.33
EM Present Value (Pv)		−7268.97	−4403.16	−2667.21	−1615.65	−978.68
Functioning (F)	−4500	−4500.00	−4545.00	−4590.45	−4636.35	−4682.72
Present Value Accumulated (NPV)		−3619.69	−6560.40	−8949.49	−10890.43	−12467.29
Present Value (NPV)		−3619.69	−2940.71	−2389.09	−1940.94	−1576.86
Energy	−200	−200.00	−400.00	−600.00	−800.00	−1000.00
Energy (NPV)		−160.88	−258.81	−312.27	−334.91	−336.74
Accumulated (FPV)		−15780.57	−22599.77	−31861.53	−43086.87	−55890.90
Maintenance (M)	−300	−300.00	−303.00	−306.03	−309.09	−312.18
Present Value (MPV)		−241.31	−196.05	−159.27	−129.40	−105.12
Accumulated (MPV)		−241.31	−437.36	−596.63	−726.03	−831.15
Non Production (NP)	0	0.00	0.00	0.00	0.00	0.00
Present Value (NPPV)		0.00	0.00	0.00	0.00	0.00
Accumulated NPPV		0.00	0.00	0.00	0.00	0.00
Accumulated Total Costs (APV)		−16021.88	−23037.13	−32458.16	−43812.90	−56722.05
Accumulated Total Costs		−5000.00	−10248.00	−15744.48	−21489.92	−27484.82
Benefit (B)	50000	44642.86	44642.86	44642.86	44642.86	44642.86
Present Value (BPV)		35909.63	28884.84	23234.27	18689.08	15033.05
Accumulated Benefits BPV		35909.63	64794.47	88028.74	106717.82	121750.87
BPV+FPV+MPV+ NPPV+IIPV		7887.76	29757.34	43570.58	50904.93	53028.82
Uniform Annual Income UAI_n (LDM)		−11519.60	−4921.18	−2786.88	−1764.99	−1185.10
Uniform Annual Income UAI_n (SDM)		−11076.15	−4542.29	−2465.02	−1493.18	−956.96
Uniform Annual Income UAI_n (EM)		−9036.79	−3033.74	−1352.65	−675.87	−358.86

(Continued)

TABLE 1.1 (*Continued*)
Data to Simulate the Uniform Annual Income Method

Time	6	7	8	9	10
Initial Investment (*II*)					
Initial Investment (*II*)					
II Present Value (IIPV) – Return	−3250.39	−2614.53	−2103.07	−1691.66	−1360.73
Life Time simulated					
Residual value of equipment (RV)					
Linear Depreciation Method (Vn)	−9117.60	−8637.20	−8156.80	−7676.40	−7196.00
LDM Present Value (Vn)	−2469.64	−1881.85	−1429.53	−1082.15	−815.98
Sum of Digits Method (Vn)	−7011.23	−6309.11	−5643.94	−5015.72	−4424.46
SDM Present Value (Vn)	−1899.10	−1374.61	−989.13	−707.07	−501.71
Exponential Method (Vn)	−2188.66	−1648.20	−1241.20	−934.71	−703.90
EM Present Value (Pv)	−592.83	−359.11	−217.53	−131.77	−79.82
Functioning (F)	−4729.55	−4776.84	−4824.61	−4872.86	−4921.58
Present Value Accumulated (NPV)	−13748.36	−14789.13	−15634.67	−16321.60	−16879.68
Present Value (NPV)	−1281.07	−1040.77	−845.54	−686.93	−558.08
Energy	−1200.00	−1400.00	−1600.00	−1800.00	−2000.00
Energy (NPV)	−325.04	−305.03	−280.41	−253.75	−226.79
Accumulated (FPV)	−69964.30	−85058.45	−100973.53	−117548.88	−134655.35
Maintenance (M)	−315.30	−318.46	−321.64	−324.86	−328.11
Present Value (MPV)	−85.40	−69.38	−56.37	−45.80	−37.21
Accumulated (MPV)	−916.56	−985.94	−1042.31	−1088.11	−1125.31
Non Production (NP)	0.00	0.00	0.00	0.00	0.00
Present Value (NPPV)	0.00	0.00	0.00	0.00	0.00
Accumulated NPPV	0.00	0.00	0.00	0.00	0.00
Accumulated Total Costs (APV)	−70880.85	−86044.40	−102015.84	−118636.99	−135780.66
Accumulated Total Costs	−33729.67	−40224.97	−46971.22	−53968.93	−61218.62
Benefit (B)	44642.86	44642.86	44642.86	44642.86	44642.86
Present Value (BPV)	12092.22	9726.69	7823.91	6293.37	5062.23
Accumulated Benefits BPV	133843.09	143569.78	151393.69	157687.05	162749.28
BPV+FPV+MPV+NPPV+IIPV	50962.23	45525.38	37377.84	27050.06	14968.62
Uniform Annual Income UAI_n (LDM)	−823.74	−585.16	−421.54	−306.37	−223.83
Uniform Annual Income UAI_n (SDM)	−633.44	−427.43	−291.67	−200.18	−137.62
Uniform Annual Income UAI_n (EM)	−197.74	−111.66	−64.14	−37.30	−21.89

(Continued)

TABLE 1.1 (*Continued*)
Data to Simulate the Uniform Annual Income Method

Time	11	12	13	14	15
Initial Investment (*II*)					
Initial Investment (*II*)					
II Present Value (IIPV) – Return	−1094.54	−880.42	−708.19	−569.65	−458.21
Life Time simulated					
Residual value of equipment (RV)					
Linear Depreciation Method (Vn)	−6715.60	−6235.20	−5754.80	−5274.40	−4794.00
LDM Present Value (Vn)	−612.54	−457.47	−339.62	−250.38	−183.06
Sum of Digits Method (Vn)	−3870.15	−3352.80	−2872.40	−2428.95	−2022.46
SDM Present Value (Vn)	−353.00	−245.99	−169.52	−115.30	−77.23
Exponential Method (Vn)	−530.08	−399.19	−300.61	−226.38	−170.48
EM Present Value (Pv)	−48.35	−29.29	−17.74	−10.75	−6.51
Functioning (F)	−4970.80	−5020.51	−5070.71	−5121.42	−5172.63
Present Value Accumulated (NPV)	−17333.07	−17701.42	−18000.67	−18243.79	−18441.30
Present Value (NPV)	−453.39	−368.35	−299.25	−243.12	−197.51
Energy	−2200.00	−2400.00	−2600.00	−2800.00	−3000.00
Energy (NPV)	−200.67	−176.08	−153.44	−132.92	−114.55
Accumulated (FPV)	−152189.09	−170066.59	−188220.70	−206597.41	−225153.26
Maintenance (M)	−331.39	−334.70	−338.05	−341.43	−344.84
Present Value (MPV)	−30.23	−24.56	−19.5	−16.21	−13.17
Accumulated (MPV)	−1155.54	−1180.09	−1200.04	−1216.25	−1229.42
Non Production (NP)	0.00	0.00	0.00	0.00	0.00
Present Value (NPPV)	0.00	0.00	0.00	0.00	0.00
Accumulated NPPV	0.00	0.00	0.00	0.00	0.00
Accumulated Total Costs (APV)	−153344.63	−171246.68	−189420.75	−207813.66	−226382.68
Accumulated Total Costs	−68720.81	−76476.01	−84484.77	−92747.62	−101265.10
Benefit (B)	44642.86	44642.86	44642.86	44642.86	44642.86
Present Value (BPV)	4071.94	3275.37	2634.63	2119.23	1704.66
Accumulated Benefits BPV	166821.22	170096.59	172731.21	174850.44	176555.10
BPV+FPV+MPV+NPPV+IIPV	1476.59	−13150.10	−28689.53	−44963.22	−61827.58
Uniform Annual Income UAI_n (LDM)	−163.92	−120.06	−87.78	−63.93	−46.29
Uniform Annual Income UAI_n (SDM)	−94.47	−64.56	−43.81	−29.44	−19.53
Uniform Annual Income UAI_n (EM)	−12.94	−7.69	−4.59	−2.74	−1.65

(Continued)

TABLE 1.1 (*Continued*)
Data to Simulate the Uniform Annual Income Method

Time	16	17	18	19	20
Initial Investment (*II*)					
Initial Investment (*II*)					
II Present Value (IIPV) – Return	−368.57	−296.47	−238.48	−191.82	−154.30
Life Time simulated					
Residual value of equipment (RV)					
Linear Depreciation Method (Vn)	−4313.60	−3833.20	−3352.80	−2872.40	−2392.00
LDM Present Value (Vn)	−132.49	−94.70	−66.63	−45.92	−30.76
Sum of Digits Method (Vn)	−1652.92	−1320.34	−1024.71	−766.03	−544.31
SDM Present Value (Vn)	−50.77	−32.62	−20.36	−12.25	−7.00
Exponential Method (Vn)	−128.38	−96.68	−72.81	−54.83	−41.29
EM Present Value (Pv)	−3.94	−2.39	−1.45	−0.88	−0.53
Functioning (F)	−5224.36	−5276.60	−5329.37	−5382.66	−5436.49
Present Value Accumulated (NPV)	−18601.76	−18732.13	−18838.04	−18924.08	−18993.99
Present Value (NPV)	−160.46	−130.36	−105.91	−86.04	−69.90
Energy	−3200.00	−3400.00	−3600.00	−3800.00	−4000.00
Energy (NPV)	−98.29	−84.00	−71.54	−60.74	−51.43
Accumulated (FPV)	−243853.31	−262669.44	−281579.02	−300563.85	−319609.27
Maintenance (M)	−348.29	−351.77	−355.29	−358.84	−362.43
Present Value (MPV)	−10.70	−8.69	−7.06	−5.74	−4.66
Accumulated (MPV)	−1240.12	−1248.81	−1255.87	−1261.61	−1266.27
Non Production (NP)	0.00	0.00	0.00	0.00	0.00
Present Value (NPPV)	0.00	0.00	0.00	0.00	0.00
Accumulated NPPV	0.00	0.00	0.00	0.00	0.00
Accumulated Total Costs (APV)	−245093.43	−263918.25	−282834.89	−301825.45	−320875.53
Accumulated Total Costs	−110037.75	−119066.13	−128350.79	−137892.30	−147691.22
Benefit (B)	44642.86	44642.86	44642.86	44642.86	44642.86
Present Value (BPV)	1371.18	1102.95	887.18	713.63	574.03
Accumulated Benefits BPV	177926.28	179029.23	179916.42	180630.05	181204.07
BPV+FPV+MPV+ NPPV+IIPV	−79167.14	−96889.02	−114918.47	−133195.41	−151671.46
Uniform Annual Income UAI_n (LDM)	−33.24	−23.62	−16.53	−11.35	−7.58
Uniform Annual Income UAI_n (SDM)	−12.74	−8.13	−5.05	−3.03	−1.72
Uniform Annual Income UAI_n (EM)	−0.99	−0.60	−0.36	−0.22	−0.13

(*Continued*)

TABLE 1.1 (*Continued*)
Data to Simulate the Uniform Annual Income Method

Time	21	22	23	24	25
Initial Investment (*II*)					
Initial Investment (*II*)					
II Present Value (IIPV) – Return	−124.11	−99.83	−80.30	−64.59	−51.96
Life Time simulated					
Residual value of equipment (RV)					
Linear Depreciation Method (Vn)	−1911.60	−1431.20	−950.80	−470.40	10.00
LDM Present Value (Vn)	−19.77	−11.91	−6.36	−2.53	0.04
Sum of Digits Method (Vn)	−359.54	−211.72	−100.86	−26.95	10.00
SDM Present Value (Vn)	−3.72	−1.76	−0.67	−0.15	0.04
Exponential Method (Vn)	−31.09	−23.42	−17.63	−13.28	−10.00
EM Present Value (Pv)	−0.32	−0.19	−0.12	−0.07	−0.04
Functioning (F)	−5490.86	−5545.76	−5601.22	−5657.23	−5713.81
Present Value Accumulated (NPV)	−19050.78	−19096.91	−19134.40	−19164.85	−19189.59
Present Value (NPV)	−56.79	−46.14	−37.48	−30.45	−24.74
Energy	−4200.00	−4400.00	−4600.00	−4800.00	−5000.00
Energy (NPV)	−43.44	−36.61	−30.78	−25.84	−21.65
Accumulated (FPV)	−338703.48	−357837.01	−377002.19	−396192.88	−415404.12
Maintenance (M)	−366.06	−369.72	−373.41	−377.15	−380.92
Present Value (MPV)	−3.79	−3.08	−2.50	−2.03	−1.65
Accumulated (MPV)	−1270.05	−1273.13	−1275.63	−1277.66	−1279.31
Non Production (NP)	0.00	0.00	0.00	0.00	0.00
Present Value (NPPV)	0.00	0.00	0.00	0.00	0.00
Accumulated NPPV	0.00	0.00	0.00	0.00	0.00
Accumulated Total Costs (APV)	−339973.54	−359110.13	−378277.81	−397470.53	−416683.42
Accumulated Total Costs	−157748.13	−168063.61	−178638.25	−189472.63	−200567.36
Benefit (B)	44642.86	44642.86	44642.86	44642.86	44642.86
Present Value (BPV)	461.73	371.41	298.75	240.31	193.30
Accumulated Benefits BPV	181665.81	182037.21	182335.96	182576.27	182769.57
BPV+FPV+MPV+NPPV+IIPV	−170307.73	−189072.92	−207941.85	−226894.26	−245913.85
Uniform Annual Income UAI_n (LDM)	−4.86	−2.92	−1.56	−0.62	0.01
Uniform Annual Income UAI_n (SDM)	−0.91	−0.43	−0.17	−0.04	0.01
Uniform Annual Income UAI_n (EM)	−0.08	−0.05	−0.03	−0.02	−0.01

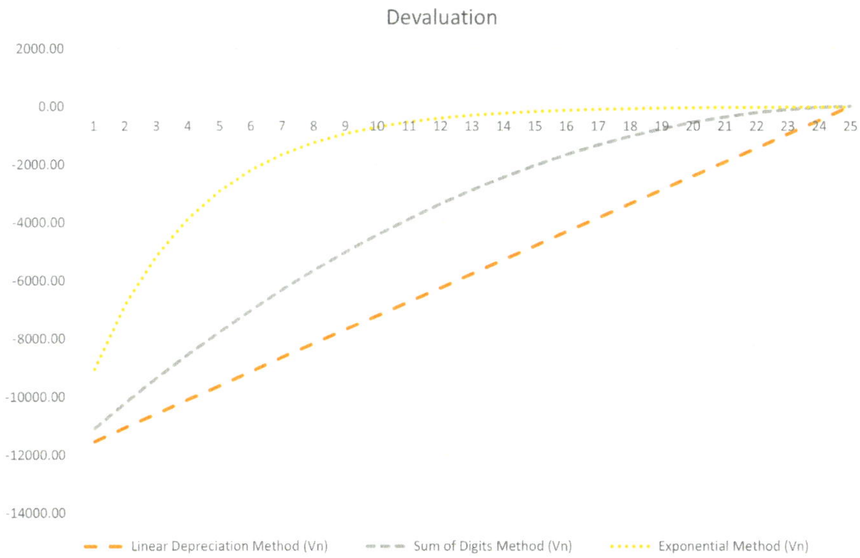

FIGURE 1.1 Curves of three devaluation methods.

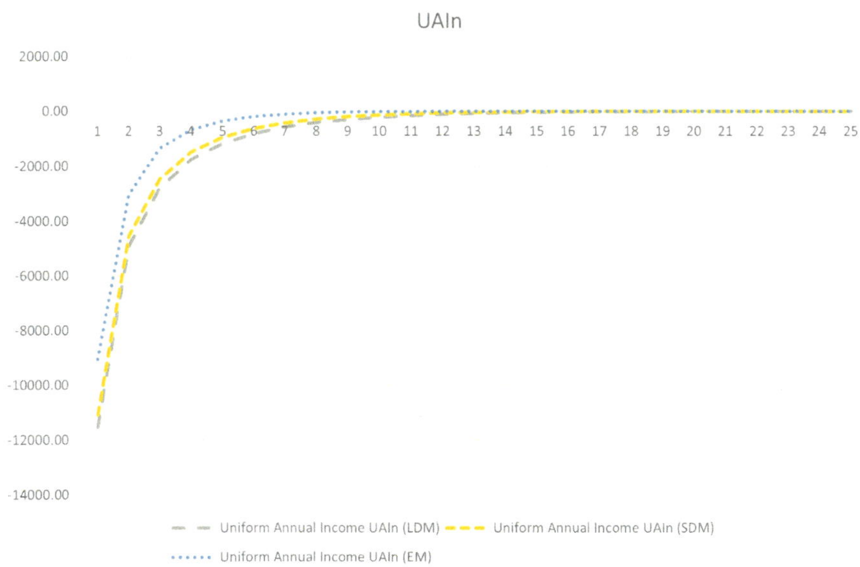

FIGURE 1.2 Curves of uniform annual income method.

1.9.2 Minimization of Total Average Cost Method

The second method simulated is the MMTAC, using the data included in Table 1.2.

The reference values for MMTAC are the same of Table 1.1, as were emphasised in the preceding section (values of the second column).

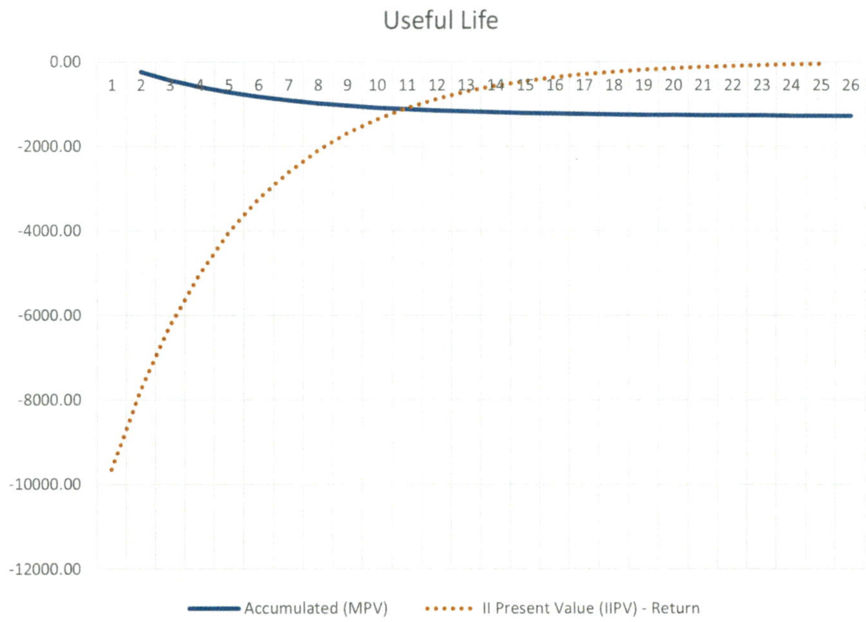

FIGURE 1.3 Useful Life.

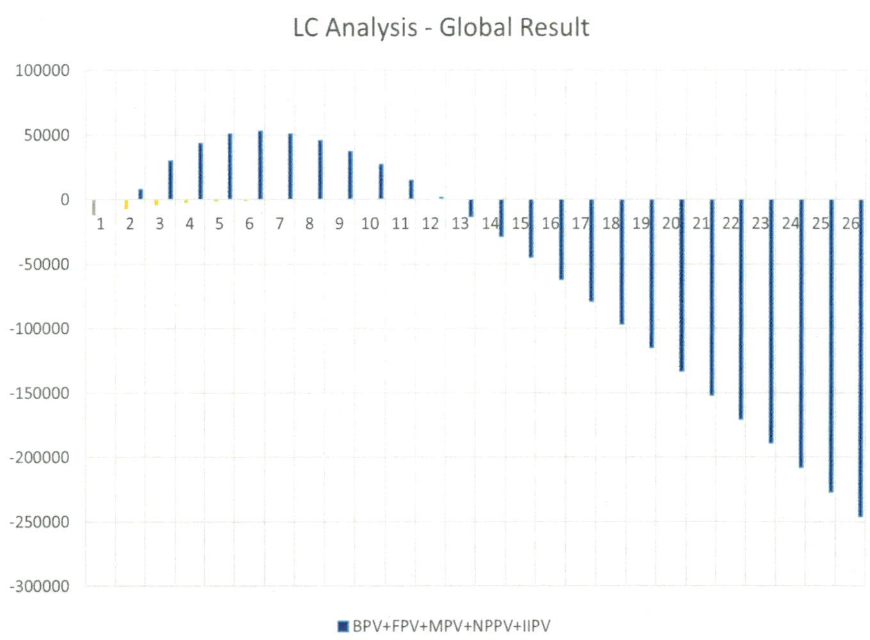

FIGURE 1.4 Global result of the Physical Asset Life Cycle.

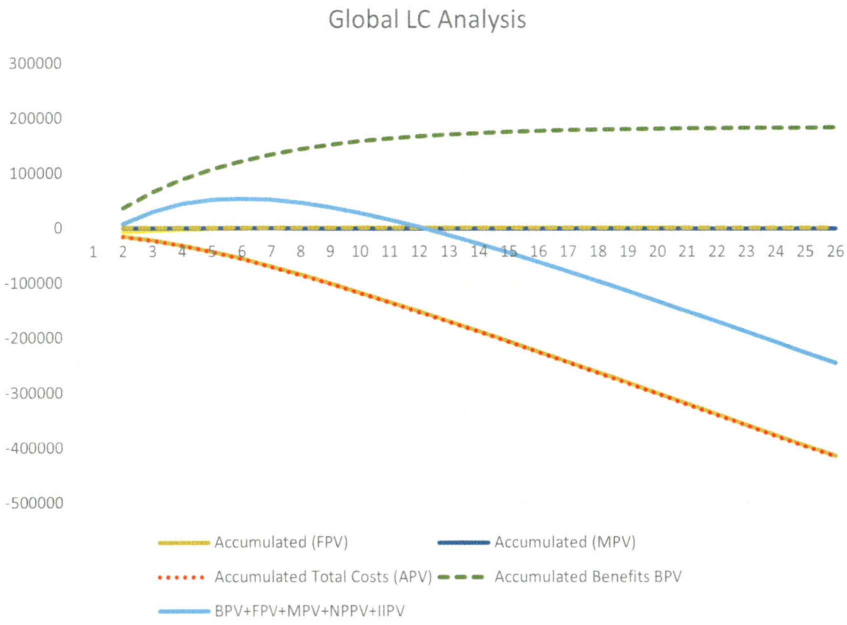

FIGURE 1.5 Global Physical Asset Life Cycle analysis.

From the simulation presented in Table 1.2, some results are emphasized:

- Figure 1.6 presents the Devaluation curves resulting from the Linear, Sum of Digits, and Exponential Methods.
- Figure 1.7 shows the results from MMTAC according to the three preceding Devaluation methods; it can be seen that the point of convergence is near the period of time considered 25 years.
- Figure 1.8 shows the Useful Life; as can be seen, the Useful Life is higher than the period of 25 years considered because it is not considered the Apparent Rate.
- Figure 1.9 shows the final balance between expenses and profits.
- Figure 1.10 shows the Global Result of the PA Life Cycle.
- It can be stated that, without Apparent Rate and with the investment and profits considered, the Life Cycle of this PA always gives a positive return. This conclusion emphasizes the importance of the Apparent Rate because it gives the context of PAs in the real world.

1.9.3 Minimization of Total Average Cost Method Reduced to Present Value

The third method simulated is the Minimization of Total Average Cost Method Reduced to Present Value (MMTAC-RPV), using the data included in Table 1.3.

TABLE 1.2
Data to Simulate the Minimization of Total Average Cost Method

Time	0	1	2	3	4	5
Initial Investment (II)	**−12000**					
Initial Investment (II)	**−12000**					
II Present Value (IIPV) - Return	−12000	−12000.00	−12000.00	−12000.00	−12000.00	−12000.00
Life Time simulated	25					
Residual value of equipment (RV)	10					
Linear Depreciation Method (Vn)	−12000	−11519.60	−11039.20	−10558.80	−10078.40	−9598.00
LDM Present Value (PV)		−11519.60	−11039.20	−10558.80	−10078.40	−9598.00
Sum of Digits Method (Vn)	−12000	−11076.15	−10189.26	−9339.32	−8526.34	−7750.31
SDM Present Value (PV)		−11076.15	−10189.26	−9339.32	−8526.34	−7750.31
Exponential Method (Vn)	−12000	−9036.79	−6805.29	−5124.83	−3859.33	−2906.33
EM Present Value (Pv)		−9036.79	−6805.29	−5124.83	−3859.33	−2906.33
Functioning (F)	−4500	−4500.00	−4545.00	−4590.45	−4636.35	−4682.72
Functioning Accumulated		−4500.00	−9045.00	−13635.45	−18271.80	−22954.52
Present Value Accumulated (NPV)		−4500.00	−9045.00	−13635.45	−18271.80	−22954.52
Present Value (NPV)		−4500.00	−4545.00	−4590.45	−4636.35	−4682.72
Energy	**−200**	−200.00	−400.00	−600.00	−800.00	−1000.00
Energy Accumulated		−200.00	−600.00	−1200.00	−2000.00	−3000.00
Energy (NPV)		−200.00	−400.00	−600.00	−800.00	−1000.00
Accumulated FPV		−16700.00	−26145.00	−40380.45	−59452.25	−83406.78
Maintenance (M)	**−300**	−300.00	−303.00	−306.03	−309.09	−312.18
Maintenance Accumulated		−300.00	−603.00	−909.03	−1218.12	−1530.30
Present Value (MPV)		−300.00	−303.00	−306.03	−309.09	−312.18
Accumulated MPV		−300.00	−603.00	−909.03	−1218.12	−1530.30
Non Production (NP)	**0**	0.00	0.00	0.00	0.00	0.00
Non Production Accumulated		0.00	0.00	0.00	0.00	0.00
Present Value (NPPV)		0.00	0.00	0.00	0.00	0.00
Accumulated NPPV		0.00	0.00	0.00	0.00	0.00
Accumulated Total Costs (APV)		−17000.00	−26748.00	−41289.48	−60670.37	−84937.08
Accumulated Total Costs		−5000.00	−10248.00	−15744.48	−21489.92	−27484.82
Benefit (B)	50000	44642.86	44642.86	44642.86	44642.86	44642.86
Benefit Accumulated		44642.86	89285.71	133928.57	178571.43	223214.29
Present Value (BPV)		44642.86	44642.86	44642.86	44642.86	44642.86
Accumulated Benefits BPV		44642.86	89285.71	133928.57	178571.43	223214.29
BPV+FPV+MPV+ NPPV+IIPV		27642.86	67037.71	106184.09	145081.50	183729.46
MTAC (LDM)		−28519.60	−16643.60	−12767.76	−10892.08	−9816.56
MTAC (SDM)		−28076.15	−16218.63	−12361.27	−10504.07	−9447.03
MTAC (EM)		−26036.79	−14526.65	−10956.44	−9337.31	−8478.23

(Continued)

TABLE 1.2 (*Continued*)
Data to Simulate the Minimization of Total Average Cost Method

Time	6	7	8	9	10
Initial Investment (II)					
Initial Investment (II)					
II Present Value (IIPV) - Return	−12000.00	−12000.00	−12000.00	−12000.00	−12000.00
Life Time simulated					
Residual value of equipment (RV)					
Linear Depreciation Method (Vn)	−9117.60	−8637.20	−8156.80	−7676.40	−7196.00
LDM Present Value (PV)	−9117.60	−8637.20	−8156.80	−7676.40	−7196.00
Sum of Digits Method (Vn)	−7011.23	−6309.11	−5643.94	−5015.72	−4424.46
SDM Present Value (PV)	−7011.23	−6309.11	−5643.94	−5015.72	−4424.46
Exponential Method (Vn)	−2188.66	−1648.20	−1241.20	−934.71	−703.90
EM Present Value (Pv)	−2188.66	−1648.20	−1241.20	−934.71	−703.90
Functioning (F)	−4729.55	−4776.84	−4824.61	−4872.86	−4921.58
Functioning Accumulated	−27684.07	−32460.91	−37285.52	−42158.37	−47079.96
Present Value Accumulated (NPV)	−27684.07	−32460.91	−37285.52	−42158.37	−47079.96
Present Value (NPV)	−4729.55	−4776.84	−4824.61	−4872.86	−4921.58
Energy	−1200.00	−1400.00	−1600.00	−1800.00	−2000.00
Energy Accumulated	−4200.00	−5600.00	−7200.00	−9000.00	−11000.00
Energy (NPV)	−1200.00	−1400.00	−1600.00	−1800.00	−2000.00
Accumulated FPV	−112290.84	−146151.75	−185037.27	−228995.64	−278075.60
Maintenance (M)	−315.30	−318.46	−321.64	−324.86	−328.11
Maintenance Accumulated	−1845.60	−2164.06	−2485.70	−2810.56	−3138.66
Present Value (MPV)	−315.30	−318.46	−321.64	−324.86	−328.11
Accumulated MPV	−1845.60	−2164.06	−2485.70	−2810.56	−3138.66
Non Production (NP)	0.00	0.00	0.00	0.00	0.00
Non Production Accumulated	0.00	0.00	0.00	0.00	0.00
Present Value (NPPV)	0.00	0.00	0.00	0.00	0.00
Accumulated NPPV	0.00	0.00	0.00	0.00	0.00
Accumulated Total Costs (APV)	−114136.45	−148315.81	−187522.97	−231806.20	−281214.26
Accumulated Total Costs	−33729.67	−40224.97	−46971.22	−53968.93	−61218.62
Benefit (B)	44642.86	44642.86	44642.86	44642.86	44642.86
Benefit Accumulated	267857.14	312500.00	357142.86	401785.71	446428.57
Present Value (BPV)	44642.86	44642.86	44642.86	44642.86	44642.86
Accumulated Benefits BPV	267857.14	312500.00	357142.86	401785.71	446428.57
BPV+FPV+MPV+ NPPV+IIPV	222127.47	260275.03	298171.64	335816.78	373209.95
MTAC (LDM)	−9141.21	−8694.60	−8391.00	−8182.81	−8041.46
MTAC (SDM)	−8790.15	−8362.01	−8076.89	−7887.18	−7764.31
MTAC (EM)	−7986.39	−7696.17	−7526.55	−7433.74	−7392.25

(*Continued*)

TABLE 1.2 (*Continued*)
Data to Simulate the Minimization of Total Average Cost Method

Time	11	12	13	14	15
Initial Investment (II)					
Initial Investment (II)					
II Present Value (IIPV) - Return	−12000.00	−12000.00	−12000.00	−12000.00	−12000.00
Life Time simulated					
Residual value of equipment (RV)					
Linear Depreciation Method (Vn)	−6715.60	−6235.20	−5754.80	−5274.40	−4794.00
LDM Present Value (PV)	−6715.60	−6235.20	−5754.80	−5274.40	−4794.00
Sum of Digits Method (Vn)	−3870.15	−3352.80	−2872.40	−2428.95	−2022.46
SDM Present Value (PV)	−3870.15	−3352.80	−2872.40	−2428.95	−2022.46
Exponential Method (Vn)	−530.08	−399.19	−300.61	−226.38	−170.48
EM Present Value (Pv)	−530.08	−399.19	−300.61	−226.38	−170.48
Functioning (F)	−4970.80	−5020.51	−5070.71	−5121.42	−5172.63
Functioning Accumulated	−52050.76	−57071.26	−62141.98	−67263.40	−72436.03
Present Value Accumulated (NPV)	−52050.76	−57071.26	−62141.98	−67263.40	−72436.03
Present Value (NPV)	−4970.80	−5020.51	−5070.71	−5121.42	−5172.63
Energy	−2200.00	−2400.00	−2600.00	−2800.00	−3000.00
Energy Accumulated	−13200.00	−15600.00	−18200.00	−21000.00	−24000.00
Energy (NPV)	−2200.00	−2400.00	−2600.00	−2800.00	−3000.00
Accumulated FPV	−332326.36	−391797.62	−456539.60	−526602.99	−602039.02
Maintenance (M)	−331.39	−334.70	−338.05	−341.43	−344.84
Maintenance Accumulated	−3470.05	−3804.75	−4142.80	−4484.23	−4829.07
Present Value (MPV)	−331.39	−334.70	−338.05	−341.43	−344.84
Accumulated MPV	−3470.05	−3804.75	−4142.80	−4484.23	−4829.07
Non Production (NP)	0.00	0.00	0.00	0.00	0.00
Non Production Accumulated	0.00	0.00	0.00	0.00	0.00
Present Value (NPPV)	0.00	0.00	0.00	0.00	0.00
Accumulated NPPV	0.00	0.00	0.00	0.00	0.00
Accumulated Total Costs (APV)	−335796.41	−395602.37	−460682.39	−531087.22	−606868.09
Accumulated Total Costs	−68720.81	−76476.01	−84484.77	−92747.62	−101265.10
Benefit (B)	44642.86	44642.86	44642.86	44642.86	44642.86
Benefit Accumulated	491071.43	535714.29	580357.14	625000.00	669642.86
Present Value (BPV)	44642.86	44642.86	44642.86	44642.86	44642.86
Accumulated Benefits BPV	491071.43	535714.29	580357.14	625000.00	669642.86
BPV+FPV+MPV+NPPV+IIPV	410350.62	447238.27	483872.37	520252.38	556377.76
MTAC (LDM)	−7948.76	−7892.60	−7864.58	−7858.72	−7870.61
MTAC (SDM)	−7690.09	−7652.40	−7642.86	−7655.47	−7685.84
MTAC (EM)	−7386.44	−7406.27	−7445.03	−7498.14	−7562.37

(*Continued*)

TABLE 1.2 (*Continued*)
Data to Simulate the Minimization of Total Average Cost Method

Time	16	17	18	19	20
Initial Investment (II)					
Initial Investment (II)					
II Present Value (IIPV) - Return	−12000.00	−12000.00	−12000.00	−12000.00	−12000.00
Life Time simulated					
Residual value of equipment (RV)					
Linear Depreciation Method (Vn)	−4313.60	−3833.20	−3352.80	−2872.40	−2392.00
LDM Present Value (PV)	−4313.60	−3833.20	−3352.80	−2872.40	−2392.00
Sum of Digits Method (Vn)	−1652.92	−1320.34	−1024.71	−766.03	−544.31
SDM Present Value (PV)	−1652.92	−1320.34	−1024.71	−766.03	−544.31
Exponential Method (Vn)	−128.38	−96.68	−72.81	−54.83	−41.29
EM Present Value (Pv)	−128.38	−96.68	−72.81	−54.83	−41.29
Functioning (F)	−5224.36	−5276.60	−5329.37	−5382.66	−5436.49
Functioning Accumulated	−77660.39	−82936.99	−88266.36	−93649.03	−99085.52
Present Value Accumulated (NPV)	−77660.39	−82936.99	−88266.36	−93649.03	−99085.52
Present Value (NPV)	−5224.36	−5276.60	−5329.37	−5382.66	−5436.49
Energy	−3200.00	−3400.00	−3600.00	−3800.00	−4000.00
Energy Accumulated	−27200.00	−30600.00	−34200.00	−38000.00	−42000.00
Energy (NPV)	−3200.00	−3400.00	−3600.00	−3800.00	−4000.00
Accumulated FPV	−682899.41	−769236.41	−861102.77	−958551.80	−1061637.32
Maintenance (M)	−348.29	−351.77	−355.29	−358.84	−362.43
Maintenance Accumulated	−5177.36	−5529.13	−5884.42	−6243.27	−6605.70
Present Value (MPV)	−348.29	−351.77	−355.29	−358.84	−362.43
Accumulated MPV	−5177.36	−5529.13	−5884.42	−6243.27	−6605.70
Non Production (NP)	0.00	0.00	0.00	0.00	0.00
Non Production Accumulated	0.00	0.00	0.00	0.00	0.00
Present Value (NPPV)	0.00	0.00	0.00	0.00	0.00
Accumulated NPPV	0.00	0.00	0.00	0.00	0.00
Accumulated Total Costs (APV)	−688076.77	−774765.54	−866987.19	−964795.07	−1068243.02
Accumulated Total Costs	−110037.75	−119066.13	−128350.79	−137892.30	−147691.22
Benefit (B)	44642.86	44642.86	44642.86	44642.86	44642.86
Benefit Accumulated	714285.71	758928.57	803571.43	848214.29	892857.14
Present Value (BPV)	44642.86	44642.86	44642.86	44642.86	44642.86
Accumulated Benefits BPV	714285.71	758928.57	803571.43	848214.29	892857.14
BPV+FPV+MPV+NPPV+IIPV	592247.96	627862.44	663220.64	698321.99	733165.92
MTAC (LDM)	−7896.96	−7935.25	−7983.53	−8040.25	−8104.16
MTAC (SDM)	−7730.67	−7787.44	−7854.19	−7929.39	−8011.78
MTAC (EM)	−7635.38	−7715.46	−7801.31	−7891.95	−7986.63

(Continued)

TABLE 1.2 (*Continued*)
Data to Simulate the Minimization of Total Average Cost Method

Time	21	22	23	24	25
Initial Investment (II)					
Initial Investment (II)					
II Present Value (IIPV) - Return	−12000.00	−12000.00	−12000.00	−12000.00	−12000.00
Life Time simulated					
Residual value of equipment (RV)					
Linear Depreciation Method (Vn)	−1911.60	−1431.20	−950.80	−470.40	10.00
LDM Present Value (PV)	−1911.60	−1431.20	−950.80	−470.40	10.00
Sum of Digits Method (Vn)	−359.54	−211.72	−100.86	−26.95	10.00
SDM Present Value (PV)	−359.54	−211.72	−100.86	−26.95	10.00
Exponential Method (Vn)	−31.09	−23.42	−17.63	−13.28	−10.00
EM Present Value (Pv)	−31.09	−23.42	−17.63	−13.28	−10.00
Functioning (F)	−5490.86	−5545.76	−5601.22	−5657.23	−5713.81
Functioning Accumulated	−104576.37	−110122.14	−115723.36	−121380.59	−127094.40
Present Value Accumulated (NPV)	−104576.37	−110122.14	−115723.36	−121380.59	−127094.40
Present Value (NPV)	−5490.86	−5545.76	−5601.22	−5657.23	−5713.81
Energy	−4200.00	−4400.00	−4600.00	−4800.00	−5000.00
Energy Accumulated	−46200.00	−50600.00	−55200.00	−60000.00	−65000.00
Energy (NPV)	−4200.00	−4400.00	−4600.00	−4800.00	−5000.00
Accumulated FPV	−1170413.69	−1284935.83	−1405259.18	−1531439.78	−1663534.17
Maintenance (M)	−366.06	−369.72	−373.41	−377.15	−380.92
Maintenance Accumulated	−6971.76	−7341.48	−7714.89	−8092.04	−8472.96
Present Value (MPV)	−366.06	−369.72	−373.41	−377.15	−380.92
Accumulated MPV	−6971.76	−7341.48	−7714.89	−8092.04	−8472.96
Non Production (NP)	0.00	0.00	0.00	0.00	0.00
Non Production Accumulated	0.00	0.00	0.00	0.00	0.00
Present Value (NPPV)	0.00	0.00	0.00	0.00	0.00
Accumulated NPPV	0.00	0.00	0.00	0.00	0.00
Accumulated Total Costs (APV)	−1177385.45	−1292277.30	−1412974.07	−1539531.82	−1672007.13
Accumulated Total Costs	−157748.13	−168063.61	−178638.25	−189472.63	−200567.36
Benefit (B)	44642.86	44642.86	44642.86	44642.86	44642.86
Benefit Accumulated	937500.00	982142.86	1026785.71	1071428.57	1116071.43
Present Value (BPV)	44642.86	44642.86	44642.86	44642.86	44642.86
Accumulated Benefits BPV	937500.00	982142.86	1026785.71	1071428.57	1116071.43
BPV+FPV+MPV+NPPV+IIPV	767751.87	802079.24	836147.47	869955.94	903504.07
MTAC (LDM)	−8174.27	−8249.76	−8329.96	−8414.29	−8502.29
MTAC (SDM)	−8100.37	−8194.33	−8293.00	−8395.82	−8502.29
MTAC (EM)	−8084.72	−8185.77	−8289.39	−8395.25	−8503.09

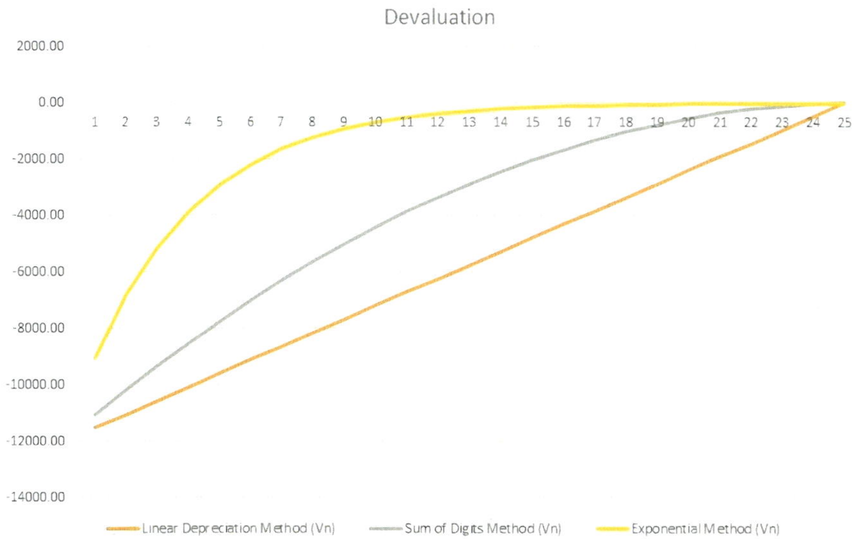

FIGURE 1.6 Curves of three devaluation methods.

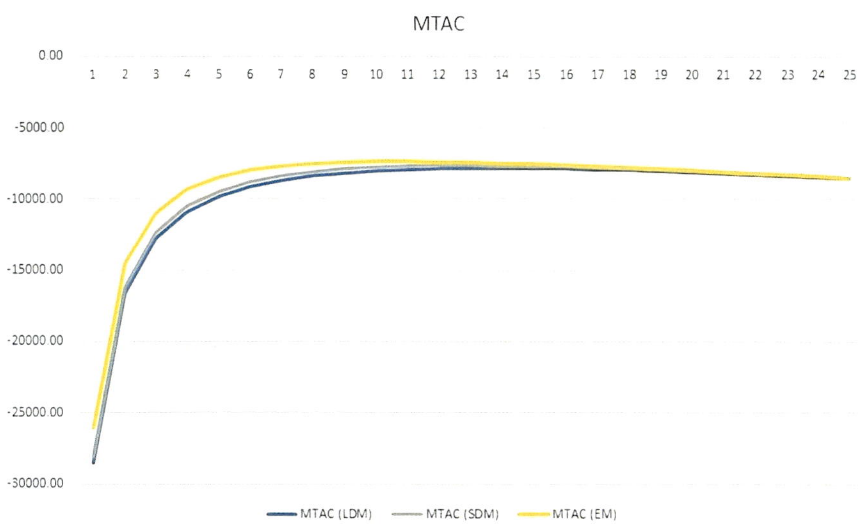

FIGURE 1.7 Curves of minimization of total average cost method.

The reference values for MMTAC-RPV are the same of Table 1.1, as were emphasized in the preceding section (values of the second column).

From the simulation presented in Table 1.3, some results are emphasized:

- Figure 1.11 presents the Devaluation curves resulting from the Linear, Sum of Digits, and Exponential Methods.

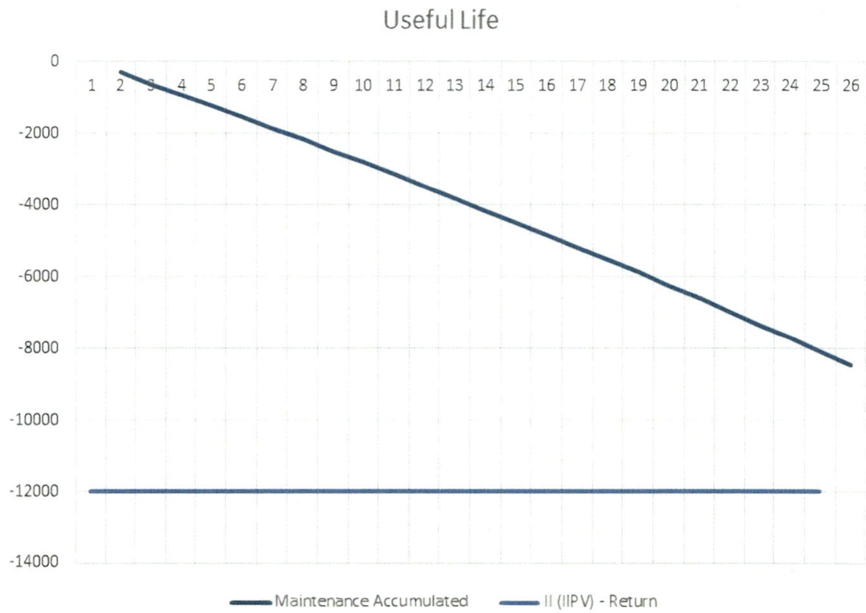

FIGURE 1.8 Useful Life.

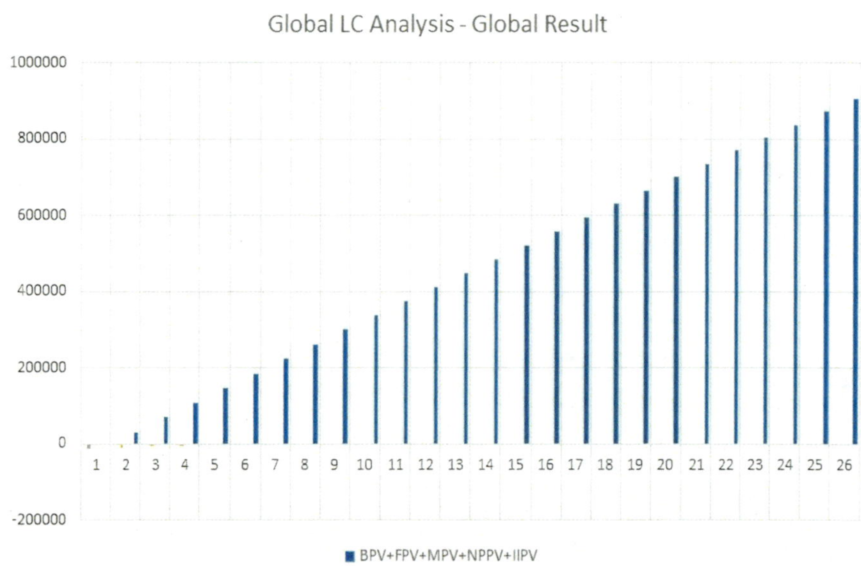

FIGURE 1.9 Global result of the Physical Asset Life Cycle.

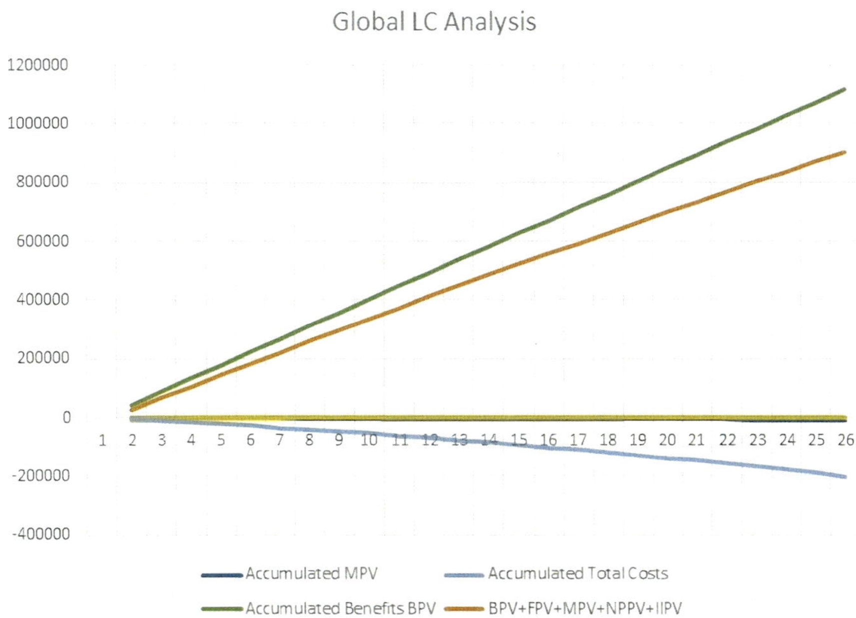

FIGURE 1.10 Global Physical Asset Life Cycle analysis.

- Figure 1.12 shows the results from MMTAC-RPV according to the three preceding Devaluation methods; as can be seen, the results do not permit to identify objectively a point of convergence.
- Figure 1.13 shows the point of Useful Life; as can be seen, the Useful Life is lower than the MMTAC.
- Figure 1.14 shows the final balance between expenses and profits.
- Figure 1.15 shows the Global Result of the PA life cycle.

TABLE 1.3
Minimization of Total Average Cost Method Reduced to Present Value

Time	0	1	2	3	4	5
Initial Investment (II)	−12000					
Initial Investment (II)	−12000					
II Present Value (IIPV) – Return	−12000	−9652.51	−7764.25	−6245.37	−5023.63	−4040.88
Life Time simulated	25					
Residual value of equipment (RV)	10					
Linear Depreciation Method (Vn)	−12000	−11519.60	−11039.20	−10558.80	−10078.40	−9598.00
LDM Present Value (Vn)		−9266.09	−7142.59	−5495.30	−4219.18	−3232.03
Sum of Digits Method (Vn)	−12000	−11076.15	−10189.26	−9339.32	−8526.34	−7750.31
SDM Present Value (Vn)		−8909.39	−6592.66	−4860.63	−3569.43	−2609.84
Exponential Method (Vn)	−12000	−9036.79	−6805.29	−5124.83	−3859.33	−2906.33
EM Present Value (Pv)		−7268.97	−4403.16	−2667.21	−1615.65	−978.68
Functioning (F)	−4500	−4500.00	−4545.00	−4590.45	−4636.35	−4682.72
Present Value Accumulated (NPV)		−3619.69	−6560.40	−8949.49	−10890.43	−12467.29
Present Value (NPV)		−3619.69	−2940.71	−2389.09	−1940.94	−1576.86
Energy (E)	−200	−200.00	−400.00	−600.00	−800.00	−1000.00
Energy (NPV)		−160.88	−258.81	−312.27	−334.91	−336.74
Accumulated FPV		−15780.57	−22599.77	−31861.53	−43086.87	−55890.90
Maintenance (M)	−300	−300.00	−303.00	−306.03	−309.09	−312.18
Present Value (MPV)		−241.31	−196.05	−159.27	−129.40	−105.12
Accumulated (MPV)		−241.31	−437.36	−596.63	−726.03	−831.15
Non Production (NP)	0	0.00	0.00	0.00	0.00	0.00
Present Value (NPPV)		0.00	0.00	0.00	0.00	0.00
Accumulated NPPV		0.00	0.00	0.00	0.00	0.00
Accumulated Total Costs (APV)		−16021.88	−23037.13	−32458.16	−43812.90	−56722.05
Accumulated Total Costs		−5000.00	−10248.00	−15744.48	−21489.92	−27484.82
Benefit (B)	50000	44642.86	44642.86	44642.86	44642.86	44642.86
Present Value (BPV)		35909.63	28884.84	23234.27	18689.08	15033.05
Accumulated Benefits BPV		35909.63	64794.47	88028.74	106717.82	121750.87
BPV+FPV+MPV+NPPV+IIPV		7887.76	29757.34	43570.58	50904.93	53028.82
MTAC-PV (LDM)		−37287.97	−21089.86	−16651.15	−15008.02	−14390.82
MTAC-PV (SDM)		−36931.27	−20814.90	−16439.60	−14845.58	−14266.38
MTAC-PV (EM)		−35290.85	−19720.15	−15708.46	−14357.14	−13940.15

(Continued)

TABLE 1.3 (*Continued*)
Minimization of Total Average Cost Method Reduced to Present Value

Time	6	7	8	9	10
Initial Investment (II)					
Initial Investment (II)					
II Present Value (IIPV) – Return	−3250.39	−2614.53	−2103.07	−1691.66	−1360.73
Life Time simulated					
Residual value of equipment (RV)					
Linear Depreciation Method (Vn)	−9117.60	−8637.20	−8156.80	−7676.40	−7196.00
LDM Present Value (Vn)	−2469.64	−1881.85	−1429.53	−1082.15	−815.98
Sum of Digits Method (Vn)	−7011.23	−6309.11	−5643.94	−5015.72	−4424.46
SDM Present Value (Vn)	−1899.10	−1374.61	−989.13	−707.07	−501.71
Exponential Method (Vn)	−2188.66	−1648.20	−1241.20	−934.71	−703.90
EM Present Value (Pv)	−592.83	−359.11	−217.53	−131.77	−79.82
Functioning (F)	−4729.55	−4776.84	−4824.61	−4872.86	−4921.58
Present Value Accumulated (NPV)	−13748.36	−14789.13	−15634.67	−16321.60	−16879.68
Present Value (NPV)	−1281.07	−1040.77	−845.54	−686.93	−558.08
Energy (E)	−1200.00	−1400.00	−1600.00	−1800.00	−2000.00
Energy (NPV)	−325.04	−305.03	−280.41	−253.75	−226.79
Accumulated FPV	−69964.30	−85058.45	−100973.53	−117548.88	−134655.35
Maintenance (M)	−315.30	−318.46	−321.64	−324.86	−328.11
Present Value (MPV)	−85.40	−69.38	−56.37	−45.80	−37.21
Accumulated (MPV)	−916.56	−985.94	−1042.31	−1088.11	−1125.31
Non Production (NP)	0.00	0.00	0.00	0.00	0.00
Present Value (NPPV)	0.00	0.00	0.00	0.00	0.00
Accumulated NPPV	0.00	0.00	0.00	0.00	0.00
Accumulated Total Costs (APV)	−70880.85	−86044.40	−102015.84	−118636.99	−135780.66
Accumulated Total Costs	−33729.67	−40224.97	−46971.22	−53968.93	−61218.62
Benefit (B)	44642.86	44642.86	44642.86	44642.86	44642.86
Present Value (BPV)	12092.22	9726.69	7823.91	6293.37	5062.23
Accumulated Benefits BPV	133843.09	143569.78	151393.69	157687.05	162749.28
BPV+FPV+MPV+NPPV+IIPV	50962.23	45525.38	37377.84	27050.06	14968.62
MTAC-PV (LDM)	−14225.08	−14275.18	−14430.67	−14635.46	−14859.66
MTAC-PV (SDM)	−14129.99	−14202.72	−14375.62	−14593.78	−14828.24
MTAC-PV (EM)	−13912.28	−14057.64	−14279.17	−14529.86	−14786.05

(*Continued*)

TABLE 1.3 (*Continued*)
Minimization of Total Average Cost Method Reduced to Present Value

Time	11	12	13	14	15
Initial Investment (II)					
Initial Investment (II)					
II Present Value (IIPV) – Return	−1094.54	−880.42	−708.19	−569.65	−458.21
Life Time simulated					
Residual value of equipment (RV)					
Linear Depreciation Method (Vn)	−6715.60	−6235.20	−5754.80	−5274.40	−4794.00
LDM Present Value (Vn)	−612.54	−457.47	−339.62	−250.38	−183.06
Sum of Digits Method (Vn)	−3870.15	−3352.80	−2872.40	−2428.95	−2022.46
SDM Present Value (Vn)	−353.00	−245.99	−169.52	−115.30	−77.23
Exponential Method (Vn)	−530.08	−399.19	−300.61	−226.38	−170.48
EM Present Value (Pv)	−48.35	−29.29	−17.74	−10.75	−6.51
Functioning (F)	−4970.80	−5020.51	−5070.71	−5121.42	−5172.63
Present Value Accumulated (NPV)	−17333.07	−17701.42	−18000.67	−18243.79	−18441.30
Present Value (NPV)	−453.39	−368.35	−299.25	−243.12	−197.51
Energy (E)	−2200.00	−2400.00	−2600.00	−2800.00	−3000.00
Energy (NPV)	−200.67	−176.08	−153.44	−132.92	−114.55
Accumulated FPV	−152189.09	−170066.59	−188220.70	−206597.41	−225153.26
Maintenance (M)	−331.39	−334.70	−338.05	−341.43	−344.84
Present Value (MPV)	−30.23	−24.56	−19.95	−16.21	−13.17
Accumulated (MPV)	−1155.54	−1180.09	−1200.04	−1216.25	−1229.42
Non Production (NP)	0.00	0.00	0.00	0.00	0.00
Present Value (NPPV)	0.00	0.00	0.00	0.00	0.00
Accumulated NPPV	0.00	0.00	0.00	0.00	0.00
Accumulated Total Costs (APV)	−153344.63	−171246.68	−189420.75	−207813.66	−226382.68
Accumulated Total Costs	−68720.81	−76476.01	−84484.77	−92747.62	−101265.10
Benefit (B)	44642.86	44642.86	44642.86	44642.86	44642.86
Present Value (BPV)	4071.94	3275.37	2634.63	2119.23	1704.66
Accumulated Benefits BPV	166821.22	170096.59	172731.21	174850.44	176555.10
BPV+FPV+MPV+NPPV+IIPV	1476.59	−13150.10	−28689.53	−44963.22	−61827.58
MTAC-PV (LDM)	−15087.02	−15308.68	−15520.03	−15718.86	−15904.38
MTAC-PV (SDM)	−15063.42	−15291.06	−15506.94	−15709.21	−15897.33
MTAC-PV (EM)	−15035.73	−15273.00	−15495.27	−15701.74	−15892.61

(*Continued*)

TABLE 1.3 (*Continued*)
Minimization of Total Average Cost Method Reduced to Present Value

Time	16	17	18	19	20
Initial Investment (II)					
Initial Investment (II)					
II Present Value (IIPV) – Return	−368.57	−296.47	−238.48	−191.82	−154.30
Life Time simulated					
Residual value of equipment (RV)					
Linear Depreciation Method (Vn)	−4313.60	−3833.20	−3352.80	−2872.40	−2392.00
LDM Present Value (Vn)	−132.49	−94.70	−66.63	−45.92	−30.76
Sum of Digits Method (Vn)	−1652.92	−1320.34	−1024.71	−766.03	−544.31
SDM Present Value (Vn)	−50.77	−32.62	−20.36	−12.25	−7.00
Exponential Method (Vn)	−128.38	−96.68	−72.81	−54.83	−41.29
EM Present Value (Pv)	−3.94	−2.39	−1.45	−0.88	−0.53
Functioning (F)	−5224.36	−5276.60	−5329.37	−5382.66	−5436.49
Present Value Accumulated (NPV)	−18601.76	−18732.13	−18838.04	−18924.08	−18993.99
Present Value (NPV)	−160.46	−130.36	−105.91	−86.04	−69.90
Energy (E)	−3200.00	−3400.00	−3600.00	−3800.00	−4000.00
Energy (NPV)	−98.29	−84.00	−71.54	−60.74	−51.43
Accumulated FPV	−243853.31	−262669.44	−281579.02	−300563.85	−319609.27
Maintenance (M)	−348.29	−351.77	−355.29	−358.84	−362.43
Present Value (MPV)	−10.70	−8.69	−7.06	−5.74	−4.66
Accumulated (MPV)	−1240.12	−1248.81	−1255.87	−1261.61	−1266.27
Non Production (NP)	0.00	0.00	0.00	0.00	0.00
Present Value (NPPV)	0.00	0.00	0.00	0.00	0.00
Accumulated NPPV	0.00	0.00	0.00	0.00	0.00
Accumulated Total Costs (APV)	−245093.43	−263918.25	−282834.89	−301825.45	−320875.53
Accumulated Total Costs	−110037.75	−119066.13	−128350.79	−137892.30	−147691.22
Benefit (B)	44642.86	44642.86	44642.86	44642.86	44642.86
Present Value (BPV)	1371.18	1102.95	887.18	713.63	574.03
Accumulated Benefits BPV	177926.28	179029.23	179916.42	180630.05	181204.07
BPV+FPV+MPV+ NPPV+IIPV	−79167.14	−96889.02	−114918.47	−133195.41	−151671.46
MTAC-PV (LDM)	−16076.62	−16236.06	−16383.42	−16519.55	−16645.31
MTAC-PV (SDM)	−16071.51	−16232.40	−16380.85	−16517.77	−16644.13
MTAC-PV (EM)	−16068.59	−16230.63	−16379.80	−16517.18	−16643.80

(*Continued*)

TABLE 1.3 (*Continued*)
Minimization of Total Average Cost Method Reduced to Present Value

Time	21	22	23	24	25
Initial Investment (II)					
Initial Investment (II)					
II Present Value (IIPV) – Return	−124.11	−99.83	−80.30	−64.59	−51.96
Life Time simulated					
Residual value of equipment (RV)					
Linear Depreciation Method (Vn)	−1911.60	−1431.20	−950.80	−470.40	10.00
LDM Present Value (Vn)	−19.77	−11.91	−6.36	−2.53	0.04
Sum of Digits Method (Vn)	−359.54	−211.72	−100.86	−26.95	10.00
SDM Present Value (Vn)	−3.72	−1.76	−0.67	−0.15	0.04
Exponential Method (Vn)	−31.09	−23.42	−17.63	−13.28	−10.00
EM Present Value (Pv)	−0.32	−0.19	−0.12	−0.07	−0.04
Functioning (F)	−5490.86	−5545.76	−5601.22	−5657.23	−5713.81
Present Value Accumulated (NPV)	−19050.78	−19096.91	−19134.40	−19164.85	−19189.59
Present Value (NPV)	−56.79	−46.14	−37.48	−30.45	−24.74
Energy (E)	−4200.00	−4400.00	−4600.00	−4800.00	−5000.00
Energy (NPV)	−43.44	−36.61	−30.78	−25.84	−21.65
Accumulated FPV	−338703.48	−357837.01	−377002.19	−396192.88	−415404.12
Maintenance (M)	−366.06	−369.72	−373.41	−377.15	−380.92
Present Value (MPV)	−3.79	−3.08	−2.50	−2.03	−1.65
Accumulated (MPV)	−1270.05	−1273.13	−1275.63	−1277.66	−1279.31
Non Production (NP)	0.00	0.00	0.00	0.00	0.00
Present Value (NPPV)	0.00	0.00	0.00	0.00	0.00
Accumulated NPPV	0.00	0.00	0.00	0.00	0.00
Accumulated Total Costs (APV)	−339973.54	−359110.13	−378277.81	−397470.53	−416683.42
Accumulated Total Costs	−157748.13	−168063.61	−178638.25	−189472.63	−200567.36
Benefit (B)	44642.86	44642.86	44642.86	44642.86	44642.86
Present Value (BPV)	461.73	371.41	298.75	240.31	193.30
Accumulated Benefits BPV	181665.81	182037.21	182335.96	182576.27	182769.57
BPV+FPV+MPV+ NPPV+IIPV	−170307.73	−189072.92	−207941.85	−226894.26	−245913.85
MTAC-PV (LDM)	−16761.59	−16869.18	−16968.88	−17061.38	−17147.34
MTAC-PV (SDM)	−16760.82	−16868.72	−16968.63	−17061.28	−17147.34
MTAC-PV (EM)	−16760.66	−16868.65	−16968.61	−17061.28	−17147.34

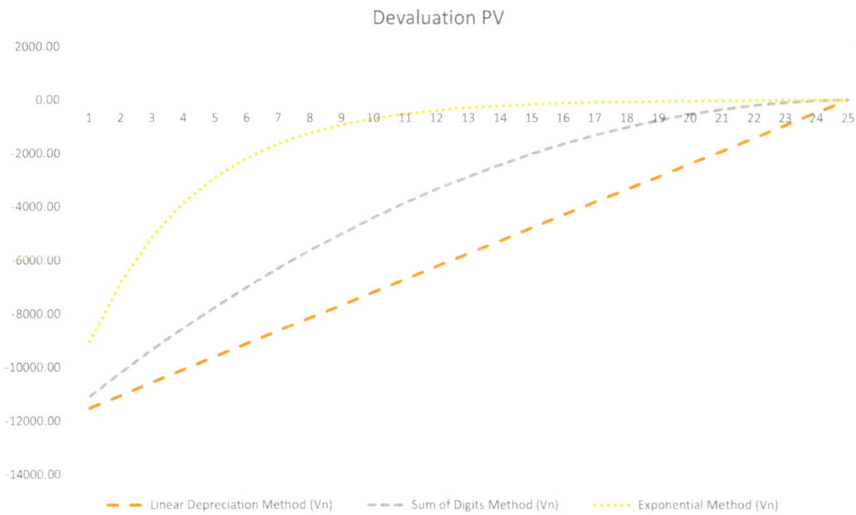

FIGURE 1.11 Curves of three devaluation methods.

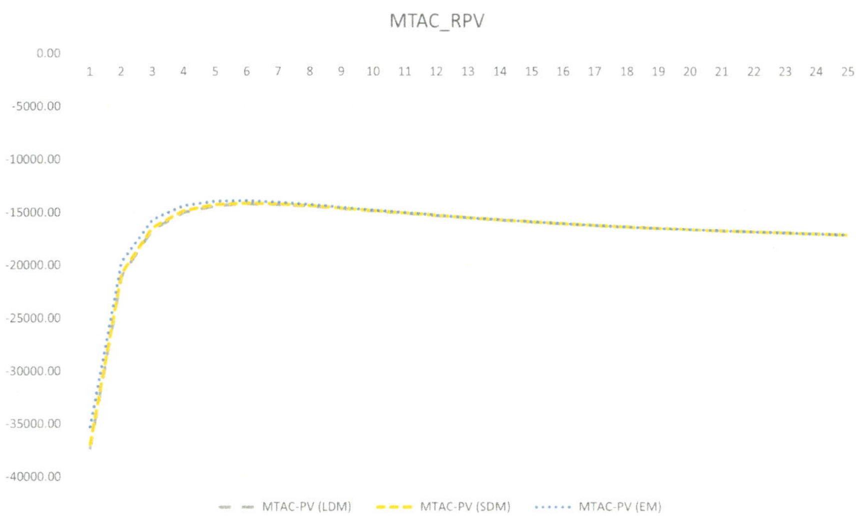

FIGURE 1.12 Curves of minimization of total average cost method reduced to present value.

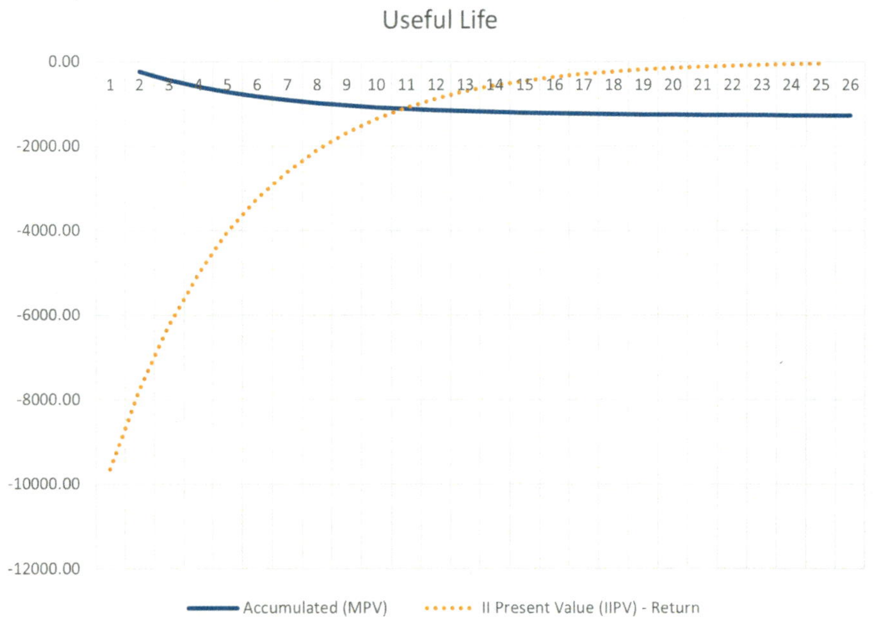

FIGURE 1.13 Useful Life.

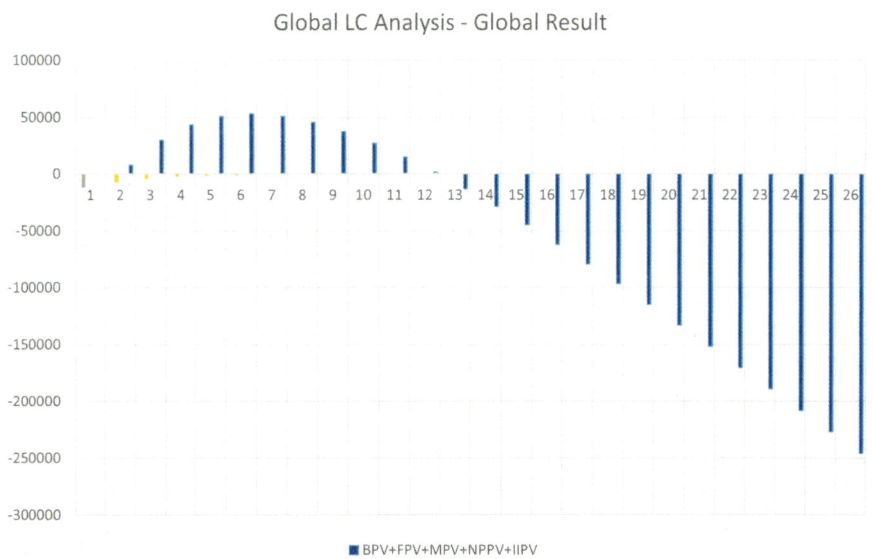

FIGURE 1.14 Global result of the Physical Asset Life Cycle.

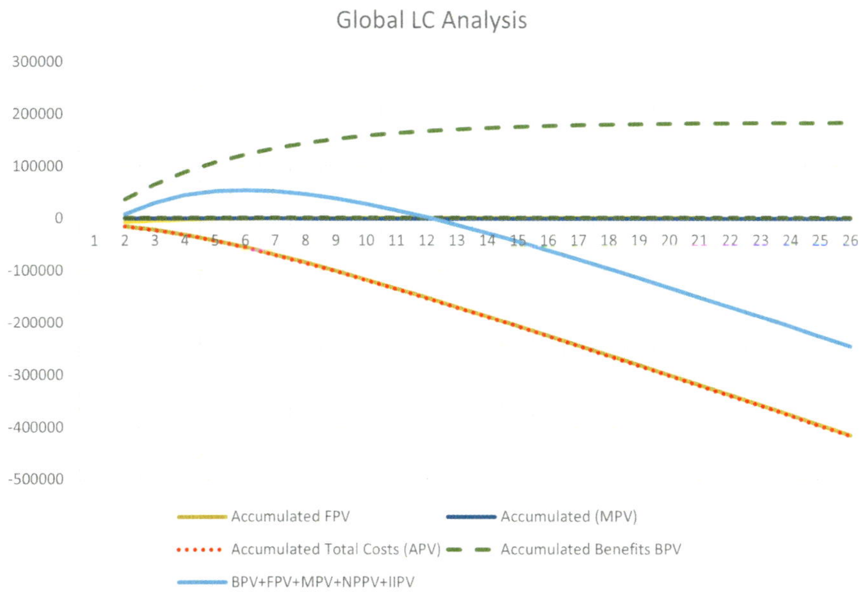

FIGURE 1.15 Global Physical Asset Life Cycle Analysis.

REFERENCES

Farinha, J. M. T. (2018) *Asset Maintenance Engineering Methodologies*. Boca Raton, FL: CRC Press.

Raposo, H. N. (2018) Modelos de Optimização do Tempo de Substituição de Autocarros de Transporte Urbano com Indexação à Frota de Reserva. PhD Thesis. Coimbra University, Portugal

Raposo, H., Farinha, J. T., Ferreira, L., & Galar, D. (2017) An integrated econometric model for bus replacement and determination of reserve fleet size based on predictive maintenance. *Eksploatacja i Niezawodnosc - Maintenance and Reliability*, *19*(3), 358–368. https://doi.org/10.17531/ein.2017.3.6

Raposo, H., Farinha, J. T., Ferreira, L., & Galar, D. (2018) Dimensioning reserve bus fleet using life cycle cost models and condition based/predictive maintenance: a case study. *Public Transport*, *10*(1), 169–190. https://doi.org/10.1007/s12469-017-0167-x

Raposo, H. D. N., Farinha, J. M. T., Ferreira, L. A., & Galar, D. (2019) Reserve fleet indexed to exogenous cost variables. *Transport*, *34*(4), 437–454. https://doi.org/10.3846/transport.2019.11079

2 Life Cycle Investment of Physical Assets

2.1 INTRODUCTION

This chapter presents a model for the Life Cycle of Physical Assets that includes the profits and expenses, called Life Cycle Investment (LCI), because it permits to evaluate the Life Cycle of Physical Asset from the point of view of investment and the respective return (Farinha *et al.*, 2020).

This chapter is structured as follows:

- First, it presents the model of a holistic approach of the Physical Assets Life Cycle;
- Second, it presents the model of the LCI, with emphasis on the relation in the maintenance policies;
- Third, it presents a case study simulation and the respective conclusions.

2.2 THE MODEL OF A HOLISTIC APPROACH OF THE PHYSICAL ASSETS LIFE CYCLE

In the traditional approach of the economic life cycle, under the concept of LCC, the benefits are considered null (Farinha, 2018). The final Global Result (GR_n) in year n can be represented by Equation 2.1:

$$GR_n = \sum_{j=0}^{n} \frac{B_j}{(1+i_A)^j} - \sum_{j=0}^{n} \frac{F_j}{(1+i_A)^j} - \sum_{j=0}^{n} \frac{M_j}{(1+i_A)^j} \tag{2.1}$$

where

i_A is the Apparent Rate

B_j is the value of benefits in year j

F_j is the value of functioning in year j

M_j is the value of maintenance in year j

From Table 2.1 some values stand out, like the following ones:

- Initial Investment (*II*) –12000
- II Present Value (IIPV) – Return –12000
- Functioning (F) –4500
- Energy (E) –200
- Maintenance (M) –300
- Non-Production (NP) 0
- Benefit (B) including IIPV 50000

 DOI: 10.1201/9781003395690-3

TABLE 2.1
Values of Investment and Functioning

Time	0	1	2	3	4	5	6	7	8	9	10	11	12
Initial Investment (*II*)	−12000												
II Present Value (IIPV) - Return	−12000	−9652.51	−7764.25	−6245.37	−5023.63	−4040.88	−3250.39	−2614.53	−2103.07	−1691.66	−1360.73	−1094.54	−880.42
Functioning (*F*)	−4500	−4500.00	−4500.00	−4500.00	−4500.00	−4500.00	−4500.00	−4500.00	−4500.00	−4500.00	−4500.00	−4500.00	−4500.00
Present Value (*PV*) Accumulated		−3619.69	−6531.28	−8873.30	−10757.16	−12272.49	−13491.38	−14471.83	−15260.48	−15894.86	−16405.13	−16815.58	−17145.74
Energy (*E*)	−200	−200.00	−200.00	−200.00	−200.00	−200.00	−200.00	−200.00	−200.00	−200.00	−200.00	−200.00	−200.00
Energy (*PV*)		−160.88	−129.40	−104.09	−83.73	−67.35	−54.17	−43.58	−35.05	−28.19	−22.68	−18.24	−14.67
Accumulated (AFPV)		−15780.57	−6660.69	−15638.07	−26478.96	−38818.79	−52364.35	−66879.76	−82175.29	−98098.34	−114526.15	−131359.97	−148520.38
Maintenance (*M*)	−300	−300.00	−303.00	−306.03	−309.09	−312.18	−315.30	−318.46	−321.64	−324.86	−328.11	−331.39	−334.70
Present Value (*PV*)		−241.31	−194.11	−156.13	−125.59	−101.02	−81.26	−65.36	−52.58	−42.29	−34.02	−27.36	−22.01
Accumulated (AMPV)	−5000	−241.31	−435.42	−591.55	−717.14	−818.17	−899.43	−964.79	−1017.37	−1059.66	−1093.68	−1121.04	−1143.05
Non-Production (NP)	0	0.00	0.00	0.00	0.00	0.00	0.00	0.00	0.00	0.00	0.00	0.00	0.00
Present Value (PV)		0.00	0.00	0.00	0.00	0.00	0.00	0.00	0.00	0.00	0.00	0.00	0.00
Accumulated (ANPPV)		0.00	0.00	0.00	0.00	0.00	0.00	0.00	0.00	0.00	0.00	0.00	0.00
Accumulated Total Costs (ATCPV)		−16021.88	−7096.11	−16229.63	−27196.10	−39636.96	−53263.78	−67844.55	−83192.66	−99158.00	−115619.82	−132481.01	−149663.43
Benefit (B) including IIPV	50000	34990.35	36878.61	38397.49	39619.23	40601.97	41392.47	42028.32	42539.79	42951.20	43282.13	43548.32	43762.44
Present Value (PV)		28145.39	23861.22	19983.88	16586.01	13672.32	11211.80	9157.04	7455.34	6054.89	4907.93	3972.10	3210.77
Accumulated Benefits (ABPV)		28145.39	52006.60	71990.48	88576.50	102248.82	113460.62	122617.65	130072.99	136127.88	141035.82	145007.92	148218.69
ATCPV + ABPV		12123.51	44910.50	55760.86	61380.40	62611.86	60196.84	54773.11	46880.33	36969.88	25415.99	12526.91	−1444.74

(Continued)

TABLE 2.1 (*Continued*)
Values of Investment and Functioning

Time	13	14	15	16	17	18	19	20	21	22	23	24	25
Initial Investment (*II*)													
II Present Value (IIPV) - Return	−708.19	−569.65	−458.21	−368.57	−296.47	−238.48	−191.82	−154.30	−124.11	−99.83	−80.30	−64.59	−51.96
Functioning (*F*)	−4500.00	−4500.00	−4500.00	−4500.00	−4500.00	−4500.00	−4500.00	−4500.00	−4500.00	−4500.00	−4500.00	−4500.00	−4500.00
Present Value (*PV*) Accumulated	−17411.31	−17624.92	−17796.75	−17934.97	−18046.15	−18135.57	−18207.51	−18265.37	−18311.91	−18349.35	−18379.47	−18403.69	−18423.17
Energy (*E*)	−200.00	−200.00	−200.00	−200.00	−200.00	−200.00	−200.00	−200.00	−200.00	−200.00	−200.00	−200.00	−200.00
Energy (*PV*)	−11.80	−9.49	−7.64	−6.14	−4.94	−3.97	−3.20	−2.57	−2.07	−1.66	−1.34	−1.08	−0.87
Accumulated (AFPV)	−165943.49	−183577.91	−201382.30	−219323.41	−237374.50	−255514.05	−273724.76	−291992.70	−310306.68	−328657.69	−347038.50	−365443.26	−383867.30
Maintenance (*M*)	−338.05	−341.43	−344.84	−348.29	−351.77	−355.29	−358.84	−362.43	−366.06	−369.72	−373.41	−377.15	−380.92
Present Value (*PV*)	−17.70	−14.24	−11.46	−9.21	−7.41	−5.96	−4.80	−3.86	−3.10	−2.50	−2.01	−1.61	−1.30
Accumulated (AMPV)	−1160.75	−1174.99	−1186.45	−1195.66	−1203.08	−1209.04	−1213.83	−1217.69	−1220.79	−1223.29	−1225.30	−1226.91	−1228.21
Non-Production (NP)	0.00	0.00	0.00	0.00	0.00	0.00	0.00	0.00	0.00	0.00	0.00	0.00	0.00
Present Value (PV)	0.00	0.00	0.00	0.00	0.00	0.00	0.00	0.00	0.00	0.00	0.00	0.00	0.00
Accumulated (ANPPV)	0.00	0.00	0.00	0.00	0.00	0.00	0.00	0.00	0.00	0.00	0.00	0.00	0.00
Accumulated Total Costs (ATCPV)	−167104.24	−184752.90	−202568.75	−220519.08	−238577.58	−256723.09	−274938.59	−293210.39	−311527.47	−329880.98	−348263.80	−366670.18	−385095.51
Benefit (B) including IIPV	43934.67	44073.21	44184.65	44274.28	44346.38	44404.38	44451.03	44488.56	44518.74	44543.02	44562.55	44578.26	44590.90
Present Value (PV)	2592.83	2092.19	1687.16	1359.86	1095.62	882.45	710.56	572.04	460.45	370.58	298.21	239.96	193.07
Accumulated Benefits (ABPV)	150811.52	152903.71	154590.87	155950.73	157046.36	157928.80	158639.37	159211.41	159671.86	160042.43	160340.65	160580.61	160773.68
ATCPV + ABPV	−16292.72	−31849.19	−47977.88	−64568.34	−81531.22	−98794.29	−116299.22	−133998.98	−151855.62	−169838.55	−187923.15	−206089.57	−224321.83

- Internal Rate of Return (r) 0.12
- Risk (R) 0.11
- Apparent Rate 0.2432
- Maintenance Increase Rate 0.01
- Availability 0.8929
- MTBF 500
- MTTR 50
- MWT 10

Figures 2.1–2.3 show the Useful Life, a Global LCI analysis, and a Global Economical Result, respectively.

The preceding presents a global approach to making the balance between expenses and profits. However, it must be considered the influence of other variables, like the maintenance policies, and the non-production among others. These aspects are discussed in the next section.

2.3 LCI CONSIDERING MAINTENANCE AND NON-PRODUCTION

The investment and costs in a general way mean different things: for example, the initial cost of a physical asset is called Investment, but the remaining values along its Life Cycle are usually called Costs. It is important to clarify these concepts, naming the initial cost as the initial investment and naming the remaining values along time

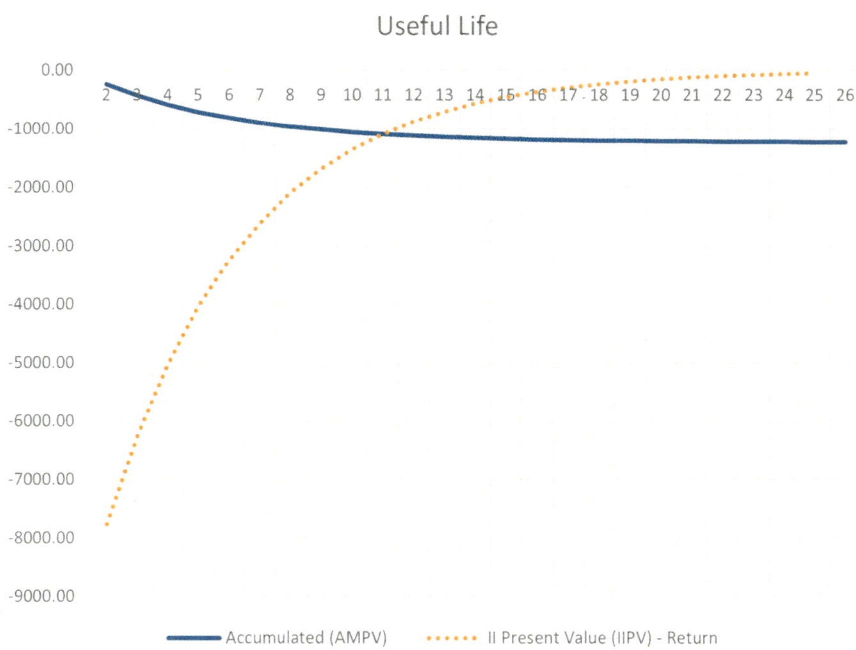

FIGURE 2.1 Useful Life.

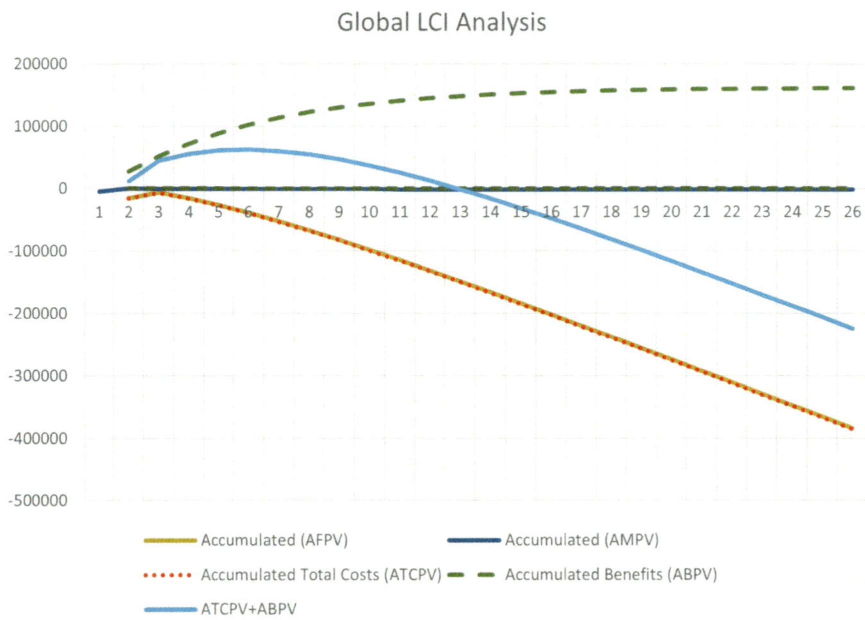

FIGURE 2.2 Global LCI analysis.

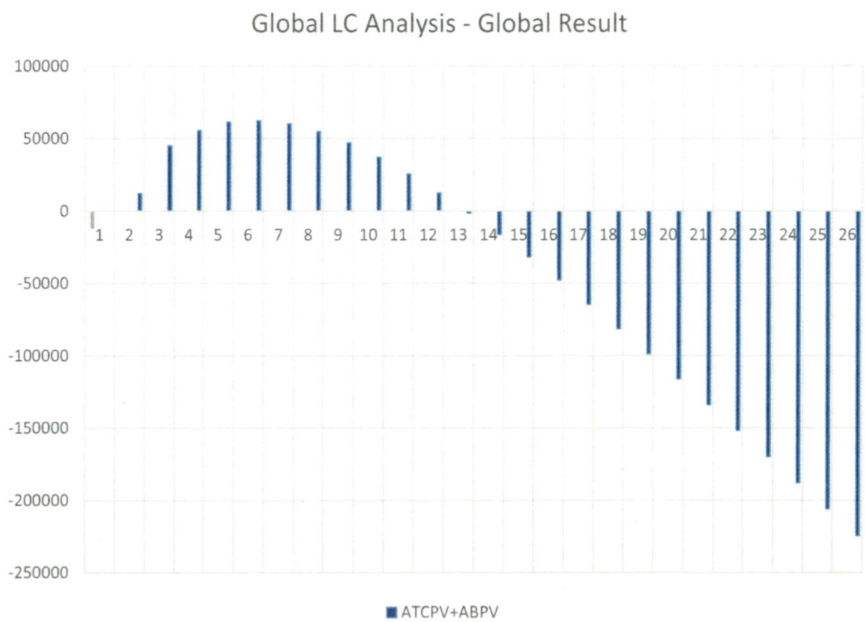

FIGURE 2.3 Global Result of LCI analysis.

(Life Cycle) as variable investments. This, in turn, suggests the need to change the acronym LCC to LCI if the asset is used for production objectives.

Some maintenance KPIs must be considered in industrial production companies because the expected productivity of the equipment depends on them, namely the Availability.

The profits (benefits) are directly related to Availability: The benefits must be calculated considering that the equipment need downtime for maintenance, and the downtime duration is directly related to the maintenance policy, measured through the Mean Time To Repair (MTTR) - the lower MTTR the maximum Availability.

Additionally, it must be taken into account the non-production time because of the downtimes and, in particular the ones caused by maintenance. These stopping times, in many companies, correspond to the most expensive because of the loss of production.

The variable that measures the maintenance downtime is the Time To Repair (TTR), generally evaluated by its mean, i.e., the MTTR. The time of good functioning is usually called the Time Between Failures (TBF), usually evaluated by its mean, i.e., the Mean Time Between Failures (MTBF). Based on these two KPIs, the Availability can be calculated using Equation 2.2:

$$A = \frac{MTBF}{MTTR + MTBF} \tag{2.2}$$

The Availability at time t is given by Equation 2.3.

$$A_t = \frac{MTBF_t}{MWT_t + MTTR_t} \tag{2.3}$$

However, when there is a fault, there may be a time interval between the fault communication and the technician intervention. This time is usually called Waiting Time (WT), generally evaluated by its mean, i.e., Mean Waiting Time (MWT). If this variable is considered in Equation 2.3, the new Availability evaluation, at time t, is done by Equation 2.4:

$$A_t = \frac{MTBF_t}{MWT_t + MTTR_t + MTBF_t} \tag{2.4}$$

Under this perspective, Equation 2.1 can be upgraded to include Availability, which is multiplied by the total benefits (B_t), because the useful time for production only happens when the physical asset has Availability. Equation 2.5 includes this new variable:

$$GR_n = \sum_{t=0}^{n} \frac{B_t * A_t}{(1+i_t)^t} - \sum_{t=0}^{n} \frac{F_t}{(1+i_t)^t} - \sum_{t=0}^{n} \frac{M_t}{(1+i_t)^t} - \sum_{t=0}^{n} \frac{N_t}{(1+i_t)^t} + \sum_{t=0}^{n} \frac{I_t}{(1+i_t)^t} \tag{2.5}$$

where

- B_t is the value of benefit in year t
- A_t is the availability in year t
- F_t is the value of functioning in year t
- M_t is the value of maintenance in year t
- N_t is the value of non-production in year t
- I_t is the value of the physical asset in year t

If variable A_t from Equation 2.4 is inserted into Equation 2.5, this results in Equation 2.6:

$$GR_n = \sum_{t=0}^{n} \frac{B_t * \dfrac{MTBF_t}{MWT_t + MTTR_t + MTBF_t}}{(1+i_t)^t} + \sum_{t=0}^{n} \frac{F_t}{(1+i_t)^t} + \sum_{t=0}^{n} \frac{M_t}{(1+i_t)^t} + \sum_{t=0}^{n} \frac{N_t}{(1+i_t)^t} + \sum_{t=0}^{n} \frac{I_t}{(1+i_t)^t} \tag{2.6}$$

In addition, if we consider the values of Non-Production related to the unavailability of the physical asset, the variable N_j may be evaluated using Equation 2.7:

$$N_t = B_t * (1 - A_t) = B_t * \left(1 - \frac{MTBF_t}{MWT_t + MTTR_t + MTBF_t}\right) \tag{2.7}$$

By substituting the variable N_t of Equation 2.7 into Equation 2.6, we get Equation 2.8.

$$GR_n = \sum_{t=0}^{n} \frac{B_t * \dfrac{MTBF_t}{MWT_t + MTTR_t + MTBF_t}}{(1+i_t)^t} + \sum_{t=0}^{n} \frac{F_t}{(1+i_t)^t} + \sum_{t=0}^{n} \frac{M_t}{(1+i_t)^t} + \sum_{t=0}^{n} \frac{B_t * \left(1 - \dfrac{MTBF_t}{MWT_t + MTTR_t + MTBF_t}\right)}{(1+i_t)^t} + \sum_{t=0}^{n} \frac{I_t}{(1+i_t)^t} \tag{2.8}$$

Equation 2.8 includes the initial investment and the variable maintenance annual investments along the Physical Asset's Life Cycle, giving the global result that a company may expect from them from an Investment perspective.

2.4 CASE STUDY SIMULATION

This section uses the model presented in the previous sections to demonstrate the importance of Availability (based on the maintenance policy) and the Non-Production in the Physical Asset LCI.

2.4.1 Case Study Simulation Considering the Availability

The first simulation considers a maintenance policy that has a lower Availability than what was considered in Table 2.1 ($A = 0.8929$). The new values considered are the following:

- Availability 0.8065
- MTBF 500
- MTTR 100
- MWT 20

TABLE 2.2
Simulation of the LCI with an Availability of 0.7463

Time	0	1	2	3	4	5	6	7	8	9	10	11	12
Initial Investment (*II*)	−12000												
II Present Value (IIPV) – Return	−12000	−9652.51	−7764.25	−6245.37	−5023.63	−4040.88	−3250.39	−2614.53	−2103.07	−1691.66	−1360.73	−1094.54	−880.42
Functioning (*F*)	−4500	−4500.00	−4500.00	−4500.00	−4500.00	−4500.00	−4500.00	−4500.00	−4500.00	−4500.00	−4500.00	−4500.00	−4500.00
Present Value (*PV*) Accumulated		−3619.69	−6531.28	−8873.30	−10757.16	−12272.49	−13491.38	−14471.83	−15260.48	−15894.86	−16405.13	−16815.58	−17145.74
Energy (*E*)	−200	−200.00	−200.00	−200.00	−200.00	−200.00	−200.00	−200.00	−200.00	−200.00	−200.00	−200.00	−200.00
Energy (*PV*)		−160.88	−129.40	−104.09	−83.73	−67.35	−54.17	−43.58	−35.05	−28.19	−22.68	−18.24	−14.67
Accumulated (AFPV)		−15780.57	−6660.69	−15638.07	−26478.96	−38818.79	−52364.35	−66879.76	−82175.29	−98098.34	−114526.15	−131359.97	−148520.38
Maintenance (*M*)	−300	−300.00	−303.00	−306.03	−309.09	−312.18	−315.30	−318.46	−321.64	−324.86	−328.11	−331.39	−334.70
Present Value (*PV*)		−241.31	−194.11	−156.13	−125.59	−101.02	−81.26	−65.36	−52.58	−42.29	−34.02	−27.36	−22.01
Accumulated (AMPV)	−5000	−241.31	−435.42	−591.55	−717.14	−818.17	−899.43	−964.79	−1017.37	−1059.66	−1093.68	−1121.04	−1143.05
Non-Production (NP)	0	0.00	0.00	0.00	0.00	0.00	0.00	0.00	0.00	0.00	0.00	0.00	0.00
Present Value (PV)		0.00	0.00	0.00	0.00	0.00	0.00	0.00	0.00	0.00	0.00	0.00	0.00
Accumulated (ANPPV)		0.00	0.00	0.00	0.00	0.00	0.00	0.00	0.00	0.00	0.00	0.00	0.00
Accumulated Total Costs (ATCPV)		−16021.88	−7096.11	−16229.63	−27196.10	−39636.96	−53263.78	−67844.55	−83192.66	−99158.00	−115619.82	−132481.01	−149663.43
Benefit (B) including IIPV	50000	30670.07	32558.34	34077.21	35298.96	36281.70	37072.19	37708.05	38219.51	38630.92	38961.85	39228.04	39442.16
Present Value (PV)		24670.26	21065.91	17735.40	14777.39	12217.51	10041.58	8215.75	6698.18	5445.86	4418.04	3578.04	2893.80
Accumulated Benefits (ABPV)		24670.26	45736.17	63471.57	78248.97	90466.47	100508.06	108723.80	115421.99	120867.84	125285.88	128863.93	131757.73
ATCPV + ABPV		8648.38	38640.07	47241.95	51052.87	50829.51	47244.28	40879.26	32229.33	21709.85	9666.06	−3617.08	−17905.70

(Continued)

TABLE 2.2 (*Continued*)
Simulation of the LCI with an Availability of 0.7463

Time	13	14	15	16	17	18	19	20	21	22	23	24	25
Initial Investment (*II*)													
II Present Value (IIPV) – Return	−708.19	−569.65	−458.21	−368.57	−296.47	−238.48	−191.82	−154.30	−124.11	−99.83	−80.30	−64.59	−51.96
Functioning (*F*)	−4500.00	−4500.00	−4500.00	−4500.00	−4500.00	−4500.00	−4500.00	−4500.00	−4500.00	−4500.00	−4500.00	−4500.00	−4500.00
Present Value (*PV*) Accumulated	−17411.31	−17624.92	−17796.75	−17934.97	−18046.15	−18135.57	−18207.51	−18265.37	−18311.91	−18349.35	−18379.47	−18403.69	−18423.17
Energy (*E*)	−200.00	−200.00	−200.00	−200.00	−200.00	−200.00	−200.00	−200.00	−200.00	−200.00	−200.00	−200.00	−200.00
Energy (*PV*)	−11.80	−9.49	−7.64	−6.14	−4.94	−3.97	−3.20	−2.57	−2.07	−1.66	−1.34	−1.08	−0.87
Accumulated (AFPV)	−165943.49	−183577.91	−201382.30	−219323.41	−237374.50	−255514.05	−273724.76	−291992.70	−310306.68	−328657.69	−347038.50	−365443.26	−383867.30
Maintenance (*M*)	−338.05	−341.43	−344.84	−348.29	−351.77	−355.29	−358.84	−362.43	−366.06	−369.72	−373.41	−377.15	−380.92
Present Value (*PV*)	−17.70	−14.24	−11.46	−9.21	−7.41	−5.96	−4.80	−3.86	−3.10	−2.50	−2.01	−1.61	−1.30
Accumulated (AMPV)	−1160.75	−1174.99	−1186.45	−1195.66	−1203.08	−1209.04	−1213.83	−1217.69	−1220.79	−1223.29	−1225.30	−1226.91	−1228.21
Non-Production (NP)	0.00	0.00	0.00	0.00	0.00	0.00	0.00	0.00	0.00	0.00	0.00	0.00	0.00
Present Value (PV)	0.00	0.00	0.00	0.00	0.00	0.00	0.00	0.00	0.00	0.00	0.00	0.00	0.00
Accumulated (ANPPV)	0.00	0.00	0.00	0.00	0.00	0.00	0.00	0.00	0.00	0.00	0.00	0.00	0.00
Accumulated Total Costs (ATCPV)	−167104.24	−184752.90	−202568.75	−220519.08	−238577.58	−256723.09	−274938.59	−293210.39	−311527.47	−329880.98	−348263.80	−366670.18	−385095.51
Benefit (B) including IIPV	39614.39	39752.93	39864.37	39954.01	40026.11	40084.11	40130.76	40168.28	40198.47	40222.75	40242.28	40257.99	40270.62
Present Value (PV)	2337.87	1887.10	1522.19	1227.17	988.89	796.59	641.50	516.49	415.77	334.63	269.30	216.70	174.37
Accumulated Benefits (ABPV)	134095.60	135982.70	137504.89	138732.06	139720.95	140517.54	141159.04	141675.53	142091.30	142425.93	142695.23	142911.94	143086.30
ATCPV + ABPV	−33008.65	−48770.21	−65063.86	−81787.02	−98856.63	−116205.55	−133779.55	−151534.86	−169436.18	−187455.06	−205568.56	−223758.24	−242009.21

Based on these values, the new data simulation is presented in Table 2.2. Figures 2.4–2.6 show the Useful Life, a Global LCI analysis, and a Global Economical Result, respectively.

As can be seen from Figures 2.5 and 2.6, with this new value of Availability, lower than what is used in Table 2.1, the positive results of the Life Cycle of the Physical Asset diminish a lot, being all the positive results under 50000.

2.4.2 Case Study Simulation Considering the Non-Production

This section uses the model presented in the previous sections to demonstrate the importance of Non-Production in the Physical Asset LCI.

The first simulation considers a maintenance policy that has a loss of production, with the following values of reference:

- Availability 0.8065
- MTBF 500
- MTTR 100
- MWT 20
- Non-Production –3000

Based on these values, the new data simulation is presented in Table 2.3. Figures 2.7–2.9 show the Useful Life, a Global LCI analysis, and a Global Economical Result, respectively.

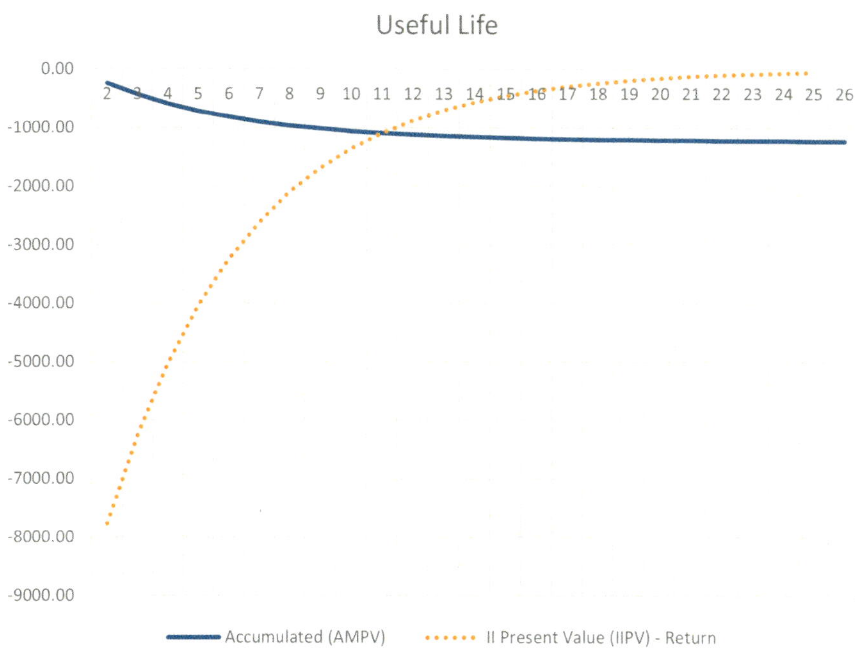

FIGURE 2.4 Useful Life.

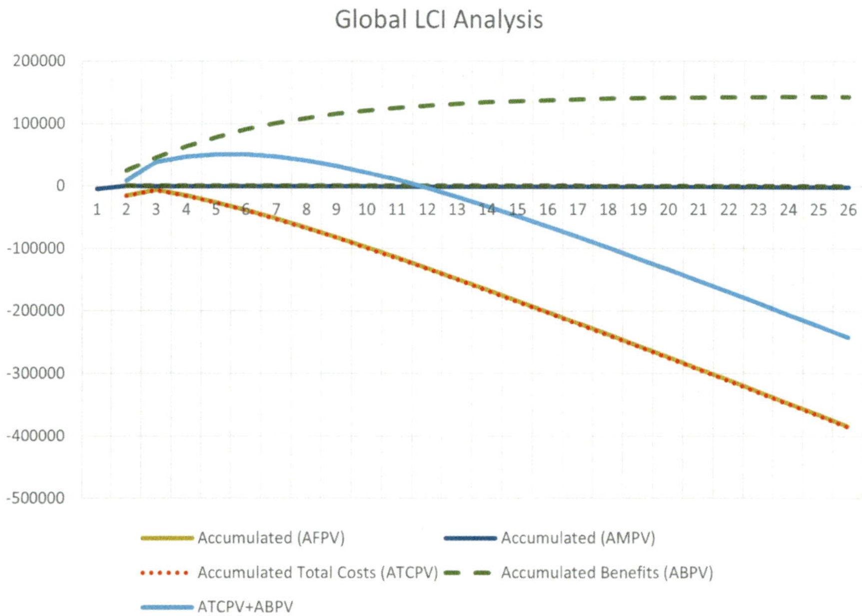

FIGURE 2.5 Global LCI analysis.

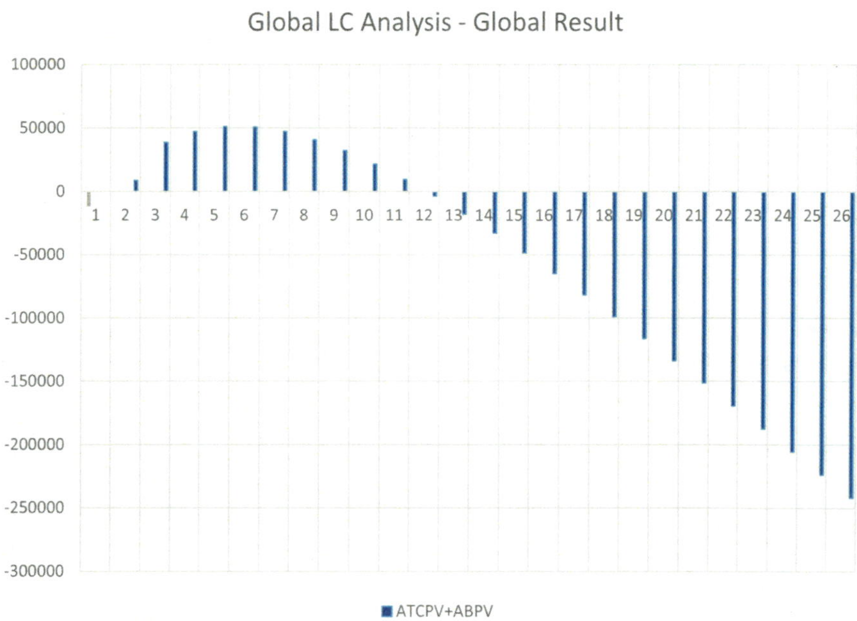

FIGURE 2.6 Global Result of LCI analysis.

TABLE 2.3
Simulation of the LCI Considering Non-Production Values

Time	0	1	2	3	4	5	6	7	8	9	10	11	12
Initial Investment (*II*)	−12000												
II Present Value (IIPV) - Return	−12000	−9652.51	−7764.25	−6245.37	−5023.63	−4040.88	−3250.39	−2614.53	−2103.07	−1691.66	−1360.73	−1094.54	−880.42
Functioning (*F*)	−4500	−4500.00	−4500.00	−4500.00	−4500.00	−4500.00	−4500.00	−4500.00	−4500.00	−4500.00	−4500.00	−4500.00	−4500.00
Present Value (*PV*) Accumulated		−3619.69	−6531.28	−8873.30	−10757.16	−12272.49	−13491.38	−14471.83	−15260.48	−15894.86	−16405.13	−16815.58	−17145.74
Energy (*E*)	−200	−200.00	−200.00	−200.00	−200.00	−200.00	−200.00	−200.00	−200.00	−200.00	−200.00	−200.00	−200.00
Energy (*PV*)		−160.88	−129.40	−104.09	−83.73	−67.35	−54.17	−43.58	−35.05	−28.19	−22.68	−18.24	−14.67
Accumulated (AFPV)		−15780.57	−6660.69	−15638.07	−26478.96	−38818.79	−52364.35	−66879.76	−82175.29	−98098.34	−114526.15	−131359.97	−148520.38
Maintenance (*M*)	−300	−300.00	−303.00	−306.03	−309.09	−312.18	−315.30	−318.46	−321.64	−324.86	−328.11	−331.39	−334.70
Present Value (*PV*)		−241.31	−194.11	−156.13	−125.59	−101.02	−81.26	−65.36	−52.58	−42.29	−34.02	−27.36	−22.01
Accumulated (AMPV)	−5000	−241.31	−435.42	−591.55	−717.14	−818.17	−899.43	−964.79	−1017.37	−1059.66	−1093.68	−1121.04	−1143.05
Non-Production (NP)	−3000	−3000.00	−3030.00	−3060.30	−3090.90	−3121.81	−3153.03	−3184.56	−3216.41	−3248.57	−3281.06	−3313.87	−3347.01
Present Value (PV)		−2413.13	−1960.47	−1592.73	−1293.96	−1051.24	−854.05	−693.85	−563.69	−457.96	−372.05	−302.26	−245.56
Accumulated (ANPPV)		−2413.13	−4373.60	−5966.33	−7260.29	−8311.53	−9165.57	−9859.42	−10423.11	−10881.07	−11253.12	−11555.38	−11800.95
Accumulated Total Costs (ATCPV)		−18435.01	−11469.71	−22195.95	−34456.39	−47948.49	−62429.35	−77703.97	−93615.77	−110039.07	−126872.94	−144036.39	−161464.38
Benefit (B) including IIPV	50000	30670.07	32558.34	34077.21	35298.96	36281.70	37072.19	37708.05	38219.51	38630.92	38961.85	39228.04	39442.16
Present Value (PV)		24670.26	21065.91	17735.40	14777.39	12217.51	10041.58	8215.75	6698.18	5445.86	4418.04	3578.04	2893.80
Accumulated Benefits (ABPV)		24670.26	45736.17	63471.57	78248.97	90466.47	100508.06	108723.80	115421.99	120867.84	125285.88	128863.93	131757.73
ATCPV + ABPV		6235.26	34266.47	41275.62	43792.58	42517.99	38078.71	31019.84	21806.22	10828.78	−1587.06	−15172.46	−29706.65

(Continued)

TABLE 2.3 (*Continued*)
Simulation of the LCI Considering Non-Production Values

Time	13	14	15	16	17	18	19	20	21	22	23	24	25
Initial Investment (*II*)													
II Present Value (IIPV) - Return	−708.19	−569.65	−458.21	−368.57	−296.47	−238.48	−191.82	−154.30	−124.11	−99.83	−80.30	−64.59	−51.96
Functioning (*F*)	−4500.00	−4500.00	−4500.00	−4500.00	−4500.00	−4500.00	−4500.00	−4500.00	−4500.00	−4500.00	−4500.00	−4500.00	−4500.00
Present Value (*PV*) Accumulated	−17411.31	−17624.92	−17796.75	−17934.97	−18046.15	−18135.57	−18207.51	−18265.37	−18311.91	−18349.35	−18379.47	−18403.69	−18423.17
Energy (*E*)	−200.00	−200.00	−200.00	−200.00	−200.00	−200.00	−200.00	−200.00	−200.00	−200.00	−200.00	−200.00	−200.00
Energy (*PV*)	−11.80	−9.49	−7.64	−6.14	−4.94	−3.97	−3.20	−2.57	−2.07	−1.66	−1.34	−1.08	−0.87
Accumulated (AFPV)	−165943.49	−183577.91	−201382.30	−219323.41	−237374.50	−255514.05	−273724.76	−291992.70	−310306.68	−328657.69	−347038.50	−365443.26	−383867.30
Maintenance (*M*)	−338.05	−341.43	−344.84	−348.29	−351.77	−355.29	−358.84	−362.43	−366.06	−369.72	−373.41	−377.15	−380.92
Present Value (*PV*)	−17.70	−14.24	−11.46	−9.21	−7.41	−5.96	−4.80	−3.86	−3.10	−2.50	−2.01	−1.61	−1.30
Accumulated (AMPV)	−1160.75	−1174.99	−1186.45	−1195.66	−1203.08	−1209.04	−1213.83	−1217.69	−1220.79	−1223.29	−1225.30	−1226.91	−1228.21
Non-Production (NP)	−3380.48	−3414.28	−3448.42	−3482.91	−3517.74	−3552.91	−3588.44	−3624.33	−3660.57	−3697.18	−3734.15	−3771.49	−3809.20
Present Value (PV)	−199.50	−162.08	−131.68	−106.98	−86.91	−70.61	−57.36	−46.60	−37.86	−30.76	−24.99	−20.30	−16.49
Accumulated (ANPPV)	−12000.45	−12162.53	−12294.20	−12401.18	−12488.09	−12558.69	−12616.05	−12662.66	−12700.52	−12731.28	−12756.27	−12776.57	−12793.06
Accumulated Total Costs (ATCPV)	−179104.69	−196915.43	−214862.95	−232920.25	−251065.66	−269281.78	−287554.64	−305873.05	−324227.99	−342612.26	−361020.06	−379446.74	−397888.57
Benefit (B) including IIPV	39614.39	39752.93	39864.37	39954.01	40026.11	40084.11	40130.76	40168.28	40198.47	40222.75	40242.28	40257.99	40270.62
Present Value (PV)	2337.87	1887.10	1522.19	1227.17	988.89	796.59	641.50	516.49	415.77	334.63	269.30	216.70	174.37
Accumulated Benefits (ABPV)	134095.60	135982.70	137504.89	138732.06	139720.95	140517.54	141159.04	141675.53	142091.30	142425.93	142695.23	142911.94	143086.30
ATCPV + ABPV	−45009.09	−60932.73	−77358.06	−94188.19	−111344.72	−128764.24	−146395.61	−164197.52	−182136.70	−200186.33	−218324.83	−236534.81	−254802.27

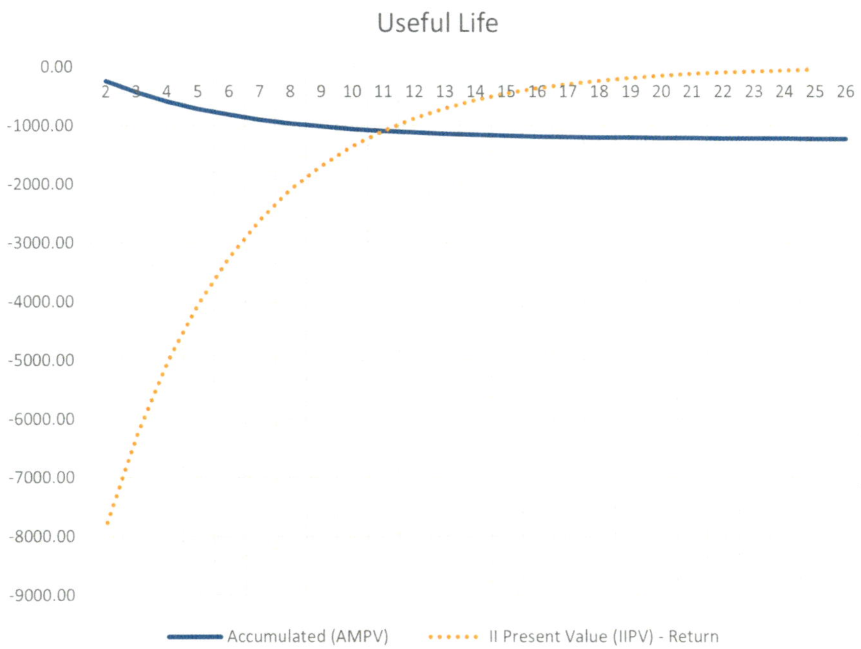

FIGURE 2.7 Useful Life.

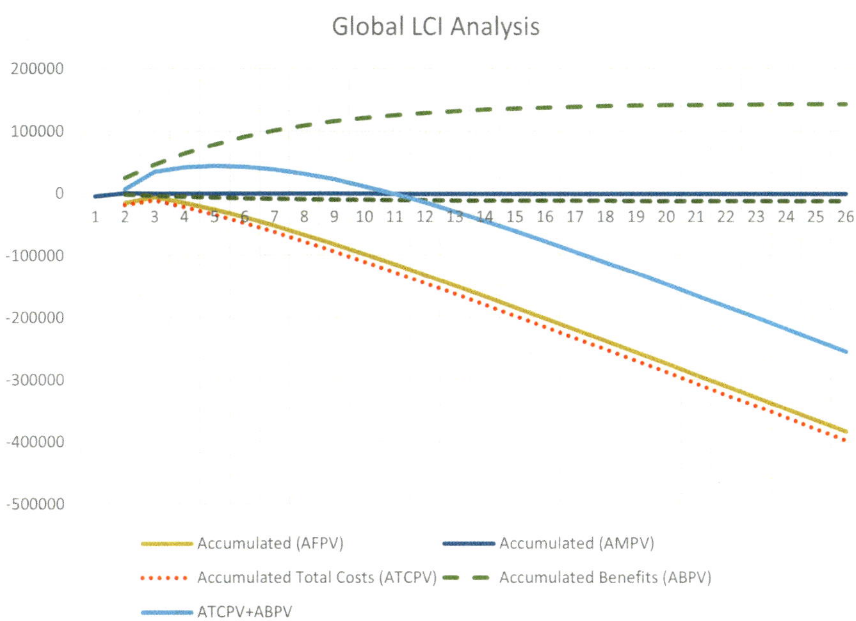

FIGURE 2.8 Global LCI analysis.

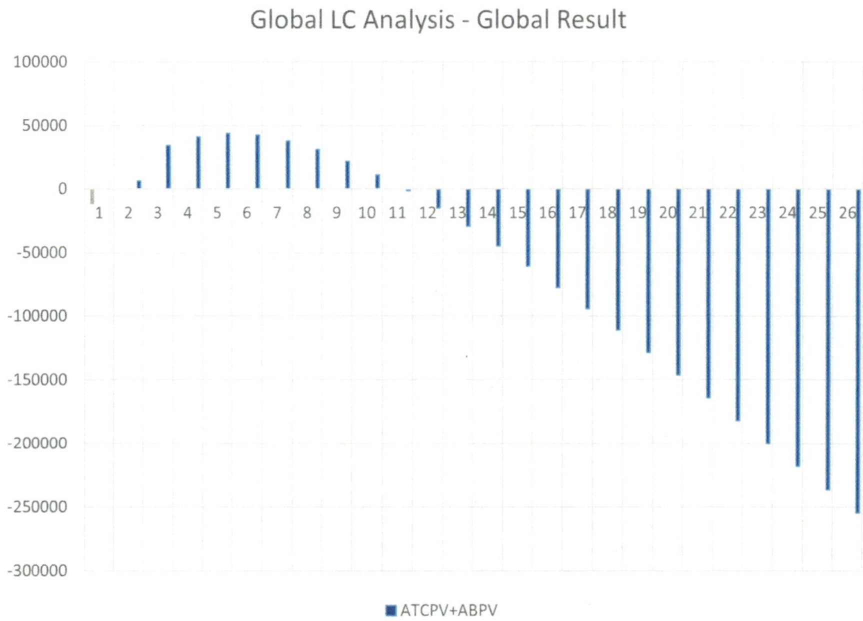

FIGURE 2.9 Global Result of LCI analysis.

As can be seen from Figures 2.8 and 2.9, with the impact of the Non-Production value, the positive results of the Life Cycle of the Physical Asset diminish by one year and the maximum positive results are all under 50000.

2.5 CONCLUSIONS

This chapter presented a Life Cycle model to make this evaluation, aiming to simulate several situations, being considered the Availability and Non-Production, but other variables would be considered.

It is proposed to change the traditional Life Cycle Cost (LCC) analysis concept to the LCI analysis. The objective is to emphasize the importance of all financial movements, namely initial investment, the variable investment related to functioning and maintenance, and also the profits resulting from the production of the Physical Assets along their Life Cycles.

The model presented shows high versatility and ability to be adequate for any type of Physical Asset.

REFERENCES

Farinha, J. T. (2018) *Asset Maintenance Engineering Methodologies*. Boca Raton, FL: CRC Press.

Farinha, J. T., Raposo, H. N., & Galar, D. (2020) Life Cycle Cost versus Life Cycle Investment – A New Approach. In *WSEAS Transactions on Systems and Control* (Vol. 15, pp. 743–753). World Scientific and Engineering Academy and Society (WSEAS). https://doi.org/10.37394/23203.2020.15.75

3 Life Cycle of Physical Assets with Recovery

3.1 INTRODUCTION

The evaluation of the Physical Asset Life Cycle may have complimentary perspectives, as was demonstrated in the previous chapters. Another perspective is to take into account the necessary money at the end of Physical Asset Life Cycle aiming to acquire a new one. This is the approach presented in this chapter, which is structured as follows:

- The next section presents the global methodology;
- Then, it presents the basic concepts to support math models;
- Next, it evaluates the behaviour of the Physical Asset value along time;
- The production economic movements are discussed;
- Next the functioning economic movements;
- Then, the global movements analysis is done;
- The Return On Investment is analyzed;
- Finally, the models' validation and discussion are presented.

3.2 METHODOLOGY

The methodology adopted in this chapter is based on math models that take into account economic variables related to Physical Assets, such as acquisition cost, functioning costs, maintenance costs, inflation rate, interest rate, profit rate and risk.

The preceding variables among others are modelled by equations that translate the Physical Asset Life Cycle functioning.

The methodology is supported in math models with a coherent development for all economic variables that converge in a global model that translates dynamically the Life Cycle Investment (LCI) behaviour, namely the relation between production levels and global functioning expenses taking into account the several rates.

The methodology is validated through the introduction of several situations of financial movements and evaluating the robustness of the results.

3.3 BASIC CONCEPTS TO SUPPORT MATH MODELS

The Life Cycle of Physical Assets analysis is based on the economic variables that permit to evaluate their Life Cycle from the perspective of the Investment Cycle. It is usual to discuss the Life Cycle Cost (LCC) of the Physical Asset. However, when an organization invests in a Physical Asset, it is because it has a perspective of a Return On Investment (ROI) in a realistic time. This means that the organization

DOI: 10.1201/9781003395690-4

is conscious about the negative monetary fluxes, which correspond to the initial investment and the functioning and maintenance expenses, necessary to permit the Physical Asset to have incomes and the final balance is positive, providing good profits. By consequence, the Life Cycle must be evaluated in this holistic perspective, which corresponds to the LCI.

The main variables that must be considered in the Physical Asset Life Cycle are the following ones:

- Acquisition cost;
- Functioning costs;
- Maintenance costs;
- Inflation rate;
- Interest rate;
- Risk rate;
- Profit rate.

Some of these variables can be decomposed into other variables, namely the functioning and maintenance costs. The remaining variables interact with the preceding and, among them, as is described next.

This section starts describing the model, step by step.

The main structural variables that influence the models are the inflation, interest, and profit rates. The relationship between the real interest and inflation rates can be described using Fisher's equation (Equation 3.1).

$$(1+i_A)=(1+r)\times(1+h)\times(1+p) \tag{3.1}$$

where

i_A is the Apparent Rate
r is the Interest Rate
h is the Inflation Rate
p is the Profit Rate

Inflation and interest rates are inversely related; it means that a higher inflation rate has a lower interest rate and vice versa. Using a higher Apparent Rate (i_A) means a lower present value of a cash flow. It means that when future cash flows are discounted with a higher i_A, the Net Present Value (NPV) of the future cash flows decreases.

Taking into account the Apparent Rate (i_A), all movements, incomes, and expenses must be actualized to their NPV, which represents the sum of all future cash flows over the entire life of a Physical Asset discounted to the present day. NPV is the estimate of the worth of future cash flows in today's terms. It tells how much the future flows will be worth if received today (Equation 3.2).

$$NPV=\sum_{t=1}^{N}\frac{F_t}{(1+i_A)^t} \tag{3.2}$$

where

NPV is the Net Present Value
F_t is the Financial Movement
i_A is the Apparent Rate

If we take into account the Initial Investment on Physical Asset (C_0) and if F_t corresponds to the profits, it must be discounted the value of C_0 in Equation 3.2, originating Equation 3.3.

$$NPV = \sum_{t=1}^{N} \frac{F_t}{(1+i_A)^t} - C_0 \quad (3.3)$$

3.4 BEHAVIOUR OF THE PHYSICAL ASSET VALUE ALONG TIME

The value of the Physical Asset along time may be seen from two perspectives: The devaluation, taking into account the Apparent Rate. The value of a new one takes into account the Apparent Rate.

The preceding perspectives are opposites and are evaluated as follows:

- The devaluation of the Physical Asset taking into account the Apparent Rate (Equation 3.4)

$$C_{0D}^{t} = C_{0D}^{t-1} \times i_A \quad (3.4)$$

where

C_{0D}^{t} is the value of the Physical Asset in time t
C_{0D}^{t-1} is the value of the Physical Asset in time $t-1$
i_A is the Apparent Rate

- The value of a new Physical Asset taking into account the Apparent Rate (Equation 3.5)

$$C_{0N}^{t} = C_{0}^{t-1} / i_A \quad (3.5)$$

Where,

C_{0N}^{t} is the value of the Physical Asset in time t
C_{0N}^{t-1} is the value of the Physical Asset in time $t-1$
i_A is the Apparent Rate

3.5 PRODUCTION ECONOMIC MOVEMENTS

Physical Assets are used to produce services and goods and, by consequence, the values of the production movements must be actualized through NPV using the value of the Initial Investment (C_0) – (Equation 3.3).

However, the value of the Physical Asset may also be evaluated considering what was discussed in the preceding section. Then, instead of considering C_0, the value with devaluation and the value of a new one may be considered.

- The calculation of NPV of the Financial Production (F_p) movements considering the devaluation of the Physical Asset taking into account the Apparent Rate is given by Equation 3.6.

$$NPV_{FpD} = \sum_{t=1}^{N} \frac{F_P}{(1+i_A)^t} - C_{0D}^t \tag{3.6}$$

where

NPV_{FpD}	is the Net Present Value of the Financial Production movement
F_P	is the Financial Production movement
i_A	is the Apparent Rate
C_{0D}^t	is the devaluation of the Physical Asset taking into account the Apparent Rate.

- The calculation of NPV of the Financial Production (F_p) movements considering the value of a new Physical Asset taking into account the Apparent Rate is given by Equation 3.7.

$$NPV_{FpN} = \sum_{t=1}^{N} \frac{F_P}{(1+i_A)^t} - C_{0N}^t \tag{3.7}$$

where

NPV_{FPN}	is the Net Present Value of the Financial Production movement
F_P	is the Financial Production movement
i_A	is the Apparent Rate
C_{0N}^t	is the value of a new Physical Asset taking into account the Apparent Rate.

The production is strongly dependent on the Physical Assets Availability, which can be calculated using Equation 3.8.

$$A = \frac{MTBF}{MTTR + MTBF} \tag{3.8}$$

where

A	is the Availability
$MTBF$	is the Mean Time Between Failures
$MTTR$	is the Mean Time To Repair

If we consider the Availability at each time instant (A_t), Equation 3.8 transforms into Equation 3.9.

$$A_t = \frac{MTBF_t}{MTTR_t + MTBF_t} \tag{3.9}$$

The Production movements (F_P) indexed to Availability (A) correspond to a product between them ($F_P \times A$) which permits to transform Equations 3.6 and 3.7 into Equations 3.10 and 3.11.

$$NPV_{FpD} = \sum_{t=1}^{N} \frac{F_P \times A}{(1+i_A)^t} - C_{0D}^t \tag{3.10}$$

$$NPV_{FpN} = \sum_{t=1}^{N} \frac{F_P \times A}{(1+i_A)^t} - C_{0N}^t \tag{3.11}$$

3.6 FUNCTIONING ECONOMIC MOVEMENTS

Aiming for the good functioning of the Physical Assets, it implies expenses, namely in energy and maintenance, among others. These financial movements also need to be actualized to their NPV. By consequence, the NPV of global functioning expenses can be evaluated by Equation 3.12.

$$NPV_{Fe} = \sum_{t=1}^{N} \frac{F_e}{(1+i_A)^t} \tag{3.12}$$

where

NPV_{Fe} is the Net Present Value of the Financial expenses movement
F_e is the Financial expense movement
i_A is the Apparent Rate.

3.7 GLOBAL MOVEMENTS ANALYSIS

Considering the approaches made in the last two sections, it can be calculated the Total NPV of investments and expenses and, again, taking into account the two opposite situations of the Physical Assets (Devaluation and New).

- The Total NPV considering the devaluation of the Physical Asset taking into account the Apparent Rate is given by Equation 3.13.

$$NPV_{TD} = NPV_{FpD} + NPV_{Fe} \tag{3.13}$$

  where

  NPV_{TD} is the Total Net Present Value considering Physical Asset Devaluation
  NPV_{FpD} is the Net Present Value of the Financial Production movement
  NPV_{Fe} is the Net Present Value of the Financial expenses movement
- The Total NPV considering the value of a new Physical Asset taking into account the Apparent Rate is given by Equation 3.14.

$$NPV_{TN} = NPV_{FpN} + NPV_{Fe} \tag{3.14}$$

where

NPV_{TN} is the Total Net Present Value considering a New Physical Asset
NPV_{FpN} is the Net Present Value of the Financial Production movement
NPV_{Fe} is the Net Present Value of the Financial expenses movement.

3.8 RETURN ON INVESTMENT

Considering the approaches of the preceding sections it is possible to evaluate the final financial movement results, what permits to see along time the balance among production, functioning and Physical Asset investments, and what represents the dynamics of the Physical Asset LCI.

This final approach permits to define the time when the ROI is reached as well as to permit supporting the decision about a renewal or an acquisition of a new Physical Asset.

Like in the preceding sections, ROI is calculated considering the two opposite situations of the Physical Assets (Devaluation and New).

- The evaluation of the ROI with Physical Asset devaluation is given by Equation 3.15.

$$ROI_D = NPV_{TF} + C_{0N}^t \tag{3.15}$$

- The evaluation of the ROI with a New Physical Asset is given by Equation 3.16.

$$ROI_N = NPV_{TF} + C_{0N}^t \tag{3.16}$$

3.9 MODELS VALIDATION AND DISCUSSION

In this section, we evaluate the robustness of the model presented in the previous section as well as, in their subsections.

We consider two situations very different aiming to evaluate if the model corresponds to what is expected in real situations.

For the first simulation (Table 3.1) the reference values are the following:

- Interest Rate (r) 0.05
- Inflation Rate (h) 0.02
- Profit Rate (p) 0.05
- Maintenance Increase Rate 0.01
- *MTBF* 500.0
- *MTTR* 60.0
- Risk (R) 0.10
- Initial Investment (II) (C_0) –12000
- F_p Financial Movement (Production) **50000**
- Functioning Costs –4500
- Energy –200
- Maintenance Costs –300

TABLE 3.1
Values of Simulation in a Window Time of 25 Years

Time (t)	0	1	2	3	4	5	6	7	8	9	10	11	12
r - real interest rate $(1+r)$		1.05	1.05	1.05	1.05	1.05	1.05	1.05	1.05	1.05	1.05	1.05	1.05
h - inflation rate $(1+h)$		1.02	1.02	1.02	1.02	1.02	1.02	1.02	1.02	1.02	1.02	1.02	1.02
p - profit rate		1.05	1.05	1.05	1.05	1.05	1.05	1.05	1.05	1.05	1.05	1.05	1.05
i - money rate (nominal interest rate)		0.12	0.12	0.12	0.12	0.12	0.12	0.12	0.12	0.12	0.12	0.12	0.12
i - Apparent Rate with Risk		0.24	0.24	0.24	0.24	0.24	0.24	0.24	0.24	0.24	0.24	0.24	0.24
$1/((1+i)\wedge t))$		0.81	0.65	0.53	0.43	0.35	0.28	0.23	0.18	0.15	0.12	0.10	0.08
Initial Investment (II) (C_0)	−12000	−12000	−12000	−12000	−12000	−12000	−12000	−12000	−12000	−12000	−12000	−12000	−12000
Initial Investment (II) (C_0) with Devaluation (NPV)	−12000	−9701	−7842	−6340	−5125	−4143	−3349	−2708	−2189	−1769	−1430	−1156	−935
Initial Investment (II) (C_0) considering New Asset (NPV)	−12000	−14844	−18362	−22714	−28097	−34757	−42994	−53184	−65789	−81381	−100669	−124528	−154042
F_p Financial Movement (Production)	50000	50000	50000	50000	50000	50000	50000	50000	50000	50000	50000	50000	50000
F_p Financial Movement (NPV)	50000	36089	29175	23585	19066	15413	12460	10073	8143	6583	5322	4302	3478
F_p Financial Movement (NPV) Accumulated	50000	36089	65264	88849	107916	123329	135789	145862	154005	160588	165909	170211	173689
Functioning Costs	−4500	−4500	−4500	−4500	−4500	−4500	−4500	−4500	−4500	−4500	−4500	−4500	−4500
Functioning Costs (NPV) Accumulated	−4500	−3638	−6579	−8956	−10878	−12432	−13688	−14703	−15524	−16187	−16724	−17157	−17508
Energy	−200	−200	−200	−200	−200	−200	−200	−200	−200	−200	−200	−200	−200
Energy Accumulated (NPV) Accumulated	−200	−162	−292	−398	−483	−553	−608	−653	−690	−719	−743	−763	−778
Maintenance Costs	−300	−300	−300	−300	−300	−300	−300	−300	−300	−300	−300	−300	−300
Maintenance Costs (NPV) Accumulated	−300	−245	−443	−603	−732	−837	−922	−990	−1045	−1090	−1126	−1155	−1179
Total Costs (NPV) Accumulated	−5000	−4083	−11304	−21063	−32874	−46342	−61152	−77044	−93813	−111291	−129340	−147853	−166740
F_T Net Present Value (NPV) (With II Devaluated)		22306	46118	61446	69917	72844	71288	66110	58003	47528	35138	21202	6015
F_T Net Present Value (With II New Asset)		17163	35598	45072	46944	42230	31643	15633	−5597	−32084	−64100	−102170	−147093

(Continued)

TABLE 3.1 (*Continued*)
Values of Simulation in a Window Time of 25 Years

Time (t)	13	14	15	16	17	18	19	20	21	22	23	24	25
r - real interest rate $(1+r)$	1.05	1.05	1.05	1.05	1.05	1.05	1.05	1.05	1.05	1.05	1.05	1.05	1.05
h - inflation rate $(1+h)$	1.02	1.02	1.02	1.02	1.02	1.02	1.02	1.02	1.02	1.02	1.02	1.02	1.02
p - profit rate	1.05	1.05	1.05	1.05	1.05	1.05	1.05	1.05	1.05	1.05	1.05	1.05	1.05
i - money rate (nominal interest rate)	0.12	0.12	0.12	0.12	0.12	0.12	0.12	0.12	0.12	0.12	0.12	0.12	0.12
i - Apparent Rate with Risk	0.24	0.24	0.24	0.24	0.24	0.24	0.24	0.24	0.24	0.24	0.24	0.24	0.24
$1/((1+i) \wedge t))$	0.06	0.05	0.04	0.03	0.03	0.02	0.02	0.01	0.01	0.01	0.01	0.01	0.00
Initial Investment (II) (C_0)	−12000	−12000	−12000	−12000	−12000	−12000	−12000	−12000	−12000	−12000	−12000	−12000	−12000
Initial Investment (II) (C_0) with Devaluation (NPV)	−756	−611	−494	−399	−323	−261	−211	−171	−138	−111	−90	−73	−59
Initial Investment (II) (C_0) considering New Asset (NPV)	−190551	−235712	−291577	−360682	−446165	−551909	−682714	−844521	−1044676	−1292270	−1598544	−1977407	−2446063
F_p Financial Movement (Production)	50000	50000	50000	50000	50000	50000	50000	50000	50000	50000	50000	50000	50000
F_p Financial Movement (NPV)	2811	2273	1837	1485	1201	971	785	634	513	415	335	271	219
F_p Financial Movement (NPV) Accumulated	176500	178773	180610	182096	183296	184267	185052	185686	186199	186613	186949	187219	187438
Functioning Costs	−4500	−4500	−4500	−4500	−4500	−4500	−4500	−4500	−4500	−4500	−4500	−4500	−4500
Functioning Costs (NPV) Accumulated	−17791	−18020	−18206	−18355	−18476	−18574	−18653	−18717	−18769	−18811	−18844	−18872	−18894
Energy	−200	−200	−200	−200	−200	−200	−200	−200	−200	−200	−200	−200	−200
Energy Accumulated (NPV) Accumulated	−791	−801	−809	−816	−821	−826	−829	−832	−834	−836	−838	−839	−840
Maintenance Costs	−300	−300	−300	−300	−300	−300	−300	−300	−300	−300	−300	−300	−300
Maintenance Costs (NPV) Accumulated	−1198	−1213	−1226	−1236	−1244	−1251	−1256	−1260	−1264	−1267	−1269	−1271	−1272
Total Costs (NPV) Accumulated	−185929	−205362	−224994	−244785	−264705	−284730	−304839	−325017	−345249	−365527	−385840	−406182	−426548
F_T Net Present Value (NPV) (With II Devaluated)	−10184	−27200	−44877	−63089	−81732	−100724	−119998	−139501	−159188	−179025	−198981	−219036	−239169
F_T Net Present Value (With II New Asset)	−199979	−262301	−335960	−423371	−527574	−652372	−802502	−983851	−1203727	−1471183	−1797436	−2196370	−2685173

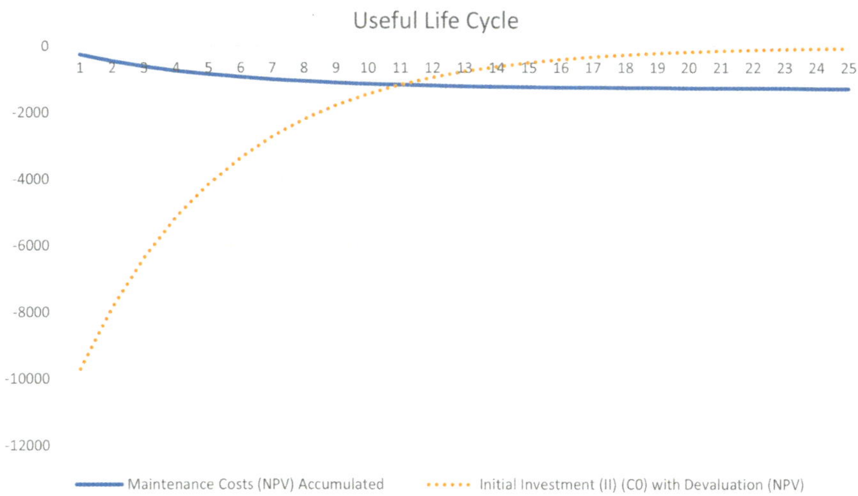

FIGURE 3.1 Useful Life Cycle.

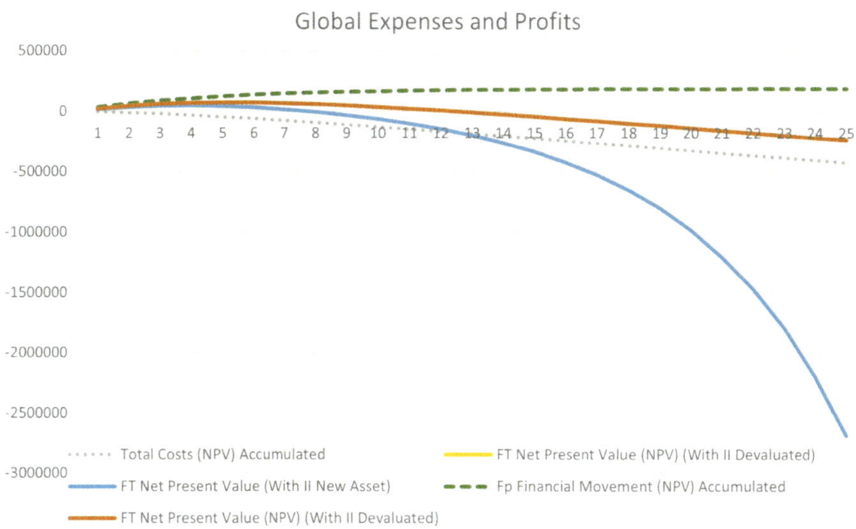

FIGURE 3.2 Global Investment and Profits.

Figure 3.1 shows the Useful Life Cycle; Figure 3.2 shows all economical movements (Expenses and Profits); Figure 3.3 shows the global results taking into account the devaluation of the Physical Asset and the opposite situation the increasing of value along time of a new one.

For the second simulation (Table 3.2) the reference values are the following:

- Interest Rate (r) 0.05
- Inflation Rate (h) 0.02
- Profit Rate (p) 0.05

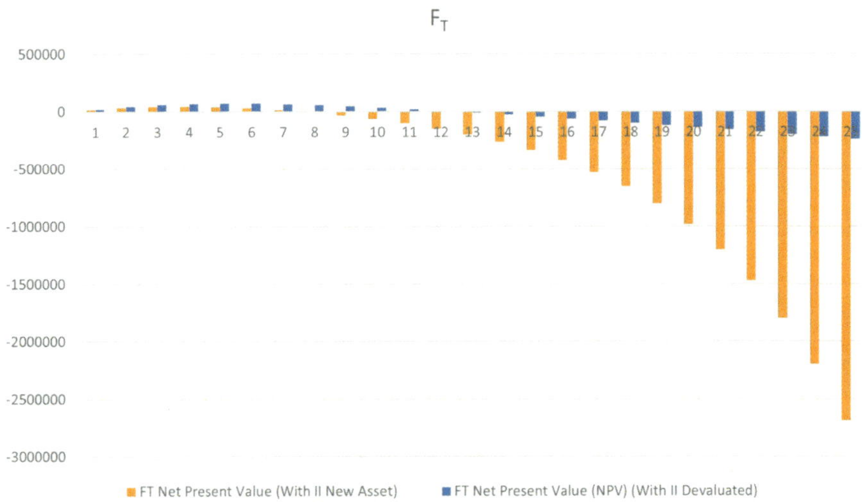

FIGURE 3.3 Final result considering devaluation and recovery of the Physical Asset.

- Maintenance Increase Rate 0.01
- *MTBF* 500.0
- *MTTR* 60.0
- Risk (*R*) 0.10
- Initial Investment (*II*) (C_0) –12000
- F_p Financial Movement (Production) **70000**
- Functioning Costs –4500
- Energy –200
- Maintenance Costs –300

The only change in relation to the first simulation was the increase of production to 70000MU.

Figure 3.4 shows the Useful Life Cycle; Figure 3.5 shows all economical movements (Expenses and Profits); Figure 3.6 shows the global results taking into account the devaluation of the Physical Asset and the opposite situation the increasing of value along time of a new one.

Figure 3.4 shows the Useful Life Cycle; Figure 3.5 shows all economical movements (expenses and profits). Figure 3.6 shows the global results taking into account the devaluation of the Physical Asset and the opposite situation the increasing of value along time of a new one.

According to the preceding simulation, it can be clearly concluded that, with the increase of the production profits, the positive result of the Life Cycle increases in both situations, Devaluations and Recovery: five years for the first situation and two years for the second. However, it can be stated that with the increasing of production (profits), maintaining the remaining variables at the same value, the final results increase, both in maximum value and the Life Cycle (four years).

TABLE 3.2
Values of Simulation in a Window Time of 25 Years

Time (t)	0	1	2	3	4	5	6	7	8	9	10	11	12
r - real interest rate $(1+r)$		1.05	1.05	1.05	1.05	1.05	1.05	1.05	1.05	1.05	1.05	1.05	1.05
h - inflation rate $(1+h)$		1.02	1.02	1.02	1.02	1.02	1.02	1.02	1.02	1.02	1.02	1.02	1.02
p - profit rate		1.05	1.05	1.05	1.05	1.05	1.05	1.05	1.05	1.05	1.05	1.05	1.05
i - money rate (nominal interest rate)		0.12	0.12	0.12	0.12	0.12	0.12	0.12	0.12	0.12	0.12	0.12	0.12
i - Apparent Rate with Risk		0.24	0.24	0.24	0.24	0.24	0.24	0.24	0.24	0.24	0.24	0.24	0.24
$1/((1+i)$ ^ $t))$		0.81	0.65	0.53	0.43	0.35	0.28	0.23	0.18	0.15	0.12	0.10	0.08
Initial Investment (II) (C_0)	−12000	−12000	−12000	−12000	−12000	−12000	−12000	−12000	−12000	−12000	−12000	−12000	−12000
Initial Investment (II) (C_0) with Devaluation (NPV)	−12000	−9701	−7842	−6340	−5125	−4143	−3349	−2708	−2189	−1769	−1430	−1156	−935
Initial Investment (II) (C_0) considering New Asset (NPV)	−12000	−14844	−18362	−22714	−28097	−34757	−42994	−53184	−65789	−81381	−100669	−124528	−154042
F_p Financial Movement (Production)	70000	70000	70000	70000	70000	70000	70000	70000	70000	70000	70000	70000	70000
F_p Financial Movement (NPV)	70000	50525	40845	33019	26693	21579	17444	14102	11400	9216	7450	6023	4869
F_p Financial Movement (NPV) Accumulated	70000	50525	91370	124389	151082	172661	190105	204207	215607	224823	232273	238296	243164
Functioning Costs	−4500	−4500	−4500	−4500	−4500	−4500	−4500	−4500	−4500	−4500	−4500	−4500	−4500
Functioning Costs (NPV) Accumulated	−4500	−3638	−6579	−8956	−10878	−12432	−13688	−14703	−15524	−16187	−16724	−17157	−17508
Energy	−200	−200	−200	−200	−200	−200	−200	−200	−200	−200	−200	−200	−200
Energy Accumulated (NPV) Accumulated	−200	−162	−292	−398	−483	−553	−608	−653	−690	−719	−743	−763	−778
Maintenance Costs	−300	−300	−300	−300	−300	−300	−300	−300	−300	−300	−300	−300	−300
Maintenance Costs (NPV) Accumulated	−300	−245	−443	−603	−732	−837	−922	−990	−1045	−1090	−1126	−1155	−1179
Total Costs (NPV) Accumulated	−5000	−4083	−11304	−21063	−32874	−46342	−61152	−77044	−93813	−111291	−129340	−147853	−166740
F_T Net Present Value (NPV) (With II Devaluated)		36742	72224	96986	113083	122175	125604	124455	119605	111763	101502	89286	75490
F_T Net Present Value (With II New Asset)		31598	61704	80612	90111	91561	85959	73978	56005	32151	2264	−34085	−77617

(Continued)

TABLE 3.2 (*Continued*)
Values of Simulation in a Window Time of 25 Years

Time (*t*)	**13**	**14**	**15**	**16**	**17**	**18**	**19**	**20**	**21**	**22**	**23**	**24**	**25**
r - real interest rate $(1+r)$	1.05	1.05	1.05	1.05	1.05	1.05	1.05	1.05	1.05	1.05	1.05	1.05	1.05
h - inflation rate $(1+h)$	1.02	1.02	1.02	1.02	1.02	1.02	1.02	1.02	1.02	1.02	1.02	1.02	1.02
p - profit rate	1.05	1.05	1.05	1.05	1.05	1.05	1.05	1.05	1.05	1.05	1.05	1.05	1.05
i - money rate (nominal interest rate)	0.12	0.12	0.12	0.12	0.12	0.12	0.12	0.12	0.12	0.12	0.12	0.12	0.12
i - Apparent Rate with Risk	0.24	0.24	0.24	0.24	0.24	0.24	0.24	0.24	0.24	0.24	0.24	0.24	0.24
$1/((1+i)^\wedge t))$	0.06	0.05	0.04	0.03	0.03	0.02	0.02	0.01	0.01	0.01	0.01	0.01	0.00
Initial Investment (II) (C_0)	−12000	−12000	−12000	−12000	−12000	−12000	−12000	−12000	−12000	−12000	−12000	−12000	−12000
Initial Investment (II) (C_0) with Devaluation (NPV)	−756	−611	−494	−399	−323	−261	−211	−171	−138	−111	−90	−73	−59
Initial Investment (II) (C_0) considering New Asset (NPV)	−190551	−235712	−291577	−360682	−446165	−551909	−682714	−844521	−1044676	−1292270	−1598544	−1977407	−2446063
F_p Financial Movement (Production)	70000	70000	70000	70000	70000	70000	70000	70000	70000	70000	70000	70000	70000
F_p Financial Movement (NPV)	3936	3182	2572	2079	1681	1359	1099	888	718	580	469	379	307
F_p Financial Movement (NPV) Accumulated	247100	250282	252854	254934	256615	257974	259072	259960	260678	261259	261728	262107	262414
Functioning Costs	−4500	−4500	−4500	−4500	−4500	−4500	−4500	−4500	−4500	−4500	−4500	−4500	−4500
Functioning Costs (NPV) Accumulated	−17791	−18020	−18206	−18355	−18476	−18574	−18653	−18717	−18769	−18811	−18844	−18872	−18894
Energy	−200	−200	−200	−200	−200	−200	−200	−200	−200	−200	−200	−200	−200
Energy Accumulated (NPV) Accumulated	−791	−801	−809	−816	−821	−826	−829	−832	−834	−836	−838	−839	−840
Maintenance Costs	−300	−300	−300	−300	−300	−300	−300	−300	−300	−300	−300	−300	−300
Maintenance Costs (NPV) Accumulated	−1198	−1213	−1226	−1236	−1244	−1251	−1256	−1260	−1264	−1267	−1269	−1271	−1272
Total Costs (NPV) Accumulated	−185929	−205362	−224994	−244785	−264705	−284730	−304839	−325017	−345249	−365527	−385840	−406182	−426548
F_T Net Present Value (NPV) (With II Devaluated)	60416	44309	27367	9750	−8413	−27017	−45978	−65227	−84709	−104379	−124202	−144148	−164193
F_T Net Present Value (With II New Asset)	−129379	−190792	−263716	−350533	−454256	−578665	−728481	−909577	−1129247	−1396538	−1722656	−2121482	−2610197

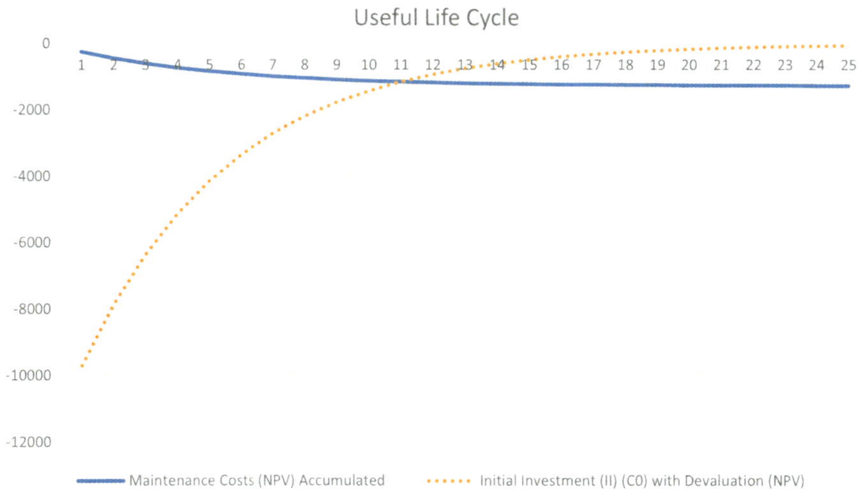

FIGURE 3.4 Useful Life Cycle.

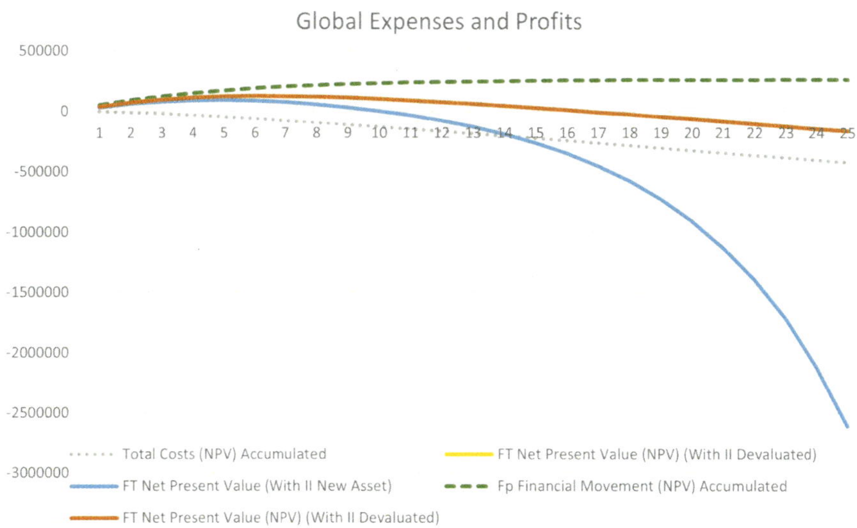

FIGURE 3.5 Global investment and profits.

3.10 CONCLUSIONS

The chapter takes the main variables indexed to an LCI, namely the value of the initial investment, the functioning expenses, including the maintenance ones, and the profits. Taking this into account, this chapter develops and implements an LCI model, considering the situation of the Physical Asset has a Devaluation given by the NPV and the situation of Recovery with the same NPV.

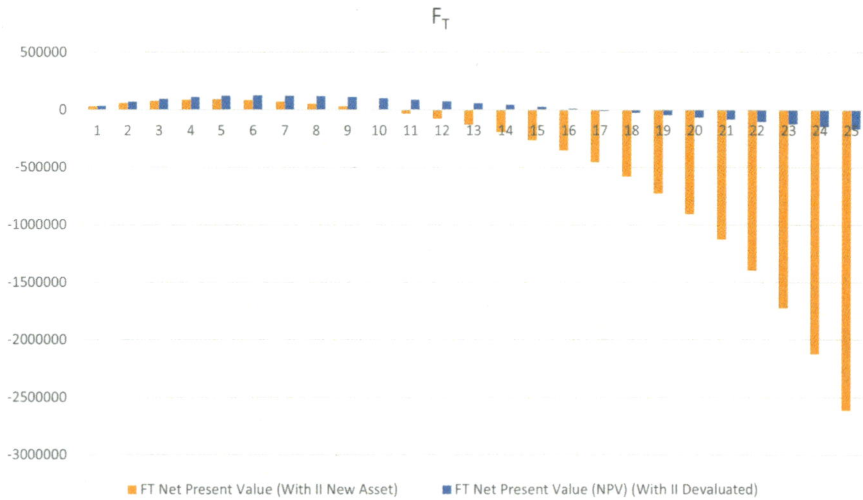

FIGURE 3.6 Final result considering devaluation and recovery of the Physical Asset.

These two scenarios, at last instance, permit organizations to consider the Physical Asset Life Cycle until it does not have interest to maintain functioning and, by consequence, to withdrawal.

But, for organizations, it is more prudent to evaluate the Physical Life Cycle taking into consideration the necessary money to acquire a new one at its Life Cycle withdrawal: this is the approach that this chapter presented and demonstrated its robustness.

PART II

Maintenance Methodologies

4 Maintenance Types and Strategies

4.1 INTRODUCTION

According to EN13306 (Maintenance terminology), the maintenance types and strategies that this standard refers are the following ones:

- Preventive maintenance
- Scheduled maintenance
- Predetermined maintenance
- Condition-based maintenance
- Predictive maintenance
- Corrective maintenance
- Remote maintenance
- Deferred maintenance
- Immediate maintenance
- On line maintenance
- On site maintenance
- Operator maintenance

The preceding can be resumed as follows:

- Maintenance
  - Preventive maintenance
    - Predetermined Maintenance
      - Scheduled
    - Condition-Based Maintenance
      - Scheduled, on request or continuous
  - Corrective maintenance
    - Deferred
    - Immediate

These are the main subjects that are discussed in this chapter, emphasizing their importance for the management of the Physical Life Cycle, namely when related to the Life Cycle models discussed in Part I of this book.

4.2 PREDETERMINED MAINTENANCE

4.2.1 Scheduled

Scheduled maintenance is carried out in accordance with an established time schedule or established number of units of use.

DOI: 10.1201/9781003395690-6

Weeks	1	2	3	4	5	6	7	8	9	10	11	12	13	14	15	16	17	18	19	20	21	22	23	24	25	26	27	28	29	30	31	32	33	34	35	36	37	38	39	40	41	42	43	44	45	46	47	48	49	50	51	52
Physical Asset																																																				

FIGURE 4.1 Example of a table to schedule maintenance by calendar.

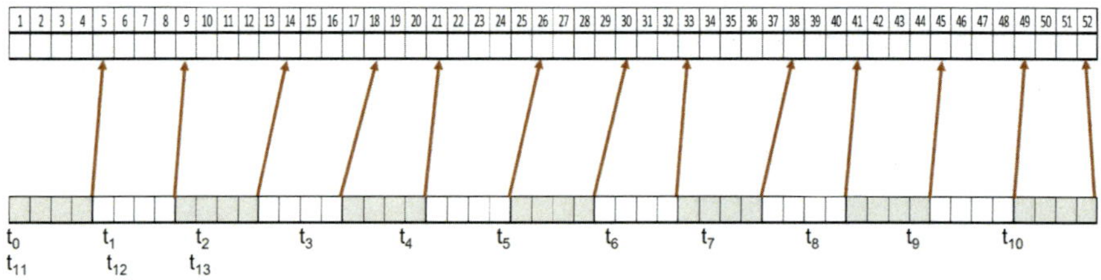

FIGURE 4.2 Example of the correspondence between a systematic variable and the calendar.

Scheduled maintenance can, according to the preceding, be done:

- By calendar;
- By Control Variable.

If the Predetermined Maintenance, Scheduled, is based on calendar then, as the name indicates, it means that is planned based on a calendar, usually with a weekly design, as illustrated in Figure 4.1.

If the Predetermined Maintenance, Scheduled, is based on number of units of use (Control Variable), even the periodicity of the intervals of number of units of use, the time intervals in calendar may not be periodic. Some examples of these type of variables are: hours of functioning, kilometres travelled, pieces manufactured, etc. (Figure 4.2).

4.2.2 Condition-Based Maintenance

Condition-based maintenance (CBM) is "carried out following a forecast derived from repeated analysis or known characteristics and evaluation of the significant parameters of the degradation of the item" (EN 13306).

The accompaniment of each condition parameter (variable) permits to evaluate the health (condition) state of the Physical Asset and, by consequence, to define the best time to make the maintenance intervention before fault.

If the CBM is "carried out following a forecast derived from repeated analysis or known characteristics and evaluation of the significant parameters of the degradation of the item" (EN 13306), it is called Predictive Maintenance. This subject is discussed later.

The significant parameters of the degradation of an item may be several: each Physical Asset must be evaluated carefully aiming to identify the most important ones. Some of the most usual are the following:

- Oil Analysis;
- Vibration Analysis;
- Ultrasonic Monitoring;
- Effluents Analysis;
- Temperature Analysis;
- Speed;
- Torque;
- Electrical Current;
- Pressure;
- Radiography;
- Power;
- Structural;
- Humidity;
- Laser Interferometry.

4.2.2.1 Oil Analysis

About the CBM each variable must be evaluated *per si*, being the oil, usually, the most difficult to evaluate, because it contains a lot of components (usually more than 20) that must be analyzed carefully. Due to the fact of these large number of components, one of the most usual techniques to prioritize the importance of each one is the Principal Component Analysis (PCA).

For example, if we are analyzing oil of Diesel engines, the most important variables that must be considered are the following (Rodrigues *et al.*, 2020): Si Content; Fe Content; Al Content; Cr Content; Sn Content; PQ Index; Ni Content; Soot; Oxidation; V Content; Cu Content; Sulfation; Mo Content; Pb Content; Fuel; Na Content; Viscosity; TBN; Nitration; Water content; Antifreeze.

The PCA is a statistical procedure that allows to summarize the data content in large tables by means of a smaller set of variables that are more relevant to the problem under discussion.

Another problem linked to oil analysis is the necessity to be evaluated offline for almost all of their variables. In fact, it is necessary to take regular oil samples and send them to a laboratory. By consequence, the results are not online, even if the delay be minimum.

4.2.2.2 Vibration Analysis

Vibration analysis is one of the strongest approaches to CBM because they exist in almost all types of Physical Assets.

The most common failures which vibration analysis can help a lot are the following ones:

- Unbalance;
- Misalignment;

- Loose components;
- Bearing defects;
- Gear failures;
- Chafing.

To analyze vibrations, the more important approach is the Fourier Analysis and the following metrics: harmonics; phase; amplitude; average value; Root Mean Square (RMS) value. The most usual way to evaluate vibrations is through Velocity and Acceleration.

The most common sensor used to monitor vibrations is the accelerometer.

Obviously, the vibration measures may be offline or online. The accelerometer is not a cheap sensor, but the Return on Investment (ROI) in this type of technology, usually is short, even using the online or offline approach.

4.2.2.3 Ultrasonic Monitoring

All equipment and most leakage problems produce a broad range of sound. The high-frequency ultrasonic components of these sounds are extremely short waves in nature. A shortwave signal tends to be fairly directional.

Ultrasonic Monitoring is a non-destructive technique based on the propagation of ultrasonic waves in an object. In most common Ultrasonic Applications, very short ultrasonic pulse waves with centre frequencies ranging from 0.1 to 50 MHz, are transmitted into objects to detect internal flaws or to characterize materials.

Ultrasonic corrosion monitoring is highly favoured for steel and other metals and alloys. It is also used to test the physical properties of concrete, wood, and composites, flaw detection and evaluation, dimensional measurements, and material characterization.

Ultrasonic monitoring usually provides the following information:

- Qualitative – with the ability to "hear" ultrasounds through a noise-isolating headphone;
- Quantitative – via intensity (dB) readings on a metre or display panel;
- Analytical – with the use of spectral analysis software to review recorded sound samples.

4.2.2.4 Effluents Analysis

Many Physical Assets have exhaust systems that emit pollutants aiming: through their measure and accompanying it is possible to know the condition of the system. These kinds of systems are chimneys, car exhaust devices, among others. In case of Diesel engines, the most usual effluents are NO_x, PM, CO_2, and noise.

4.2.2.5 Temperature Analysis

The temperature analysis is the easiest and one ore the most popular CBM techniques. Additionally, the sensors of temperature are not expensive, which facilitates their implementation.

This approach is used commonly offline and online.

When temperature analysis is made offline, they are used, usually, thermography equipment that, simultaneously measures the temperature and make thermographic images of the Physical Asset under study. This means that it caught not only the temperature but also the thermographic image; this last aspect needs a lot of research aiming to help making prediction about the health Physical Asset based on temperature values, complemented to images.

4.2.2.6 Speed

The rotational speed measurement is a very important variable for condition monitoring, namely for speed control and supervision of rotation equipment, such as generators and, in general, rotating equipment.

The rotating speed measurement method can be classified into digital and analogue categories. The analogue tachometer gives as output a voltage or a current signal that is proportional to the speed.

There are several types of tachometers based on different mechanisms, such as optical, electrical, and magnetic induction.

The speed sensors usually work with vibration sensors to correlate vibrations to rotating speed and shaft angular position.

4.2.2.7 Torque

The force that makes every shaft rotate is called torque, and it is directly proportional to the power. This means that when it is necessary to maintain a certain operating speed in the process, and the load increases, the motor must deliver more power, which is achieved by increasing the torque.

Torque is the rotational analogue of linear force. It is also referred to as the moment of force. It describes the rate of change of angular momentum that would be imparted to an isolated body.

The mechanical shaft power is proportional to the product of torque and speed: torque measurement is a critical parameter in mechanical efficiency measurements.

The torque evaluation in condition monitoring permits, among other conclusions, to identify: excessive load; equipment performance such as slippage; deterioration trending; and transient events.

4.2.2.8 Electrical Current

Electrical current analysis is a powerful way to evaluate the condition of electrical motors. The electrical condition monitoring techniques include: checking out the overall circuitry of an electric motor, covering the resistance of the windings and the integrity of the insulation; evaluating resistance, inductance, impedance, capacitance, and phase angle; checking current and voltage unbalance, distortion, rotor conditions, power levels, and power factor; the motor current signature analysis, used for identifying developing problems inside motors, particularly the development of defects in the rotor cage structure, such as rotor bar cracking; the model-based voltage and current systems to identify the distortions of the motor current that have not been caused by distortions of the voltage power supply; partial discharge testing, used to detect small electrical discharges that take place within insulation what may signify insulation deterioration.

The applications for electrical condition monitoring techniques include motors and their driven equipment, generators and their prime movers, transformers, switchgear, inverters, supply sources, etc.

The most usual motor in the industry is the squirrel cage induction motor and, for these, two of the most frequently used current monitoring methods used in condition monitoring are: the wavelet analysis-based monitoring; and Park's vector approach. A wavelet is an oscillation similar to that recorded by a seismograph or a heart monitor, in which amplitude starts out at zero, increases and then decreases back to zero. When any type of fault occurs, the three-phased current phase current system becomes unbalanced, which results in an elliptic representation of the Park's current vector.

4.2.2.9 Pressure

Pressure is the force applied perpendicularly to the surface of an object per unit area over which that force is distributed.

The weight of the atmosphere pushing down on each unit area of Earth's surface constitutes the atmospheric pressure. In International System units, the pressure is measured in Pascal; one Pascal is equal to one Newton per square metre; the atmospheric pressure is close to 100,000 Pascal.

Pressure sensors, for example, are used to monitor hydraulically controlled processes, while measuring the occurring pressure, and translate the condition within the control loop.

Another example is fluid pressure, which is a very important variable in condition monitoring. Pressure monitoring is the second most common variable measured after temperature. Pressure measurement applications range from simple set point monitoring to ensure sufficiently high or low pressure levels, to continuous monitoring as part of a complex automation system. The amplitude of applications that require pressure measurement explains the diversity of sensing products on the market.

4.2.2.10 Radiography

The radiography is a non-destructive method. Imaging allows technicians to inspect internal defects, such as corrosion in sintered parts and flaws in welding. The main advantage is that it can be used in all types of materials, provided technicians protect themselves.

X-ray radiography helps to identify the surface and subsurface flaws caused by problems such as stress; corrosion; fatigue; poor or incomplete welds; and trapped gases. It can be used to locate semiconductor faults and loose wires. The technique produces a radiograph by passing X-rays through opaque materials and producing an image of those materials. The film exposed to X-rays captures a shadow image. The darkest is where the object is thinnest or absorbs the least radiation. This technique can be used to analyze welds, steel structures, plastic structures, and metallic wear components of engines, compressors, gearboxes, pumps, and shafts.

4.2.2.11 Power

Power monitors give early warnings to maintenance teams when there is a problem with almost of electrical devices. Power monitors measure fluctuations in current, voltage, frequency, and energy consumption, all of which can point to premature degradation on the Physical Asset.

Power monitoring is one of the condition-monitoring techniques to prevent unplanned downtime. Beyond detecting power problems that could lead to outages, a power monitoring solution plays an important role in data centre challenges, namely improving energy efficiency and supporting better capacity planning. For example, power monitoring permits ensuring the system availability, efficient operations. Power monitoring is an essential tool for protecting and optimizing a data centre environment.

Power monitoring is achieved by using power monitors, sensors, and metres to monitor devices, including Uninterruptable Power Supply (UPS) and Power Distribution Units (PDU). Utility sources, generators, branch circuits, and other critical infrastructure can also be monitored. Nowadays UPS and PDUs usually are always connected, meaning they have built-in monitoring and communication capabilities.

Power monitors, sensors, and metres, whether built-in or freestanding, work together with power monitoring software to provide data centre managers with the reliability of data.

4.2.2.12 Structural

Structural Health Monitoring (SHM) implies the analysis of a system over time using periodically sampled response measurements to monitor changes to the material and geometric properties of physical structures such as dams, bridges, and planes.

After extreme events, such as earthquakes or blast loading, SHM is a technique used for condition monitoring and to provide information regarding the integrity of the structure.

The SHM process involves several types of sensors and methods, the data acquisition, transfer and storage hardware and software. Measurements permit to detect any degradation or damage that may occur in a system or indirectly allow to predict the state of the system.

The SHM process implies to identify features in the acquired data that allows to identify damages in the structure. The damage accumulation testing may involve induced-damage testing, fatigue testing, corrosion growth, or temperature cycling to accumulate certain types of damage in an accelerated fashion.

4.2.2.13 Humidity

Humidity sensors are electronic devices enabling to measure the humidity and convert the data into a corresponding electrical signal that can be used for condition monitoring. The humidity sensors are called hygrometers.

Humidity sensors are categorized based on their size and their functionality. There are two different methods to measure humidity: Absolute Humidity (AH) and Relative Humidity. The type that measures absolute humidity utilizes two probes: one of them measures dry nitrogen and the other measures the environment's air; with these two types of data, the sensor can calculate AH. The relative humidity compares the instant humidity at a given temperature with the highest amount of the air's humidity at the exact same temperature.

Humidity sensors are used, for example, as surface-mounted or flush-mounted devices to monitor the humidity of indoor spaces, or outdoor spaces, often in combination with temperature sensors.

Duct humidity sensors are intended for the exact detection of air humidity (and temperature) in air ducts or fans, like in refrigeration, air conditioning, ventilation, or clean room technology.

4.2.2.14 Laser Interferometry

Laser interferometers measure changes to calculate displacement based on laser-generated wavelengths. They are used in condition monitoring to identify surface and subsurface defects, like corrosion and cavities. Interferometry includes laser ultrasonic, strain mapping, electronic speckle pattern interferometry, and digital holography.

Interferometry is a technique that uses the *interference* of superimposed waves to extract information, using electromagnetic waves. It is an important technique in the fields of fibre optics, surface profile, mechanical stress/strain measurement, velocimetry, and optometry, making holograms.

Interferometers are widely used in industry for the measurement of microscopic displacements, refractive index changes, and surface irregularities.

4.2.2.14.1 Sensors

There are three types of sensors:

- Wired sensors
- Wireless sensors
- Contactless sensors

4.2.2.14.1.1 Wired Sensors Wired sensors are devices used to monitor the condition of equipment and machinery using cables or wires to transmit data to a central monitoring system.

Wired sensors link directly the sensor to the device that is receiving the input. The wired sensors are very reliable, having a long life cycle. However, the wired systems require a large amount of physical space to install. This problem increases in size as more and more sensors are used. But, as a big advantage, wired sensors have high predictable data transfer time.

Wired sensors are reliable systems for CBM, as they directly link the sensor to the device that is receiving the data. This means that wired sensors are also the most durable systems that do not need to be replaced often. However, it should be noted that wired systems require a large amount of space and some maintenance. This burden increases in size as more sensors are used based on the application that the sensor is designed for.

Wired sensors offer several advantages, namely over wireless sensors, including:

- Reliability – Wired sensors are typically more reliable than wireless sensors, as they do not rely on wireless signals that can be disrupted by interference or other factors;
- Stability – Wired sensors offer greater stability than wireless sensors, as they do not suffer from signal dropouts;
- Accuracy – Wired sensors typically offer greater precision than wireless sensors, as they can transmit data more reliably and with greater resolution.

The advantages and disadvantages of wired sensors are the following:

- Advantages:
  - Generally, they are more reliable and stable than wireless sensors, as they don't rely on wireless signals that can be disrupted by interference or other factors;
  - Generally, they offer greater accuracy and precision than wireless sensors, as they can transmit data more reliably and with greater resolution;
  - Generally, they are less expensive than wireless ones, requiring fewer components.
- Disadvantages:
  - They can be more difficult to install than wireless sensors, as they require cables or wires to transmit data;
  - They can be limited in placement and location, requiring a physical connection to a signal source;
  - They can be affected by cable or wire damage or disconnection, resulting in inaccurate data;
  - They may not be suitable for remote or hard-to-reach locations.

4.2.2.14.1.2 Wireless Sensors Wireless sensors are devices used to monitor the condition of equipment without cables. They use wireless technology, such as Wi-Fi, Bluetooth, or cellular networks to transmit data from the sensor to a central monitoring system or a cloud-based platform. Wireless sensors can be battery-powered or use energy harvesting technology to power themselves, eliminating the need for wiring or external power sources.

Wireless sensors offer several advantages over wired sensors, including:

- Flexibility – Wireless sensors can be installed in almost any location, regardless of accessibility, which allows businesses to monitor equipment in areas that are difficult or impossible to reach with wired sensors.
- Ease of installation – Since wireless sensors don't require wiring, installation is quicker and easier. This can help reduce downtime and improve efficiency during installation.
- Scalability – Wireless sensors can be added or removed easily, making it simple to scale up or down a condition monitoring system as needed.
- Cost-effectiveness – In some cases, wireless sensors can be more cost-effective than wired sensors, especially when installation costs are factored in.

The advantages and disadvantages of wireless sensors are the following:

- Advantages:
  - They are not much expensive to install and easy to maintain;
  - They allow for a greater level of flexibility, making it possible to adapt a sensor system to the needs of the user aiming for a CBM policy;
  - Greater flexibility and ease of installation compared to wired sensors;
  - Can be installed in hard-to-reach or remote areas, making them ideal for monitoring equipment in challenging environments;

  - They can be added or removed easily, making them a scalable solution for condition monitoring;
  - They can transmit data in real time, providing businesses with up-to-date information about the condition of their equipment.
- Disadvantages:
  - They are more expensive than wired ones due to the required components;
  - They can be affected by interference or signal loss, which can result in inaccurate data;
  - They may be subject to interference from other devices such as radios and cell phones;
  - The speed at which the data is transferred depends on the location of the receiving device relative to the sensor;
  - Typically require a power source, such as batteries, which can be a maintenance issue;
  - They may have a shorter range than wired sensors, making them less suitable for larger facilities or outdoor environments.

About contactless sensors, it implies to distinguish between sensors themselves and the contactless systems.

4.2.2.14.1.3 Contactless Sensors Usually, the sensors are physically linked to equipment in a mechanical way. However, many sensors don't need to have physical contact with the piece of equipment or the device that must be monitored. Next are presented several types of these types of sensors.

4.2.2.14.1.3.1 Air Sensing or Gauging Technology A flow of pressurized air is used to measure the dimensions of parts or distances such as diameters or tapers. The technology relies on the principle that as the distance between the air gauge and the object being measured decreases, the flow rate drops while the air pressure increases. Similarly, increasing the separation distance increases the flow rate and decreases pressure. These changes can be measured and used to establish measured values of distance for parts.

4.2.2.14.1.3.2 Hall Effect Sensor Technology Hall effect sensors use a thin, rectangular piece of p-type semiconductor material to which a constant Direct Current (DC) voltage is applied. When the sensor material is brought in proximity to an external magnetic field, the lines of magnetic flux exert a net force on the charge carriers in the sensor, resulting in a deposition of like charge carriers (electrons or holes) to accumulate on opposite sides of the semiconductor material. This charge accumulation results in an electrical potential difference to develop between these sides of the material, which can be measured, and which is known as the Hall voltage.

4.2.2.14.1.3.3 Ultrasonic Sensor Technology An ultrasonic sensor makes use of high-frequency sound waves as a detection or sensing mechanism. By measuring the time that it takes for the sound to make a round trip from the sensor to the target and back again, the distance to the target can be easily determined.

4.2.2.14.1.3.4 Photonic Sensor Technology Photonic or optical sensors use light energy directed through fibres as a means of determining the displacement from or the distance to the target object by measuring the intensity of the light that is reflected off the target. Photonic-based sensors have the advantage of being unaffected by the presence of Electromagnetic Interference (EMI) emissions or high voltage, but the characteristics of the target's surface finish, which determine its reflectivity, can impact the sensor operation.

4.2.2.14.1.3.5 Capacitive Sensor Technology Capacitive sensors rely on detecting a change in capacitance to provide information about the movement or position of a target. A capacitor is a device that has the ability to store energy in an electric field between two plates that are known as electrodes. With capacitive sensors, the sensor functions as one plate of the capacitor with the target functioning as the other plate. If a fixed frequency Alternating Current (AC) is applied to the sensor, the amplitude of the AC voltage will provide a proportional measure of the distance between the sensor and target.

4.2.2.14.1.3.6 Inductive Sensor Technology Inductive sensors make use of magnetic fields generated in coils to detect motion or the position of a target. One common type of inductive sensor is a Linear Variable Differential Transformer (LVDT). A set of coils, one primary and two secondaries are contained in the LVDT. When the primary coil is energized with an AC voltage, induced Electric and Magnetic Fields (EMF) are triggered in each of the secondary coils. By measuring the voltage difference between the two secondary coils, the sensor can establish the movement of the target and determine its position.

Another type of inductive sensor technology makes use of Eddy currents and is used when targets are conductive. An Eddy current sensor uses an AC applied to a coil to generate an alternating magnetic field. When the conductive target approaches the sensor, the magnetic field induces currents in the conductive target, called Eddy currents. These currents cause the generation of a secondary magnetic field which opposes the primary field of the sensor. The field interaction can be measured and used as an indicator of the distance that the sensor is from the target.

4.2.2.14.1.3.7 Laser Displacement Sensor Technology Laser displacement sensors, also called laser triangulation sensors, are a non-contact sensor technology that uses the triangulation of a reflected light beam to establish the movement or position of a target across a measurement range. A semiconductor laser generates a light beam that is sent through a transmitting lens towards the target. Light reflected off the target passes through a receiving lens in the sensor and is focused onto a detector element such as a Charge Coupled Device (CCD) array. As the target changes its position relative to the sensor, the angle of the reflected light will change which results in that light focusing on a different part of the detector element. The change in light position can be analyzed and used to establish the position of the target.

4.2.2.14.1.3.8 Non-contact Sensor Uses for Thermal Measurement Resistance Temperature Detectors (RTD) and thermocouples are common types of contact sensors that can be utilized. But there are a number of non-contact sensor types that can be used to perform measurement of temperatures. These include radiation thermometers, thermal imagers, optical pyrometers, and fibre optic temperature sensors.

4.2.2.14.1.3.9 Radiation Thermometers Radiation thermometers gauge temperature based on the radiation released from an object. An object's ability to release radiation is called its emissivity. Sensors in this category include spot measuring devices that can produce 1D and 2D temperature readings, as well as thermal imaging thermometers which can display temperature data as a 2D image. These types of visual representations are useful in terms of industrial processes because they can help identify potential problems or inconsistencies.

4.2.2.14.1.3.10 Thermal Imagers A thermal imager can measure a two-dimensional space, in essence providing an accurate picture of both the source of radiation and the space around it. Thermal imagers can be used to locate areas in a cord that are overheating, as well as by firefighters to locate people amid smoke and fire. The devices can also be used to locate heat leaks in buildings with weak insulation.

4.2.2.14.1.3.11 Optical Pyrometers Optical pyrometers receive such a meaningful name because they can gauge temperatures that are too bright to see with the naked eye. A pyrometer has two parts: an optical system and a detector. To conduct a proper temperature reading, the optical system works to focus the chosen region of thermal radiation onto the detector, which, in turn, translates data into a readable temperature. This particular type of measuring device is especially helpful for measuring the temperature of moving objects, or objects that cannot be touched.

4.2.2.14.1.3.12 Fibre Optic Temperature Sensors Many fibre optic temperature sensors are merely variations of radiation thermometers. Relatively simple in design, the fibre optic features an active sensing device attached to a system that processes the radiation data and produces a temperature reading. They are extremely useful in automotive applications because they can set a temperature limit signal for engines.

4.2.2.14.1.4 Contactless Systems The contactless (non-contact) systems detect and measure a physical property and records or otherwise, without necessary contact with the subject of measurement.

Contactless systems, generally rely on technologies that are based on electrical, magnetic, optical, sonic, or other principles, not depending on physical contact or mechanical movement to obtain readings.

Nowadays, there are contactless systems, that are much more than sensors, which permit to evaluate the condition of the Physical Asset without any physical contact.

Some examples of these systems are the High-Speed Cameras, that can manage three main types of vibration analysis:

- Frequency analysis – permits to determine how often a vibration occurs over a period of time;
- Amplitude analysis – permits to determine to measure the magnitude, or strength, of a vibration;
- Phase analysis – permits to determine the timing of a vibration in relation to other vibrations.

In the field of bridge monitoring, the contactless solution can be, for example:

- Profile scanning – the Terrestrial Laser Scanners (TLS), also known as ground-based LIDAR, enable the digitization of the entire environment in a 360° panorama in the form of a 3D point cloud; during the scanning process, a high-frequency rotating mirror deflects the laser beam, and the TLS, additionally, rotates around its standing axis; this sequential acquisition method produces a high-resolution point cloud of the visible environment.
- Microwave interferometry – it is a useful technique for understanding the development and propagation of detonation waves. The velocity of the front can be determined directly with the dielectric constant of the explosive and the instantaneous phase difference of the reflected microwave signal from the detonation front.

About the preceding subjects, some additional references must be taken into account, like Fraden (2015) and Nyce (2023).

Additionally, it must be noted that nowadays there are more and more devices with sensors with multi-functions, i.e., devices that measure several variables simultaneously, like vibration, temperature, humidity, air pressure, among others.

4.3 CORRECTIVE MAINTENANCE

According to EN13306 (Maintenance terminology), Corrective Maintenance can be of two types:

- Deferred
- Immediate.

If the fault in Physical Asset is not too urgent, in many situations it is possible to defer a little the intervention and to prepare the intervention making a previous diagnosis and preparing the corresponding resources to solve the problem. Sometimes it is possible to manage this delay until the next scheduled intervention.

If the fault needs a very urgent intervention, the Corrective Maintenance intervention must be Immediate.

However, in both preceding situations, a Working Order (WO) must be launched, filled and, at last, closed. Only if all WO closed their circuits, it is possible to evaluate the maintenance performance with rigour.

4.4 ENTERPRISE ASSET MANAGEMENT SYSTEMS

Traditionally, the information systems for maintenance are called Computerized Maintenance Management System (CMMS) and are focused on the main aspects aiming to manage maintenance, that are (Farinha, 2018):

- Physical Assets dossier;
- Customers;
- Suppliers;

- Technicians;
- Tools;
- Spare Parts;
- Work Orders (WO);
- Intervention Requests;
- Fault Diagnosis;
- Maintenance Plans;
- Maintenance Contracts.

Nowadays, it is important to manage all points of Life Cycle of the Physical Assets and, for this, are being added to the CMMS Life Cycle models that permit to evaluate the Life Cycle from the design to withdrawal. The three first chapters of this book propose models to answer this, and, the Part III of this presents several approaches of them, namely when implementing several standards.

A CMMS that includes Life Cycle models to evaluate and accompany the Physical Life Cycle is called an Enterprise Asset Management (EAM) system.

An EAM system, as referred, has some additional *moduli*, namely the following:

- Acquisition;
- Life Cycle Evaluation;
- Withdrawal.

These preceding *moduli* must be supported in Life Cycle models and approaches like the ones presented in this book.

The author of this book developed an EAM called SMIT (Integrated Modular System of Terology) – Figure 4.3.

4.5 MAINTENANCE PERFORMANCE

The evaluation of the Maintenance Types and Strategies is made through Key Performance Indicators (KPI). The Standard EN 15341 (Maintenance - Maintenance Key Performance Indicators) permits any Organization to choose the most adequate

FIGURE 4.3 Integrated Modular System of Terology.

KPI that fits its profile. Additionally, each Organization must pay attention to other similar companies aiming to compare its performance based on the same set of KPIs.

The Standard EN 1541 proposes three types of KPI:

- Technical
- Organizational
- Economic.

The final result is a *Scoreboard* or a "Cockpit Chart" with the most relevant KPI that permits to monitor internally the Organization and also when compared to the main competitors.

REFERENCES

Farinha, J. M. T. (2018) *Asset Maintenance Engineering Methodologies*. Boca Raton, FL: CRC Press.

Rodrigues, J., Cost, I., Farinha, J. T., Mendes, M., & Margalho, L. (2020). Predicting motor oil condition using artificial neural networks and principal component analysis. *Eksploatacja i Niezawodnosc – Maintenance and Reliability*, *22*(3), 440–448. doi:10.17531/ein.2020.3.6

Fraden, J. (2015) *Handbook of Modern Sensors: Physics, Designs, and Applications* (English Edition, N. 5 edition). Springer International Publishing Switzerland. ISBN: 3319307673

Nyce, D. (2023) *Understanding Position Sensors* (N. 1 edition). CRC Press. ISBN-13: 978-1032436999

5 Maintenance Algorithms

5.1 INTRODUCTION

The algorithms presented in synthesis in this chapter are the most usual in condition-based maintenance (CBM) and in predictive maintenance (PdM), both based on stochastic and in artificial intelligence approaches.

The next sections of this chapter are the following:

- Stochastic algorithms;
- Artificial intelligence algorithms.

5.2 STOCHASTIC ALGORITHMS

The book "Time Series Analysis: Forecasting and Control addresses the analysis of Time Series" (Box *et al.*, 2015) presents important applications of the methodologies and the integrated moving averages and regressive author models were presented, namely Auto-Regressive Integrated Moving Average (ARIMA), as well as extensions of these models. It also presented model properties and criteria for selecting them, estimating parameters, and how to validate the model.

According to Makridakis *et al.* (1997), there are five basic steps in any forecasting task for which quantitative data are available:

1. Problem definition
    - The definition of the problem involves a deep understanding of how the forecasts will be used, what requires the forecasts, and how the forecasting function fits within the organization.
2. Gathering information
    - There are two kinds of information available: statistical data, and the accumulated judgement and expertise of key personnel. Both kinds of information must be selected. It is necessary to collect historical data of the items of interest, namely the historical data aiming to construct a model which can be used for forecasting.
3. Preliminary (exploratory) analysis
    - Usually, the process starts by graphing the data for visual inspection. Then they are made some simple descriptive statistics (e.g., mean, standard deviation, minimum, maximum, and percentiles) associated with each set of data. The purpose in all cases at this stage is to get a feel for the data – Are there consistent patterns? Is there a significant trend? Is the seasonality important? Are there any outliers that need to be explained? How strong are the relationships among the variables available for analysis?

DOI: 10.1201/9781003395690-7

4. Choosing and fitting models
   - Each model is itself an artificial construct. It is based on a set of assumptions that, usually, involve one or more parameters which must be "fitted" using the historical data. They can be used in stochastic or artificial intelligence models.
5. Using and evaluating a forecasting model
   - Once a model has been selected and its parameters estimated, the model is ready to make forecasts. A forecasting assignment is not complete until the model fits the data. The performance of the model can only be evaluated after the data for the forecast period have become available.

When exploring the data, it is important to visualize the data through graphs. The basic features of the data visualization usually include patterns and the anomalous data can be easily seen through graphs. Many times the graphs suggest possible explanations for the data behaviour.

The most usual graphs are the following ones:

- Time plots and time series patterns (Figure 5.1).
- Scatterplots (Figure 5.2).
- Boxplots (Figure 5.3).

The most usual stochastic algorithms are the following:

- Moving Average;
- Exponential Smoothing;

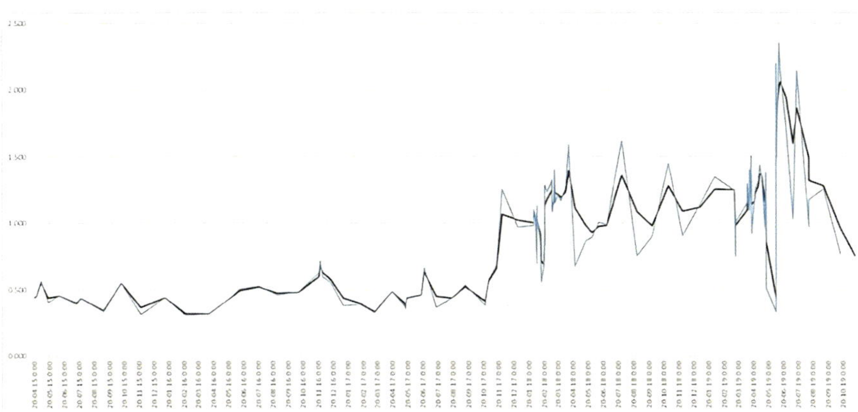

FIGURE 5.1 Time plot example.

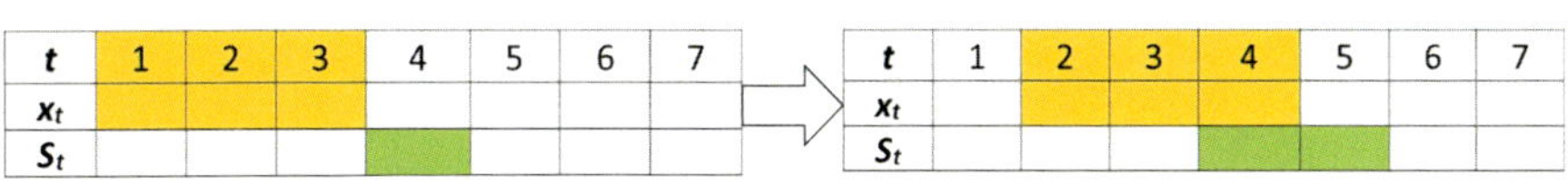

FIGURE 5.2 Scatterplot example.

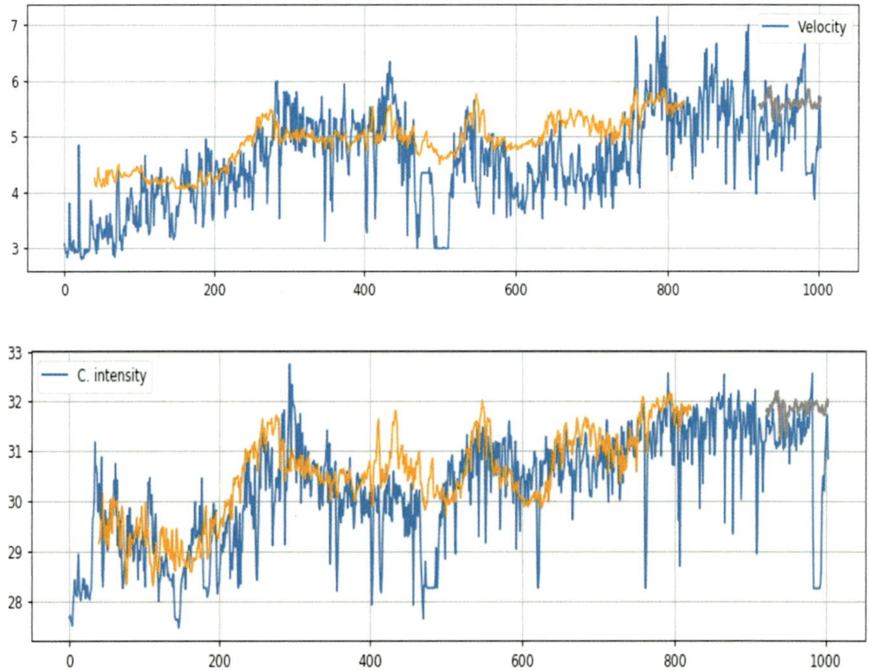

FIGURE 5.3 Boxplot example.

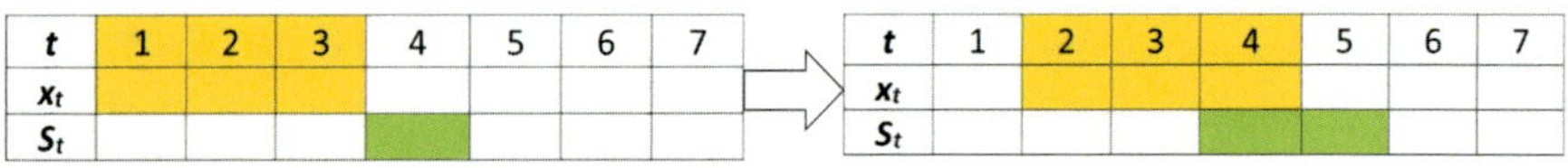

FIGURE 5.4 Moving average graphical movement.

- Auto Regressive Moving Average (ARMA);
- Autoregressive Integrated Moving Average (ARIMA);
- Seasonal Autoregressive Integrated Moving Average (SARIMA).

Next is a brief description of each one of the preceding algorithms; about these subjects see also Farinha (2018).

5.2.1 Moving Average

The Moving Average method makes the forecast for the next period from the average of *n* previous periods. Figure 5.4 shows the sliding window that corresponds to the concept of Moving Average.

In Figure 5.4,

- t is the period of the series
- x_t is the actual value
- S_t is the forecast for the next period

The Equation for calculating the value of the next period is given by Equation 5.1:

$$S_{t+1} = \frac{\sum_{i=t-n}^{i=t} X_i}{n} \tag{5.1}$$

being

- S_{t+1} the forecast for the next period
- X_i the actual value recorded for each one of the n periods before, between $i = t - n$ and $i = t$

5.2.2 Exponential Smoothing

Based on moving average, it appears that Exponential Smoothing (ES) is a technique for smoothing time series data using the exponential window function. Whereas in the simple moving average, the past observations are weighted equally, exponential functions are used to assign exponentially decreasing weights over time.

ES is a time series forecasting method for univariate data that can be extended to support data with a systematic trend or seasonal component, being given by Equation 5.2.

$$S_{t+1} = \alpha_t \times x_t + (1 - \alpha_t) \times S_t \tag{5.2}$$

$$0 \leq \alpha \leq 1$$

being

- S_{t+1} the forecast for the next period
- x_t the actual value in the present time
- S_t the forecasted value for the present time
- α the smoothing parameter

If the time series has a tendency to forecast, the second-order ES (Equation 5.3) can be used:

$$S_{t+m} = \alpha_t + b_t \times m \tag{5.3}$$

being

- S_{t+m} is the expected value of the 2nd order for period m

where

$$S_t' = \alpha_t \times x_{t-1} + (1 - \alpha_t) \times S_{t-1}'$$

$$S_t'' = \alpha \times S_t' + (1 - \alpha_t) \times S_{t-1}''$$

$$a_t = 2 \times S_t' - S_t''$$

$$b_t = \frac{\alpha}{1-\alpha}\left(S_t' - S_t''\right)$$

$$b_t = \frac{\alpha}{1-\alpha}\left(S_t' - S_t''\right)$$

$$0 \leq \alpha \leq 1$$

being

- $S_t^{'}$ is the first second-order forecast for the next period
- $S_t^{''}$ is the second second-order forecast for the next period
- a_t is the coefficient to support the calculation of the second-order forecast
- b_t is the coefficient to support the calculation of the second-order forecast - second smoothing

In the preceding algorithms, namely the ES of first order, the value of the smoothing parameter (α) must be decided according to the time series behaviour: for a stable series, the value of α ought to be low and inversely.

An alternative to a fixed smoothing parameter, a self-adaptive α can be used given by Equation 5.4:

$$\alpha_t = \left|\frac{A_t}{B_t}\right| \tag{5.4}$$

$$S_{t+1} = \alpha_t \times x_t + (1-\alpha_t) \times S_t$$

$$A_t = \beta \times E_t + (1-\beta) \times A_{t-1}$$

$$M_t = \beta \times |E_t| + (1-\beta) \times M_{t-1}$$

$$E_t = X_t - S_t$$

$$0 \leq \alpha \leq 1$$

$$0 \leq \beta \leq 1$$

5.2.3 Autoregressive Moving Average

Autoregressive Moving Average (ARMA) models provide a description of a (weakly) stationary stochastic process. The series is assumed stationary in the mean and in the variance.

An ARMA process consists of two models: An Autoregression (AR) model and a Moving Average (MA) model. Compared with the pure AR and MA models, ARMA models provide an effective linear model of stationary time series since they are capable of modelling the unknown process with the minimum number of parameters:

- Autoregression (AR) – means a model that shows a changing variable that regresses on its own lagged, or prior, values;
- Moving Average (MA) – incorporates the dependency between an observation and a residual error from a moving average model applied to lagged observations.

An ARMA model of order p is defined by Equation 5.5:

$$Y_t = c + \theta_1 \times Y_{t-1} + \theta_2 \times Y_{t-2} + \cdots + \theta_p \times Y_{t-p} + \epsilon_t \tag{5.5}$$

where

- p is the order of the process
- c is a constant
- θ_i are the parameters of the model
- ε_t is the noise

5.2.4 Autoregressive Integrated Moving Average

ARIMA is a statistical analysis model that uses time series data to either understand the data set or to predict future trends.

Data values are replaced by the difference between the data values and the previous values, being ARIMA the result of:

- AR – Autoregression - equation terms created based on past data points;
- I – Integration or differencing - accounting for overall "trend" in the data;
- MA – Moving Average - equation terms of error or noise based on past data points.

An ARIMA model is characterized by three terms (p, d, q):

- p is the order of the AR process;
- q is the order of the MA process;
- d is the number of differencing required to make the time series stationary.

The equation for the simplest case of ARIMA is given by Equation 5.6:

$$Y_t' = c + \theta_1 \times Y_{t-1}' + \cdots + \theta_p \times Y_{t-p}' + \theta_1 \times \epsilon_{t-1} + \cdots + \theta_p \times \epsilon_{t-q} + \epsilon_t \tag{5.6}$$

where

- c is a constant
- θ_i are the parameters of the model
- ε_t is the error

5.2.5 Seasonal Autoregressive Integrated Moving Average

When ARMA models are extended to non-stationary series by allowing differencing of the data series, they are called Autoregressive Integrated Moving Average (ARIMA) models (Box *et al.*, 2008).

Seasonal Autoregressive Integrated Moving Average (SARIMA), or Seasonal ARIMA, is an extension of ARIMA, which explicitly supports univariate time series data with a seasonal component.

A seasonal ARIMA model is formed by including additional seasonal terms in the ARIMA model: the seasonal part consists of terms that are very similar to the non-seasonal components of the model, but they involve regression of the seasonal period (Equation 5.7).

$$Y_t = c + \alpha_1 \times Y_{t-1} + \cdots + \alpha_p \times Y_{t-p} + \theta_1 \times \epsilon_{t-1} + \cdots + \theta_q \times \epsilon_{t-q} + \varnothing_1 \times Y_{t-1} + \cdots + \varnothing_P \times Y_{t-P} + \varphi_1 \times \varepsilon_{t-1} + \cdots + \varphi_Q \times \varepsilon_{t-Q} + \epsilon_t \tag{5.7}$$

5.3 ARTIFICIAL INTELLIGENCE ALGORITHMS

Instead of the methods presented in the preceding section, artificial intelligence (AI) can be used to forecast new maintenance interventions.

Generally, AI is the knowledge field that is concerned with the development of techniques that allow computers to act in a way that looks like an intelligent organism, as a human brain (Raynor, 1999).

Nowadays Industries take advantage of some technological tools that provide them with significant advantages, such as the State of The Art of nonlinear temporal Machine Learning (ML) methods, based on evolutionary algorithms, genetic algorithms, Artificial Neural Networks (ANN), and other AI techniques; these ones have been considered necessary to achieve automated decision-making based on databases (Gubbi *et al.*, 2013). For example, Big Data analysis, AI and ML, and the Internet of Things (IoT) allow for real-time predictions of manufacturing equipment, making possible to predict equipment's faults on time allowing to launch a Work Order before the fault occurs; this permits to plan the maintenance procedures and resources, like the technicians and spare parts (Chen *et al.*, 2019; Farinha, 2018).

Within the scope of the development of AI, there has been a promising development with a high degree of success, regarding the applications of neural networks in the detection of defects in power systems.

The intelligent behaviour of an ANN comes from the interactions among the processing units of the network. An artificial neuron consists of input signals (x_i), a set of real value weights (w_i), an activation level $\left(\sum w_i x_i\right)$, and a threshold function (f) (Ozel *et al.*, 2008). The processing units make the operations only on their local data, which are the inputs that are received by their connections (Figure 5.5).

Bansal *et al.* (2004) presented a new real-time predictive maintenance system for machine systems based on neural networks. Ghaboussi and Joghataie (1995) and Bruneo and De Vita (2019) emphasize the relevance of ANNs for predictive maintenance.

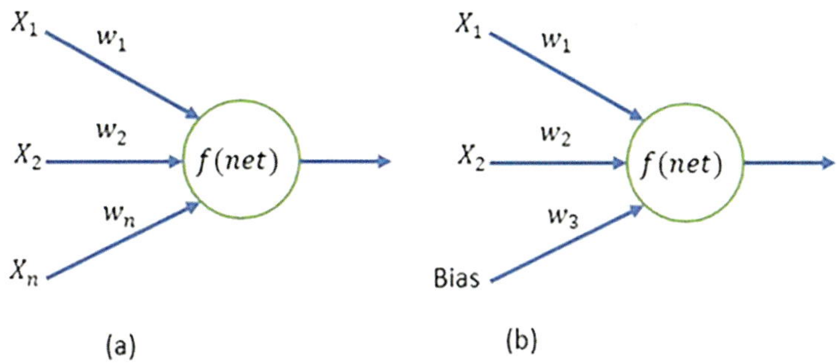

FIGURE 5.5 (a) An artificial neuron and (b) the perceptron net.

MLP and Recurrent Networks: Multilayer Perceptron (MLP) neural networks were used with success for predicting and diagnosing pump failures, showing the promising results with different types of failures (Wang & Hope, 2004; Kittisupakorn *et al.*, 2009; Yasaka *et al.*, 2018; Krizhevsky *et al.*, 2012). According to Ni and Wang (2000) and Partovi and Anandarajan (2002), neural networks have a high prediction accuracy. Neural networks have been an important aid in decision-making (Fonseca *et al.*, 2003).

About Recurrent Neural Networks (RNN), the Long Short Term Memory (LSTM)-based models presented good performance, both in time series classification and in prediction tasks (Guo *et al.*, 2017). An LSTM network is useful to solve nonlinear problems due to its nonlinear processing capacity (Liu *et al.*, 2020).

Sakalle *et al.* (2021) used an LSTM network to recognize a number of emotions in brain waves. The results obtained with this algorithm were better when compared to the other models referred in the study. The same approach was used in predictive maintenance for high-speed rail power equipment (Wang *et al.*, 2020). There are some architectures that have a good ability to predict univariate or multivariate temporal series with LSTM and Gated Recurrent Units (GRU) networks (Park *et al.*, 2018; Essien & Giannetti, 2020; Soloway & Haley, 1996).

The models that use RNN are usually suitable for time-series data. Hochreiter and Schmidhuber (1997) proposed an LSTM that demonstrated extraordinary performance in several sequence-centric tasks, such as handwriting recognition (Sak *et al.*, 2014; Dahl *et al.*, 2012), in auditory speech demonstration, (Alharbi *et al.*, 2018; Zaremba *et al.*, 2015), in dialect modelling (Luong *et al.*, 2015) in dialect translation. Besides these areas, networks were also used in predicting heart failure (Lasko *et al.*, 2013).

Deep Learning: Recently, Deep Learning strategies have been used with success in different areas (Bengio, 2009). Vincent *et al.* (2008) demonstrate that Deep Neural Networks can outperform other methods in voice recognition. A similar approach was also used in audio processing (Lee *et al.*, 2009).

Yasaka *et al.* (2018) use Deep Learning with a Convolutional Neural Network (CNN) to obtain a high performance in image recognition. The images themselves can be used in a learning process with this technique, and feature extraction prior to the learning process is not necessary. Other studies in the field of computer vision include: (Kriegeskorte & Golan, 2019; Zhang *et al.*, 2016).

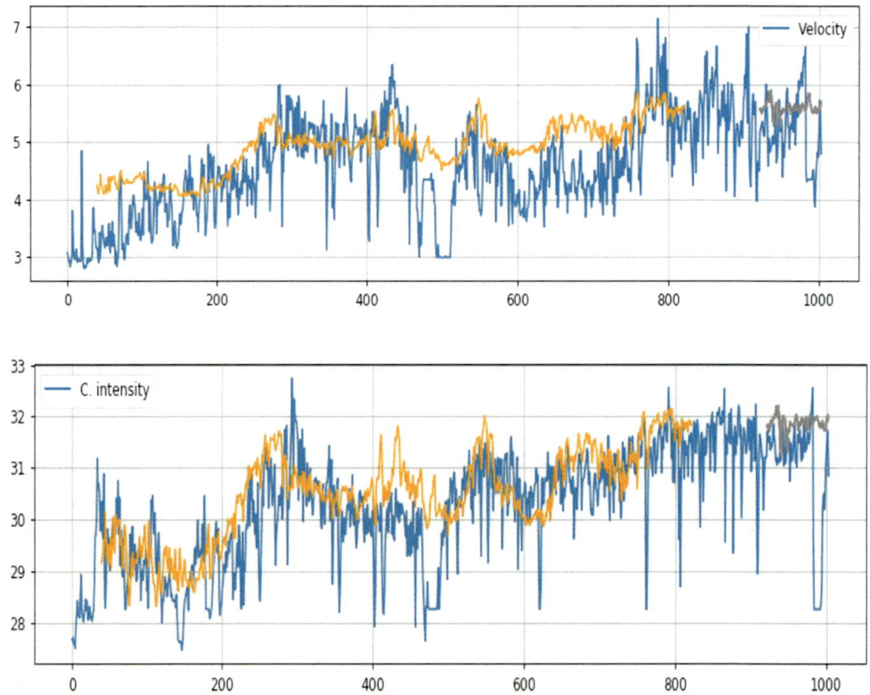

FIGURE 5.6 Results from training and test of algorithms of AI.

Krizhevsky *et al.* (2012) showed good results in image processing, employing a layered pre-training technique. The analysis shows that a large deep CNN can achieve excellent results in challenging data collection using supervised learning. This study also demonstrates how important is the amount of convolutional layers to achieve good results. There is a need for an exhaustive exploration of the types of layers, size, transfer functions, and other hyperparameters (Mateus *et al.*, 2021).

Figure 5.6 shows some results from training and test of algorithms of AI.

5.3.1 Some Predictive Algorithms

This section presents some of the most common predictive algorithms based on AI.

5.3.1.1 Artificial Neural Network

An ANN is based on a collection of connected units or nodes called artificial neurons, which try to model the neurons like a biological brain. Each connection, like the synapses in a biological brain, can transmit a signal to other neurons. An artificial neuron receives "signals" and processes them and the signal neurons connected to it. The "signal" at a connection is a real number, being the output of each neuron computed by some nonlinear function of the sum of its inputs. The connections are called *edges*. Neurons and edges typically have a *weight* that adjusts as learning proceeds. The *weight* increases or decreases the strength of the signal at a connection.

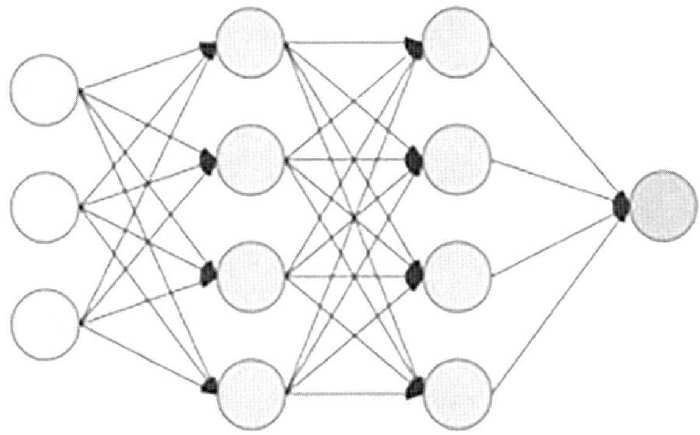

FIGURE 5.7 Artificial neural network.

Neurons may have a threshold such that a signal is sent only if the aggregate signal overtakes that threshold.

Typically, neurons are aggregated into layers. Different layers may perform different transformations on their inputs. Signals travel from the first layer (the input layer), to the last layer (the output layer), possibly after traversing the layer's multiple times (Figure 5.7).

5.3.1.2 Recurrent Neural Network

A RNN is a class of ANNs where connections between nodes can create a cycle, allowing output from some nodes to affect subsequent input to the same nodes. This allows it to exhibit temporal dynamic behaviour. Derived from feedforward neural networks, RNN can use their internal state (memory) to process variable length sequences of inputs. This makes them applicable to tasks such as unsegmented, connected handwriting recognition or speech recognition (Figure 5.8).

5.3.1.3 Random Forest

A Random Forest consists of multiple random decision trees. Two types of randomnesses are built into the trees: First, each tree is built on a random sample from the original data: Second, at each tree node, a subset of features are randomly selected to generate the best split.

Random Forest is an algorithm used for classification, regression, and feature selection.

In a Random Forest, each decision tree is built independently by selecting a random sample of features and a random subset of data points from the training set. The trees are then trained using their data sets and the predictions are made by multiple trees that are combined to reach the final decision (Figure 5.9).

5.3.1.4 Long Short-Term Memory

LSTM is an ANN used in the fields of AI and Deep Learning (DL). Unlike standard feedforward neural networks, LSTM has feedback connections, such a RNN can process not only single data points (such as images) but also entire sequences

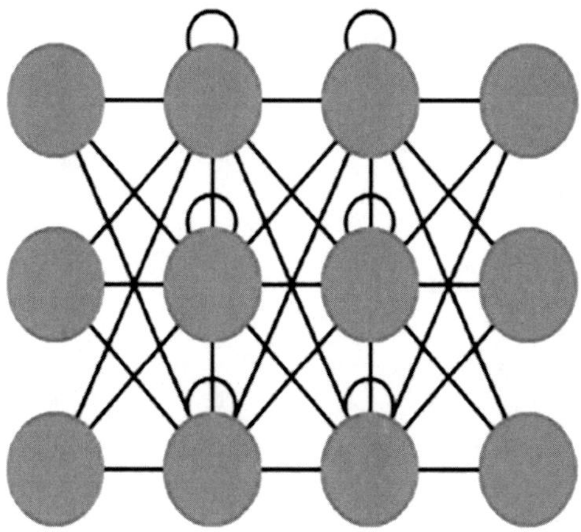

FIGURE 5.8 Recurrent neural network.

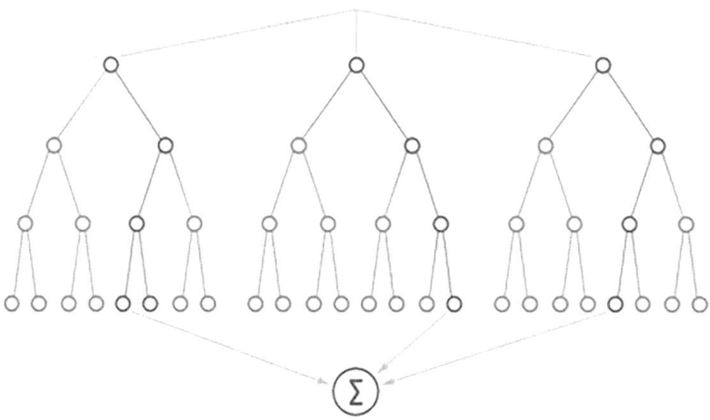

FIGURE 5.9 Random forest.

of data (such as speech or video). This characteristic makes LSTM networks ideal for processing and predicting data. For example, LSTM is applicable to tasks such as unsegmented, connected handwriting recognition, speech recognition, machine translation, speech activity detection, robot control, video games, and healthcare.

LSTM has a chain structure that contains four neural networks and different memory blocks called cells.

Information is retained by the cells and the memory manipulations are done by the gates. There are three gates (Figure 5.10):

- Forget Gate;
- Input gate;
- Output gate.

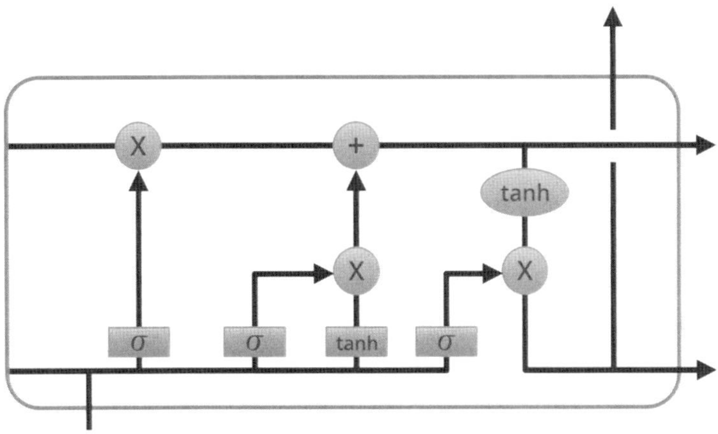

FIGURE 5.10 LSTM.

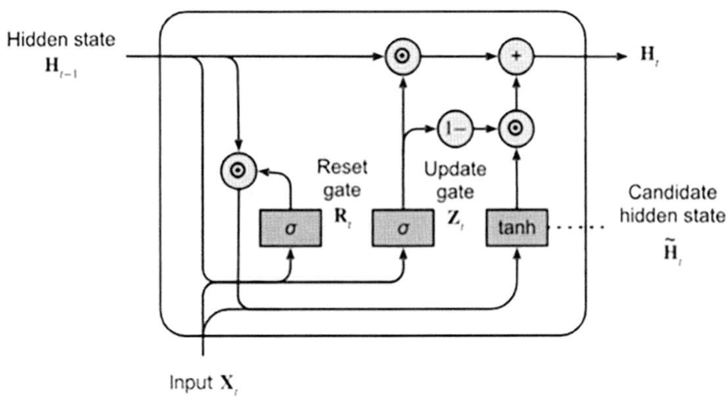

FIGURE 5.11 GRU.

5.3.1.5 Gated Recurrent Unit

GRU is a gating mechanism in RNN. The GRU is like a LSTM with a forget gate, but has fewer parameters than LSTM, as it lacks an output gate. GRU's performance on certain tasks of polyphonic music modelling, speech signal modelling, and natural language processing was found to be similar to that of LSTM (Figure 5.11).

5.3.1.6 Seq2Seq

Sequence to Sequence (Seq2Seq) is a model used in a sequence of prediction tasks, such as language modelling and machine translation. The idea is to use one LSTM, the *encoder*, to read the input sequence one-time step at a time, to obtain a large fixed dimensional vector representation (a context vector) and, then to use another LSTM, the *decoder*, to extract the output sequence from that vector. The second LSTM is essentially a RNN language model except that it is conditioned on the input sequence (Figure 5.12).

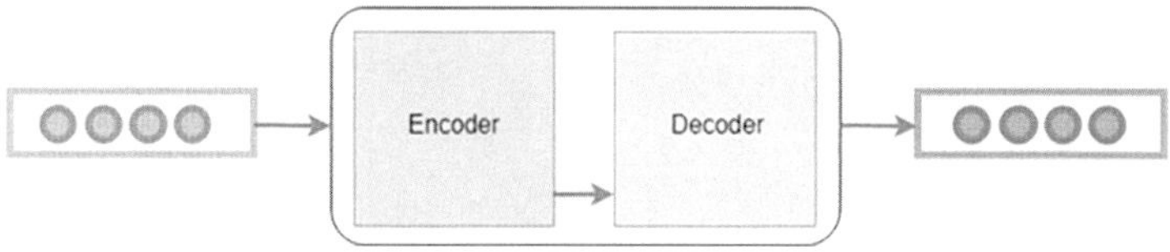

FIGURE 5.12 Seq2Seq.

REFERENCES

Alharbi, R., Magdy, W., Darwish, K., AbdelAli, A., & Mubarak, H. (2018) Part-of-speech tagging for Arabic Gulf dialect using Bi-LSTM. *Proceedings of the Eleventh International Conference on Language Resources and Evaluation (LREC 2018). European Language Resources Association (ELRA),* 2018-05. Miyazaki, Japan. https://aclanthology.org/L18-1620.pdf

Bansal, D., Evans, D. J., & Jones, B. (2004) A real-time predictive maintenance system for machine systems. *International Journal of Machine Tools and Manufacture, 44*, 759–766. https://doi.org/10.1016/j.ijmachtools.2004.02.004

Bengio, Y. (2009) *Learning Deep Architectures for AI.* Hanover, MA: Now Publishers Inc. https://doi.org/10.1561/2200000006

Box, G. E. P., Jenkins, G. M., Reinsel, G. C., & Ljung, G. M. (2015) *Time Series Analysis: Forecasting and Control.* Hoboken, NJ: John Wiley & Sons.

Bruneo, D. & De Vita, F. (2019) On the use of LSTM networks for predictive maintenance in smart industries. *2019 IEEE International Conference on Smart Computing (SMARTCOMP),* Washington, DC, USA, pp. 241–248. https://doi.org/10.1109/SMARTCOMP.2019.00059. https://ieeexplore.ieee.org/document/8784003

Chen, J., Gusikhin, O., Finkenstaedt, W., & Liu, Y.-N. (2019) Maintenance, repair, and operations parts inventory management in the era of industry 4.0. *IFAC-PapersOnLine, 52*(13), 171–176. https://doi.org/10.1016/j.ifacol.2019.11.171

Dahl, G., Yu, D., Deng, L., & Acero, A. (2012) Context-dependent pre-trained deep neural networks for large-vocabulary speech recognition. *IEEE Transactions on Audio, Speech, and Language Processing, 20*, 30–42. https://doi.org/10.1109/TASL.2011.2134090

Essien, A. & Giannetti, C. (2020) A deep learning model for smart manufacturing using convolutional LSTM neural network autoencoders. *IEEE Transactions on Industrial Informatics, 16*, 6069–6078. https://doi.org/10.1109/TII.2020.2967556

Farinha, J. T. (2018) *Asset Maintenance Engineering Methodologies* (1st Edition). Boca Raton, FL: CRC Press. https://doi.org/10.1201/9781315232867. https://www.taylorfrancis.com/books/mono/10.1201/9781315232867/asset-maintenance-engineering-methodologies-jos%C3%A9-manuel-torres-farinha

Fonseca, D., Navaresse, D., & Moynihan, G. (2003) Simulation metamodeling through artificial neural networks. *Engineering Applications of Artificial Intelligence, 16*, 177–183. https://doi.org/10.1016/S0952-1976(03)00043-5

Ghaboussi, J. & Joghataie, A. (1995) Active control of structures using neural networks. *Journal of Engineering Mechanics*, 121, 555–567. https://doi.org/10.1061/(ASCE)0733-9399(1995)121:4(555)

Gubbi, J., Buyya, R., Marusic, S., & Palaniswami, M. (2013). Internet of Things (IoT): a vision, architectural elements, and future directions. *Future Generation Computer Systems, 29*(7), 1645–1660. https://doi.org/10.1016/j.future.2013.01.010

Guo, Y., Wu, Z., & Ji, Y. (2017) A hybrid deep representation learning model for time series classification and prediction. *2017 3rd International Conference on Big Data Computing and Communications (BIGCOM),* Chengdu, China, pp. 226–231. https://doi.org/10.1109/BIGCOM.2017.13. https://ieeexplore.ieee.org/document/8113070

Hochreiter, S. & Schmidhuber, J. (1997) Long short-term memory. *Neural Computation, 9*, 1735–1780. https://doi.org/10.1162/neco.1997.9.8.1735

Kittisupakorn, P., Thitiyasook, P., Hussain, M., & Daosud, W. (2009) Neural network based model predictive control for a steel pickling process. *Journal of Process Control, 19*, 579–590. https://doi.org/10.1016/j.jprocont.2008.09.003

Kriegeskorte, N. & Golan, T. (2019) Neural network models and deep learning. *Current Biology Magazine 29*, R225–R240. https://doi.org/10.1016/j.cub.2019.02.034

Krizhevsky, A., Sutskever, I., & Hinton, G. E. (2012) Imagenet classification with deep convolutional neural networks. *Advances in Neural Information Processing Systems 2012, 25*, 1097–1105.

Lasko, T. A., Denny, J. C., & Levy, M. A. (2013) Computational phenotype discovery using unsupervised feature learning over noisy, sparse, and irregular clinical data. *PLOS One, 8*, e66341. https://doi.org/10.1371/journal.pone.0066341

Lee, H., Pham, P., Largman, Y., & Ng, A. (2009) Unsupervised feature learning for audio classification using convolutional deep belief networks. *Advances in Neural Information Processing Systems, 22*, 1096–1104.

Liu, Y., Duan, W., Huang, L., Duan, S., & Ma, X. (2020) The input vector space optimization for LSTM deep learning model in real-time prediction of ship motions. *Ocean Engineering, 213*, 107681. https://doi.org/10.1016/j.oceaneng.2020.107681

Luong, M. T., Sutskever, I., Le, Q. V., Vinyals, O., & Zaremba, W. (2015) Addressing the rare word problem in neural machine translation, 2015, [arXiv:cs.CL/1410.8206]

Makridakis, S. G., Wheelwright, S. C., & Hyndman, R. J. (1997) *Forecasting: Methods and Applications* (3rd Edition). New York: Wiley.

Mateus, B. C., Mendes, M., Farinha, J. T., & Cardoso, A. M. (2021) Anticipating future behavior of an industrial press using LSTM networks. *Applied Sciences, 11*(13), 6101. https://doi.org/10.3390/app11136101

Ni, H. G. & Wang, J. Z. (2000) Prediction of compressive strength of concrete by neural networks. *Cement and Concrete Research, 30*, 1245–1250. https://doi.org/10.1016/S0008-8846(00)00345-8

Ozel, Y., Guney, I., & Arca, E. (2008). Neural network solution to the cogeneration system by using coal. *Proceedings of the 12th WSEAS International Conference on CIRCUITS, Heraklion*, Greece, July 22–24, pp 279–283.

Park, S. H., Kim, B., Kang, C. M., Chung, C. C., & Choi, J.W. (2018) Sequence-to-sequence prediction of vehicle trajectory via LSTM encoder-decoder architecture. *2018 IEEE Intelligent Vehicles Symposium (IV)*. IEEE, pp. 1672–1678. https://doi.org/10.1109/IVS.2018.8500658

Partovi, F. Y. & Anandarajan, M. (2002) Classifying inventory using an artificial neural network approach. *Computers & Industrial Engineering, 41*, 389–404. https://doi.org/10.1016/S0360-8352(01)00064-X.

Raynor, W. J. (1999). *The International Dictionary of Artificial Intelligence* (Library ed). New York: Glenlake Pub. Co.; Fitzroy Dearborn Pub. https://doi.org/10.4324/9781315074108

Sak, H., Senior, A. W., & Beaufays, F. (2014) Long short-term memory recurrent neural network architectures for large scale acoustic modeling. ArXiv,volume={abs/1402.1128}

Sakalle, A., Tomar, P., Bhardwaj, H., Acharya, D., & Bhardwaj, A. (2021) A LSTM based deep learning network for recognizing emotions using wireless brainwave driven system. *Expert Systems with Applications, 173*, 114516. https://doi.org/10.1016/j.eswa.2020.114516

Soloway, D. & Haley, P. J. (1996) Neural generalized predictive control. *Proceedings of the 1996 {IEEE} International Symposium on Intelligent Control*. IEEE, 1996, pp. 277–282. https://doi.org/10.1109/ISIC.1996.556214

Vincent, P., Larochelle, H., Bengio, Y., & Manzagol, P. A. (2008) Extracting and composing robust features with denoising autoencoders. *Proceedings of the 25th International Conference on Machine Learning*. Association for Computing Machinery, ICML'08, Helsinki Finland, pp. 1096–1103. https://doi.org/10.1145/1390156.1390294

Wang, L. & Hope, A. D. (2004) Fault diagnosis: bearing fault diagnosis using multi-layer neural networks. *Insight - Non-Destructive Testing and Condition Monitoring*, *46*, 451–455. https://doi.org/10.1784/insi.46.8.451.39377

Wang, Q., Bu, S., & He, Z. (2020) Achieving predictive and proactive maintenance for high-speed railway power equipmentwith LSTM-RNN. *IEEE Transactions on Industrial Informatics*, *16*, 6509–6517. https://doi.org/10.1109/TII.2020.2966033

Yasaka, K., Akai, H., Kunimatsu, A., Kiryu, S., & Abe, O. (2018) Deep learning with convolutional neural network in radiology. *Japanese Journal of Radiology*, *36*, 257–272. https://doi.org/10.1007/s11604-018-0726-3

Zaremba, W., Sutskever, I., & Vinyals, O. (2015) Recurrent neural network regularization, 2015, [arXiv:cs.NE/1409.2329]

Zhang, L., Yang, F., Daniel Zhang, Y., & Zhu, Y. J. (2016) Road crack detection using deep convolutional neural network. *2016 IEEE International Conference on Image Processing (ICIP)*, Phoenix, AZ, USA, pp. 3708–3712. https://doi.org/10.1109/ICIP.2016.7533052

PART III

Standards

6 Maintenance Standards

6.1 INTRODUCTION

This chapter presents a main approach about the maintenance standards, namely the NP 4492, structured in the following perspectives and sections:

- Characterization of Maintenance Standard NP 4492, that presents the main guidelines of this standard and other norms related with this;
- Diagnosis of the Maintenance State, that permits to evaluate which are the changes necessary to do in the company aiming to have a state-of-the-art maintenance organization;
- Diagnosis to implement the standard NP 4492 that permits to evaluate which are the changes necessary to do in the organization aiming to implement it;
- Economical implications of implementing Maintenance Standards, that permit to evaluate the economic implications resulting from that implementation, according to the Life Cycle models presented in the first three chapters of this book.

6.2 CHARACTERIZATION OF MAINTENANCE STANDARD NP 4492

The standard NP4492 has the constituent standards the following ones:

- NP 4483:2009 - Guide for the implementation of the maintenance management system;
- NP EN 13269 - Maintenance – Instructions for preparing maintenance contracts;
- NP EN 13306 - Maintenance terminology;
- NP EN ISO 9000 - Quality management systems – Fundamentals and vocabulary;
- NP EN 13460 - Maintenance – Documentation for maintenance;
- NP EN 15341 - Maintenance – Maintenance Performance Indicators (KPI);
- CEN/TR 15628 - Maintenance – Qualification of Maintenance personnel.

To implement the Norm NP 4492, it is necessary to have:

- A Service provision manual

And to accomplish the following requirements:

1. From the organization;
2. Service offer;

DOI: 10.1201/9781003395690-9

3. Human resources;
4. Material resources;
5. Proposal elaboration;
6. Purchases of goods, equipment and services;
7. Management of maintenance service contracts and management control;
8. Quality program;
9. Provider preparation, planning and control;
10. Engineering studies;
11. Materials and parts management;
12. Safety, health and environment;
13. Audits;
14. Performance Indicators – Control KPI.

To apply the standard, it is desirable that each company make a diagnosis of the state of its internal organization aiming to prepare its implementation. The model proposed in this book is based in the model presented by Farinha (2018) and that is partially reproduced here to help the reader more easily follow its application.

The referred model corresponds to the diagnosis of the maintenance state with emphasis on the maintenance organisation, as is described in the next section and subsections.

6.3 DIAGNOSIS OF THE MAINTENANCE STATE

Either the introduction of integrated maintenance management systems in a company and or the need of reorganizing its maintenance department, namely to implement new administrative, organizational and methodological solutions, implies that previously be made an adequate diagnosis of the present state of its situation.

To manage this problem, it is particularly useful to use internal audit methods to diagnose the state of maintenance. These methods must be used initially to aid the company's reorganization. After this phase, they should be carried out periodically in order to assess the situation of the maintenance organization and correct deviations from the plan defined.

Diagnosing the maintenance department permits to identify the nonconforming aspects to correct and the new solutions to be implemented. Using this methodology, the reorganization can be implemented consistently and according to the most appropriate model that corresponds to the company profile.

The diagnosis method for assessing the maintenance state, outlined in this section, takes the data from the current maintenance status as inputs and generates a suite of reports, graphs, and indicators to facilitate the implementation of a reorganization. This will enable a more efficient organization and management as well as to implement new approaches for the organization of the maintenance department.

The diagnosis method also permit a permanent monitoring of the maintenance state, as well as the identification, at any time, of any deviations and therefore the definition of corrections to ensure proper consistency between the conceptual model of the organization and the one effectively implemented.

6.3.1 Holistic Diagnosis Model

The diagnosis model presented here was called Holistic Diagnosis Model (HDM) comprising three main components:

1. Basic organization;
2. Transversal concepts;
3. Management models.

The diagnosis method of the maintenance state is based on the use of a sequence of questionnaires whose evaluation of the responses places the situation of the maintenance department state in relation to reference landmarks of each subject analysed.

This diagnosis method was designed to be a practical tool for direct use by the responsible of each organization on assessing the maintenance management models and also systematic measures for their improvement. Additionally, it permits to evaluate the efficiency of the utilization of their Physical Assets.

The diagnosis of the maintenance state aims to support the implementation of an evolutionary step of this activity, providing it with innovative tools designed to accomplish the best international practices or, in other words, the state of the art. This way is essentially based on two vectors that guide throughout their application:

1. The first is the introduction of new tools essential to the processing of several data for an adequate maintenance management of Physical Assets, with the objective of obtaining an effective management of their Life Cycle Cost;
2. The second is the quality assurance systems through the potential subsequent certification by the appropriate general and specific standards, namely the maintenance ones. To reach this, it is required that maintenance function better interpret their actions, their operation, and their management.

The diagnosis model developed may be easily implemented through common tools, like Excel program, or yet programmed through a software language. It is important to have a friendly interface in order to help the user to easily understand and interact with the diagnosis model.

For the application of the surveys it was considered its suitability for any company and activity sector, having the concern do not individualize or limit the model that covers three aspects (Table 6.1):

1. Organization basis;
2. Transversal concepts;
3. Management models.

On the organization basis, seven questionnaires were considered:

1. Technical Asset Management;
2. Maintenance of 1st level;

TABLE 6.1
Sheds and Stadiums of the Questionnaires

Groups	Phases
Organization basis	1. Technical asset management
	2. Maintenance of 1st level
	3. Planning and security
	4. Data base
	5. Maintenance works
	6. Spare-parts
	7. Cost analysis
Transversal concepts	1. RAMS analysis
	2. RCM
	3. RBM
	4. Transversal tools
Management models	1. 5S
	2. TPM
	3. LEAN maintenance

3. Planning and security;
4. Data base;
5. Maintenance works;
6. Spare parts;
7. Cost analysis.

About the transversal concepts, the following four questionnaires were considered:

1. RAMS analysis;
2. RCM;
3. RBM;
4. Transversal tools.

Finally, on the part of Management Models, three questionnaires were used:

1. 5S;
2. TPM;
3. LEAN maintenance.

The final diagnosis is structured on the following questionnaires:

1. Technical asset management;
2. 1st level maintenance;
3. Planning and security;
4. Data base;
5. Maintenance works (interventions);

6. Spare parts;
7. Cost analysis.
8. RAMS analysis;
9. RCM;
10. RBM;
11. Transversal tools;
12. 5S;
13. TPM;
14. LEAN maintenance.

This method for diagnosing the maintenance state is based on a sequence of the referred precedent questionnaires whose evaluation of the responses places the position of the maintenance department referenced to landmarks defined in the diagnosis model. The methodology is based on the following phases:

- Data collection – questionnaires responses;
- Analysis of the data collected;
- Establishment of an improvement action plan.

6.3.2 Questionnaires

The methodology here presented is based on 14 questionnaires, called "Diagnosis Sheets" to be completed by the company maintenance manager or by external consultants. Each questionnaire is accompanied by an explanatory record about each question to answer to some doubt that may appear. Table 6.2 shows the 14 sheets that

TABLE 6.2
The 14 Stages of the Diagnostic Model with the Respective Highest and Lowest Scores

Stadiums	Activity	Maximum Score	Minimum Score
1	Management of physical assets	17	9.2
2	1st level maintenance	16	7.2
3	Planning and security	18	8.6
4	Data Base	12	5.0
5	Maintenance works (interventions)	16	6.9
6	Spare parts	16	8.2
7	Cost analysis	14	6.8
8	RAMS analysis	18	9.5
9	RCM	14	6.4
10	RBM	14	7.0
11	Transversal tools	12	5.4
12	5S	18	10.8
13	TPM	20	12.0
14	LEAN maintenance	21	10.6

constitute several questionnaires in which the diagnosis model with the respective highest and lowest scores is supported.

In each questionnaire, there are several questions with five possible answers, which are the following:

1. *Always* (it verified always);
2. *Almost always* (not always found);
3. *Sometimes* (sometimes it is found);
4. *Almost never* (seldom verified);
5. *Never* (never verified).

According to the type of answers, the final punctuation is evaluated having into consideration the limits referred in Table 6.2:

- The maximum score is reached when all responses are *Always* in all questions, being equal to the number of the total questions of each questionnaire;
- The minimum score depends on the type of questionnaire, according to the column and colours of each one. Each column has an evaluation of 1, 0.7, 0.5, 0.3, and 0. The colours are green, yellow, orange, and red.

Figure 6.1 shows a questionnaire number 1, in which it can be seen the global approach that is repeated in all questionnaires. Obviously, each one has its own specificities according to their own theme, like the first one exemplified.

However, it is necessary to consider the questions under the business context. For example, if the company subcontracts all maintenance services, it will not be able to answer questions related to the implementation of new internal works. In this case, it should not be marked by any response. If they are not answered more than 25% of the questions in each questionnaire, then it is not taken into account when evaluating the final assessment. It is considered that it does not met the minimum conditions for a sustained evaluation.

6.3.3 The Explanation Sheets

In order to minimize doubts about the content and understanding of the questions in each questionnaire, each one is accompanied by an explanatory sheet that allows, question by question, to understand the question and its several choices.

With the objective to exemplify this, Table 6.3 shows the first question of the diagnostic Sheet 12, about 5S. As is illustrated in it, the exemplified question is given an interpretation to the case of the answer, which is True or False, and also a suggestion for improvement.

Diagnosis sheets are identified at the top, through the stage number (1–14), by the term “Diagnosis sheet nº”.

It follows an intermediate zone where there is a grid with the questions and the columns reserved to their response. Each line begins with the indication of a number associated to each statement that consists of three or four digits. The first represents

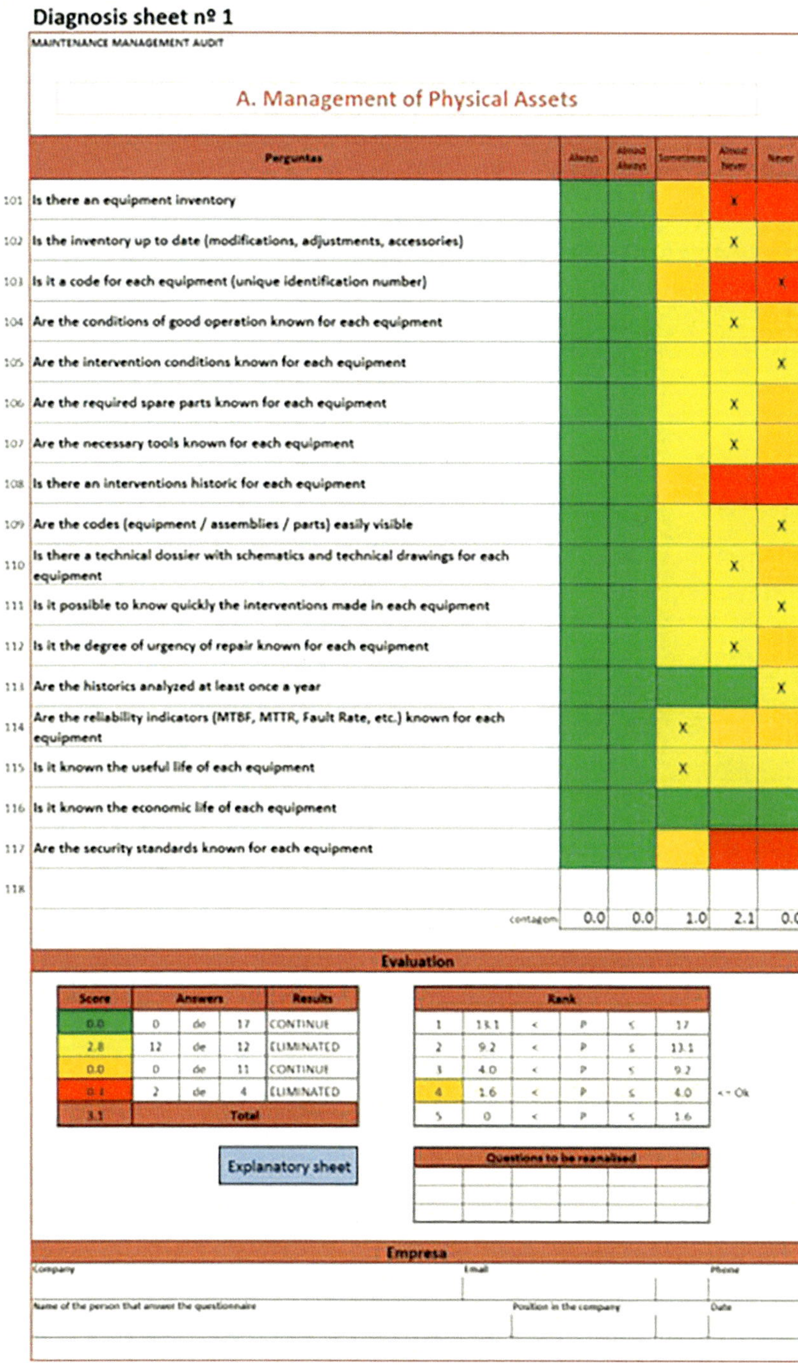
Diagnosis sheet nº 1

MAINTENANCE MANAGEMENT AUDIT

A. Management of Physical Assets

	Perguntas	Always	Almost Always	Sometimes	Almost Never	Never
101	Is there an equipment inventory				X	
102	Is the inventory up to date (modifications, adjustments, accessories)				X	
103	Is it a code for each equipment (unique identification number)					X
104	Are the conditions of good operation known for each equipment				X	
105	Are the intervention conditions known for each equipment					X
106	Are the required spare parts known for each equipment				X	
107	Are the necessary tools known for each equipment				X	
108	Is there an interventions historic for each equipment					
109	Are the codes (equipment / assemblies / parts) easily visible					X
110	Is there a technical dossier with schematics and technical drawings for each equipment				X	
111	Is it possible to know quickly the interventions made in each equipment					X
112	Is it the degree of urgency of repair known for each equipment				X	
113	Are the historics analyzed at least once a year					X
114	Are the reliability indicators (MTBF, MTTR, Fault Rate, etc.) known for each equipment			X		
115	Is it known the useful life of each equipment			X		
116	Is it known the economic life of each equipment					
117	Are the security standards known for each equipment					
118						
	contagem	0.0	0.0	1.0	2.1	0.0

Evaluation

Score	Answers			Results
0.0	0	de	17	CONTINUE
2.8	12	de	12	ELIMINATED
0.0	0	de	11	CONTINUE
[illegible]	2	de	4	ELIMINATED
3.1	Total			

Rank						
1	13.1	<	P	≤	17	
2	9.2	<	P	≤	13.1	
3	4.0	<	P	≤	9.2	
4	1.6	<	P	≤	4.0	<- Ok
5	0	<	P	≤	1.6	

Explanatory sheet

Questions to be reanalised

Empresa

Company	Email	Phone

Name of the person that answer the questionnaire	Position in the company	Date

FIGURE 6.1 Diagnostic sheet nº 1.

TABLE 6.3
Questionnaire 12/Question 1 – Two Explanations and Improvement Suggestions

1201	Separate the equipment, tools and materials useful from useless, eliminating the unnecessary ones (Seiri)
Explanation about the True	**Explanation about the False**
It separates the useful from the useless, eliminating unnecessary. The work begins to be putting things in order to use only what is really necessary and applicable. So, it is important to have only the necessary, in appropriate quantity and controlled to facilitate operations.	Do not separate the useful from the useless. It does not eliminate the unnecessary.
Improvement Suggestion	
Separating the useful from the useless, eliminating unnecessary. It is important to have the necessary, in adequate quantity and controlled to facilitate the operations. The work begins to be putting things in order, and only to use what is really necessary and applicable. It can also be interpreted with Sense of Use, Storage, Organization and Selection. It is essential to know how to separate and classify the useful objects from the unnecessary ones as follows: what is always used is put near the workplace; what is used almost always is put near the workplace; what is occasionally used is put a little away from the work place; what is rarely used, but necessary, is placed separately in a particular place; what is unnecessary is retired, sold or eliminated, because it occupies space needed to the working objects. Advantages: it reduces the need and the expense of space, stock, storage, transportation and insurance; it facilitates internal transport, physical arrangement, and production control; it avoids the purchase of materials and components in duplicate, and the damage of materials or stored products; it increases the productivity of equipment and people involved; it brings greater sense of humanization, organization, economy, less physical fatigue and greater ease of operation; it reduces risks of accidental use of these materials by staff. All team members must be able to distinguish the useful from the useless, what it is really necessary and what is not.	

the number of the questionnaire and the remaining two identify the number of the question and is indexed to the field "Explanatory sheet".

The final answers may have five possible options, as can be seen in the example presented in Figure 6.1:

1. Always;
2. Almost Always;
3. Sometimes;
4. Almost Never;
5. Never.

Finally, the bottom is reserved for the determination:

- Of the obtained score;
- Of the resulting rank categories;
- Of the achieved elimination criteria.

6.3.4 Organization and Analysis of Information Collected

Depending on the responses from the questionnaire, that interprets the current maintenance state, each diagnostic record reaches a certain score that is classified in a Rank with five levels:

1. Level 1
2. Level 2
3. Level 3
4. Level 4
5. Level 5

The most positive situation has Level 1 that is synonymous of very good positioning in relation to the highest value of the rank. Level 2 translates a good position. The negative situations are the next levels: Level 3 indicates that there are aspects to be improved in the organization; Level 4 reflects poor performance of the maintenance department, indicating that it must be carried on a broad and deep intervention to reorganize it; Level 5 means that a bad or inexistent maintenance organization.

Each questionnaire and stage must achieve a minimum score in order to sustain the position of the next stage. That is, the company cannot adequately ensure the implementation of the issues of a certain stage without the previous stage having reached a certain position that is considered positive. In practice, it is established that the company should reach the threshold of the third category as the minimum sustainability to ensure the implementation of the next stage.

6.3.5 Elimination Criteria

For every possibility of response to the several questions, a degree of importance is assigned, working as a criterion for elimination, classified in four colours – green, yellow, orange, and red – according to the interpretation given in Table 6.4.

It is considered as an elimination criterion if the company does not respond to a question of critical importance or exceeds the maximum number of allowable responses of exceptional importance.

The final evaluation of the questionnaires permits to determine the status of the maintenance department about the organization and management made over the equipment and facilities of the organization.

It can be seen in several questionnaires that the columns relating to responses like "always" or "almost always" are the most desirable possibilities of response, so they have always the "green" colour of the elimination criterion. However, if the answer is negative, in this case, this is not a critical point.

For each question answered in the negative options, i.e., "Sometimes" or "Almost Never" or "Never", the model automatically prepares a report of weak points (responses obtained in orange areas) or reports of critical points (responses obtained in red zones).

TABLE 6.4
Criteria Importance of Responses in the Position of State Maintenance

Green	**Adequate response** This answer is always desirable.
Yellow	**Inadequate response** Only some answers ought to be this way and the company should improve them.
Orange	**Exceptional response** Few answers should be of this type, although these responses are not qualifiers, the company must improve them as soon as possible.
Red	**Critical response** The company should never have this kind of response, being the first to be reviewed.

6.3.6 Elimination Grid

Qualifying grids are no more than coloured cells in red, orange, yellow, and green that are part of the grid of the questionnaire of each diagnosis questionnaire. The answers can be critical, exceptional, inadequate, and adequate, respectively, allowing to identify if the company, in each question, is eliminated or not, according to the process described above.

The process of assigning colours according to the criterions of elimination has resulted from the importance of how each question contributes to the maintenance state diagnosis and, by consequence, for its implications to the maintenance department reorganization.

The score (R) obtained by the company results from Equation 6.1:

$$R = \sum R_S + \sum R_{QS} + \sum R_{AV} + \sum R_{QN} \tag{6.1}$$

where

R_S corresponds to the answer Always
R_{QS} corresponds to the answer Almost Always
R_{AV} corresponds to the answer Sometimes
R_{QN} corresponds to the answer almost Never

The score gives rise to the category achieved by the company in each stage or questionnaire.

Table 6.5 shows an example of the application of the elimination criteria to the possibilities of answer.

6.3.7 Establishment of an Improvement Action Plan

The final stage of diagnosis allows to establish or propose an improvement action plan in the audited company. After the completion and evaluation of the scores of questionnaires, the model produces several positioning graphics, called "Radar

TABLE 6.5
Application of the Elimination Criteria to the Possibilities of Answer

1	Always	Almost Always	Sometimes	Almost Never	Never	The answer to the question is not fundamental to make the maintenance state diagnosis and the implementation of a new management model. A negative answer to this type of question only contributes to the positioning in a category of poor technical maintenance management.
	G	G	G	G	G	
2	Always	Almost Always	Sometimes	Almost Never	Never	The answer to the question is not fundamental to make the maintenance state diagnosis and the implementation of a new management model. However, if the company responds "Never", it is desirable to improve this positioning as soon as possible.
	G	G	G	G	Y	
3	Always	Almost Always	Sometimes	Almost Never	Never	The answer to the question is not fundamental to make the maintenance state diagnosis and the implementation of a new management model. However, if the company responds "Never", or "Almost Never" it is desirable to improve this positioning as soon as possible.
	G	G	G	Y	Y	
4	Always	Almost Always	Sometimes	Almost Never	Never	The answer to the question is not fundamental to make the maintenance state diagnosis and the implementation of a new management model. However, if the company responds "Never", or "Almost Never" it is desirable to improve this positioning as soon as possible.
	G	G	Y	Y	Y	
5	Always	Almost Always	Sometimes	Almost Never	Never	The answer to the question is important to make the maintenance state diagnosis and consequent implementation of a new management model. However, a negative positioning of the company on the various questions of this type may put into question an improvement program and should therefore contribute to the elimination of the stage under analysis.
	G	G	Y	Y	O	

(Continued)

TABLE 6.5 (*Continued*)
Application of the Elimination Criteria to the Possibilities of Answer

6	Always G	Almost Always G	Sometimes Y	Almost Never O	Never O	The answer to the question is important to make the maintenance state diagnosis and consequent implementation of a new management model. However, a negative positioning of the company on the various questions of this type may put into question an improvement program and should therefore contribute to the elimination of the stage under analysis.
7	Always G	Almost Always G	Sometimes O	Almost Never O	Never O	The answer to the question is important to make the maintenance state diagnosis and consequent implementation of a new management model. However, a negative positioning of the company on the various questions of this type may put into question an improvement program and should therefore contribute to the elimination of the stage under analysis.
8	Always G	Almost Always G	Sometimes O	Almost Never O	Never R	The answer to the question is important to make the maintenance state diagnosis and consequent implementation of a new management model. It is not possible to develop any improvement program if the company answers "Never".
9	Always G	Almost Always G	Sometimes O	Almost Never R	Never R	The answer to the question is important to make the maintenance state diagnosis and consequent implementation of a new management model. It is not possible to develop any improvement program if the company answers "Never" or "Almost Never".
10	Always G	Almost Always G	Sometimes R	Almost Never R	Never R	The answer to the question is important to make the maintenance state diagnosis and consequent implementation of a new management model. It is only possible to develop an improvement program if the company already applies the subject of the question.

Notes: G = green, Y = yellow, O = orange, R = red

Maps" that allow graphically view the positioning of the company and the most critical questionnaires that must be taken into account to improve the organization.

The diagnostic model automatically produces six radar maps about the company's positioning radar:

1. General position of the maintenance state;
2. General position of the maintenance state by each specific subject;
3. Position on the organization base;
4. Position on the transversal concepts;
5. Position on the transversal concepts and management models;
6. Position in the management models.

Additionally, to the radar maps positioning, several reports can be printed that help to make a characterization of the maintenance state, such as the following:

- The position report;
- The critical points report;
- The fragile points report;
- The weaknesses points report.

The position report, as its name indicates, has the objective to put the organization maintenance state in its several stages. It takes the diagnostic model making an immediate diagnosis of the general maintenance state, namely through the radar maps. The other reports allow obtaining the punctuations according to their designations (critical, fragile, and weaknesses), as well as suggestions for improvement for each referred point. Figure 6.2 shows an example of a Radar Map.

6.4 DIAGNOSIS TO IMPLEMENT THE STANDARD NP 4492

According to what was referred in the beginning of the previous section, to implement Standard NP 4492 it is necessary to have:

- A Service provision manual.

And to accomplish the following Requirements:

1. From the Organization;
2. Service Offer;
3. Human Resources;
4. Material resources;
5. Proposal elaboration;
6. Purchases of goods, equipment, and services;
7. Management of maintenance service contracts and management control;
8. Quality program;

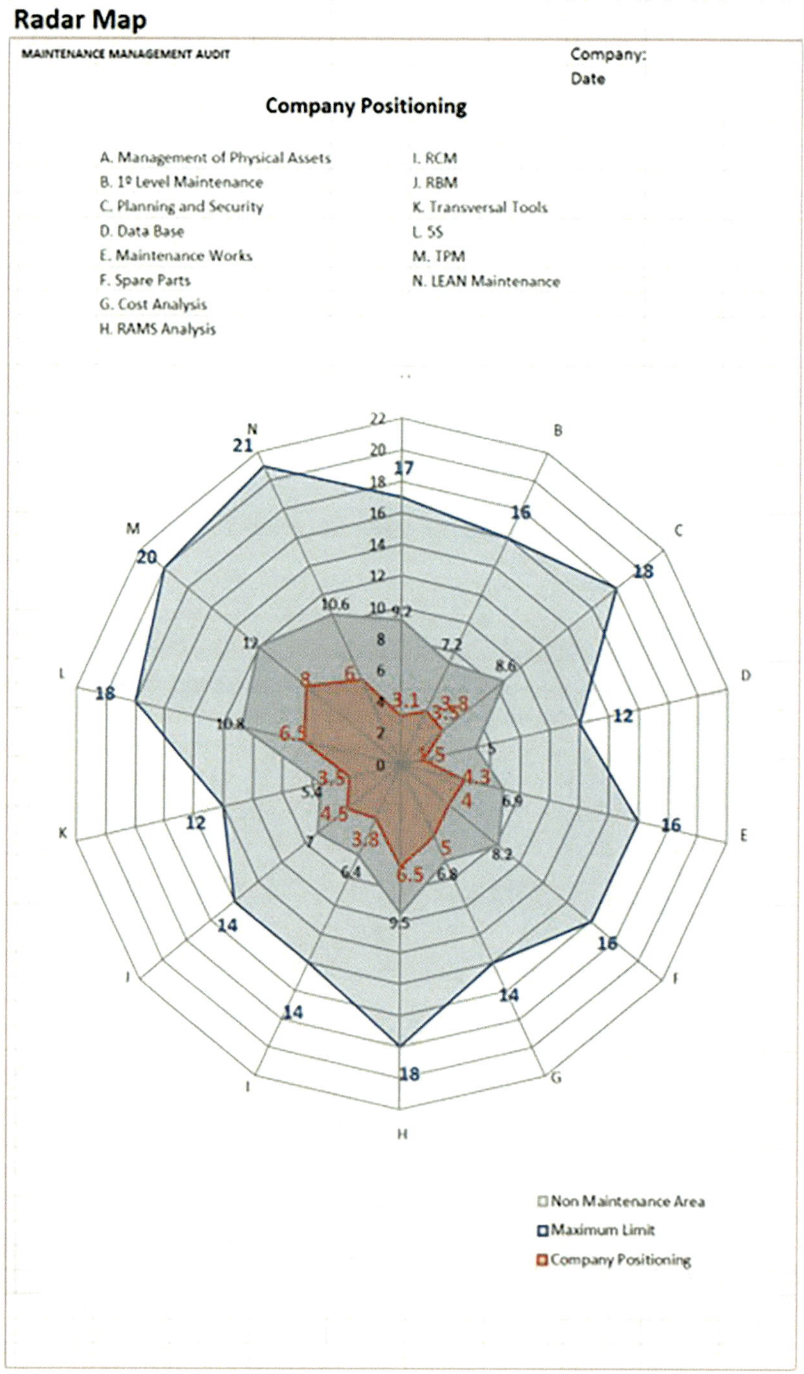

FIGURE 6.2 Radar map.

9. Provider preparation, planning, and control;
10. Engineering studies;
11. Materials and parts management;
12. Safety, health, and environment;
13. Audits;
14. Performance Indicators - Control KPI.

Then, the diagnosis model proposed to evaluate the Organization state aiming to implement the standard NP 4492 might have 15 questionnaires corresponding to the preceding requisites and the provision manual. Figure 6.3 illustrates the sheet about Service Offer.

6.5 ECONOMIC IMPLICATIONS OF IMPLEMENTING MAINTENANCE STANDARDS

This section presents the use of Life Cycle models described in Chapter 1 taking into account the implication of implementing policy and maintenance norms, aiming to support quantitatively the decision of the responsible.

6.5.1 Life Cycle Cost of Physical Assets

This section discusses the application of the first Life Cycle model presented in this book that corresponds to the Physical Asset Life Cycle Cost, which is the most traditional approach. They have used the three devaluation methods presented in Chapter 1:

- Linear depreciation method;
- Sum of the digits method;
- Exponential method.

The three methods for determining the economic cycle of equipment replacement, also referred in Chapter 1:

- Uniform Annual Income Method (UAI);
- Minimization of Total Average Cost Method (MTAC);
- MMTAC Reduced to Present Value (MTAC-RPV).

However, even though the simulations presented in the next sub-sections are traditional, called Life Cycle Cost (LCC), they add to them the profits that result from the use of Physical Assets, which increases a lot the quality of evaluation about the situations that can happen in day-to-day companies.

The simulations ought to always follow the same approach, that is, to compare values, aiming to demonstrate how the methods presented in this book can help the manager to decide if the implementation of the standards discussed in this book adds value to the organization or not.

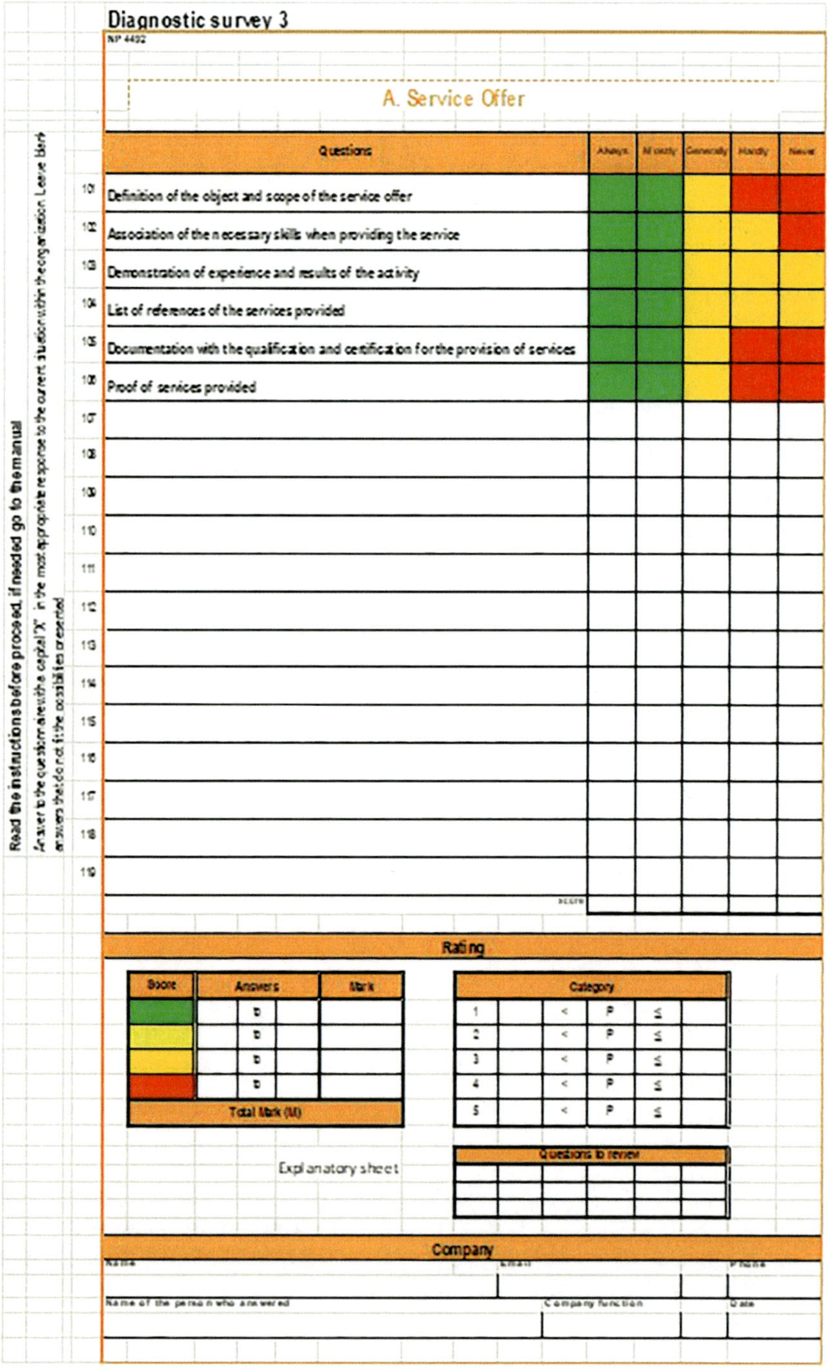

FIGURE 6.3 Diagnostic sheet referring to requisite Service Offer.

However, because this approach is so extensive and involves three methods, the simulations are made by changing only one variable; other situations may be simulated, namely changing the proposed variables.

6.5.1.1 Uniform Annual Income Method

The first simulation uses the following values:

- Initial Investment (II) –12000
- Residual value of equipment (RV) 10
- Functioning (F) –4500
- Energy –200
- Maintenance (M) –300
- Non-Production (NP) 0
- Benefit (B) 50000
- Internal Rate of Return (r) 0.12
- Risk (*R*) 0.11
- Apparent Rate 0.2432
- Maintenance Increase Rate 0.01
- Availability 0.8929
- MTBF 500
- MTTR 50
- MWT 10

The remaining values are presented in Table 6.6, and the graphical results are in Figures 6.4–6.7.

As referred in the preceding introduction of this section, the evaluation of the introduction of standard norms and maintenance policies can be simulated, primarily through the variation of MTBF and MTTR. They may be used other indicators, but may be not so intuitive.

The second simulation uses the following values (the MTBF is increased and MTTR decreases as a result of the Maintenance Standards):

- Initial Investment (II) –12000
- Residual value of equipment (RV) 10
- Functioning (*F*) –4500
- Energy –200
- Maintenance (*M*) –300
- Non-Production (NP) 0
- Benefit (*B*) 50000
- Internal Rate of Return (*r*) 0.12
- Risk (*R*) 0.11
- Apparent Rate 0.2432
- Maintenance Increase Rate 0.01
- Availability 0.8929
- **MTBF** **1000**
- **MTTR** **25**
- MWT 10

TABLE 6.6
Values Used for Simulation Based on UAI Method Related to Maintenance Standard – First Simulation

Time	0	1	2	3	4	5
Initial Investment (II)	−12000					
Initial Investment (II)	−12000					
II Present Value (IIPV) - Return	−12000	−9652.51	−7764.25	−6245.37	−5023.63	−4040.88
Life Time simulated	25					
Residual value of equipment (RV)	10					
Linear Depreciation Method (Vn)	−12000	−11519.60	−11039.20	−10558.80	−10078.40	−9598.00
LDM Present Value (Vn)		−9266.09	−7142.59	−5495.30	−4219.18	−3232.03
Sum of Digits Method (Vn)	−12000	−11076.15	−10189.26	−9339.32	−8526.34	−7750.31
SDM Present Value (Vn)		−8909.39	−6592.66	−4860.63	−3569.43	−2609.84
Exponential Method (Vn)	−12000	−9036.79	−6805.29	−5124.83	−3859.33	−2906.33
EM Present Value (Pv)		−7268.97	−4403.16	−2667.21	−1615.65	−978.68
Functioning (F)	−4500	−4500.00	−4545.00	−4590.45	−4636.35	−4682.72
Present Value Accumulated (NPV)		−3619.69	−6560.40	−8949.49	−10890.43	−12467.29
Present Value (NPV)		−3619.69	−2940.71	−2389.09	−1940.94	−1576.86
Energy	−200	−200.00	−400.00	−600.00	−800.00	−1000.00
Energy (NPV)		−160.88	−258.81	−312.27	−334.91	−336.74
Accumulated (FPV)		−15780.57	−22599.77	−31861.53	−43086.87	−55890.90
Maintenance (M)	−300	−300.00	−303.00	−306.03	−309.09	−312.18
Present Value (MPV)		−241.31	−196.05	−159.27	−129.40	−105.12
Accumulated (MPV)		−241.31	−437.36	−596.63	−726.03	−831.15
Non Production (NP)	0	0.00	0.00	0.00	0.00	0.00
Present Value (NPPV)		0.00	0.00	0.00	0.00	0.00
Accumulated NPPV		0.00	0.00	0.00	0.00	0.00
Accumulated Total Costs (APV)		−16021.88	−23037.13	−32458.16	−43812.90	−56722.05
Accumulated Total Costs		−5000.00	−10248.00	−15744.48	−21489.92	−27484.82
Benefit (B)	50000	44642.86	44642.86	44642.86	44642.86	44642.86
Present Value (BPV)		35909.63	28884.84	23234.27	18689.08	15033.05
Accumulated Benefits BPV		35909.63	64794.47	88028.74	106717.82	121750.87
BPV + FPV + MPV + NPPV + IIPV		7887.76	29757.34	43570.58	50904.93	53028.82
Uniform Annual Income UAIn (LDM)		−11519.60	−4921.18	−2786.88	−1764.99	−1185.10
Uniform Annual Income UAIn (SDM)		−11076.15	−4542.29	−2465.02	−1493.18	−956.96
Uniform Annual Income UAIn (EM)		−9036.79	−3033.74	−1352.65	−675.87	−358.86

(Continued)

TABLE 6.6 (*Continued*)
Values Used for Simulation Based on UAI Method Related to Maintenance Standard – First Simulation

Time	6	7	8	9	10
Initial Investment (II)					
Initial Investment (II)					
II Present Value (IIPV) - Return	−3250.39	−2614.53	−2103.07	−1691.66	−1360.73
Life Time simulated					
Residual value of equipment (RV)					
Linear Depreciation Method (Vn)	−9117.60	−8637.20	−8156.80	−7676.40	−7196.00
LDM Present Value (Vn)	−2469.64	−1881.85	−1429.53	−1082.15	−815.98
Sum of Digits Method (Vn)	−7011.23	−6309.11	−5643.94	−5015.72	−4424.46
SDM Present Value (Vn)	−1899.10	−1374.61	−989.13	−707.07	−501.71
Exponential Method (Vn)	−2188.66	−1648.20	−1241.20	−934.71	−703.90
EM Present Value (Pv)	−592.83	−359.11	−217.53	−131.77	−79.82
Functioning (F)	−4729.55	−4776.84	−4824.61	−4872.86	−4921.58
Present Value Accumulated (NPV)	−13748.36	−14789.13	−15634.67	−16321.60	−16879.68
Present Value (NPV)	−1281.07	−1040.77	−845.54	−686.93	−558.08
Energy	−1200.00	−1400.00	−1600.00	−1800.00	−2000.00
Energy (NPV)	−325.04	−305.03	−280.41	−253.75	−226.79
Accumulated (FPV)	−69964.30	−85058.45	−100973.53	−117548.88	−134655.35
Maintenance (M)	−315.30	−318.46	−321.64	−324.86	−328.11
Present Value (MPV)	−85.40	−69.38	−56.37	−45.80	−37.21
Accumulated (MPV)	−916.56	−985.94	−1042.31	−1088.11	−1125.31
Non Production (NP)	0.00	0.00	0.00	0.00	0.00
Present Value (NPPV)	0.00	0.00	0.00	0.00	0.00
Accumulated NPPV	0.00	0.00	0.00	0.00	0.00
Accumulated Total Costs (APV)	−70880.85	−86044.40	−102015.84	−118636.99	−135780.66
Accumulated Total Costs	−33729.67	−40224.97	−46971.22	−53968.93	−61218.62
Benefit (B)	44642.86	44642.86	44642.86	44642.86	44642.86
Present Value (BPV)	12092.22	9726.69	7823.91	6293.37	5062.23
Accumulated Benefits BPV	133843.09	143569.78	151393.69	157687.05	162749.28
BPV + FPV + MPV + NPPV + IIPV	50962.23	45525.38	37377.84	27050.06	14968.62
Uniform Annual Income UAIn (LDM)	−823.74	−585.16	−421.54	−306.37	−223.83
Uniform Annual Income UAIn (SDM)	−633.44	−427.43	−291.67	−200.18	−137.62
Uniform Annual Income UAIn (EM)	−197.74	−111.66	−64.14	−37.30	−21.89

(*Continued*)

TABLE 6.6 (*Continued*)
Values Used for Simulation Based on UAI Method Related to Maintenance Standard – First Simulation

Time	11	12	13	14	15
Initial Investment (II)					
Initial Investment (II)					
II Present Value (IIPV) - Return	−1094.54	−880.42	−708.19	−569.65	−458.21
Life Time simulated					
Residual value of equipment (RV)					
Linear Depreciation Method (Vn)	−6715.60	−6235.20	−5754.80	−5274.40	−4794.00
LDM Present Value (Vn)	−612.54	−457.47	−339.62	−250.38	−183.06
Sum of Digits Method (Vn)	−3870.15	−3352.80	−2872.40	−2428.95	−2022.46
SDM Present Value (Vn)	−353.00	−245.99	−169.52	−115.30	−77.23
Exponential Method (Vn)	−530.08	−399.19	−300.61	−226.38	−170.48
EM Present Value (Pv)	−48.35	−29.29	−17.74	−10.75	−6.51
Functioning (F)	−4970.80	−5020.51	−5070.71	−5121.42	−5172.63
Present Value Accumulated (NPV)	−17333.07	−17701.42	−18000.67	−18243.79	−18441.30
Present Value (NPV)	−453.39	−368.35	−299.25	−243.12	−197.51
Energy	−2200.00	−2400.00	−2600.00	−2800.00	−3000.00
Energy (NPV)	−200.67	−176.08	−153.44	−132.92	−114.55
Accumulated (FPV)	−152189.09	−170066.59	−188220.70	−206597.41	−225153.26
Maintenance (M)	−331.39	−334.70	−338.05	−341.43	−344.84
Present Value (MPV)	−30.23	−24.56	−19.95	−16.21	−13.17
Accumulated (MPV)	−1155.54	−1180.09	−1200.04	−1216.25	−1229.42
Non Production (NP)	0.00	0.00	0.00	0.00	0.00
Present Value (NPPV)	0.00	0.00	0.00	0.00	0.00
Accumulated NPPV	0.00	0.00	0.00	0.00	0.00
Accumulated Total Costs (APV)	−153344.63	−171246.68	−189420.75	−207813.66	−226382.68
Accumulated Total Costs	−68720.81	−76476.01	−84484.77	−92747.62	−101265.10
Benefit (B)	44642.86	44642.86	44642.86	44642.86	44642.86
Present Value (BPV)	4071.94	3275.37	2634.63	2119.23	1704.66
Accumulated Benefits BPV	166821.22	170096.59	172731.21	174850.44	176555.10
BPV + FPV + MPV + NPPV + IIPV	1476.59	−13150.10	−28689.53	−44963.22	−61827.58
Uniform Annual Income UAIn (LDM)	−163.92	−120.06	−87.78	−63.93	−46.29
Uniform Annual Income UAIn (SDM)	−94.47	−64.56	−43.81	−29.44	−19.53
Uniform Annual Income UAIn (EM)	−12.94	−7.69	−4.59	−2.74	−1.65

(Continued)

TABLE 6.6 (*Continued*)
Values Used for Simulation Based on UAI Method Related to Maintenance Standard – First Simulation

Time	16	17	18	19	20
Initial Investment (II)					
Initial Investment (II)					
II Present Value (IIPV) - Return	–368.57	–296.47	–238.48	–191.82	–154.30
Life Time simulated					
Residual value of equipment (RV)					
Linear Depreciation Method (Vn)	–4313.60	–3833.20	–3352.80	–2872.40	–2392.00
LDM Present Value (Vn)	–132.49	–94.70	–66.63	–45.92	–30.76
Sum of Digits Method (Vn)	–1652.92	–1320.34	–1024.71	–766.03	–544.31
SDM Present Value (Vn)	–50.77	–32.62	–20.36	–12.25	–7.00
Exponential Method (Vn)	–128.38	–96.68	–72.81	–54.83	–41.29
EM Present Value (Pv)	–3.94	–2.39	–1.45	–0.88	–0.53
Functioning (F)	–5224.36	–5276.60	–5329.37	–5382.66	–5436.49
Present Value Accumulated (NPV)	–18601.76	–18732.13	–18838.04	–18924.08	–18993.99
Present Value (NPV)	–160.46	–130.36	–105.91	–86.04	–69.90
Energy	–3200.00	–3400.00	–3600.00	–3800.00	–4000.00
Energy (NPV)	–98.29	–84.00	–71.54	–60.74	–51.43
Accumulated (FPV)	–243853.31	–262669.44	–281579.02	–300563.85	–319609.27
Maintenance (M)	–348.29	–351.77	–355.29	–358.84	–362.43
Present Value (MPV)	–10.70	–8.69	–7.06	–5.74	–4.66
Accumulated (MPV)	–1240.12	–1248.81	–1255.87	–1261.61	–1266.27
Non Production (NP)	0.00	0.00	0.00	0.00	0.00
Present Value (NPPV)	0.00	0.00	0.00	0.00	0.00
Accumulated NPPV	0.00	0.00	0.00	0.00	0.00
Accumulated Total Costs (APV)	–245093.43	–263918.25	–282834.89	–301825.45	–320875.53
Accumulated Total Costs	–110037.75	–119066.13	–128350.79	–137892.30	–147691.22
Benefit (B)	44642.86	44642.86	44642.86	44642.86	44642.86
Present Value (BPV)	1371.18	1102.95	887.18	713.63	574.03
Accumulated Benefits BPV	177926.28	179029.23	179916.42	180630.05	181204.07
BPV + FPV + MPV + NPPV + IIPV	–79167.14	–96889.02	–114918.47	–133195.41	–151671.46
Uniform Annual Income UAIn (LDM)	–33.24	–23.62	–16.53	–11.35	–7.58
Uniform Annual Income UAIn (SDM)	–12.74	–8.13	–5.05	–3.03	–1.72
Uniform Annual Income UAIn (EM)	–0.99	–0.60	–0.36	–0.22	–0.13

(*Continued*)

TABLE 6.6 (*Continued*)
Values Used for Simulation Based on UAI Method Related to Maintenance Standard – First Simulation

Time	21	22	23	24	25
Initial Investment (II)					
Initial Investment (II)					
II Present Value (IIPV) - Return	−124.11	−99.83	−80.30	−64.59	−51.96
Life Time simulated					
Residual value of equipment (RV)					
Linear Depreciation Method (Vn)	−1911.60	−1431.20	−950.80	−470.40	10.00
LDM Present Value (Vn)	−19.77	−11.91	−6.36	−2.53	0.04
Sum of Digits Method (Vn)	−359.54	−211.72	−100.86	−26.95	10.00
SDM Present Value (Vn)	−3.72	−1.76	−0.67	−0.15	0.04
Exponential Method (Vn)	−31.09	−23.42	−17.63	−13.28	−10.00
EM Present Value (Pv)	−0.32	−0.19	−0.12	−0.07	−0.04
Functioning (F)	−5490.86	−5545.76	−5601.22	−5657.23	−5713.81
Present Value Accumulated (NPV)	−19050.78	−19096.91	−19134.40	−19164.85	−19189.59
Present Value (NPV)	−56.79	−46.14	−37.48	−30.45	−24.74
Energy	−4200.00	−4400.00	−4600.00	−4800.00	−5000.00
Energy (NPV)	−43.44	−36.61	−30.78	−25.84	−21.65
Accumulated (FPV)	−338703.48	−357837.01	−377002.19	−396192.88	−415404.12
Maintenance (M)	−366.06	−369.72	−373.41	−377.15	−380.92
Present Value (MPV)	−3.79	−3.08	−2.50	−2.03	−1.65
Accumulated (MPV)	−1270.05	−1273.13	−1275.63	−1277.66	−1279.31
Non Production (NP)	0.00	0.00	0.00	0.00	0.00
Present Value (NPPV)	0.00	0.00	0.00	0.00	0.00
Accumulated NPPV	0.00	0.00	0.00	0.00	0.00
Accumulated Total Costs (APV)	−339973.54	−359110.13	−378277.81	−397470.53	−416683.42
Accumulated Total Costs	−157748.13	−168063.61	−178638.25	−189472.63	−200567.36
Benefit (B)	44642.86	44642.86	44642.86	44642.86	44642.86
Present Value (BPV)	461.73	371.41	298.75	240.31	193.30
Accumulated Benefits BPV	181665.81	182037.21	182335.96	182576.27	182769.57
BPV + FPV + MPV + NPPV + IIPV	−170307.73	−189072.92	−207941.85	−226894.26	−245913.85
Uniform Annual Income UAIn (LDM)	−4.86	−2.92	−1.56	−0.62	0.01
Uniform Annual Income UAIn (SDM)	−0.91	−0.43	−0.17	−0.04	0.01
Uniform Annual Income UAIn (EM)	−0.08	−0.05	−0.03	−0.02	−0.01

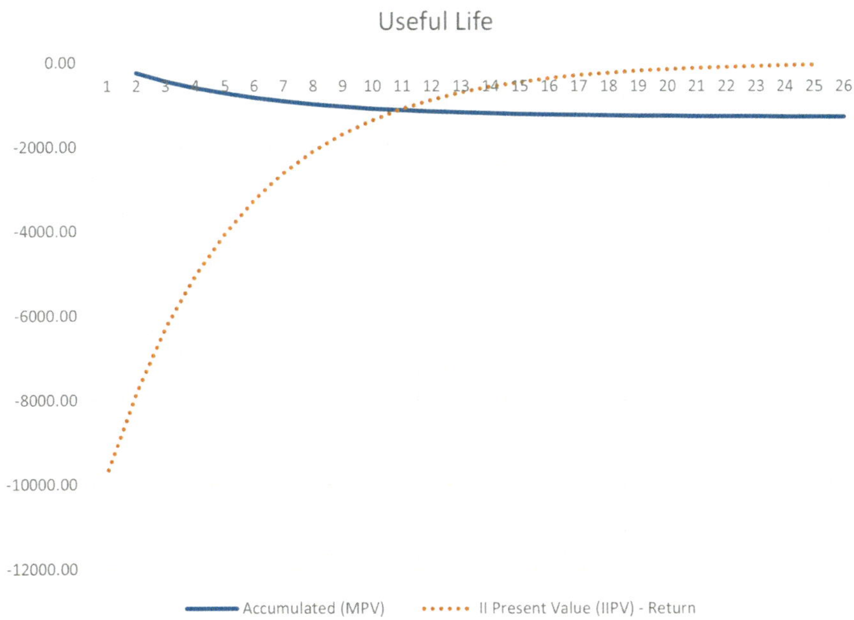

FIGURE 6.4 Physical asset useful life.

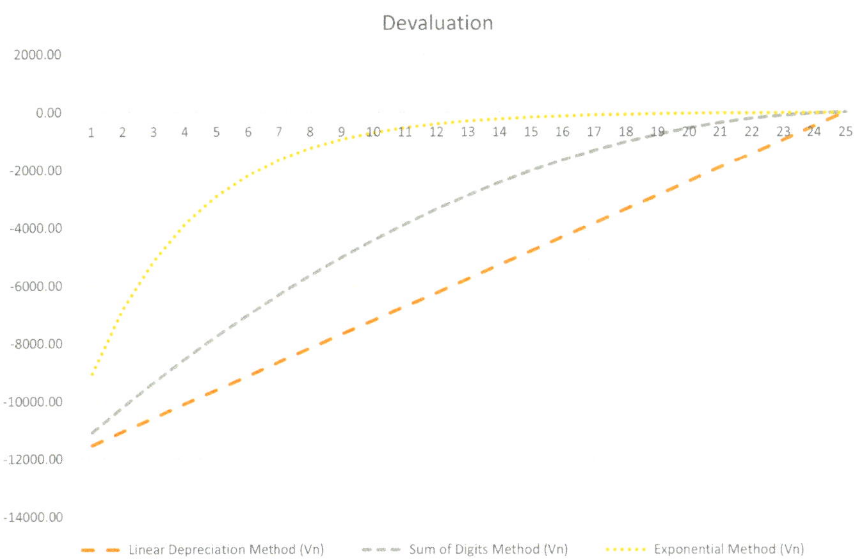

FIGURE 6.5 The three devaluation methods.

The remaining values are presented in Table 6.7, and the graphical results are shown in Figures 6.8–6.11.

As can be seen, namely comparing Figures 6.7 and 6.11, with the increasing of the maintenance management quality helped by the maintenance standards, it is getting one more positive year, but the positive incomes increase a lot.

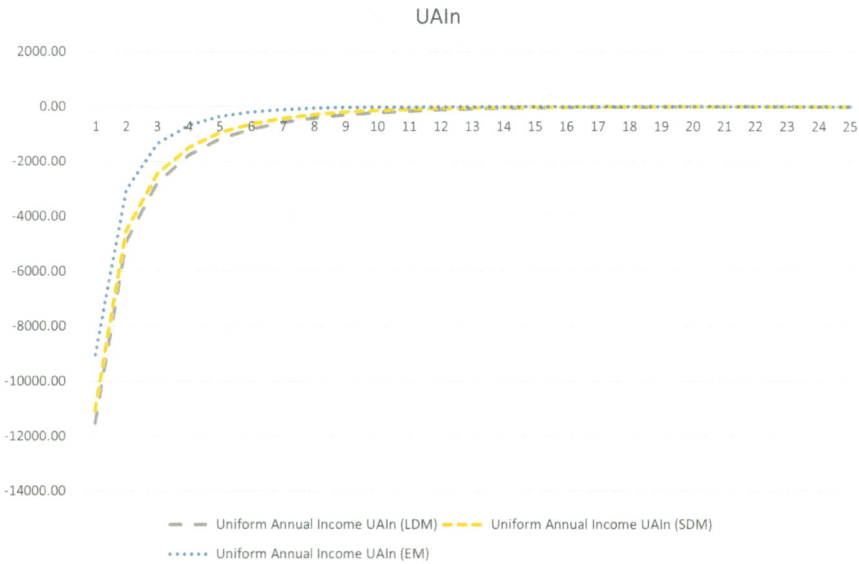

FIGURE 6.6 The UAI results from the three devaluation methods.

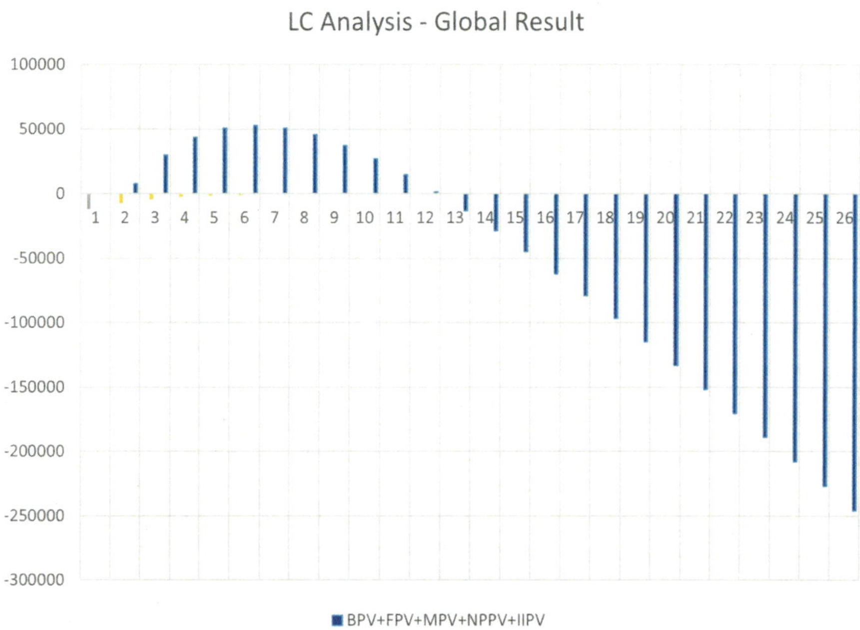

FIGURE 6.7 Life cycle analysis global result including profits.

TABLE 6.7
Values Used for Simulation based on UAI Method Related to Maintenance Standard – Second Simulation

Time	0	1	2	3	4	5
Initial Investment (II)	−12000					
Initial Investment (II)	−12000					
II Present Value (IIPV) - Return	−12000	−9652.51	−7764.25	−6245.37	−5023.63	−4040.88
Life Time simulated	25					
Residual value of equipment (RV)	10					
Linear Depreciation Method (Vn)	−12000	−11519.60	−11039.20	−10558.80	−10078.40	−9598.00
LDM Present Value (Vn)		−9266.09	−7142.59	−5495.30	−4219.18	−3232.03
Sum of Digits Method (Vn)	−12000	−11076.15	−10189.26	−9339.32	−8526.34	−7750.31
SDM Present Value (Vn)		−8909.39	−6592.66	−4860.63	−3569.43	−2609.84
Exponential Method (Vn)	−12000	−9036.79	−6805.29	−5124.83	−3859.33	−2906.33
EM Present Value (Pv)		−7268.97	−4403.16	−2667.21	−1615.65	−978.68
Functioning (*F*)	−4500	−4500.00	−4545.00	−4590.45	−4636.35	−4682.72
Present Value Accumulated (NPV)		−3619.69	−6560.40	−8949.49	−10890.43	−12467.29
Present Value (NPV)		−3619.69	−2940.71	−2389.09	−1940.94	−1576.86
Energy	−200	−200.00	−400.00	−600.00	−800.00	−1000.00
Energy (NPV)		−160.88	−258.81	−312.27	−334.91	−336.74
Accumulated (FPV)		−15780.57	−22599.77	−31861.53	−43086.87	−55890.90
Maintenance (*M*)	−300	−300.00	−303.00	−306.03	−309.09	−312.18
Present Value (MPV)		−241.31	−196.05	−159.27	−129.40	−105.12
Accumulated (MPV)		−241.31	−437.36	−596.63	−726.03	−831.15
Non Production (NP)	0	0.00	0.00	0.00	0.00	0.00
Present Value (NPPV)		0.00	0.00	0.00	0.00	0.00
Accumulated NPPV		0.00	0.00	0.00	0.00	0.00
Accumulated Total Costs (APV)		−16021.88	−23037.13	−32458.16	−43812.90	−56722.05
Accumulated Total Costs		−5000.00	−10248.00	−15744.48	−21489.92	−27484.82
Benefit (*B*)	50000	48309.18	48309.18	48309.18	48309.18	48309.18
Present Value (BPV)		38858.73	31257.03	25142.40	20223.93	16267.64
Accumulated Benefits BPV		38858.73	70115.76	95258.16	115482.09	131749.73
BPV + FPV + MPV + NPPV + IIPV		10836.86	35078.63	50799.99	59669.19	63027.68
Uniform Annual Income UAIn (LDM)		−11519.60	−4921.18	−2786.88	−1764.99	−1185.10
Uniform Annual Income UAIn (SDM)		−11076.15	−4542.29	−2465.02	−1493.18	−956.96
Uniform Annual Income UAIn (EM)		−9036.79	−3033.74	−1352.65	−675.87	−358.86

(Continued)

TABLE 6.7 (*Continued*)
Values Used for Simulation based on UAI Method Related to Maintenance Standard – Second Simulation

Time	6	7	8	9	10
Initial Investment (II)					
Initial Investment (II)					
II Present Value (IIPV) - Return	−3250.39	−2614.53	−2103.07	−1691.66	−1360.73
Life Time simulated					
Residual value of equipment (RV)					
Linear Depreciation Method (Vn)	−9117.60	−8637.20	−8156.80	−7676.40	−7196.00
LDM Present Value (Vn)	−2469.64	−1881.85	−1429.53	−1082.15	−815.98
Sum of Digits Method (Vn)	−7011.23	−6309.11	−5643.94	−5015.72	−4424.46
SDM Present Value (Vn)	−1899.10	−1374.61	−989.13	−707.07	−501.71
Exponential Method (Vn)	−2188.66	−1648.20	−1241.20	−934.71	−703.90
EM Present Value (Pv)	−592.83	−359.11	−217.53	−131.77	−79.82
Functioning (*F*)	−4729.55	−4776.84	−4824.61	−4872.86	−4921.58
Present Value Accumulated (NPV)	−13748.36	−14789.13	−15634.67	−16321.60	−16879.68
Present Value (NPV)	−1281.07	−1040.77	−845.54	−686.93	−558.08
Energy	−1200.00	−1400.00	−1600.00	−1800.00	−2000.00
Energy (NPV)	−325.04	−305.03	−280.41	−253.75	−226.79
Accumulated (FPV)	−69964.30	−85058.45	−100973.53	−117548.88	−134655.35
Maintenance (*M*)	−315.30	−318.46	−321.64	−324.86	−328.11
Present Value (MPV)	−85.40	−69.38	−56.37	−45.80	−37.21
Accumulated (MPV)	−916.56	−985.94	−1042.31	−1088.11	−1125.31
Non Production (NP)	0.00	0.00	0.00	0.00	0.00
Present Value (NPPV)	0.00	0.00	0.00	0.00	0.00
Accumulated NPPV	0.00	0.00	0.00	0.00	0.00
Accumulated Total Costs (APV)	−70880.85	−86044.40	−102015.84	−118636.99	−135780.66
Accumulated Total Costs	−33729.67	−40224.97	−46971.22	−53968.93	−61218.62
Benefit (*B*)	48309.18	48309.18	48309.18	48309.18	48309.18
Present Value (BPV)	13085.30	10525.50	8466.46	6810.21	5477.97
Accumulated Benefits BPV	144835.03	155360.53	163826.99	170637.20	176115.17
BPV + FPV + MPV + NPPV + IIPV	61954.18	57316.13	49811.14	40000.21	28334.51
Uniform Annual Income UAIn (LDM)	−823.74	−585.16	−421.54	−306.37	−223.83
Uniform Annual Income UAIn (SDM)	−633.44	−427.43	−291.67	−200.18	−137.62
Uniform Annual Income UAIn (EM)	−197.74	−111.66	−64.14	−37.30	−21.89

(*Continued*)

TABLE 6.7 (*Continued*)
Values Used for Simulation based on UAI Method Related to Maintenance Standard – Second Simulation

Time	11	12	13	14	15
Initial Investment (II)					
Initial Investment (II)					
II Present Value (IIPV) - Return	−1094.54	−880.42	−708.19	−569.65	−458.21
Life Time simulated					
Residual value of equipment (RV)					
Linear Depreciation Method (Vn)	−6715.60	−6235.20	−5754.80	−5274.40	−4794.00
LDM Present Value (Vn)	−612.54	−457.47	−339.62	−250.38	−183.06
Sum of Digits Method (Vn)	−3870.15	−3352.80	−2872.40	−2428.95	−2022.46
SDM Present Value (Vn)	−353.00	−245.99	−169.52	−115.30	−77.23
Exponential Method (Vn)	−530.08	−399.19	−300.61	−226.38	−170.48
EM Present Value (Pv)	−48.35	−29.29	−17.74	−10.75	−6.51
Functioning (*F*)	−4970.80	−5020.51	−5070.71	−5121.42	−5172.63
Present Value Accumulated (NPV)	−17333.07	−17701.42	−18000.67	−18243.79	−18441.30
Present Value (NPV)	−453.39	−368.35	−299.25	−243.12	−197.51
Energy	−2200.00	−2400.00	−2600.00	−2800.00	−3000.00
Energy (NPV)	−200.67	−176.08	−153.44	−132.92	−114.55
Accumulated (FPV)	−152189.09	−170066.59	−188220.70	−206597.41	−225153.26
Maintenance (*M*)	−331.39	−334.70	−338.05	−341.43	−344.84
Present Value (MPV)	−30.23	−24.56	−19.95	−16.21	−13.17
Accumulated (MPV)	−1155.54	−1180.09	−1200.04	−1216.25	−1229.42
Non Production (NP)	0.00	0.00	0.00	0.00	0.00
Present Value (NPPV)	0.00	0.00	0.00	0.00	0.00
Accumulated NPPV	0.00	0.00	0.00	0.00	0.00
Accumulated Total Costs (APV)	−153344.63	−171246.68	−189420.75	−207813.66	−226382.68
Accumulated Total Costs	−68720.81	−76476.01	−84484.77	−92747.62	−101265.10
Benefit (*B*)	48309.18	48309.18	48309.18	48309.18	48309.18
Present Value (BPV)	4406.35	3544.36	2851.00	2293.27	1844.65
Accumulated Benefits BPV	180521.51	184065.87	186916.87	189210.14	191054.79
BPV + FPV + MPV + NPPV + IIPV	15176.89	819.19	−14503.88	−30603.52	−47327.89
Uniform Annual Income UAIn (LDM)	−163.92	−120.06	−87.78	−63.93	−46.29
Uniform Annual Income UAIn (SDM)	−94.47	−64.56	−43.81	−29.44	−19.53
Uniform Annual Income UAIn (EM)	−12.94	−7.69	−4.59	−2.74	−1.65

(*Continued*)

TABLE 6.7 (*Continued*)
Values Used for Simulation based on UAI Method Related to Maintenance Standard – Second Simulation

Time	16	17	18	19	20
Initial Investment (II)					
Initial Investment (II)					
II Present Value (IIPV) - Return	−368.57	−296.47	−238.48	−191.82	−154.30
Life Time simulated					
Residual value of equipment (RV)					
Linear Depreciation Method (Vn)	−4313.60	−3833.20	−3352.80	−2872.40	−2392.00
LDM Present Value (Vn)	−132.49	−94.70	−66.63	−45.92	−30.76
Sum of Digits Method (Vn)	−1652.92	−1320.34	−1024.71	−766.03	−544.31
SDM Present Value (Vn)	−50.77	−32.62	−20.36	−12.25	−7.00
Exponential Method (Vn)	−128.38	−96.68	−72.81	−54.83	−41.29
EM Present Value (Pv)	−3.94	−2.39	−1.45	−0.88	−0.53
Functioning (*F*)	−5224.36	−5276.60	−5329.37	−5382.66	−5436.49
Present Value Accumulated (NPV)	−18601.76	−18732.13	−18838.04	−18924.08	−18993.99
Present Value (NPV)	−160.46	−130.36	−105.91	−86.04	−69.90
Energy	−3200.00	−3400.00	−3600.00	−3800.00	−4000.00
Energy (NPV)	−98.29	−84.00	−71.54	−60.74	−51.43
Accumulated (FPV)	−243853.31	−262669.44	−281579.02	−300563.85	−319609.27
Maintenance (*M*)	−348.29	−351.77	−355.29	−358.84	−362.43
Present Value (MPV)	−10.70	−8.69	−7.06	−5.74	−4.66
Accumulated (MPV)	−1240.12	−1248.81	−1255.87	−1261.61	−1266.27
Non Production (NP)	0.00	0.00	0.00	0.00	0.00
Present Value (NPPV)	0.00	0.00	0.00	0.00	0.00
Accumulated NPPV	0.00	0.00	0.00	0.00	0.00
Accumulated Total Costs (APV)	−245093.43	−263918.25	−282834.89	−301825.45	−320875.53
Accumulated Total Costs	−110037.75	−119066.13	−128350.79	−137892.30	−147691.22
Benefit (*B*)	48309.18	48309.18	48309.18	48309.18	48309.18
Present Value (BPV)	1483.79	1193.53	960.05	772.24	621.17
Accumulated Benefits BPV	192538.59	193732.12	194692.16	195464.40	196085.57
BPV + FPV + MPV + NPPV + IIPV	−64554.84	−82186.13	−100142.73	−118361.06	−136789.97
Uniform Annual Income UAIn (LDM)	−33.24	−23.62	−16.53	−11.35	−7.58
Uniform Annual Income UAIn (SDM)	−12.74	−8.13	−5.05	−3.03	−1.72
Uniform Annual Income UAIn (EM)	−0.99	−0.60	−0.36	−0.22	−0.13

(Continued)

TABLE 6.7 (*Continued*)
Values Used for Simulation based on UAI Method Related to Maintenance Standard – Second Simulation

Time	21	22	23	24	25
Initial Investment (II)					
Initial Investment (II)					
II Present Value (IIPV) - Return	−124.11	−99.83	−80.30	−64.59	−51.96
Life Time simulated					
Residual value of equipment (RV)					
Linear Depreciation Method (Vn)	−1911.60	−1431.20	−950.80	−470.40	10.00
LDM Present Value (Vn)	−19.77	−11.91	−6.36	−2.53	0.04
Sum of Digits Method (Vn)	−359.54	−211.72	−100.86	−26.95	10.00
SDM Present Value (Vn)	−3.72	−1.76	−0.67	−0.15	0.04
Exponential Method (Vn)	−31.09	−23.42	−17.63	−13.28	−10.00
EM Present Value (Pv)	−0.32	−0.19	−0.12	−0.07	−0.04
Functioning (*F*)	−5490.86	−5545.76	−5601.22	−5657.23	−5713.81
Present Value Accumulated (NPV)	−19050.78	−19096.91	−19134.40	−19164.85	−19189.59
Present Value (NPV)	−56.79	−46.14	−37.48	−30.45	−24.74
Energy	−4200.00	−4400.00	−4600.00	−4800.00	−5000.00
Energy (NPV)	−43.44	−36.61	−30.78	−25.84	−21.65
Accumulated (FPV)	−338703.48	−357837.01	−377002.19	−396192.88	−415404.12
Maintenance (*M*)	−366.06	−369.72	−373.41	−377.15	−380.92
Present Value (MPV)	−3.79	−3.08	−2.50	−2.03	−1.65
Accumulated (MPV)	−1270.05	−1273.13	−1275.63	−1277.66	−1279.31
Non Production (NP)	0.00	0.00	0.00	0.00	0.00
Present Value (NPPV)	0.00	0.00	0.00	0.00	0.00
Accumulated NPPV	0.00	0.00	0.00	0.00	0.00
Accumulated Total Costs (APV)	−339973.54	−359110.13	−378277.81	−397470.53	−416683.42
Accumulated Total Costs	−157748.13	−168063.61	−178638.25	−189472.63	−200567.36
Benefit (*B*)	48309.18	48309.18	48309.18	48309.18	48309.18
Present Value (BPV)	499.65	401.91	323.29	260.04	209.17
Accumulated Benefits BPV	196585.22	196987.13	197310.42	197570.46	197779.63
BPV + FPV + MPV + NPPV + IIPV	−155388.32	−174123.00	−192967.40	−211900.07	−230903.79
Uniform Annual Income UAIn (LDM)	−4.86	−2.92	−1.56	−0.62	0.01
Uniform Annual Income UAIn (SDM)	−0.91	−0.43	−0.17	−0.04	0.01
Uniform Annual Income UAIn (EM)	−0.08	−0.05	−0.03	−0.02	−0.01

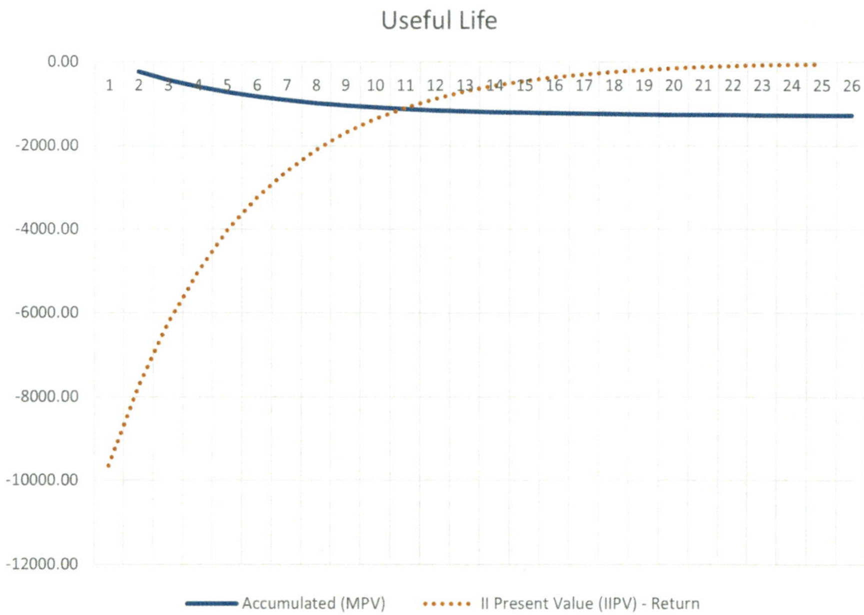

FIGURE 6.8 Physical asset useful life.

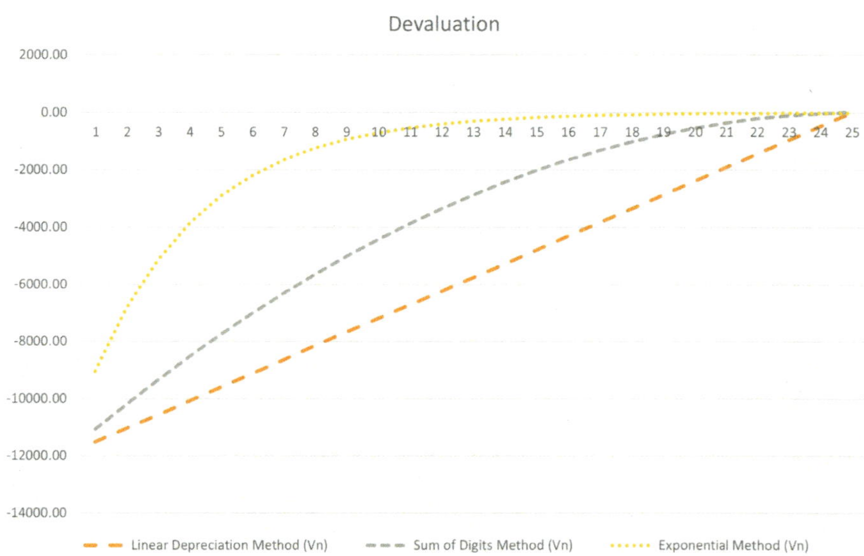

FIGURE 6.9 The three devaluation methods.

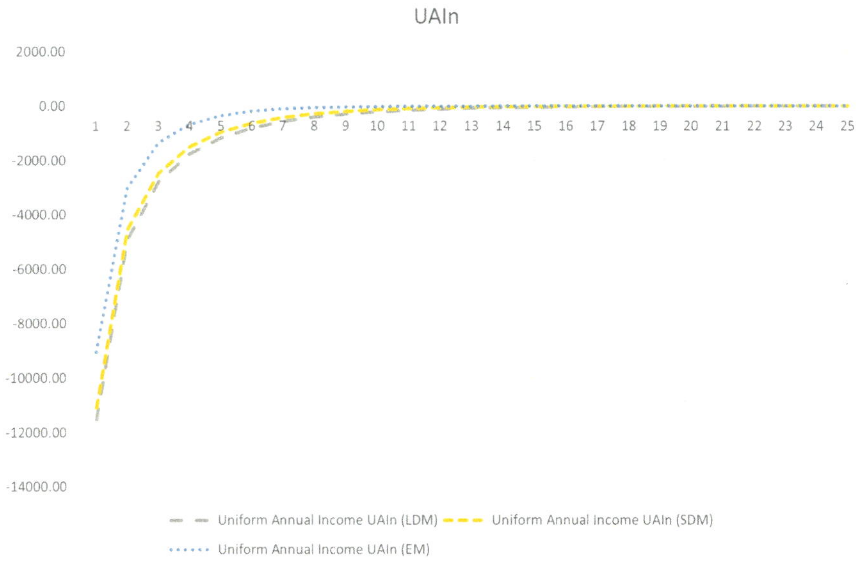

FIGURE 6.10 The UAI results from the three devaluation methods.

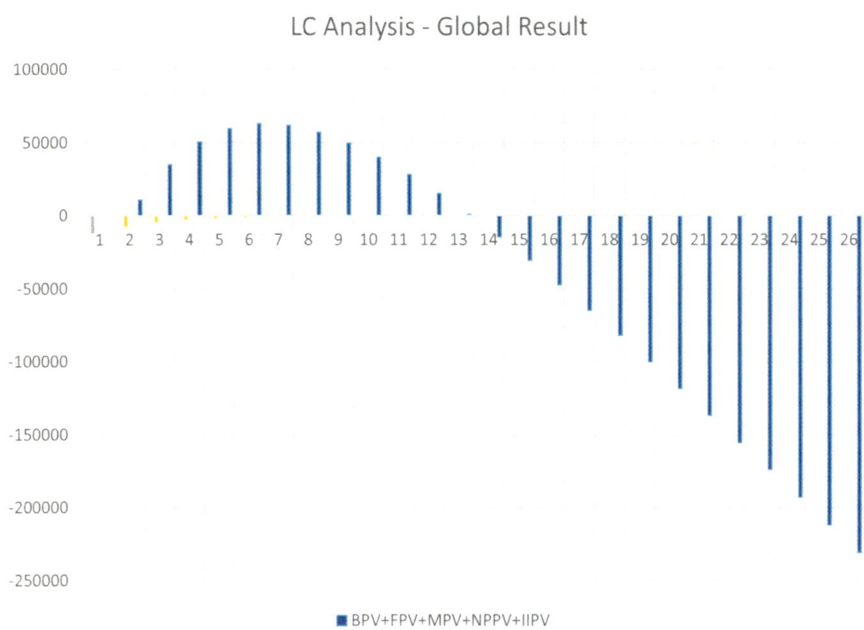

FIGURE 6.11 Life cycle analysis global result including profits.

6.5.1.2 Minimization of Total Average Cost Method (MMTAC)

The first simulation uses the following values:

- Initial Investment (II) –12000
- Life Time simulated 25
- Residual value of equipment (RV) 10
- Functioning (*F*) –4500
- Energy –200
- Maintenance (*M*) –300
- Non-Production (NP) 0
- Benefit (*B*) 50000
- Internal Rate of Return (*r*) 0.0
- Risk (*R*) 0.0
- Apparent Rate 0.0
- Maintenance Increase Rate 0.01
- Availability 0.8929
- MTBF **500**
- MTTR **50**
- MWT 10

The remaining values are presented in Table 6.8 and the graphical results are shown in Figures 6.12–6.15.

As referred in the preceding introduction of this section, the evaluation of the introduction of standard norms and maintenance policies can be simulated, primarily through the variation of MTBF and MTTR. They may be used other indicators, but may be not so intuitive.

The second simulation uses the following values (the MTBF is increased and MTTR decreases as a result of the Maintenance Standards):

- Initial Investment (II) –12000
- Life Time simulated 25
- Residual value of equipment (RV) 10
- Functioning (*F*) –4500
- Energy –200
- Maintenance (*M*) –300
- Non-Production (NP) 0
- Benefit (*B*) 50000
- Internal Rate of Return (*r*) 0.0
- Risk (*R*) 0.0
- Apparent Rate 0.0
- Maintenance Increase Rate 0.01
- Availability 0.9662
- MTBF **1000**
- MTTR **25**
- MWT 10

TABLE 6.8
Values Used for Simulation based on MMTAC Method Related to Maintenance Standard – First Simulation

Time	0	1	2	3	4	5
Initial Investment (II)	−12000					
Initial Investment (II)	−12000					
II (IIPV) - Return	−12000	−12000.00	−12000.00	−12000.00	−12000.00	−12000.00
Useful Life	25					
Residual value of equipment (RV)	10					
Linear Depreciation Method (Vn)	−12000	−11519.60	−11039.20	−10558.80	−10078.40	−9598.00
LDM Present Value (PV)		−11519.60	−11039.20	−10558.80	−10078.40	−9598.00
Sum of Digits Method (Vn)	−12000	−11076.15	−10189.26	−9339.32	−8526.34	−7750.31
SDM Present Value (PV)		−11076.15	−10189.26	−9339.32	−8526.34	−7750.31
Exponential Method (Vn)	−12000	−9036.79	−6805.29	−5124.83	−3859.33	−2906.33
EM Present Value (Pv)		−9036.79	−6805.29	−5124.83	−3859.33	−2906.33
Functioning (F)	−4500	−4500.00	−4545.00	−4590.45	−4636.35	−4682.72
Functioning Accumulated (FA)		−4500.00	−9045.00	−13635.45	−18271.80	−22954.52
Present Value Accumulated (NPV)		−4500.00	−9045.00	−13635.45	−18271.80	−22954.52
Present Value (NPV)		−4500.00	−4545.00	−4590.45	−4636.35	−4682.72
Energy	−200	−200.00	−400.00	−600.00	−800.00	−1000.00
Energy Accumulated (EA)		−200.00	−600.00	−1200.00	−2000.00	−3000.00
Energy (NPV)		−200.00	−400.00	−600.00	−800.00	−1000.00
Accumulated Functioning (FPV)		−16700.00	−26145.00	−40380.45	−59452.25	−83406.78
Maintenance (M)	−300	−300.00	−303.00	−306.03	−309.09	−312.18
Maintenane Accumulated (MA)		−300.00	−603.00	−909.03	−1218.12	−1530.30
Present Value (MPV)		−300.00	−303.00	−306.03	−309.09	−312.18
Accumulated MPV		−300.00	−603.00	−909.03	−1218.12	−1530.30
Non Production (NP)	0	0.00	0.00	0.00	0.00	0.00
Non Production Accumulated (NPA)		0.00	0.00	0.00	0.00	0.00
Present Value (NPPV)		0.00	0.00	0.00	0.00	0.00
Accumulated NPPV		0.00	0.00	0.00	0.00	0.00
Accumulated Total Costs (APV)		−17000.00	−26748.00	−41289.48	−60670.37	−84937.08
Accumulated Total Costs		−5000.00	−10248.00	−15744.48	−21489.92	−27484.82
Benefit (B)	50000	44642.86	44642.86	44642.86	44642.86	44642.86
Benefits Accumulated (BA)		44642.86	89285.71	133928.57	178571.43	223214.29
Present Value (BPV)		44642.86	44642.86	44642.86	44642.86	44642.86
Accumulated Benefits BPV		44642.86	89285.71	133928.57	178571.43	223214.29
BPV+FPV+MPV+ NPPV+IIPV		27642.86	67037.71	106184.09	145081.50	183729.46
MTAC (LDM)		−5480.40	−11208.80	−17185.68	−23411.52	−29886.82
MTAC (SDM)		−5923.85	−12058.74	−18405.16	−24963.59	−31734.52
MTAC (EM)		−7963.21	−15442.71	−22619.65	−29630.59	−36578.49

(Continued)

TABLE 6.8 (*Continued*)
Values Used for Simulation based on MMTAC Method Related to Maintenance Standard – First Simulation

Time	6	7	8	9	10
Initial Investment (II)					
Initial Investment (II)					
II (IIPV) - Return	−12000.00	−12000.00	−12000.00	−12000.00	−12000.00
Useful Life					
Residual value of equipment (RV)					
Linear Depreciation Method (Vn)	−9117.60	−8637.20	−8156.80	−7676.40	−7196.00
LDM Present Value (PV)	−9117.60	−8637.20	−8156.80	−7676.40	−7196.00
Sum of Digits Method (Vn)	−7011.23	−6309.11	−5643.94	−5015.72	−4424.46
SDM Present Value (PV)	−7011.23	−6309.11	−5643.94	−5015.72	−4424.46
Exponential Method (Vn)	−2188.66	−1648.20	−1241.20	−934.71	−703.90
EM Present Value (Pv)	−2188.66	−1648.20	−1241.20	−934.71	−703.90
Functioning (F)	−4729.55	−4776.84	−4824.61	−4872.86	−4921.58
Functioning Accumulated (FA)	−27684.07	−32460.91	−37285.52	−42158.37	−47079.96
Present Value Accumulated (NPV)	−27684.07	−32460.91	−37285.52	−42158.37	−47079.96
Present Value (NPV)	−4729.55	−4776.84	−4824.61	−4872.86	−4921.58
Energy	−1200.00	−1400.00	−1600.00	−1800.00	−2000.00
Energy Accumulated (EA)	−4200.00	−5600.00	−7200.00	−9000.00	−11000.00
Energy (NPV)	−1200.00	−1400.00	−1600.00	−1800.00	−2000.00
Accumulated Functioning (FPV)	−112290.84	−146151.75	−185037.27	−228995.64	−278075.60
Maintenance (M)	−315.30	−318.46	−321.64	−324.86	−328.11
Maintenane Accumulated (MA)	−1845.60	−2164.06	−2485.70	−2810.56	−3138.66
Present Value (MPV)	−315.30	−318.46	−321.64	−324.86	−328.11
Accumulated MPV	−1845.60	−2164.06	−2485.70	−2810.56	−3138.66
Non Production (NP)	0.00	0.00	0.00	0.00	0.00
Non Production Accumulated (NPA)	0.00	0.00	0.00	0.00	0.00
Present Value (NPPV)	0.00	0.00	0.00	0.00	0.00
Accumulated NPPV	0.00	0.00	0.00	0.00	0.00
Accumulated Total Costs (APV)	−114136.45	−148315.81	−187522.97	−231806.20	−281214.26
Accumulated Total Costs	−33729.67	−40224.97	−46971.22	−53968.93	−61218.62
Benefit (B)	44642.86	44642.86	44642.86	44642.86	44642.86
Benefits Accumulated (BA)	267857.14	312500.00	357142.86	401785.71	446428.57
Present Value (BPV)	44642.86	44642.86	44642.86	44642.86	44642.86
Accumulated Benefits BPV	267857.14	312500.00	357142.86	401785.71	446428.57
BPV+FPV+MPV+ NPPV+IIPV	222127.47	260275.03	298171.64	335816.78	373209.95
MTAC (LDM)	−36612.07	−43587.77	−50814.42	−58292.53	−66022.62
MTAC (SDM)	−38718.44	−45915.86	−53327.28	−60953.21	−68794.16
MTAC (EM)	−43541.02	−50576.77	−57730.01	−65034.22	−72514.72

(*Continued*)

TABLE 6.8 (*Continued*)
Values Used for Simulation based on MMTAC Method Related to Maintenance Standard – First Simulation

Time	11	12	13	14	15
Initial Investment (II)					
Initial Investment (II)					
II (IIPV) - Return	−12000.00	−12000.00	−12000.00	−12000.00	−12000.00
Useful Life					
Residual value of equipment (RV)					
Linear Depreciation Method (Vn)	−6715.60	−6235.20	−5754.80	−5274.40	−4794.00
LDM Present Value (PV)	−6715.60	−6235.20	−5754.80	−5274.40	−4794.00
Sum of Digits Method (Vn)	−3870.15	−3352.80	−2872.40	−2428.95	−2022.46
SDM Present Value (PV)	−3870.15	−3352.80	−2872.40	−2428.95	−2022.46
Exponential Method (Vn)	−530.08	−399.19	−300.61	−226.38	−170.48
EM Present Value (Pv)	−530.08	−399.19	−300.61	−226.38	−170.48
Functioning (F)	−4970.80	−5020.51	−5070.71	−5121.42	−5172.63
Functioning Accumulated (FA)	−52050.76	−57071.26	−62141.98	−67263.40	−72436.03
Present Value Accumulated (NPV)	−52050.76	−57071.26	−62141.98	−67263.40	−72436.03
Present Value (NPV)	−4970.80	−5020.51	−5070.71	−5121.42	−5172.63
Energy	−2200.00	−2400.00	−2600.00	−2800.00	−3000.00
Energy Accumulated (EA)	−13200.00	−15600.00	−18200.00	−21000.00	−24000.00
Energy (NPV)	−2200.00	−2400.00	−2600.00	−2800.00	−3000.00
Accumulated Functioning (FPV)	−332326.36	−391797.62	−456539.60	−526602.99	−602039.02
Maintenance (M)	−331.39	−334.70	−338.05	−341.43	−344.84
Maintenane Accumulated (MA)	−3470.05	−3804.75	−4142.80	−4484.23	−4829.07
Present Value (MPV)	−331.39	−334.70	−338.05	−341.43	−344.84
Accumulated MPV	−3470.05	−3804.75	−4142.80	−4484.23	−4829.07
Non Production (NP)	0.00	0.00	0.00	0.00	0.00
Non Production Accumulated (NPA)	0.00	0.00	0.00	0.00	0.00
Present Value (NPPV)	0.00	0.00	0.00	0.00	0.00
Accumulated NPPV	0.00	0.00	0.00	0.00	0.00
Accumulated Total Costs (APV)	−335796.41	−395602.37	−460682.39	−531087.22	−606868.09
Accumulated Total Costs	−68720.81	−76476.01	−84484.77	−92747.62	−101265.10
Benefit (B)	44642.86	44642.86	44642.86	44642.86	44642.86
Benefits Accumulated (BA)	491071.43	535714.29	580357.14	625000.00	669642.86
Present Value (BPV)	44642.86	44642.86	44642.86	44642.86	44642.86
Accumulated Benefits BPV	491071.43	535714.29	580357.14	625000.00	669642.86
BPV+FPV+MPV+ NPPV+IIPV	410350.62	447238.27	483872.37	520252.38	556377.76
MTAC (LDM)	−74005.21	−82240.81	−90729.97	−99473.22	−108471.10
MTAC (SDM)	−76850.65	−85123.21	−93612.37	−102318.67	−111242.64
MTAC (EM)	−80190.73	−88076.83	−96184.16	−104521.24	−113094.62

(*Continued*)

TABLE 6.8 (*Continued*)
Values Used for Simulation based on MMTAC Method Related to Maintenance Standard – First Simulation

Time	16	17	18	19	20
Initial Investment (II)					
Initial Investment (II)					
II (IIPV) - Return	−12000.00	−12000.00	−12000.00	−12000.00	−12000.00
Useful Life					
Residual value of equipment (RV)					
Linear Depreciation Method (Vn)	−4313.60	−3833.20	−3352.80	−2872.40	−2392.00
LDM Present Value (PV)	−4313.60	−3833.20	−3352.80	−2872.40	−2392.00
Sum of Digits Method (Vn)	−1652.92	−1320.34	−1024.71	−766.03	−544.31
SDM Present Value (PV)	−1652.92	−1320.34	−1024.71	−766.03	−544.31
Exponential Method (Vn)	−128.38	−96.68	−72.81	−54.83	−41.29
EM Present Value (Pv)	−128.38	−96.68	−72.81	−54.83	−41.29
Functioning (F)	−5224.36	−5276.60	−5329.37	−5382.66	−5436.49
Functioning Accumulated (FA)	−77660.39	−82936.99	−88266.36	−93649.03	−99085.52
Present Value Accumulated (NPV)	−77660.39	−82936.99	−88266.36	−93649.03	−99085.52
Present Value (NPV)	−5224.36	−5276.60	−5329.37	−5382.66	−5436.49
Energy	−3200.00	−3400.00	−3600.00	−3800.00	−4000.00
Energy Accumulated (EA)	−27200.00	−30600.00	−34200.00	−38000.00	−42000.00
Energy (NPV)	−3200.00	−3400.00	−3600.00	−3800.00	−4000.00
Accumulated Functioning (FPV)	−682899.41	−769236.41	−861102.77	−958551.80	−1061637.32
Maintenance (M)	−348.29	−351.77	−355.29	−358.84	−362.43
Maintenane Accumulated (MA)	−5177.36	−5529.13	−5884.42	−6243.27	−6605.70
Present Value (MPV)	−348.29	−351.77	−355.29	−358.84	−362.43
Accumulated MPV	−5177.36	−5529.13	−5884.42	−6243.27	−6605.70
Non Production (NP)	0.00	0.00	0.00	0.00	0.00
Non Production Accumulated (NPA)	0.00	0.00	0.00	0.00	0.00
Present Value (NPPV)	0.00	0.00	0.00	0.00	0.00
Accumulated NPPV	0.00	0.00	0.00	0.00	0.00
Accumulated Total Costs (APV)	−688076.77	−774765.54	−866987.19	−964795.07	−1068243.02
Accumulated Total Costs	−110037.75	−119066.13	−128350.79	−137892.30	−147691.22
Benefit (B)	44642.86	44642.86	44642.86	44642.86	44642.86
Benefits Accumulated (BA)	714285.71	758928.57	803571.43	848214.29	892857.14
Present Value (BPV)	44642.86	44642.86	44642.86	44642.86	44642.86
Accumulated Benefits BPV	714285.71	758928.57	803571.43	848214.29	892857.14
BPV+FPV+MPV+ NPPV+IIPV	592247.96	627862.44	663220.64	698321.99	733165.92
MTAC (LDM)	−117724.15	−127232.93	−136997.99	−147019.90	−157299.22
MTAC (SDM)	−120384.83	−129745.79	−139326.08	−149126.27	−159146.91
MTAC (EM)	−121909.37	−130969.45	−140277.98	−149837.47	−159649.93

(*Continued*)

TABLE 6.8 (*Continued*)
Values Used for Simulation based on MMTAC Method Related to Maintenance Standard – First Simulation

Time	21	22	23	24	25
Initial Investment (II)					
Initial Investment (II)					
II (IIPV) - Return	−12000.00	−12000.00	−12000.00	−12000.00	−12000.00
Useful Life					
Residual value of equipment (RV)					
Linear Depreciation Method (Vn)	−1911.60	−1431.20	−950.80	−470.40	10.00
LDM Present Value (PV)	−1911.60	−1431.20	−950.80	−470.40	10.00
Sum of Digits Method (Vn)	−359.54	−211.72	−100.86	−26.95	10.00
SDM Present Value (PV)	−359.54	−211.72	−100.86	−26.95	10.00
Exponential Method (Vn)	−31.09	−23.42	−17.63	−13.28	−10.00
EM Present Value (Pv)	−31.09	−23.42	−17.63	−13.28	−10.00
Functioning (F)	−5490.86	−5545.76	−5601.22	−5657.23	−5713.81
Functioning Accumulated (FA)	−104576.37	−110122.14	−115723.36	−121380.59	−127094.40
Present Value Accumulated (NPV)	−104576.37	−110122.14	−115723.36	−121380.59	−127094.40
Present Value (NPV)	−5490.86	−5545.76	−5601.22	−5657.23	−5713.81
Energy	−4200.00	−4400.00	−4600.00	−4800.00	−5000.00
Energy Accumulated (EA)	−46200.00	−50600.00	−55200.00	−60000.00	−65000.00
Energy (NPV)	−4200.00	−4400.00	−4600.00	−4800.00	−5000.00
Accumulated Functioning (FPV)	−1170413.69	−1284935.83	−1405259.18	−1531439.78	−1663534.17
Maintenance (M)	−366.06	−369.72	−373.41	−377.15	−380.92
Maintenane Accumulated (MA)	−6971.76	−7341.48	−7714.89	−8092.04	−8472.96
Present Value (MPV)	−366.06	−369.72	−373.41	−377.15	−380.92
Accumulated MPV	−6971.76	−7341.48	−7714.89	−8092.04	−8472.96
Non Production (NP)	0.00	0.00	0.00	0.00	0.00
Non Production Accumulated (NPA)	0.00	0.00	0.00	0.00	0.00
Present Value (NPPV)	0.00	0.00	0.00	0.00	0.00
Accumulated NPPV	0.00	0.00	0.00	0.00	0.00
Accumulated Total Costs (APV)	−1177385.45	−1292277.30	−1412974.07	−1539531.82	−1672007.13
Accumulated Total Costs	−157748.13	−168063.61	−178638.25	−189472.63	−200567.36
Benefit (B)	44642.86	44642.86	44642.86	44642.86	44642.86
Benefits Accumulated (BA)	937500.00	982142.86	1026785.71	1071428.57	1116071.43
Present Value (BPV)	44642.86	44642.86	44642.86	44642.86	44642.86
Accumulated Benefits BPV	937500.00	982142.86	1026785.71	1071428.57	1116071.43
BPV+FPV+MPV+NPPV+IIPV	767751.87	802079.24	836147.47	869955.94	903504.07
MTAC (LDM)	−167836.53	−178632.41	−189687.45	−201002.23	−212577.36
MTAC (SDM)	−169388.59	−179851.89	−190537.39	−201445.68	−212577.36
MTAC (EM)	−169717.04	−180040.20	−190620.62	−201459.35	−212557.36

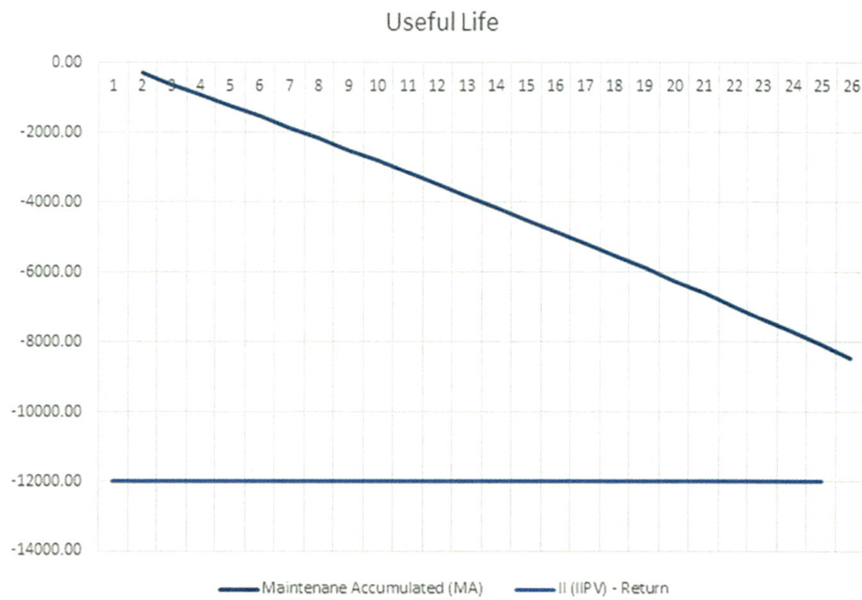

FIGURE 6.12 Physical asset useful life.

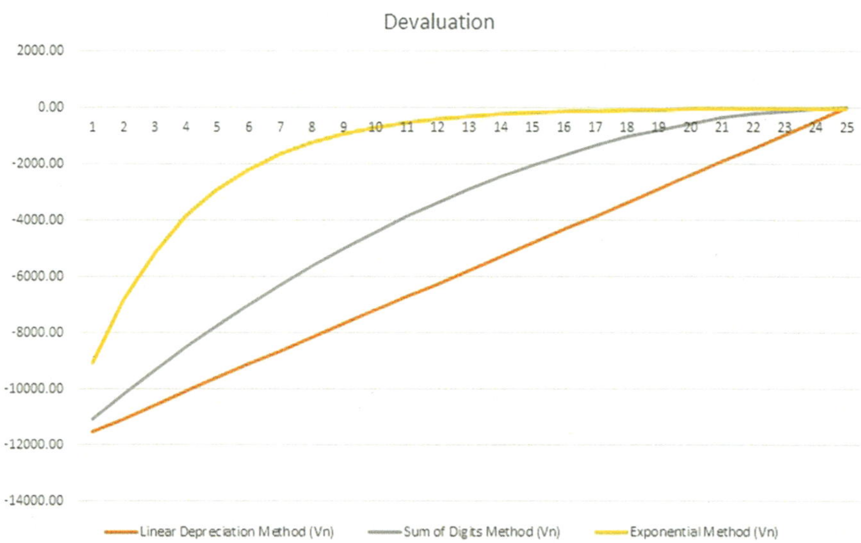

FIGURE 6.13 The three devaluation methods.

The remaining values are presented in Table 6.9, and the graphical results are shown in Figures 6.16–6.19.

As can be seen, namely comparing Figures 6.15 and 6.19, with the increasing of the maintenance management quality helped by the maintenance standards, translated in

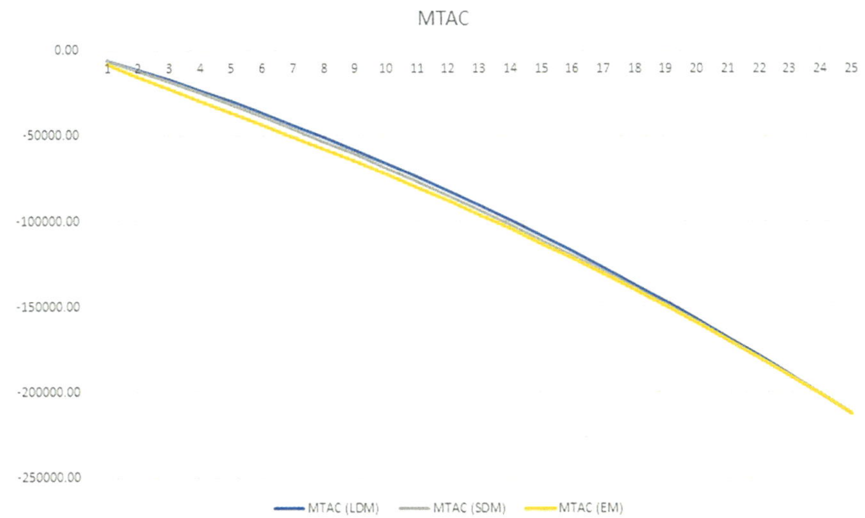

FIGURE 6.14 The MMTAC results from the three Devaluation methods.

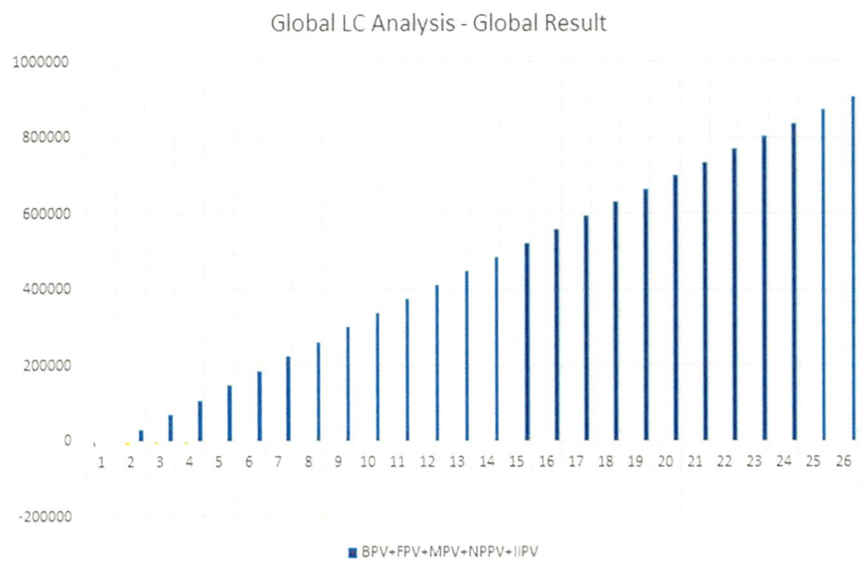

FIGURE 6.15 Life cycle analysis global result including profits.

a higher value of the Availability, the Global Result that can be seen in Figure 6.19 is higher than the results shown in Figure 6.15. However, this is in a scenario where the Apparent Rate is null, what does not happen in a real world.

TABLE 6.9
Values Used for Simulation based on MMTAC Method Related to Maintenance Standard – Second Simulation

Time	0	1	2	3	4	5
Initial Investment (II)	−12000					
Initial Investment (II)	−12000					
II (IIPV) - Return	−12000	−12000.00	−12000.00	−12000.00	−12000.00	−12000.00
Useful Life	25					
Residual value of equipment (RV)	10					
Linear Depreciation Method (Vn)	−12000	−11519.60	−11039.20	−10558.80	−10078.40	−9598.00
LDM Present Value (PV)		−11519.60	−11039.20	−10558.80	−10078.40	−9598.00
Sum of Digits Method (Vn)	−12000	−11076.15	−10189.26	−9339.32	−8526.34	−7750.31
SDM Present Value (PV)		−11076.15	−10189.26	−9339.32	−8526.34	−7750.31
Exponential Method (Vn)	−12000	−9036.79	−6805.29	−5124.83	−3859.33	−2906.33
EM Present Value (Pv)		−9036.79	−6805.29	−5124.83	−3859.33	−2906.33
Functioning (*F*)	−4500	−4500.00	−4545.00	−4590.45	−4636.35	−4682.72
Functioning Accumulated (FA)		−4500.00	−9045.00	−13635.45	−18271.80	−22954.52
Present Value Accumulated (NPV)		−4500.00	−9045.00	−13635.45	−18271.80	−22954.52
Present Value (NPV)		−4500.00	−4545.00	−4590.45	−4636.35	−4682.72
Energy	−200	−200.00	−400.00	−600.00	−800.00	−1000.00
Energy Accumulated (EA)		−200.00	−600.00	−1200.00	−2000.00	−3000.00
Energy (NPV)		−200.00	−400.00	−600.00	−800.00	−1000.00
Accumulated Functioning (FPV)		−16700.00	−26145.00	−40380.45	−59452.25	−83406.78
Maintenance (*M*)	−300	−300.00	−303.00	−306.03	−309.09	−312.18
Maintenane Accumulated (MA)		−300.00	−603.00	−909.03	−1218.12	−1530.30
Present Value (MPV)		−300.00	−303.00	−306.03	−309.09	−312.18
Accumulated MPV		−300.00	−603.00	−909.03	−1218.12	−1530.30
Non Production (NP)	0	0.00	0.00	0.00	0.00	0.00
Non Production Accumulated (NPA)		0.00	0.00	0.00	0.00	0.00
Present Value (NPPV)		0.00	0.00	0.00	0.00	0.00
Accumulated NPPV		0.00	0.00	0.00	0.00	0.00
Accumulated Total Costs (APV)		−17000.00	−26748.00	−41289.48	−60670.37	−84937.08
Accumulated Total Costs		−5000.00	−10248.00	−15744.48	−21489.92	−27484.82
Benefit (*B*)	50000	48309.18	48309.18	48309.18	48309.18	48309.18
Benefits Accumulated (BA)		48309.18	96618.36	144927.54	193236.71	241545.89
Present Value (BPV)		48309.18	48309.18	48309.18	48309.18	48309.18
Accumulated Benefits BPV		48309.18	96618.36	144927.54	193236.71	241545.89
BPV + FPV + MPV + NPPV + IIPV		31309.18	74370.36	117183.06	159746.79	202061.07
MTAC (LDM)		−5480.40	−11208.80	−17185.68	−23411.52	−29886.82
MTAC (SDM)		−5923.85	−12058.74	−18405.16	−24963.59	−31734.52
MTAC (EM)		−7963.21	−15442.71	−22619.65	−29630.59	−36578.49

(*Continued*)

TABLE 6.9 (*Continued*)
Values Used for Simulation based on MMTAC Method Related to Maintenance Standard – Second Simulation

Time	6	7	8	9	10
Initial Investment (II)					
Initial Investment (II)					
II (IIPV) - Return	−12000.00	−12000.00	−12000.00	−12000.00	−12000.00
Useful Life					
Residual value of equipment (RV)					
Linear Depreciation Method (Vn)	−9117.60	−8637.20	−8156.80	−7676.40	−7196.00
LDM Present Value (PV)	−9117.60	−8637.20	−8156.80	−7676.40	−7196.00
Sum of Digits Method (Vn)	−7011.23	−6309.11	−5643.94	−5015.72	−4424.46
SDM Present Value (PV)	−7011.23	−6309.11	−5643.94	−5015.72	−4424.46
Exponential Method (Vn)	−2188.66	−1648.20	−1241.20	−934.71	−703.90
EM Present Value (Pv)	−2188.66	−1648.20	−1241.20	−934.71	−703.90
Functioning (*F*)	−4729.55	−4776.84	−4824.61	−4872.86	−4921.58
Functioning Accumulated (FA)	−27684.07	−32460.91	−37285.52	−42158.37	−47079.96
Present Value Accumulated (NPV)	−27684.07	−32460.91	−37285.52	−42158.37	−47079.96
Present Value (NPV)	−4729.55	−4776.84	−4824.61	−4872.86	−4921.58
Energy	−1200.00	−1400.00	−1600.00	−1800.00	−2000.00
Energy Accumulated (EA)	−4200.00	−5600.00	−7200.00	−9000.00	−11000.00
Energy (NPV)	−1200.00	−1400.00	−1600.00	−1800.00	−2000.00
Accumulated Functioning (FPV)	−112290.84	−146151.75	−185037.27	−228995.64	−278075.60
Maintenance (*M*)	−315.30	−318.46	−321.64	−324.86	−328.11
Maintenane Accumulated (MA)	−1845.60	−2164.06	−2485.70	−2810.56	−3138.66
Present Value (MPV)	−315.30	−318.46	−321.64	−324.86	−328.11
Accumulated MPV	−1845.60	−2164.06	−2485.70	−2810.56	−3138.66
Non Production (NP)	0.00	0.00	0.00	0.00	0.00
Non Production Accumulated (NPA)	0.00	0.00	0.00	0.00	0.00
Present Value (NPPV)	0.00	0.00	0.00	0.00	0.00
Accumulated NPPV	0.00	0.00	0.00	0.00	0.00
Accumulated Total Costs (APV)	−114136.45	−148315.81	−187522.97	−231806.20	−281214.26
Accumulated Total Costs	−33729.67	−40224.97	−46971.22	−53968.93	−61218.62
Benefit (*B*)	48309.18	48309.18	48309.18	48309.18	48309.18
Benefits Accumulated (BA)	289855.07	338164.25	386473.43	434782.61	483091.79
Present Value (BPV)	48309.18	48309.18	48309.18	48309.18	48309.18
Accumulated Benefits BPV	289855.07	338164.25	386473.43	434782.61	483091.79
BPV + FPV + MPV + NPPV + IIPV	244125.40	285939.28	327502.21	368813.68	409873.17
MTAC (LDM)	−36612.07	−43587.77	−50814.42	−58292.53	−66022.62
MTAC (SDM)	−38718.44	−45915.86	−53327.28	−60953.21	−68794.16
MTAC (EM)	−43541.02	−50576.77	−57730.01	−65034.22	−72514.72

(*Continued*)

TABLE 6.9 (*Continued*)
Values Used for Simulation based on MMTAC Method Related to Maintenance Standard – Second Simulation

Time	11	12	13	14	15
Initial Investment (II)					
Initial Investment (II)					
II (IIPV) - Return	−12000.00	−12000.00	−12000.00	−12000.00	−12000.00
Useful Life					
Residual value of equipment (RV)					
Linear Depreciation Method (Vn)	−6715.60	−6235.20	−5754.80	−5274.40	−4794.00
LDM Present Value (PV)	−6715.60	−6235.20	−5754.80	−5274.40	−4794.00
Sum of Digits Method (Vn)	−3870.15	−3352.80	−2872.40	−2428.95	−2022.46
SDM Present Value (PV)	−3870.15	−3352.80	−2872.40	−2428.95	−2022.46
Exponential Method (Vn)	−530.08	−399.19	−300.61	−226.38	−170.48
EM Present Value (Pv)	−530.08	−399.19	−300.61	−226.38	−170.48
Functioning (*F*)	−4970.80	−5020.51	−5070.71	−5121.42	−5172.63
Functioning Accumulated (FA)	−52050.76	−57071.26	−62141.98	−67263.40	−72436.03
Present Value Accumulated (NPV)	−52050.76	−57071.26	−62141.98	−67263.40	−72436.03
Present Value (NPV)	−4970.80	−5020.51	−5070.71	−5121.42	−5172.63
Energy	−2200.00	−2400.00	−2600.00	−2800.00	−3000.00
Energy Accumulated (EA)	−13200.00	−15600.00	−18200.00	−21000.00	−24000.00
Energy (NPV)	−2200.00	−2400.00	−2600.00	−2800.00	−3000.00
Accumulated Functioning (FPV)	−332326.36	−391797.62	−456539.60	−526602.99	−602039.02
Maintenance (*M*)	−331.39	−334.70	−338.05	−341.43	−344.84
Maintenane Accumulated (MA)	−3470.05	−3804.75	−4142.80	−4484.23	−4829.07
Present Value (MPV)	−331.39	−334.70	−338.05	−341.43	−344.84
Accumulated MPV	−3470.05	−3804.75	−4142.80	−4484.23	−4829.07
Non Production (NP)	0.00	0.00	0.00	0.00	0.00
Non Production Accumulated (NPA)	0.00	0.00	0.00	0.00	0.00
Present Value (NPPV)	0.00	0.00	0.00	0.00	0.00
Accumulated NPPV	0.00	0.00	0.00	0.00	0.00
Accumulated Total Costs (APV)	−335796.41	−395602.37	−460682.39	−531087.22	−606868.09
Accumulated Total Costs	−68720.81	−76476.01	−84484.77	−92747.62	−101265.10
Benefit (*B*)	48309.18	48309.18	48309.18	48309.18	48309.18
Benefits Accumulated (BA)	531400.97	579710.14	628019.32	676328.50	724637.68
Present Value (BPV)	48309.18	48309.18	48309.18	48309.18	48309.18
Accumulated Benefits BPV	531400.97	579710.14	628019.32	676328.50	724637.68
BPV + FPV + MPV + NPPV + IIPV	450680.16	491234.13	531534.55	571580.88	611372.58
MTAC (LDM)	−74005.21	−82240.81	−90729.97	−99473.22	−108471.10
MTAC (SDM)	−76850.65	−85123.21	−93612.37	−102318.67	−111242.64
MTAC (EM)	−80190.73	−88076.83	−96184.16	−104521.24	−113094.62

(*Continued*)

TABLE 6.9 (*Continued*)
Values Used for Simulation based on MMTAC Method Related to Maintenance Standard – Second Simulation

Time	16	17	18	19	20
Initial Investment (II)					
Initial Investment (II)					
II (IIPV) - Return	−12000.00	−12000.00	−12000.00	−12000.00	−12000.00
Useful Life					
Residual value of equipment (RV)					
Linear Depreciation Method (Vn)	−4313.60	−3833.20	−3352.80	−2872.40	−2392.00
LDM Present Value (PV)	−4313.60	−3833.20	−3352.80	−2872.40	−2392.00
Sum of Digits Method (Vn)	−1652.92	−1320.34	−1024.71	−766.03	−544.31
SDM Present Value (PV)	−1652.92	−1320.34	−1024.71	−766.03	−544.31
Exponential Method (Vn)	−128.38	−96.68	−72.81	−54.83	−41.29
EM Present Value (Pv)	−128.38	−96.68	−72.81	−54.83	−41.29
Functioning (*F*)	−5224.36	−5276.60	−5329.37	−5382.66	−5436.49
Functioning Accumulated (FA)	−77660.39	−82936.99	−88266.36	−93649.03	−99085.52
Present Value Accumulated (NPV)	−77660.39	−82936.99	−88266.36	−93649.03	−99085.52
Present Value (NPV)	−5224.36	−5276.60	−5329.37	−5382.66	−5436.49
Energy	−3200.00	−3400.00	−3600.00	−3800.00	−4000.00
Energy Accumulated (EA)	−27200.00	−30600.00	−34200.00	−38000.00	−42000.00
Energy (NPV)	−3200.00	−3400.00	−3600.00	−3800.00	−4000.00
Accumulated Functioning (FPV)	−682899.41	−769236.41	−861102.77	−958551.80	−1061637.32
Maintenance (*M*)	−348.29	−351.77	−355.29	−358.84	−362.43
Maintenane Accumulated (MA)	−5177.36	−5529.13	−5884.42	−6243.27	−6605.70
Present Value (MPV)	−348.29	−351.77	−355.29	−358.84	−362.43
Accumulated MPV	−5177.36	−5529.13	−5884.42	−6243.27	−6605.70
Non Production (NP)	0.00	0.00	0.00	0.00	0.00
Non Production Accumulated (NPA)	0.00	0.00	0.00	0.00	0.00
Present Value (NPPV)	0.00	0.00	0.00	0.00	0.00
Accumulated NPPV	0.00	0.00	0.00	0.00	0.00
Accumulated Total Costs (APV)	−688076.77	−774765.54	−866987.19	−964795.07	−1068243.02
Accumulated Total Costs	−110037.75	−119066.13	−128350.79	−137892.30	−147691.22
Benefit (*B*)	48309.18	48309.18	48309.18	48309.18	48309.18
Benefits Accumulated (BA)	772946.86	821256.04	869565.22	917874.40	966183.57
Present Value (BPV)	48309.18	48309.18	48309.18	48309.18	48309.18
Accumulated Benefits BPV	772946.86	821256.04	869565.22	917874.40	966183.57
BPV + FPV + MPV + NPPV + IIPV	650909.11	690189.91	729214.43	767982.10	806492.36
MTAC (LDM)	−117724.15	−127232.93	−136997.99	−147019.90	−157299.22
MTAC (SDM)	−120384.83	−129745.79	−139326.08	−149126.27	−159146.91
MTAC (EM)	−121909.37	−130969.45	−140277.98	−149837.47	−159649.93

(Continued)

TABLE 6.9 (*Continued*)
Values Used for Simulation based on MMTAC Method Related to Maintenance Standard – Second Simulation

Time	21	22	23	24	25
Initial Investment (II)					
Initial Investment (II)					
II (IIPV) - Return	−12000.00	−12000.00	−12000.00	−12000.00	−12000.00
Useful Life					
Residual value of equipment (RV)					
Linear Depreciation Method (Vn)	−1911.60	−1431.20	−950.80	−470.40	10.00
LDM Present Value (PV)	−1911.60	−1431.20	−950.80	−470.40	10.00
Sum of Digits Method (Vn)	−359.54	−211.72	−100.86	−26.95	10.00
SDM Present Value (PV)	−359.54	−211.72	−100.86	−26.95	10.00
Exponential Method (Vn)	−31.09	−23.42	−17.63	−13.28	−10.00
EM Present Value (Pv)	−31.09	−23.42	−17.63	−13.28	−10.00
Functioning (*F*)	−5490.86	−5545.76	−5601.22	−5657.23	−5713.81
Functioning Accumulated (FA)	−104576.37	−110122.14	−115723.36	−121380.59	−127094.40
Present Value Accumulated (NPV)	−104576.37	−110122.14	−115723.36	−121380.59	−127094.40
Present Value (NPV)	−5490.86	−5545.76	−5601.22	−5657.23	−5713.81
Energy	−4200.00	−4400.00	−4600.00	−4800.00	−5000.00
Energy Accumulated (EA)	−46200.00	−50600.00	−55200.00	−60000.00	−65000.00
Energy (NPV)	−4200.00	−4400.00	−4600.00	−4800.00	−5000.00
Accumulated Functioning (FPV)	−1170413.69	−1284935.83	−1405259.18	−1531439.78	−1663534.17
Maintenance (*M*)	−366.06	−369.72	−373.41	−377.15	−380.92
Maintenane Accumulated (MA)	−6971.76	−7341.48	−7714.89	−8092.04	−8472.96
Present Value (MPV)	−366.06	−369.72	−373.41	−377.15	−380.92
Accumulated MPV	−6971.76	−7341.48	−7714.89	−8092.04	−8472.96
Non Production (NP)	0.00	0.00	0.00	0.00	0.00
Non Production Accumulated (NPA)	0.00	0.00	0.00	0.00	0.00
Present Value (NPPV)	0.00	0.00	0.00	0.00	0.00
Accumulated NPPV	0.00	0.00	0.00	0.00	0.00
Accumulated Total Costs (APV)	−1177385.45	−1292277.30	−1412974.07	−1539531.82	−1672007.13
Accumulated Total Costs	−157748.13	−168063.61	−178638.25	−189472.63	−200567.36
Benefit (*B*)	48309.18	48309.18	48309.18	48309.18	48309.18
Benefits Accumulated (BA)	1014492.75	1062801.93	1111111.11	1159420.29	1207729.47
Present Value (BPV)	48309.18	48309.18	48309.18	48309.18	48309.18
Accumulated Benefits BPV	1014492.75	1062801.93	1111111.11	1159420.29	1207729.47
BPV + FPV + MPV + NPPV + IIPV	844744.62	882738.32	920472.86	957947.66	995162.11
MTAC (LDM)	−167836.53	−178632.41	−189687.45	−201002.23	−212577.36
MTAC (SDM)	−169388.59	−179851.89	−190537.39	−201445.68	−212577.36
MTAC (EM)	−169717.04	−180040.20	−190620.62	−201459.35	−212557.36

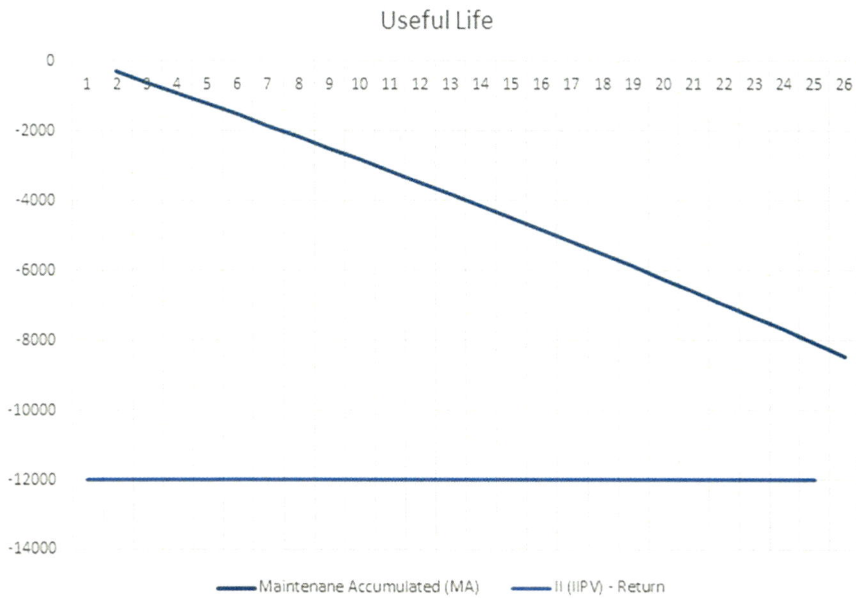

FIGURE 6.16 Physical asset useful life.

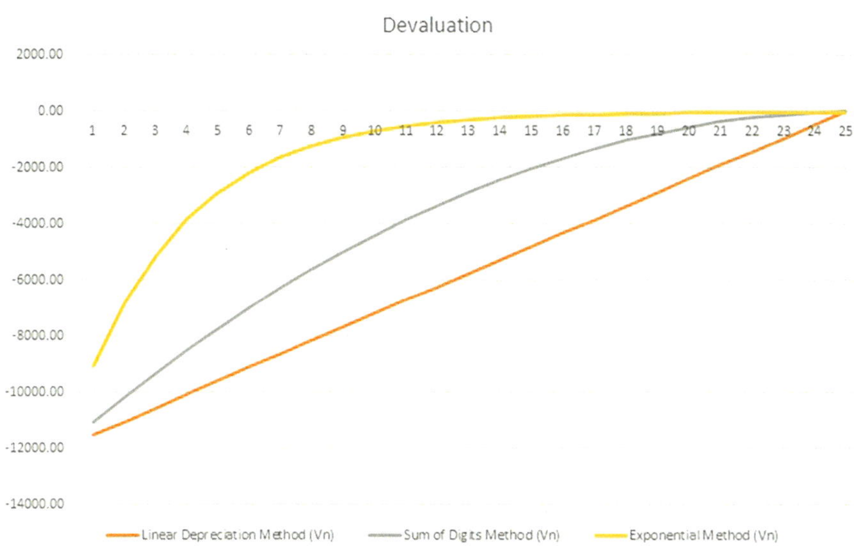

FIGURE 6.17 The three devaluation methods.

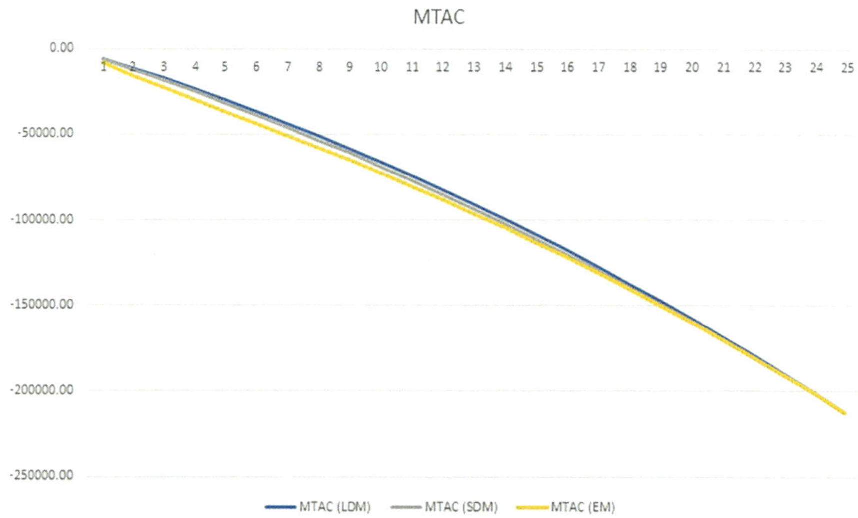

FIGURE 6.18 The MMTAC results from the three devaluation methods.

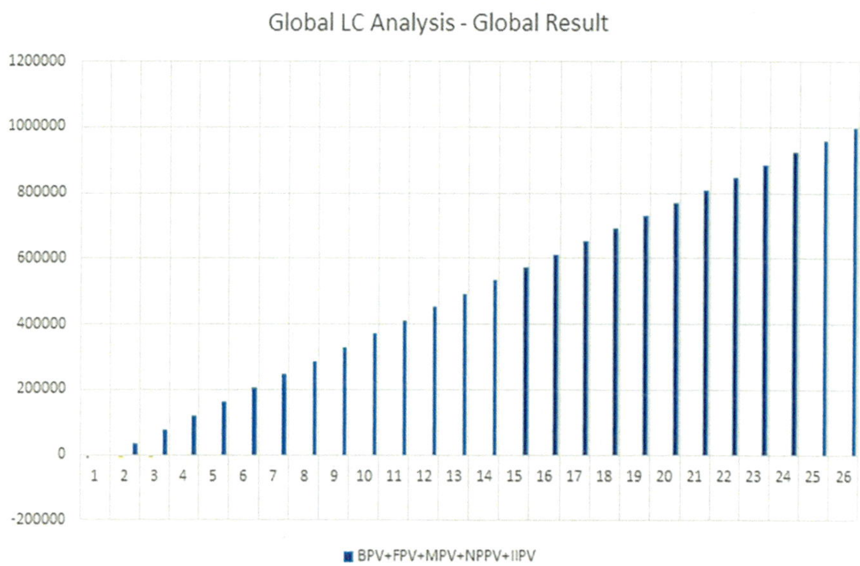

FIGURE 6.19 Life cycle analysis global result including profits.

6.5.1.3 MMTAC Reduced to Present Value

The first simulation uses the following values:

- Initial Investment (II) –12000
- Life Time simulated 25

- Residual value of equipment (RV) 10
- Functioning (*F*) –4500
- Energy (*E*) –200
- Maintenance (*M*) –300
- Non-Production (NP) 0
- Benefit (*B*) 50000
- Internal Rate of Return (*r*) 0.12
- Risk (*R*) 0.11
- Apparent Rate 0.2432
- Maintenance Increase Rate 0.01
- Availability 0.8929
- MTBF **500**
- MTTR **50**
- MWT 10

The remaining values are presented in Table 6.10, and the graphical results are shown in Figures 6.20–6.23.

As referred in the preceding introduction on this section, the evaluation of the introduction of standard norms and maintenance policies can be simulated, primarily, through the variation of MTBF and MTTR. They may be used other indicators, but may be not so intuitive.

The second simulation uses the following values (the MTBF is increased and MTTR decreases as a result of the Maintenance Standards):

- Initial Investment (II) –12000
- Life Time simulated 25
- Residual value of equipment (RV) 10
- Functioning (*F*) –4500
- Energy (*E*) –200
- Maintenance (*M*) –300
- Non Production (NP) 0
- Benefit (*B*) 50000
- Internal Rate of Return (*r*) 0.12
- Risk (*R*) 0.11
- Apparent Rate 0.2432
- Maintenance Increase Rate 0.01
- Availability 0.8929
- MTBF **1000**
- MTTR **25**
- MWT 10

The remaining values are presented in Table 6.11, and the graphical results are shown in Figures 6.24–6.27.

As can be seen, namely comparing Figures 6.23 and 6.27, with the increasing of the maintenance management quality helped by the maintenance standards, it gets one more positive year, but the positive incomes increase a lot.

TABLE 6.10
Values Used for Simulation based on MMTAC-RPV Method Related to Maintenance Standard – First Simulation

Time	0	1	2	3	4	5
Initial Investment (II)	−12000					
Initial Investment (II)	−12000					
II Present Value (IIPV) - Return	−12000	−9652.51	−7764.25	−6245.37	−5023.63	−4040.88
Life Time simulated	25					
Residual value of equipment (RV)	10					
Linear Depreciation Method (Vn)	−12000	−11519.60	−11039.20	−10558.80	−10078.40	−9598.00
LDM Present Value (Vn)		−9266.09	−7142.59	−5495.30	−4219.18	−3232.03
Sum of Digits Method (Vn)	−12000	−11076.15	−10189.26	−9339.32	−8526.34	−7750.31
SDM Present Value (Vn)		−8909.39	−6592.66	−4860.63	−3569.43	−2609.84
Exponential Method (Vn)	−12000	−9036.79	−6805.29	−5124.83	−3859.33	−2906.33
EM Present Value (Pv)		−7268.97	−4403.16	−2667.21	−1615.65	−978.68
Functioning (*F*)	−4500	−4500.00	−4545.00	−4590.45	−4636.35	−4682.72
Present Value Acummulated (NPV)		−3619.69	−6560.40	−8949.49	−10890.43	−12467.29
Present Value (NPV)		−3619.69	−2940.71	−2389.09	−1940.94	−1576.86
Energy (*E*)	−200	−200.00	−400.00	−600.00	−800.00	−1000.00
Energy (NPV)		−160.88	−258.81	−312.27	−334.91	−336.74
Accumulated FPV		−15780.57	−22599.77	−31861.53	−43086.87	−55890.90
Maintenance (*M*)	−300	−300.00	−303.00	−306.03	−309.09	−312.18
Present Value (MPV)		−241.31	−196.05	−159.27	−129.40	−105.12
Accumulated (MPV)		−241.31	−437.36	−596.63	−726.03	−831.15
Non Production (NP)	0	0.00	0.00	0.00	0.00	0.00
Present Value (NPPV)		0.00	0.00	0.00	0.00	0.00
Accumulated NPPV		0.00	0.00	0.00	0.00	0.00
Accumulated Total Costs (APV)		−16021.88	−23037.13	−32458.16	−43812.90	−56722.05
Accumulated Total Costs		−5000.00	−10248.00	−15744.48	−21489.92	−27484.82
Benefit (*B*)	50000	44642.86	44642.86	44642.86	44642.86	44642.86
Present Value (BPV)		35909.63	28884.84	23234.27	18689.08	15033.05
Accumulated Benefits BPV		35909.63	64794.47	88028.74	106717.82	121750.87
BPV + FPV + MPV + NPPV + IIPV		7887.76	29757.34	43570.58	50904.93	53028.82
MTAC-PV (LDM)		−37287.97	−21089.86	−16651.15	−15008.02	−14390.82
MTAC-PV (SDM)		−36931.27	−20814.90	−16439.60	−14845.58	−14266.38
MTAC-PV (EM)		−35290.85	−19720.15	−15708.46	−14357.14	−13940.15

(*Continued*)

TABLE 6.10 (*Continued*)
Values Used for Simulation based on MMTAC-RPV Method Related to Maintenance Standard – First Simulation

Time	6	7	8	9	10
Initial Investment (II)					
Initial Investment (II)					
II Present Value (IIPV) - Return	−3250.39	−2614.53	−2103.07	−1691.66	−1360.73
Life Time simulated					
Residual value of equipment (RV)					
Linear Depreciation Method (Vn)	−9117.60	−8637.20	−8156.80	−7676.40	−7196.00
LDM Present Value (Vn)	−2469.64	−1881.85	−1429.53	−1082.15	−815.98
Sum of Digits Method (Vn)	−7011.23	−6309.11	−5643.94	−5015.72	−4424.46
SDM Present Value (Vn)	−1899.10	−1374.61	−989.13	−707.07	−501.71
Exponential Method (Vn)	−2188.66	−1648.20	−1241.20	−934.71	−703.90
EM Present Value (Pv)	−592.83	−359.11	−217.53	−131.77	−79.82
Functioning (*F*)	−4729.55	−4776.84	−4824.61	−4872.86	−4921.58
Present Value Acummulated (NPV)	−13748.36	−14789.13	−15634.67	−16321.60	−16879.68
Present Value (NPV)	−1281.07	−1040.77	−845.54	−686.93	−558.08
Energy (*E*)	−1200.00	−1400.00	−1600.00	−1800.00	−2000.00
Energy (NPV)	−325.04	−305.03	−280.41	−253.75	−226.79
Accumulated FPV	−69964.30	−85058.45	−100973.53	−117548.88	−134655.35
Maintenance (*M*)	−315.30	−318.46	−321.64	−324.86	−328.11
Present Value (MPV)	−85.40	−69.38	−56.37	−45.80	−37.21
Accumulated (MPV)	−916.56	−985.94	−1042.31	−1088.11	−1125.31
Non Production (NP)	0.00	0.00	0.00	0.00	0.00
Present Value (NPPV)	0.00	0.00	0.00	0.00	0.00
Accumulated NPPV	0.00	0.00	0.00	0.00	0.00
Accumulated Total Costs (APV)	−70880.85	−86044.40	−102015.84	−118636.99	−135780.66
Accumulated Total Costs	−33729.67	−40224.97	−46971.22	−53968.93	−61218.62
Benefit (*B*)	44642.86	44642.86	44642.86	44642.86	44642.86
Present Value (BPV)	12092.22	9726.69	7823.91	6293.37	5062.23
Accumulated Benefits BPV	133843.09	143569.78	151393.69	157687.05	162749.28
BPV + FPV + MPV + NPPV + IIPV	50962.23	45525.38	37377.84	27050.06	14968.62
MTAC-PV (LDM)	−14225.08	−14275.18	−14430.67	−14635.46	−14859.66
MTAC-PV (SDM)	−14129.99	−14202.72	−14375.62	−14593.78	−14828.24
MTAC-PV (EM)	−13912.28	−14057.64	−14279.17	−14529.86	−14786.05

(*Continued*)

TABLE 6.10 (*Continued*)
Values Used for Simulation based on MMTAC-RPV Method Related to Maintenance Standard – First Simulation

Time	11	12	13	14	15
Initial Investment (II)					
Initial Investment (II)					
II Present Value (IIPV) - Return	−1094.54	−880.42	−708.19	−569.65	−458.21
Life Time simulated					
Residual value of equipment (RV)					
Linear Depreciation Method (Vn)	−6715.60	−6235.20	−5754.80	−5274.40	−4794.00
LDM Present Value (Vn)	−612.54	−457.47	−339.62	−250.38	−183.06
Sum of Digits Method (Vn)	−3870.15	−3352.80	−2872.40	−2428.95	−2022.46
SDM Present Value (Vn)	−353.00	−245.99	−169.52	−115.30	−77.23
Exponential Method (Vn)	−530.08	−399.19	−300.61	−226.38	−170.48
EM Present Value (Pv)	−48.35	−29.29	−17.74	−10.75	−6.51
Functioning (*F*)	−4970.80	−5020.51	−5070.71	−5121.42	−5172.63
Present Value Acummulated (NPV)	−17333.07	−17701.42	−18000.67	−18243.79	−18441.30
Present Value (NPV)	−453.39	−368.35	−299.25	−243.12	−197.51
Energy (*E*)	−2200.00	−2400.00	−2600.00	−2800.00	−3000.00
Energy (NPV)	−200.67	−176.08	−153.44	−132.92	−114.55
Accumulated FPV	−152189.09	−170066.59	−188220.70	−206597.41	−225153.26
Maintenance (*M*)	−331.39	−334.70	−338.05	−341.43	−344.84
Present Value (MPV)	−30.23	−24.56	−19.95	−16.21	−13.17
Accumulated (MPV)	−1155.54	−1180.09	−1200.04	−1216.25	−1229.42
Non Production (NP)	0.00	0.00	0.00	0.00	0.00
Present Value (NPPV)	0.00	0.00	0.00	0.00	0.00
Accumulated NPPV	0.00	0.00	0.00	0.00	0.00
Accumulated Total Costs (APV)	−153344.63	−171246.68	−189420.75	−207813.66	−226382.68
Accumulated Total Costs	−68720.81	−76476.01	−84484.77	−92747.62	−101265.10
Benefit (*B*)	44642.86	44642.86	44642.86	44642.86	44642.86
Present Value (BPV)	4071.94	3275.37	2634.63	2119.23	1704.66
Accumulated Benefits BPV	166821.22	170096.59	172731.21	174850.44	176555.10
BPV + FPV + MPV + NPPV + IIPV	1476.59	−13150.10	−28689.53	−44963.22	−61827.58
MTAC-PV (LDM)	−15087.02	−15308.68	−15520.03	−15718.86	−15904.38
MTAC-PV (SDM)	−15063.42	−15291.06	−15506.94	−15709.21	−15897.33
MTAC-PV (EM)	−15035.73	−15273.00	−15495.27	−15701.74	−15892.61

(*Continued*)

TABLE 6.10 (*Continued*)
Values Used for Simulation based on MMTAC-RPV Method Related to Maintenance Standard – First Simulation

Time	16	17	18	19	20
Initial Investment (II)					
Initial Investment (II)					
II Present Value (IIPV) - Return	−368.57	−296.47	−238.48	−191.82	−154.30
Life Time simulated					
Residual value of equipment (RV)					
Linear Depreciation Method (Vn)	−4313.60	−3833.20	−3352.80	−2872.40	−2392.00
LDM Present Value (Vn)	−132.49	−94.70	−66.63	−45.92	−30.76
Sum of Digits Method (Vn)	−1652.92	−1320.34	−1024.71	−766.03	−544.31
SDM Present Value (Vn)	−50.77	−32.62	−20.36	−12.25	−7.00
Exponential Method (Vn)	−128.38	−96.68	−72.81	−54.83	−41.29
EM Present Value (Pv)	−3.94	−2.39	−1.45	−0.88	−0.53
Functioning (*F*)	−5224.36	−5276.60	−5329.37	−5382.66	−5436.49
Present Value Acummulated (NPV)	−18601.76	−18732.13	−18838.04	−18924.08	−18993.99
Present Value (NPV)	−160.46	−130.36	−105.91	−86.04	−69.90
Energy (*E*)	−3200.00	−3400.00	−3600.00	−3800.00	−4000.00
Energy (NPV)	−98.29	−84.00	−71.54	−60.74	−51.43
Accumulated FPV	−243853.31	−262669.44	−281579.02	−300563.85	−319609.27
Maintenance (*M*)	−348.29	−351.77	−355.29	−358.84	−362.43
Present Value (MPV)	−10.70	−8.69	−7.06	−5.74	−4.66
Accumulated (MPV)	−1240.12	−1248.81	−1255.87	−1261.61	−1266.27
Non Production (NP)	0.00	0.00	0.00	0.00	0.00
Present Value (NPPV)	0.00	0.00	0.00	0.00	0.00
Accumulated NPPV	0.00	0.00	0.00	0.00	0.00
Accumulated Total Costs (APV)	−245093.43	−263918.25	−282834.89	−301825.45	−320875.53
Accumulated Total Costs	−110037.75	−119066.13	−128350.79	−137892.30	−147691.22
Benefit (*B*)	44642.86	44642.86	44642.86	44642.86	44642.86
Present Value (BPV)	1371.18	1102.95	887.18	713.63	574.03
Accumulated Benefits BPV	177926.28	179029.23	179916.42	180630.05	181204.07
BPV + FPV + MPV + NPPV + IIPV	−79167.14	−96889.02	−114918.47	−133195.41	−151671.46
MTAC-PV (LDM)	−16076.62	−16236.06	−16383.42	−16519.55	−16645.31
MTAC-PV (SDM)	−16071.51	−16232.40	−16380.85	−16517.77	−16644.13
MTAC-PV (EM)	−16068.59	−16230.63	−16379.80	−16517.18	−16643.80

(*Continued*)

TABLE 6.10 (*Continued*)
Values Used for Simulation based on MMTAC-RPV Method Related to Maintenance Standard – First Simulation

Time	21	22	23	24	25
Initial Investment (II)					
Initial Investment (II)					
II Present Value (IIPV) - Return	−124.11	−99.83	−80.30	−64.59	−51.96
Life Time simulated					
Residual value of equipment (RV)					
Linear Depreciation Method (Vn)	−1911.60	−1431.20	−950.80	−470.40	10.00
LDM Present Value (Vn)	−19.77	−11.91	−6.36	−2.53	0.04
Sum of Digits Method (Vn)	−359.54	−211.72	−100.86	−26.95	10.00
SDM Present Value (Vn)	−3.72	−1.76	−0.67	−0.15	0.04
Exponential Method (Vn)	−31.09	−23.42	−17.63	−13.28	−10.00
EM Present Value (Pv)	−0.32	−0.19	−0.12	−0.07	−0.04
Functioning (*F*)	−5490.86	−5545.76	−5601.22	−5657.23	−5713.81
Present Value Acummulated (NPV)	−19050.78	−19096.91	−19134.40	−19164.85	−19189.59
Present Value (NPV)	−56.79	−46.14	−37.48	−30.45	−24.74
Energy (*E*)	−4200.00	−4400.00	−4600.00	−4800.00	−5000.00
Energy (NPV)	−43.44	−36.61	−30.78	−25.84	−21.65
Accumulated FPV	−338703.48	−357837.01	−377002.19	−396192.88	−415404.12
Maintenance (*M*)	−366.06	−369.72	−373.41	−377.15	−380.92
Present Value (MPV)	−3.79	−3.08	−2.50	−2.03	−1.65
Accumulated (MPV)	−1270.05	−1273.13	−1275.63	−1277.66	−1279.31
Non Production (NP)	0.00	0.00	0.00	0.00	0.00
Present Value (NPPV)	0.00	0.00	0.00	0.00	0.00
Accumulated NPPV	0.00	0.00	0.00	0.00	0.00
Accumulated Total Costs (APV)	−339973.54	−359110.13	−378277.81	−397470.53	−416683.42
Accumulated Total Costs	−157748.13	−168063.61	−178638.25	−189472.63	−200567.36
Benefit (*B*)	44642.86	44642.86	44642.86	44642.86	44642.86
Present Value (BPV)	461.73	371.41	298.75	240.31	193.30
Accumulated Benefits BPV	181665.81	182037.21	182335.96	182576.27	182769.57
BPV + FPV + MPV + NPPV + IIPV	−170307.73	−189072.92	−207941.85	−226894.26	−245913.85
MTAC-PV (LDM)	−16761.59	−16869.18	−16968.88	−17061.38	−17147.34
MTAC-PV (SDM)	−16760.82	−16868.72	−16968.63	−17061.28	−17147.34
MTAC-PV (EM)	−16760.66	−16868.65	−16968.61	−17061.28	−17147.34

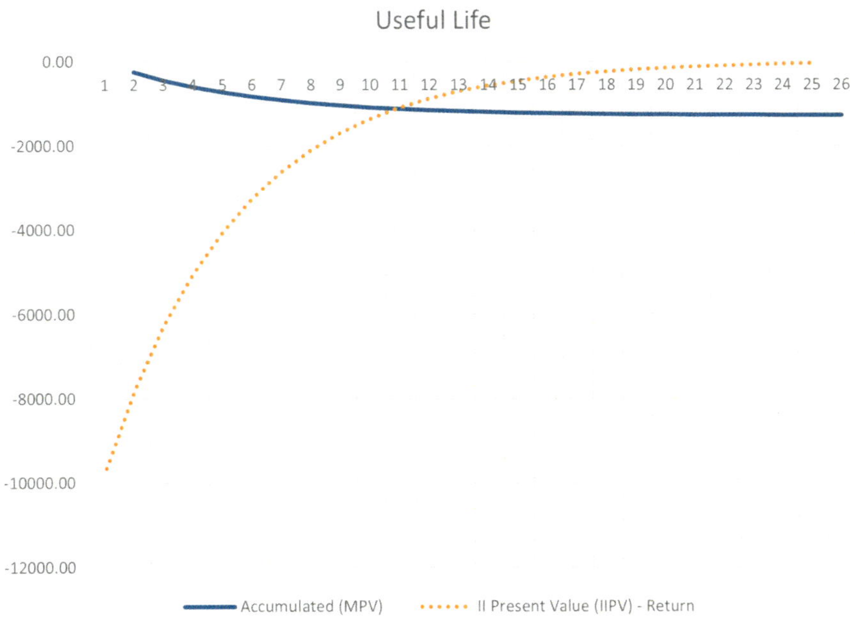

FIGURE 6.20 Physical asset useful life.

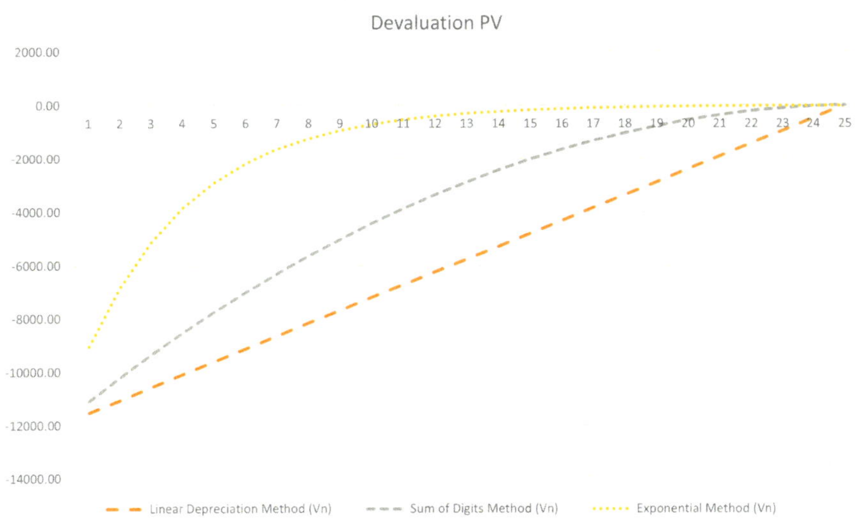

FIGURE 6.21 The three devaluation methods.

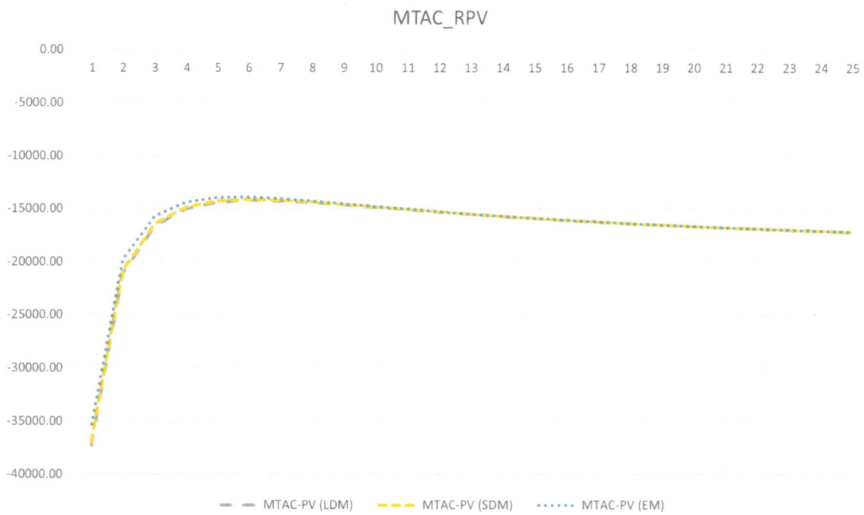

FIGURE 6.22 The MMTAC-RPV results from the three devaluation methods.

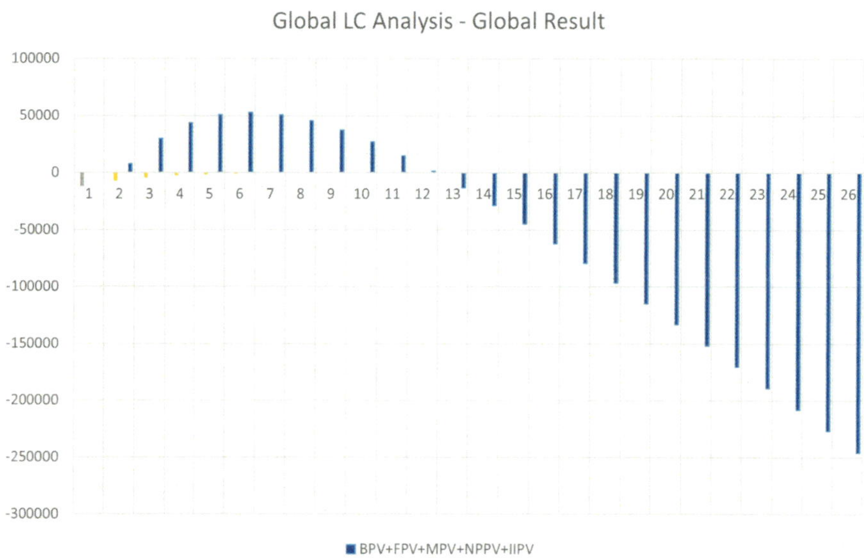

FIGURE 6.23 Life cycle analysis global result including profits.

These results from the three methods are close, which reinforces a lot the robustness of the models under discussion.

6.5.2 Life Cycle Investment of Physical Assets

When a company decides to implement a new maintenance policy, namely CBM, Predictive Maintenance (PdM) or other, it is necessary to evaluate which is the best option, quantitatively supported.

TABLE 6.11
Values Used for Simulation based on MMTAC-RPV Method Related to Maintenance Standard – Second Simulation

Time	0	1	2	3	4	5
Initial Investment (II)	−12000					
Initial Investment (II)	−12000					
II Present Value (IIPV) - Return	−12000	−9652.51	−7764.25	−6245.37	−5023.63	−4040.88
Life Time simulated	25					
Residual value of equipment (RV)	10					
Linear Depreciation Method (Vn)	−12000	−11519.60	−11039.20	−10558.80	−10078.40	−9598.00
LDM Present Value (Vn)		−9266.09	−7142.59	−5495.30	−4219.18	−3232.03
Sum of Digits Method (Vn)	−12000	−11076.15	−10189.26	−9339.32	−8526.34	−7750.31
SDM Present Value (Vn)		−8909.39	−6592.66	−4860.63	−3569.43	−2609.84
Exponential Method (Vn)	−12000	−9036.79	−6805.29	−5124.83	−3859.33	−2906.33
EM Present Value (Pv)		−7268.97	−4403.16	−2667.21	−1615.65	−978.68
Functioning (*F*)	−4500	−4500.00	−4545.00	−4590.45	−4636.35	−4682.72
Present Value Acummulated (NPV)		−3619.69	−6560.40	−8949.49	−10890.43	−12467.29
Present Value (NPV)		−3619.69	−2940.71	−2389.09	−1940.94	−1576.86
Energy (*E*)	−200	−200.00	−400.00	−600.00	−800.00	−1000.00
Energy (NPV)		−160.88	−258.81	−312.27	−334.91	−336.74
Accumulated FPV		−15780.57	−22599.77	−31861.53	−43086.87	−55890.90
Maintenance (*M*)	−300	−300.00	−303.00	−306.03	−309.09	−312.18
Present Value (MPV)		−241.31	−196.05	−159.27	−129.40	−105.12
Accumulated (MPV)		−241.31	−437.36	−596.63	−726.03	−831.15
Non Production (NP)	0	0.00	0.00	0.00	0.00	0.00
Present Value (NPPV)		0.00	0.00	0.00	0.00	0.00
Accumulated NPPV		0.00	0.00	0.00	0.00	0.00
Accumulated Total Costs (APV)		−16021.88	−23037.13	−32458.16	−43812.90	−56722.05
Accumulated Total Costs		−5000.00	−10248.00	−15744.48	−21489.92	−27484.82
Benefit (*B*)	50000	48309.18	48309.18	48309.18	48309.18	48309.18
Present Value (BPV)		38858.73	31257.03	25142.40	20223.93	16267.64
Accumulated Benefits BPV		38858.73	70115.76	95258.16	115482.09	131749.73
BPV + FPV + MPV + NPPV + IIPV		10836.86	35078.63	50799.99	59669.19	63027.68
MTAC-PV (LDM)		−37287.97	−21089.86	−16651.15	−15008.02	−14390.82
MTAC-PV (SDM)		−36931.27	−20814.90	−16439.60	−14845.58	−14266.38
MTAC-PV (EM)		−35290.85	−19720.15	−15708.46	−14357.14	−13940.15

(*Continued*)

TABLE 6.11 (*Continued*)
Values Used for Simulation based on MMTAC-RPV Method Related to Maintenance Standard – Second Simulation

Time	6	7	8	9	10
Initial Investment (II)					
Initial Investment (II)					
II Present Value (IIPV) - Return	−3250.39	−2614.53	−2103.07	−1691.66	−1360.73
Life Time simulated					
Residual value of equipment (RV)					
Linear Depreciation Method (Vn)	−9117.60	−8637.20	−8156.80	−7676.40	−7196.00
LDM Present Value (Vn)	−2469.64	−1881.85	−1429.53	−1082.15	−815.98
Sum of Digits Method (Vn)	−7011.23	−6309.11	−5643.94	−5015.72	−4424.46
SDM Present Value (Vn)	−1899.10	−1374.61	−989.13	−707.07	−501.71
Exponential Method (Vn)	−2188.66	−1648.20	−1241.20	−934.71	−703.90
EM Present Value (Pv)	−592.83	−359.11	−217.53	−131.77	−79.82
Functioning (*F*)	−4729.55	−4776.84	−4824.61	−4872.86	−4921.58
Present Value Acummulated (NPV)	−13748.36	−14789.13	−15634.67	−16321.60	−16879.68
Present Value (NPV)	−1281.07	−1040.77	−845.54	−686.93	−558.08
Energy (*E*)	−1200.00	−1400.00	−1600.00	−1800.00	−2000.00
Energy (NPV)	−325.04	−305.03	−280.41	−253.75	−226.79
Accumulated FPV	−69964.30	−85058.45	−100973.53	−117548.88	−134655.35
Maintenance (*M*)	−315.30	−318.46	−321.64	−324.86	−328.11
Present Value (MPV)	−85.40	−69.38	−56.37	−45.80	−37.21
Accumulated (MPV)	−916.56	−985.94	−1042.31	−1088.11	−1125.31
Non Production (NP)	0.00	0.00	0.00	0.00	0.00
Present Value (NPPV)	0.00	0.00	0.00	0.00	0.00
Accumulated NPPV	0.00	0.00	0.00	0.00	0.00
Accumulated Total Costs (APV)	−70880.85	−86044.40	−102015.84	−118636.99	−135780.66
Accumulated Total Costs	−33729.67	−40224.97	−46971.22	−53968.93	−61218.62
Benefit (*B*)	48309.18	48309.18	48309.18	48309.18	48309.18
Present Value (BPV)	13085.30	10525.50	8466.46	6810.21	5477.97
Accumulated Benefits BPV	144835.03	155360.53	163826.99	170637.20	176115.17
BPV + FPV + MPV + NPPV + IIPV	61954.18	57316.13	49811.14	40000.21	28334.51
MTAC-PV (LDM)	−14225.08	−14275.18	−14430.67	−14635.46	−14859.66
MTAC-PV (SDM)	−14129.99	−14202.72	−14375.62	−14593.78	−14828.24
MTAC-PV (EM)	−13912.28	−14057.64	−14279.17	−14529.86	−14786.05

(*Continued*)

TABLE 6.11 (*Continued*)
Values Used for Simulation based on MMTAC-RPV Method Related to Maintenance Standard – Second Simulation

Time	11	12	13	14	15
Initial Investment (II)					
Initial Investment (II)					
II Present Value (IIPV) - Return	−1094.54	−880.42	−708.19	−569.65	−458.21
Life Time simulated					
Residual value of equipment (RV)					
Linear Depreciation Method (Vn)	−6715.60	−6235.20	−5754.80	−5274.40	−4794.00
LDM Present Value (Vn)	−612.54	−457.47	−339.62	−250.38	−183.06
Sum of Digits Method (Vn)	−3870.15	−3352.80	−2872.40	−2428.95	−2022.46
SDM Present Value (Vn)	−353.00	−245.99	−169.52	−115.30	−77.23
Exponential Method (Vn)	−530.08	−399.19	−300.61	−226.38	−170.48
EM Present Value (Pv)	−48.35	−29.29	−17.74	−10.75	−6.51
Functioning (*F*)	−4970.80	−5020.51	−5070.71	−5121.42	−5172.63
Present Value Acummulated (NPV)	−17333.07	−17701.42	−18000.67	−18243.79	−18441.30
Present Value (NPV)	−453.39	−368.35	−299.25	−243.12	−197.51
Energy (*E*)	−2200.00	−2400.00	−2600.00	−2800.00	−3000.00
Energy (NPV)	−200.67	−176.08	−153.44	−132.92	−114.55
Accumulated FPV	−152189.09	−170066.59	−188220.70	−206597.41	−225153.26
Maintenance (*M*)	−331.39	−334.70	−338.05	−341.43	−344.84
Present Value (MPV)	−30.23	−24.56	−19.95	−16.21	−13.17
Accumulated (MPV)	−1155.54	−1180.09	−1200.04	−1216.25	−1229.42
Non Production (NP)	0.00	0.00	0.00	0.00	0.00
Present Value (NPPV)	0.00	0.00	0.00	0.00	0.00
Accumulated NPPV	0.00	0.00	0.00	0.00	0.00
Accumulated Total Costs (APV)	−153344.63	−171246.68	−189420.75	−207813.66	−226382.68
Accumulated Total Costs	−68720.81	−76476.01	−84484.77	−92747.62	−101265.10
Benefit (*B*)	48309.18	48309.18	48309.18	48309.18	48309.18
Present Value (BPV)	4406.35	3544.36	2851.00	2293.27	1844.65
Accumulated Benefits BPV	180521.51	184065.87	186916.87	189210.14	191054.79
BPV + FPV + MPV + NPPV + IIPV	15176.89	819.19	−14503.88	−30603.52	−47327.89
MTAC-PV (LDM)	−15087.02	−15308.68	−15520.03	−15718.86	−15904.38
MTAC-PV (SDM)	−15063.42	−15291.06	−15506.94	−15709.21	−15897.33
MTAC-PV (EM)	−15035.73	−15273.00	−15495.27	−15701.74	−15892.61

(*Continued*)

TABLE 6.11 (*Continued*)
Values Used for Simulation based on MMTAC-RPV Method Related to Maintenance Standard – Second Simulation

Time	16	17	18	19	20
Initial Investment (II)					
Initial Investment (II)					
II Present Value (IIPV) - Return	−368.57	−296.47	−238.48	−191.82	−154.30
Life Time simulated					
Residual value of equipment (RV)					
Linear Depreciation Method (Vn)	−4313.60	−3833.20	−3352.80	−2872.40	−2392.00
LDM Present Value (Vn)	−132.49	−94.70	−66.63	−45.92	−30.76
Sum of Digits Method (Vn)	−1652.92	−1320.34	−1024.71	−766.03	−544.31
SDM Present Value (Vn)	−50.77	−32.62	−20.36	−12.25	−7.00
Exponential Method (Vn)	−128.38	−96.68	−72.81	−54.83	−41.29
EM Present Value (Pv)	−3.94	−2.39	−1.45	−0.88	−0.53
Functioning (*F*)	−5224.36	−5276.60	−5329.37	−5382.66	−5436.49
Present Value Acummulated (NPV)	−18601.76	−18732.13	−18838.04	−18924.08	−18993.99
Present Value (NPV)	−160.46	−130.36	−105.91	−86.04	−69.90
Energy (*E*)	−3200.00	−3400.00	−3600.00	−3800.00	−4000.00
Energy (NPV)	−98.29	−84.00	−71.54	−60.74	−51.43
Accumulated FPV	−243853.31	−262669.44	−281579.02	−300563.85	−319609.27
Maintenance (*M*)	−348.29	−351.77	−355.29	−358.84	−362.43
Present Value (MPV)	−10.70	−8.69	−7.06	−5.74	−4.66
Accumulated (MPV)	−1240.12	−1248.81	−1255.87	−1261.61	−1266.27
Non Production (NP)	0.00	0.00	0.00	0.00	0.00
Present Value (NPPV)	0.00	0.00	0.00	0.00	0.00
Accumulated NPPV	0.00	0.00	0.00	0.00	0.00
Accumulated Total Costs (APV)	−245093.43	−263918.25	−282834.89	−301825.45	−320875.53
Accumulated Total Costs	−110037.75	−119066.13	−128350.79	−137892.30	−147691.22
Benefit (*B*)	48309.18	48309.18	48309.18	48309.18	48309.18
Present Value (BPV)	1483.79	1193.53	960.05	772.24	621.17
Accumulated Benefits BPV	192538.59	193732.12	194692.16	195464.40	196085.57
BPV + FPV + MPV + NPPV + IIPV	−64554.84	−82186.13	−100142.73	−118361.06	−136789.97
MTAC-PV (LDM)	−16076.62	−16236.06	−16383.42	−16519.55	−16645.31
MTAC-PV (SDM)	−16071.51	−16232.40	−16380.85	−16517.77	−16644.13
MTAC-PV (EM)	−16068.59	−16230.63	−16379.80	−16517.18	−16643.80

(*Continued*)

TABLE 6.11 (*Continued*)
Values Used for Simulation based on MMTAC-RPV Method Related to Maintenance Standard – Second Simulation

Time	21	22	23	24	25
Initial Investment (II)					
Initial Investment (II)					
II Present Value (IIPV) - Return	−124.11	−99.83	−80.30	−64.59	−51.96
Life Time simulated					
Residual value of equipment (RV)					
Linear Depreciation Method (Vn)	−1911.60	−1431.20	−950.80	−470.40	10.00
LDM Present Value (Vn)	−19.77	−11.91	−6.36	−2.53	0.04
Sum of Digits Method (Vn)	−359.54	−211.72	−100.86	−26.95	10.00
SDM Present Value (Vn)	−3.72	−1.76	−0.67	−0.15	0.04
Exponential Method (Vn)	−31.09	−23.42	−17.63	−13.28	−10.00
EM Present Value (Pv)	−0.32	−0.19	−0.12	−0.07	−0.04
Functioning (*F*)	−5490.86	−5545.76	−5601.22	−5657.23	−5713.81
Present Value Acummulated (NPV)	−19050.78	−19096.91	−19134.40	−19164.85	−19189.59
Present Value (NPV)	−56.79	−46.14	−37.48	−30.45	−24.74
Energy (*E*)	−4200.00	−4400.00	−4600.00	−4800.00	−5000.00
Energy (NPV)	−43.44	−36.61	−30.78	−25.84	−21.65
Accumulated FPV	−338703.48	−357837.01	−377002.19	−396192.88	−415404.12
Maintenance (*M*)	−366.06	−369.72	−373.41	−377.15	−380.92
Present Value (MPV)	−3.79	−3.08	−2.50	−2.03	−1.65
Accumulated (MPV)	−1270.05	−1273.13	−1275.63	−1277.66	−1279.31
Non Production (NP)	0.00	0.00	0.00	0.00	0.00
Present Value (NPPV)	0.00	0.00	0.00	0.00	0.00
Accumulated NPPV	0.00	0.00	0.00	0.00	0.00
Accumulated Total Costs (APV)	−339973.54	−359110.13	−378277.81	−397470.53	−416683.42
Accumulated Total Costs	−157748.13	−168063.61	−178638.25	−189472.63	−200567.36
Benefit (*B*)	48309.18	48309.18	48309.18	48309.18	48309.18
Present Value (BPV)	499.65	401.91	323.29	260.04	209.17
Accumulated Benefits BPV	196585.22	196987.13	197310.42	197570.46	197779.63
BPV + FPV + MPV + NPPV + IIPV	−155388.32	−174123.00	−192967.40	−211900.07	−230903.79
MTAC-PV (LDM)	−16761.59	−16869.18	−16968.88	−17061.38	−17147.34
MTAC-PV (SDM)	−16760.82	−16868.72	−16968.63	−17061.28	−17147.34
MTAC-PV (EM)	−16760.66	−16868.65	−16968.61	−17061.28	−17147.34

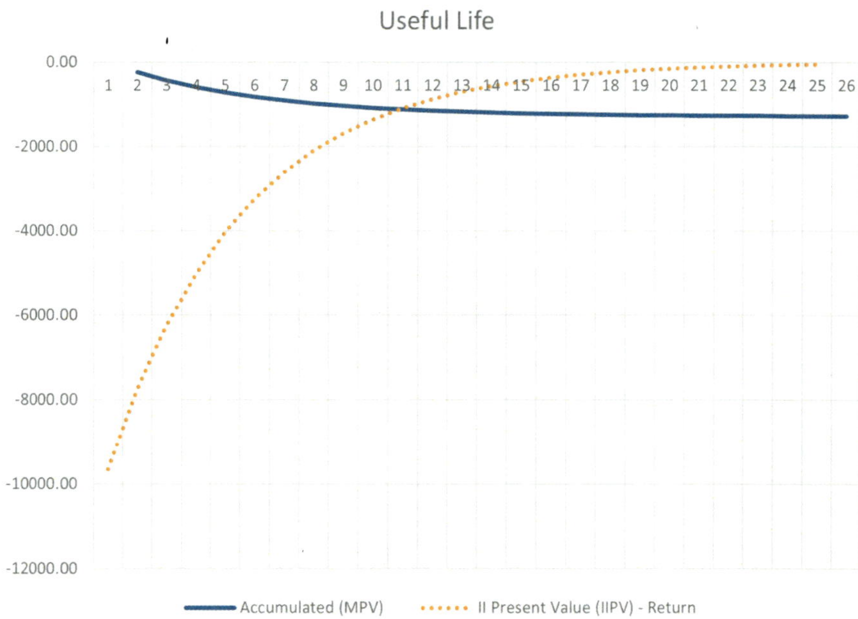

FIGURE 6.24 Physical asset useful life.

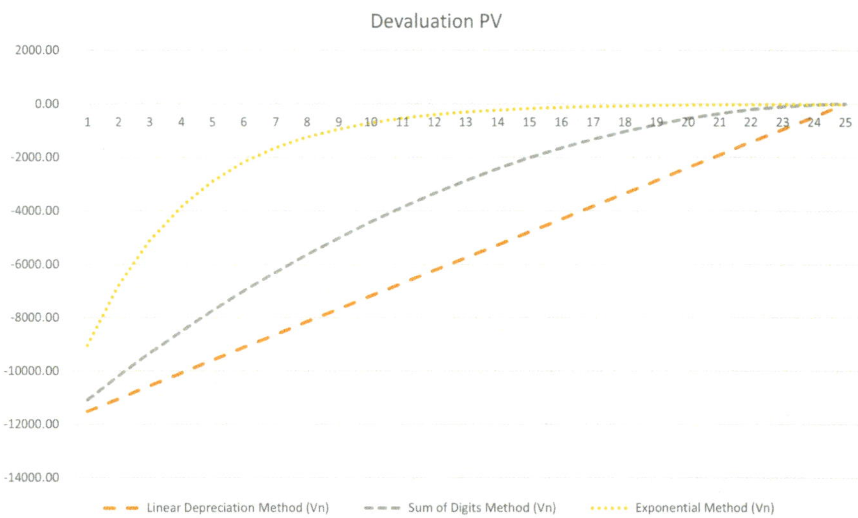

FIGURE 6.25 The three devaluation methods.

Because the main subject discussed in this chapter is the Maintenance Standard NP 4492, the MTBF, MTTR, and Availability to evaluate how the Life Cycle behaviours according to the variation of these parameters are used, which corresponds to a variation of the maintenance policies.

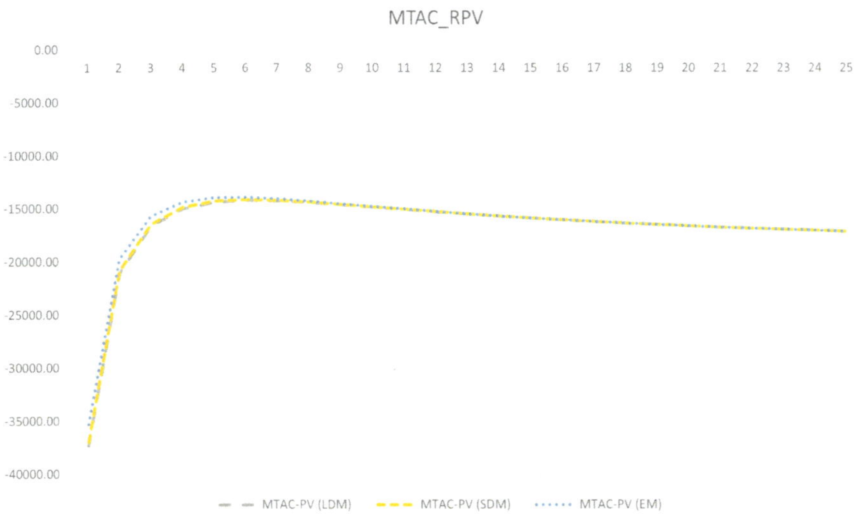

FIGURE 6.26 The MMTAC results from the three devaluation methods.

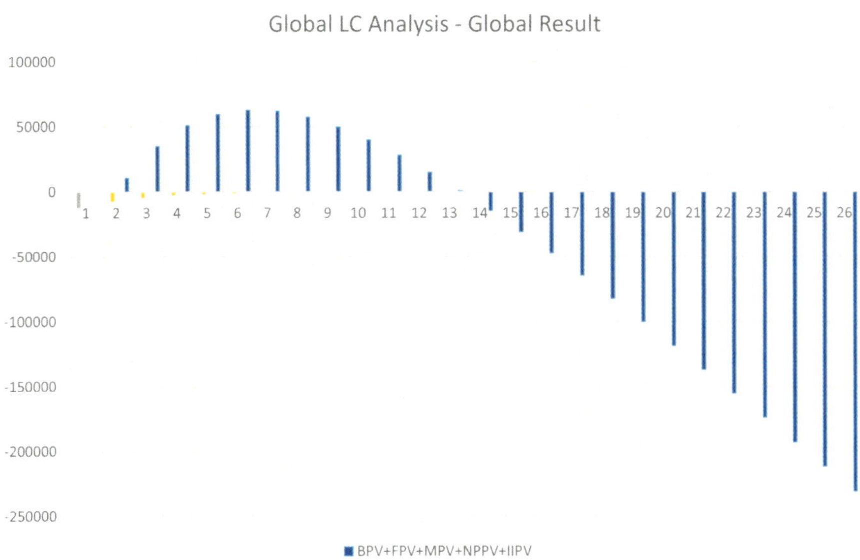

FIGURE 6.27 Life cycle analysis global result including profits.

The way to demonstrate if the return of implementing the standard is interesting or not is done, in this section, using the model of Life Cycle Investment, presented in Chapter 2.

The first simulation uses the following values:

- Initial Investment (II) –12000
- Life Time simulated 25

- Residual value of equipment (RV) 10
- Functioning (*F*) –4500
- Energy (*E*) –200
- Maintenance (*M*) –300
- Non-Production (NP) 0
- Benefit (*B*) 50000
- Internal Rate of Return (*r*) 0.12
- Risk (*R*) 0.11
- Apparent Rate 0.2432
- Maintenance Increase Rate 0.01
- Availability 0.8929
- MTBF **500**
- MTTR **50**
- MWT 10

The remaining values are presented in Table 6.12, and the graphical results are shown in Figures 6.28–6.30.

As referred in the preceding introduction of this section, the evaluation of the introduction of standard norms and maintenance policies can be simulated, primarily, through the variation of MTBF and MTTR. They may be used other indicators, but may be not so intuitive.

The second simulation uses the following values (the MTBF is increased and MTTR decreases as a result of the Maintenance Standards implementation):

- Initial Investment (II) –12000
- Life Time simulated 25
- Residual value of equipment (RV) 10
- Functioning (*F*) –4500
- Energy (*E*) –200
- Maintenance (*M*) –300
- Non-Production (*NP*) 0
- Benefit (*B*) 50000
- Internal Rate of Return (*r*) 0.12
- Risk (*R*) 0.11
- Apparent Rate 0.2432
- Maintenance Increase Rate 0.01
- Availability 0.8929
- MTBF **1000**
- MTTR **25**
- MWT 10

The remaining values are presented in Table 6.13, and the graphical results are shown in Figures 6.31–6.33.

As can be seen, when comparing Figures 6.30 and 6.33, with the increasing of the maintenance management quality helped by the maintenance standards, it gets one more positive year, but the positive incomes increase a lot.

TABLE 6.12
Values Used for Simulation based on Life Cycle Investment – First Simulation

Time	0	1	2	3	4	5	6	7	8	9	10	11	12	13
Initial Investment (II)	−12000													
II Present Value (IIPV) - Return	−12000	−9652.51	−7764.25	−6245.37	−5023.63	−4040.88	−3250.39	−2614.53	−2103.07	−1691.66	−1360.73	−1094.54	−880.42	−708.19
Functioning (*F*)	−4500	−4500.00	−4500.00	−4500.00	−4500.00	−4500.00	−4500.00	−4500.00	−4500.00	−4500.00	−4500.00	−4500.00	−4500.00	−4500.00
Present Value (PV) Accumulated		−3619.69	−6531.28	−8873.30	−10757.16	−12272.49	−13491.38	−14471.83	−15260.48	−15894.86	−16405.13	−16815.58	−17145.74	−17411.31
Energy (*E*)	−200	−200.00	−200.00	−200.00	−200.00	−200.00	−200.00	−200.00	−200.00	−200.00	−200.00	−200.00	−200.00	−200.00
Energy (PV)		−160.88	−129.40	−104.09	−83.73	−67.35	−54.17	−43.58	−35.05	−28.19	−22.68	−18.24	−14.67	−11.80
Accumulated (AFPV)		−15780.57	−6660.69	−15638.07	−26478.96	−38818.79	−52364.35	−66879.76	−82175.29	−98098.34	−114526.15	−131359.97	−148520.38	−165943.49
Maintenance (*M*)	−300	−300.00	−303.00	−306.03	−309.09	−312.18	−315.30	−318.46	−321.64	−324.86	−328.11	−331.39	−334.70	−338.05
Present Value (PV)		−241.31	−194.11	−156.13	−125.59	−101.02	−81.26	−65.36	−52.58	−42.29	−34.02	−27.36	−22.01	−17.70
Accumulated (AMPV)	−5000	−241.31	−435.42	−591.55	−717.14	−818.17	−899.43	−964.79	−1017.37	−1059.66	−1093.68	−1121.04	−1143.05	−1160.75
Non Production (NP)	0	0.00	0.00	0.00	0.00	0.00	0.00	0.00	0.00	0.00	0.00	0.00	0.00	0.00
Present Value (PV)		0.00	0.00	0.00	0.00	0.00	0.00	0.00	0.00	0.00	0.00	0.00	0.00	0.00
Accumulated (ANPPV)		0.00	0.00	0.00	0.00	0.00	0.00	0.00	0.00	0.00	0.00	0.00	0.00	0.00
Accumulated Total Costs (ATCPV)		−16021.88	−7096.11	−16229.63	−27196.10	−39636.96	−53263.78	−67844.55	−83192.66	−99158.00	−115619.82	−132481.01	−149663.43	−167104.24
Benefit (*B*) including IIPV	50000	34990.35	36878.61	38397.49	39619.23	40601.97	41392.47	42028.32	42539.79	42951.20	43282.13	43548.32	43762.44	43934.67
Present Value (PV)		28145.39	23861.22	19983.88	16586.01	13672.32	11211.80	9157.04	7455.34	6054.89	4907.93	3972.10	3210.77	2592.83
Accumulated Benefits (ABPV)		28145.39	52006.60	71990.48	88576.50	102248.82	113460.62	122617.65	130072.99	136127.88	141035.82	145007.92	148218.69	150811.52
ATCPV + ABPV		12123.51	44910.50	55760.86	61380.40	62611.86	60196.84	54773.11	46880.33	36969.88	25415.99	12526.91	−1444.74	−16292.72

(Continued)

TABLE 6.12 (*Continued*)
Values Used for Simulation based on Life Cycle Investment – First Simulation

Time	14	15	16	17	18	19	20	21	22	23	24	25
Initial Investment (II)												
II Present Value (IIPV) - Return	−569.65	−458.21	−368.57	−296.47	−238.48	−191.82	−154.30	−124.11	−99.83	−80.30	−64.59	−51.96
Functioning (*F*)	−4500.00	−4500.00	−4500.00	−4500.00	−4500.00	−4500.00	−4500.00	−4500.00	−4500.00	−4500.00	−4500.00	−4500.00
Present Value (PV) Accumulated	−17624.92	−17796.75	−17934.97	−18046.15	−18135.57	−18207.51	−18265.37	−18311.91	−18349.35	−18379.47	−18403.69	−18423.17
Energy (*E*)	−200.00	−200.00	−200.00	−200.00	−200.00	−200.00	−200.00	−200.00	−200.00	−200.00	−200.00	−200.00
Energy (PV)	−9.49	−7.64	−6.14	−4.94	−3.97	−3.20	−2.57	−2.07	−1.66	−1.34	−1.08	−0.87
Accumulated (AFPV)	−183577.91	−201382.30	−219323.41	−237374.50	−255514.05	−273724.76	−291992.70	−310306.68	−328657.69	−347038.50	−365443.26	−383867.30
Maintenance (*M*)	−341.43	−344.84	−348.29	−351.77	−355.29	−358.84	−362.43	−366.06	−369.72	−373.41	−377.15	−380.92
Present Value (PV)	−14.24	−11.46	−9.21	−7.41	−5.96	−4.80	−3.86	−3.10	−2.50	−2.01	−1.61	−1.30
Accumulated (AMPV)	−1174.99	−1186.45	−1195.66	−1203.08	−1209.04	−1213.83	−1217.69	−1220.79	−1223.29	−1225.30	−1226.91	−1228.21
Non Production (NP)	0.00	0.00	0.00	0.00	0.00	0.00	0.00	0.00	0.00	0.00	0.00	0.00
Present Value (PV)	0.00	0.00	0.00	0.00	0.00	0.00	0.00	0.00	0.00	0.00	0.00	0.00
Accumulated (ANPPV)	0.00	0.00	0.00	0.00	0.00	0.00	0.00	0.00	0.00	0.00	0.00	0.00
Accumulated Total Costs (ATCPV)	−184752.90	−202568.75	−220519.08	−238577.58	−256723.09	−274938.59	−293210.39	−311527.47	−329880.98	−348263.80	−366670.18	−385095.51
Benefit (*B*) including IIPV	44073.21	44184.65	44274.28	44346.38	44404.38	44451.03	44488.56	44518.74	44543.02	44562.55	44578.26	44590.90
Present Value (PV)	2092.19	1687.16	1359.86	1095.62	882.45	710.56	572.04	460.45	370.58	298.21	239.96	193.07
Accumulated Benefits (ABPV)	152903.71	154590.87	155950.73	157046.36	157928.80	158639.37	159211.41	159671.86	160042.43	160340.65	160580.61	160773.68
ATCPV + ABPV	−31849.19	−47977.88	−64568.34	−81531.22	−98794.29	−116299.22	−133998.98	−151855.62	−169838.55	−187923.15	−206089.57	−224321.83

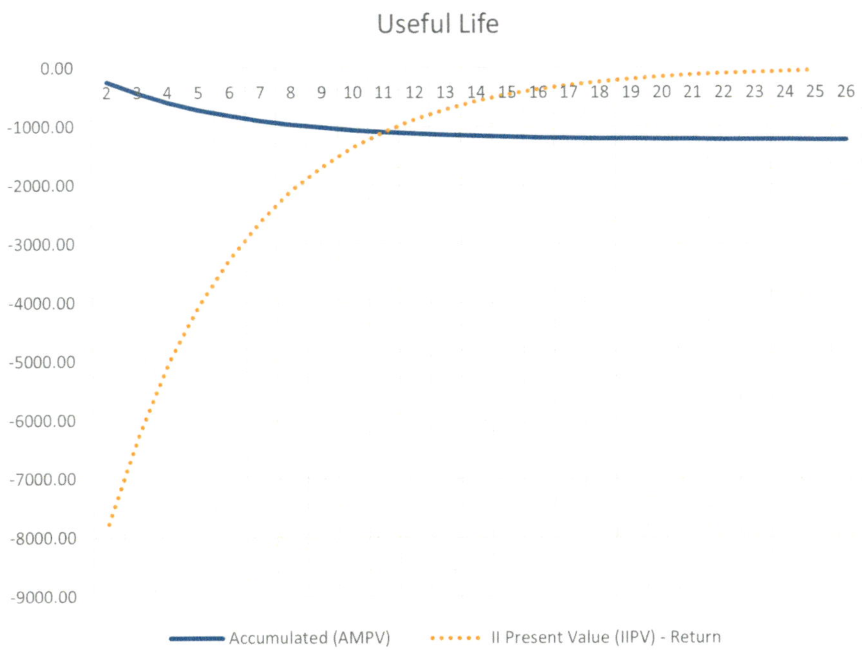

FIGURE 6.28 Physical asset useful life.

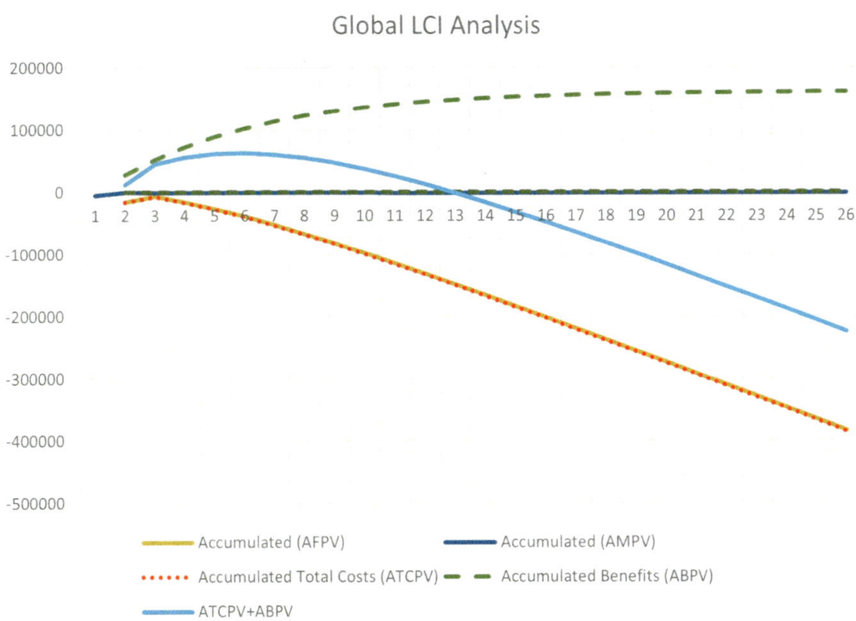

FIGURE 6.29 Global LCI analysis.

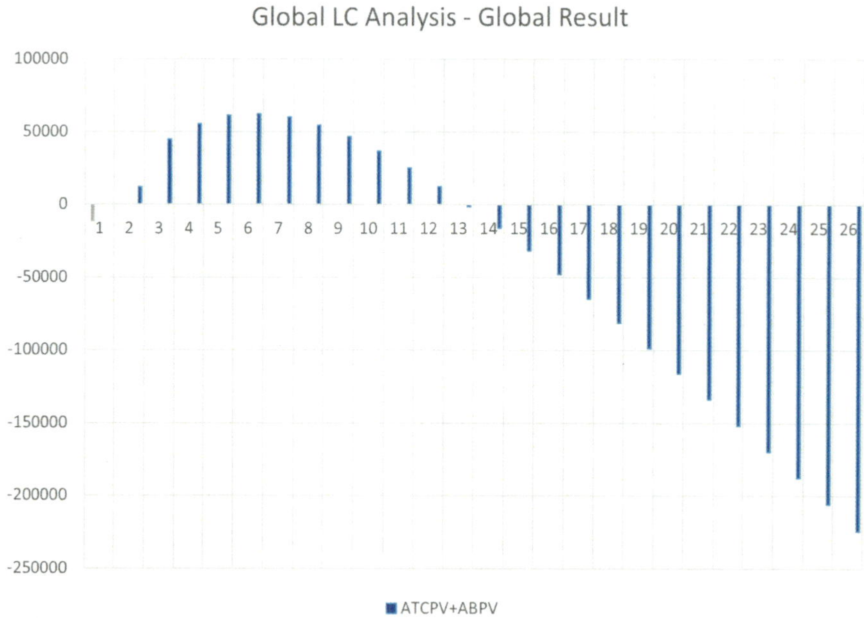

FIGURE 6.30 Life cycle analysis global result.

6.5.3 Life Cycle of Physical Assets with Recovery

Considering the same arguments of the preceding sub-section and using the Life Cycle model of Physical Asset Life Cycle based with Recovery, the result of using this model, for a first scenario, is illustrated in Table 6.14.

Other relevant values considered are the following ones:

• Interest Rate (*r*)	0.05
• Inflation Rate (*h*)	0.02
• Profit Rate (*p*)	0.05
• Maintenance Increase Rate	0.01
• **MTBF**	**500**
• **MTTR**	**60**
• Risk (*R*)	0.10
• Time (*t*)	0
• Initial Investment (II) (C_0)	–12000
• F_p Financial Movement (Production)	70000
• Functioning Costs	–4500
• Energy	–200
• Maintenance Costs	–300

The graphical results are presented in Figures 6.34–6.36.

For the second scenario, other relevant values considered are the following ones:

TABLE 6.13
Values Used for Simulation based on Life Cycle Investment – Second Simulation

Time	0	1	2	3	4	5	6	7	8	9	10	11	12	13
Initial Investment (II)	−12000													
II Present Value (IIPV) - Return	−12000	−9652.51	−7764.25	−6245.37	−5023.63	−4040.88	−3250.39	−2614.53	−2103.07	−1691.66	−1360.73	−1094.54	−880.42	−708.19
Functioning (*F*)	−4500	−4500.00	−4500.00	−4500.00	−4500.00	−4500.00	−4500.00	−4500.00	−4500.00	−4500.00	−4500.00	−4500.00	−4500.00	−4500.00
Present Value (PV) Accumulated		−3619.69	−6531.28	−8873.30	−10757.16	−12272.49	−13491.38	−14471.83	−15260.48	−15894.86	−16405.13	−16815.58	−17145.74	−17411.31
Energy (*E*)	−200	−200.00	−200.00	−200.00	−200.00	−200.00	−200.00	−200.00	−200.00	−200.00	−200.00	−200.00	−200.00	−200.00
Energy (PV)		−160.88	−129.40	−104.09	−83.73	−67.35	−54.17	−43.58	−35.05	−28.19	−22.68	−18.24	−14.67	−11.80
Accumulated (AFPV)		−15780.57	−6660.69	−15638.07	−26478.96	−38818.79	−52364.35	−66879.76	−82175.29	−98098.34	−114526.15	−131359.97	−148520.38	−165943.49
Maintenance (*M*)	−300	−300.00	−303.00	−306.03	−309.09	−312.18	−315.30	−318.46	−321.64	−324.86	−328.11	−331.39	−334.70	−338.05
Present Value (PV)		−241.31	−194.11	−156.13	−125.59	−101.02	−81.26	−65.36	−52.58	−42.29	−34.02	−27.36	−22.01	−17.70
Accumulated (AMPV)	−5000	−241.31	−435.42	−591.55	−717.14	−818.17	−899.43	−964.79	−1017.37	−1059.66	−1093.68	−1121.04	−1143.05	−1160.75
Non Production (NP)	0	0.00	0.00	0.00	0.00	0.00	0.00	0.00	0.00	0.00	0.00	0.00	0.00	0.00
Present Value (PV)		0.00	0.00	0.00	0.00	0.00	0.00	0.00	0.00	0.00	0.00	0.00	0.00	0.00
Accumulated (ANPPV)		0.00	0.00	0.00	0.00	0.00	0.00	0.00	0.00	0.00	0.00	0.00	0.00	0.00
Accumulated Total Costs (ATCPV)		−16021.88	−7096.11	−16229.63	−27196.10	−39636.96	−53263.78	−67844.55	−83192.66	−99158.00	−115619.82	−132481.01	−149663.43	−167104.24
Benefit (*B*) including IIPV	50000	38656.67	40544.93	42063.81	43285.55	44268.30	45058.79	45694.65	46206.11	46617.52	46948.45	47214.64	47428.76	47600.99
Present Value (PV)		31094.49	26233.40	21892.01	18120.87	14906.92	12204.88	9955.85	8097.88	6571.74	5323.67	4306.51	3479.76	2809.20
Accumulated Benefits (ABPV)		31094.49	57327.89	79219.90	97340.76	112247.68	124452.56	134408.41	142506.29	149078.03	154401.70	158708.21	162187.97	164997.18
ATCPV + ABPV		15072.61	50231.78	62990.27	70144.66	72610.72	71188.78	66563.86	59313.63	49920.03	38781.87	26227.20	12524.54	−2107.07

(Continued)

TABLE 6.13 (*Continued*)
Values Used for Simulation based on Life Cycle Investment – Second Simulation

Time	14	15	16	17	18	19	20	21	22	23	24	25
Initial Investment (II)												
II Present Value (IIPV) - Return	−569.65	−458.21	−368.57	−296.47	−238.48	−191.82	−154.30	−124.11	−99.83	−80.30	−64.59	−51.96
Functioning (*F*)	−4500.00	−4500.00	−4500.00	−4500.00	−4500.00	−4500.00	−4500.00	−4500.00	−4500.00	−4500.00	−4500.00	−4500.00
Present Value (PV) Accumulated	−17624.92	−17796.75	−17934.97	−18046.15	−18135.57	−18207.51	−18265.37	−18311.91	−18349.35	−18379.47	−18403.69	−18423.17
Energy (*E*)	−200.00	−200.00	−200.00	−200.00	−200.00	−200.00	−200.00	−200.00	−200.00	−200.00	−200.00	−200.00
Energy (PV)	−9.49	−7.64	−6.14	−4.94	−3.97	−3.20	−2.57	−2.07	−1.66	−1.34	−1.08	−0.87
Accumulated (AFPV)	−183577.91	−201382.30	−219323.41	−237374.50	−255514.05	−273724.76	−291992.70	−310306.68	−328657.69	−347038.50	−365443.26	−383867.30
Maintenance (*M*)	−341.43	−344.84	−348.29	−351.77	−355.29	−358.84	−362.43	−366.06	−369.72	−373.41	−377.15	−380.92
Present Value (PV)	−14.24	−11.46	−9.21	−7.41	−5.96	−4.80	−3.86	−3.10	−2.50	−2.01	−1.61	−1.30
Accumulated (AMPV)	−1174.99	−1186.45	−1195.66	−1203.08	−1209.04	−1213.83	−1217.69	−1220.79	−1223.29	−1225.30	−1226.91	−1228.21
Non Production (NP)	0.00	0.00	0.00	0.00	0.00	0.00	0.00	0.00	0.00	0.00	0.00	0.00
Present Value (PV)	0.00	0.00	0.00	0.00	0.00	0.00	0.00	0.00	0.00	0.00	0.00	0.00
Accumulated (ANPPV)	0.00	0.00	0.00	0.00	0.00	0.00	0.00	0.00	0.00	0.00	0.00	0.00
Accumulated Total Costs (ATCPV)	−184752.90	−202568.75	−220519.08	−238577.58	−256723.09	−274938.59	−293210.39	−311527.47	−329880.98	−348263.80	−366670.18	−385095.51
Benefit (*B*) including IIPV	47739.53	47850.97	47940.60	48012.71	48070.70	48117.36	48154.88	48185.06	48209.34	48228.87	48244.58	48257.22
Present Value (PV)	2266.23	1827.16	1472.47	1186.20	955.31	769.17	619.18	498.37	401.08	322.75	259.70	208.95
Accumulated Benefits (ABPV)	167263.41	169090.56	170563.04	171749.24	172704.55	173473.72	174092.90	174591.27	174992.35	175315.10	175574.79	175783.74
ATCPV + ABPV	−17489.50	−33478.19	−49956.04	−66828.34	−84018.54	−101464.87	−119117.49	−136936.20	−154888.64	−172948.70	−191095.38	−209311.77

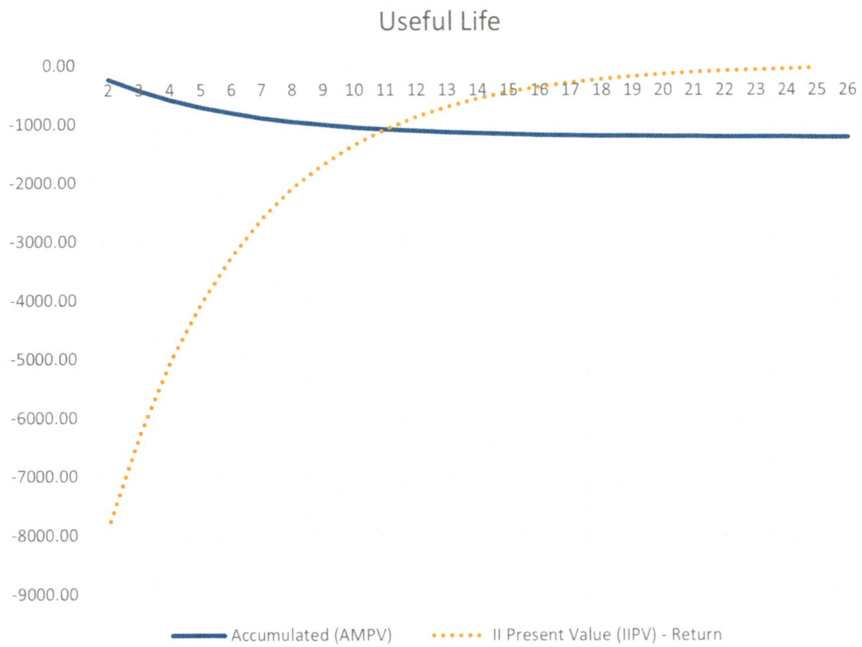

FIGURE 6.31 Physical asset useful life.

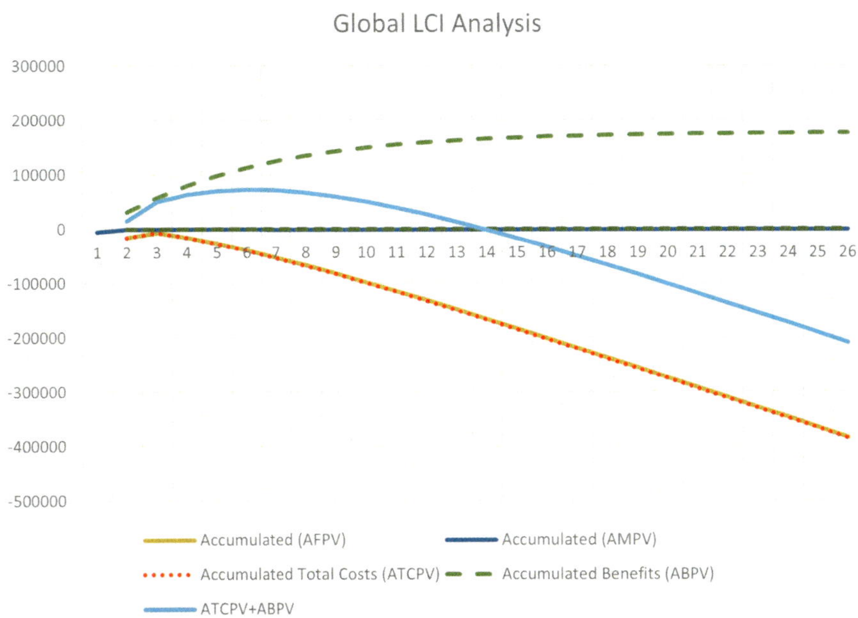

FIGURE 6.32 Global LCI analysis.

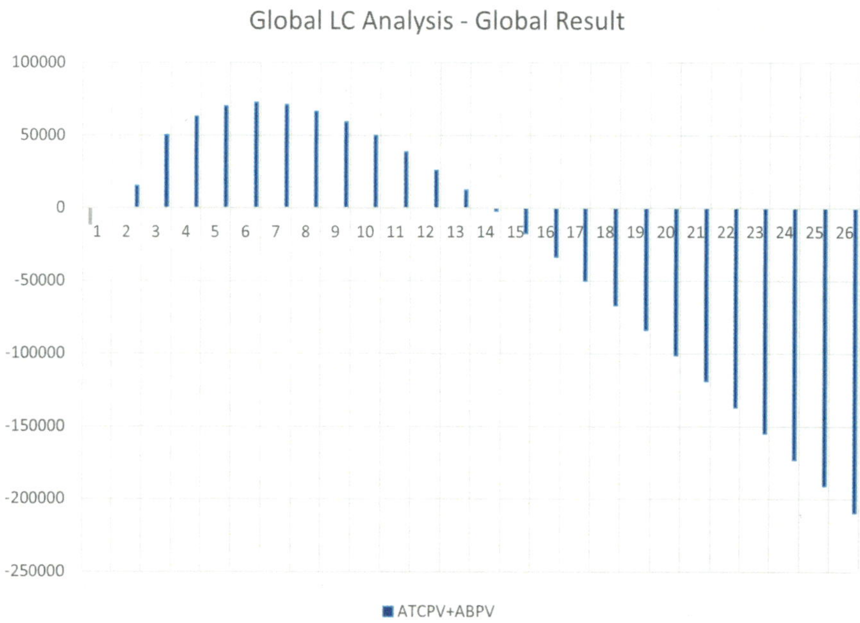

FIGURE 6.33 Life cycle analysis global result.

- Interest Rate (r) 0.05
- Inflation Rate (h) 0.02
- Profit Rate (p) 0.05
- Maintenance Increase Rate 0.01
- MTBF **1000**
- MTTR **30**
- Risk (R) 0.10
- Time (t) 0
- Initial Investment (II) (C_0) –12000
- F_p Financial Movement (Production) 70000
- Functioning Costs –4500
- Energy –200
- Maintenance Costs –300

The quantitative results are presented in Table 6.15, and the graphical results are presented in Figures 6.37–6.39.

Comparing Figures 6.36 and 6.39, it can be seen that both Life Cycles, with Devaluation and Recovery increases one year and the positive results are higher.

TABLE 6.14
Values Used for Simulation based on Life Cycle with Recovery – First Simulation

Time (t)	0	1	2	3	4	5	6	7	8	9	10	11	12	13
r - real interest rate $(1+r)$		1.05	1.05	1.05	1.05	1.05	1.05	1.05	1.05	1.05	1.05	1.05	1.05	1.05
h - inflation rate $(1+h)$		1.02	1.02	1.02	1.02	1.02	1.02	1.02	1.02	1.02	1.02	1.02	1.02	1.02
p - profit rate		1.05	1.05	1.05	1.05	1.05	1.05	1.05	1.05	1.05	1.05	1.05	1.05	1.05
i - money rate (nominal interest rate)		0.12	0.12	0.12	0.12	0.12	0.12	0.12	0.12	0.12	0.12	0.12	0.12	0.12
i - Apparent Rate with Risk		0.24	0.24	0.24	0.24	0.24	0.24	0.24	0.24	0.24	0.24	0.24	0.24	0.24
$1/((1+i)\wedge t))$		0.81	0.65	0.53	0.43	0.35	0.28	0.23	0.18	0.15	0.12	0.10	0.08	0.06
Initial Investment (II) (C_0)	−12000	−12000	−12000	−12000	−12000	−12000	−12000	−12000	−12000	−12000	−12000	−12000	−12000	−12000
Initial Investment (II) (C_0) with Devaluation (NPV)	−12000	−9701	−7842	−6340	−5125	−4143	−3349	−2708	−2189	−1769	−1430	−1156	−935	−756
Initial Investment (II) (C_0) considering NEW Asset (NPV)	−12000	−14844	−18362	−22714	−28097	−34757	−42994	−53184	−65789	−81381	−100669	−124528	−154042	−190551
Fp Financial Movement (Production)	70000	70000	70000	70000	70000	70000	70000	70000	70000	70000	70000	70000	70000	70000
Fp Financial Movement (NPV)	70000	50525	40845	33019	26693	21579	17444	14102	11400	9216	7450	6023	4869	3936
Fp Financial Movement (NPV) Accumulated	70000	50525	91370	124389	151082	172661	190105	204207	215607	224823	232273	238296	243164	247100
Functioning Costs	−4500	−4500	−4500	−4500	−4500	−4500	−4500	−4500	−4500	−4500	−4500	−4500	−4500	−4500
Functioning Costs (NPV) Accumulated	−4500	−3638	−6579	−8956	−10878	−12432	−13688	−14703	−15524	−16187	−16724	−17157	−17508	−17791
Energy	−200	−200	−200	−200	−200	−200	−200	−200	−200	−200	−200	−200	−200	−200
Energy Accumulated (NPV) Accumulated	−200	−162	−292	−398	−483	−553	−608	−653	−690	−719	−743	−763	−778	−791
Maintenance Costs	−300	−300	−300	−300	−300	−300	−300	−300	−300	−300	−300	−300	−300	−300
Maintenance Costs (NPV) Accumulated	−300	−245	−443	−603	−732	−837	−922	−990	−1045	−1090	−1126	−1155	−1179	−1198
Total Costs (NPV) Accumulated	−5000	−4083	−11304	−21063	−32874	−46342	−61152	−77044	−93813	−111291	−129340	−147853	−166740	−185929
FT Net Present Value (NPV) (With II Devaluated)		36742	72224	96986	113083	122175	125604	124455	119605	111763	101502	89286	75490	60416
FT Net Present Value (With II New Asset)		31598	61704	80612	90111	91561	85959	73978	56005	32151	2264	−34085	−77617	−129379

(Continued)

TABLE 6.14 (*Continued*)
Values Used for Simulation based on Life Cycle with Recovery – First Simulation

Time (t)	14	15	16	17	18	19	20	21	22	23	24	25
r - real interest rate $(1+r)$	1.05	1.05	1.05	1.05	1.05	1.05	1.05	1.05	1.05	1.05	1.05	1.05
h - inflation rate $(1+h)$	1.02	1.02	1.02	1.02	1.02	1.02	1.02	1.02	1.02	1.02	1.02	1.02
p - profit rate	1.05	1.05	1.05	1.05	1.05	1.05	1.05	1.05	1.05	1.05	1.05	1.05
i - money rate (nominal interest rate)	0.12	0.12	0.12	0.12	0.12	0.12	0.12	0.12	0.12	0.12	0.12	0.12
i - Apparent Rate with Risk	0.24	0.24	0.24	0.24	0.24	0.24	0.24	0.24	0.24	0.24	0.24	0.24
$1/((1+i)^\wedge t))$	0.05	0.04	0.03	0.03	0.02	0.02	0.01	0.01	0.01	0.01	0.01	0.00
Initial Investment (II) (C_0)	−12000	−12000	−12000	−12000	−12000	−12000	−12000	−12000	−12000	−12000	−12000	−12000
Initial Investment (II) (C_0) with Devaluation (NPV)	−611	−494	−399	−323	−261	−211	−171	−138	−111	−90	−73	−59
Initial Investment (II) (C_0) considering NEW Asset (NPV)	−235712	−291577	−360682	−446165	−551909	−682714	−844521	−1044676	−1292270	−1598544	−1977407	−2446063
Fp Financial Movement (Production)	70000	70000	70000	70000	70000	70000	70000	70000	70000	70000	70000	70000
Fp Financial Movement (NPV)	3182	2572	2079	1681	1359	1099	888	718	580	469	379	307
Fp Financial Movement (NPV) Accumulated	250282	252854	254934	256615	257974	259072	259960	260678	261259	261728	262107	262414
Functioning Costs	−4500	−4500	−4500	−4500	−4500	−4500	−4500	−4500	−4500	−4500	−4500	−4500
Functioning Costs (NPV) Accumulated	−18020	−18206	−18355	−18476	−18574	−18653	−18717	−18769	−18811	−18844	−18872	−18894
Energy	−200	−200	−200	−200	−200	−200	−200	−200	−200	−200	−200	−200
Energy Accumulated (NPV) Accumulated	−801	−809	−816	−821	−826	−829	−832	−834	−836	−838	−839	−840
Maintenance Costs	−300	−300	−300	−300	−300	−300	−300	−300	−300	−300	−300	−300
Maintenance Costs (NPV) Accumulated	−1213	−1226	−1236	−1244	−1251	−1256	−1260	−1264	−1267	−1269	−1271	−1272
Total Costs (NPV) Accumulated	−205362	−224994	−244785	−264705	−284730	−304839	−325017	−345249	−365527	−385840	−406182	−426548
FT Net Present Value (NPV) (With II Devaluated)	44309	27367	9750	−8413	−27017	−45978	−65227	−84709	−104379	−124202	−144148	−164193
FT Net Present Value (With II New Asset)	−190792	−263716	−350533	−454256	−578665	−728481	−909577	−1129247	−1396538	−1722656	−2121482	−2610197

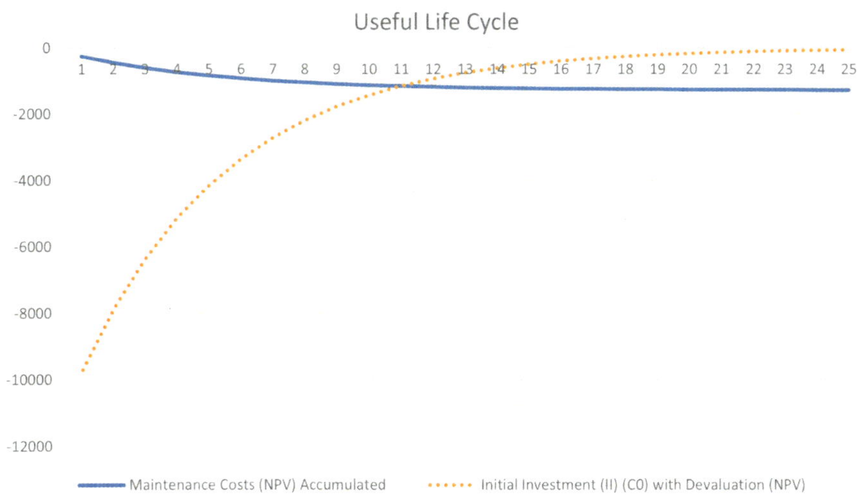

FIGURE 6.34 Useful life cycle.

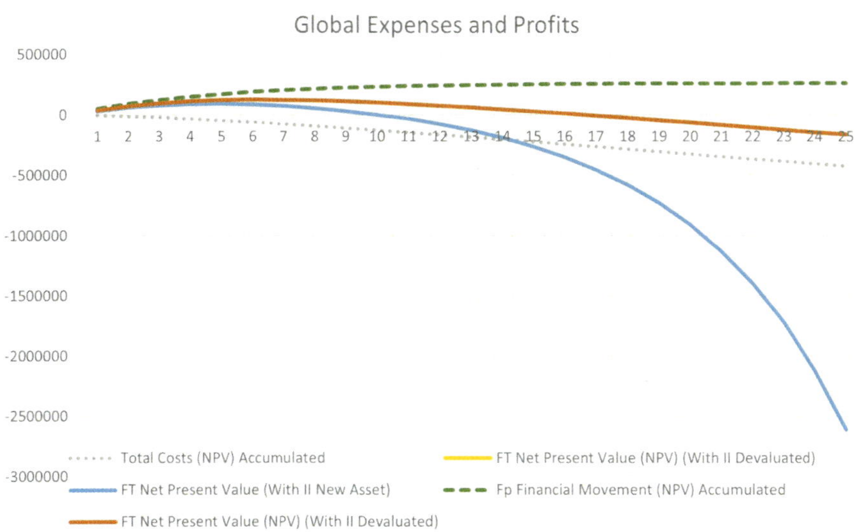

FIGURE 6.35 Global investments and profits.

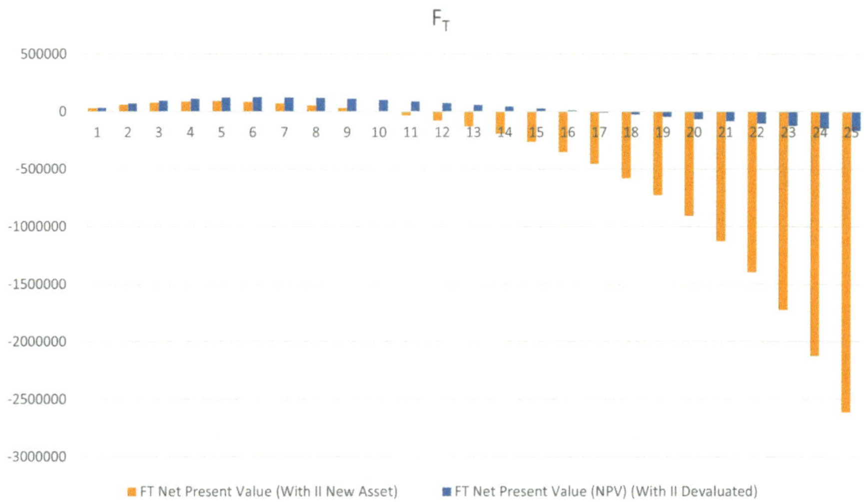

FIGURE 6.36 Results considering the acquisition of a new asset with devaluation and recovery.

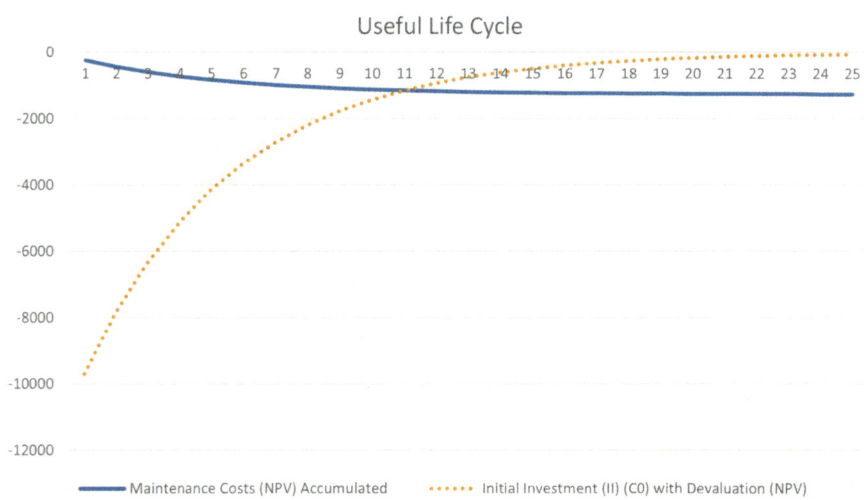

FIGURE 6.37 Useful life cycle.

TABLE 6.15
Values Used for Simulation based on Life Cycle with Recovery – Second Simulation

Time (t)	0	1	2	3	4	5	6	7	8	9	10	11	12	13
r - real interest rate (1 + r)		1.05	1.05	1.05	1.05	1.05	1.05	1.05	1.05	1.05	1.05	1.05	1.05	1.05
h - inflation rate ($1 + h$)		1.02	1.02	1.02	1.02	1.02	1.02	1.02	1.02	1.02	1.02	1.02	1.02	1.02
p - profit rate		1.05	1.05	1.05	1.05	1.05	1.05	1.05	1.05	1.05	1.05	1.05	1.05	1.05
i - money rate (nominal interest rate)		0.12	0.12	0.12	0.12	0.12	0.12	0.12	0.12	0.12	0.12	0.12	0.12	0.12
i - Apparent Rate with Risk		0.24	0.24	0.24	0.24	0.24	0.24	0.24	0.24	0.24	0.24	0.24	0.24	0.24
$1/((1 + i) \wedge t))$		0.81	0.65	0.53	0.43	0.35	0.28	0.23	0.18	0.15	0.12	0.10	0.08	0.06
Initial Investment (II) (C_0)	−12000	−12000	−12000	−12000	−12000	−12000	−12000	−12000	−12000	−12000	−12000	−12000	−12000	−12000
Initial Investment (II) (C_0) with Devaluation (NPV)	−12000	−9701	−7842	−6340	−5125	−4143	−3349	−2708	−2189	−1769	−1430	−1156	−935	−756
Initial Investment (II) (C_0) considering NEW Asset (NPV)	−12000	−14844	−18362	−22714	−28097	−34757	−42994	−53184	−65789	−81381	−100669	−124528	−154042	−190551
Fp Financial Movement (Production)	70000	70000	70000	70000	70000	70000	70000	70000	70000	70000	70000	70000	70000	70000
Fp Financial Movement (NPV)	70000	54940	44414	35904	29025	23464	18968	15334	12396	10021	8101	6549	5294	4280
Fp Financial Movement (NPV) Accumulated	70000	54940	99354	135258	164283	187747	206716	222050	234446	244467	252569	259118	264412	268692
Functioning Costs	−4500	−4500	−4500	−4500	−4500	−4500	−4500	−4500	−4500	−4500	−4500	−4500	−4500	−4500
Functioning Costs (NPV) Accumulated	−4500	−3638	−6579	−8956	−10878	−12432	−13688	−14703	−15524	−16187	−16724	−17157	−17508	−17791
Energy	−200	−200	−200	−200	−200	−200	−200	−200	−200	−200	−200	−200	−200	−200
Energy Accumulated (NPV) Accumulated	−200	−162	−292	−398	−483	−553	−608	−653	−690	−719	−743	−763	−778	−791
Maintenance Costs	−300	−300	−300	−300	−300	−300	−300	−300	−300	−300	−300	−300	−300	−300
Maintenance Costs (NPV) Accumulated	−300	−245	−443	−603	−732	−837	−922	−990	−1045	−1090	−1126	−1155	−1179	−1198
Total Costs (NPV) Accumulated	−5000	−4083	−11304	−21063	−32874	−46342	−61152	−77044	−93813	−111291	−129340	−147853	−166740	−185929
FT Net Present Value (NPV) (With II Devaluated)		41156	80207	107855	126285	137262	142215	142298	138444	131407	121798	110108	96737	82007
FT Net Present Value (With II New Asset)		36013	69687	91481	103312	106648	102570	91822	74844	51796	22559	−13263	−56370	−107788

(*Continued*)

TABLE 6.15 (*Continued*)
Values Used for Simulation based on Life Cycle with Recovery – Second Simulation

Time (t)	14	15	16	17	18	19	20	21	22	23	24	25
r - real interest rate (1 + r)	1.05	1.05	1.05	1.05	1.05	1.05	1.05	1.05	1.05	1.05	1.05	1.05
h - inflation rate $(1 + h)$	1.02	1.02	1.02	1.02	1.02	1.02	1.02	1.02	1.02	1.02	1.02	1.02
p - profit rate	1.05	1.05	1.05	1.05	1.05	1.05	1.05	1.05	1.05	1.05	1.05	1.05
i - money rate (nominal interest rate)	0.12	0.12	0.12	0.12	0.12	0.12	0.12	0.12	0.12	0.12	0.12	0.12
i - Apparent Rate with Risk	0.24	0.24	0.24	0.24	0.24	0.24	0.24	0.24	0.24	0.24	0.24	0.24
$1/((1 + i)$ ^ $t))$	0.05	0.04	0.03	0.03	0.02	0.02	0.01	0.01	0.01	0.01	0.01	0.00
Initial Investment (II) (C_0)	–12000	–12000	–12000	–12000	–12000	–12000	–12000	–12000	–12000	–12000	–12000	–12000
Initial Investment (II) (C_0) with Devaluation (NPV)	–611	–494	–399	–323	–261	–211	–171	–138	–111	–90	–73	–59
Initial Investment (II) (C_0) considering NEW Asset (NPV)	–235712	–291577	–360682	–446165	–551909	–682714	–844521	–1044676	–1292270	–1598544	–1977407	–2446063
Fp Financial Movement (Production)	70000	70000	70000	70000	70000	70000	70000	70000	70000	70000	70000	70000
Fp Financial Movement (NPV)	3460	2797	2261	1828	1478	1195	966	781	631	510	412	333
Fp Financial Movement (NPV) Accumulated	272152	274949	277210	279038	280515	281710	282675	283456	284087	284597	285010	285343
Functioning Costs	–4500	–4500	–4500	–4500	–4500	–4500	–4500	–4500	–4500	–4500	–4500	–4500
Functioning Costs (NPV) Accumulated	–18020	–18206	–18355	–18476	–18574	–18653	–18717	–18769	–18811	–18844	–18872	–18894
Energy	–200	–200	–200	–200	–200	–200	–200	–200	–200	–200	–200	–200
Energy Accumulated (NPV) Accumulated	–801	–809	–816	–821	–826	–829	–832	–834	–836	–838	–839	–840
Maintenance Costs	–300	–300	–300	–300	–300	–300	–300	–300	–300	–300	–300	–300
Maintenance Costs (NPV) Accumulated	–1213	–1226	–1236	–1244	–1251	–1256	–1260	–1264	–1267	–1269	–1271	–1272
Total Costs (NPV) Accumulated	–205362	–224994	–244785	–264705	–284730	–304839	–325017	–345249	–365527	–385840	–406182	–426548
FT Net Present Value (NPV) (With II Devaluated)	66178	49461	32025	14010	–4476	–23340	–42512	–61931	–81551	–101333	–121245	–141264
FT Net Present Value (With II New Asset)	–168923	–241622	–328257	–431833	–556124	–705844	–886862	–1106470	–1373709	–1699787	–2098580	–2587268

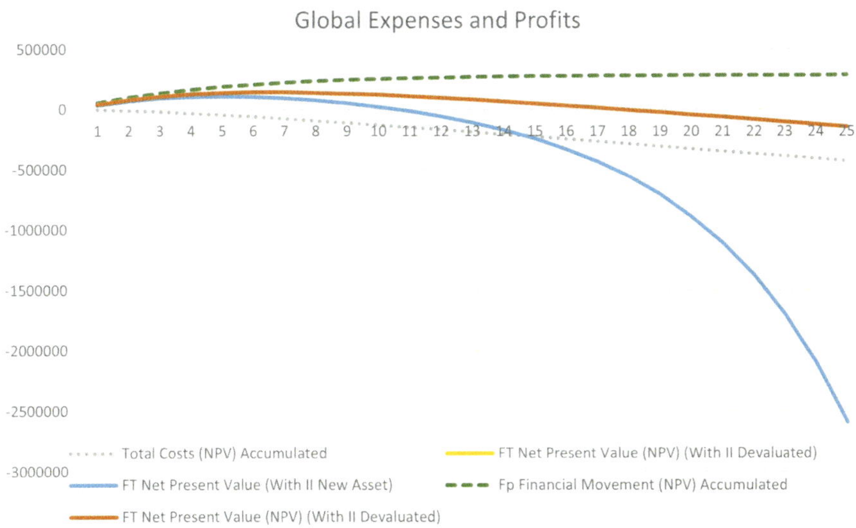

FIGURE 6.38 Global investments and profits.

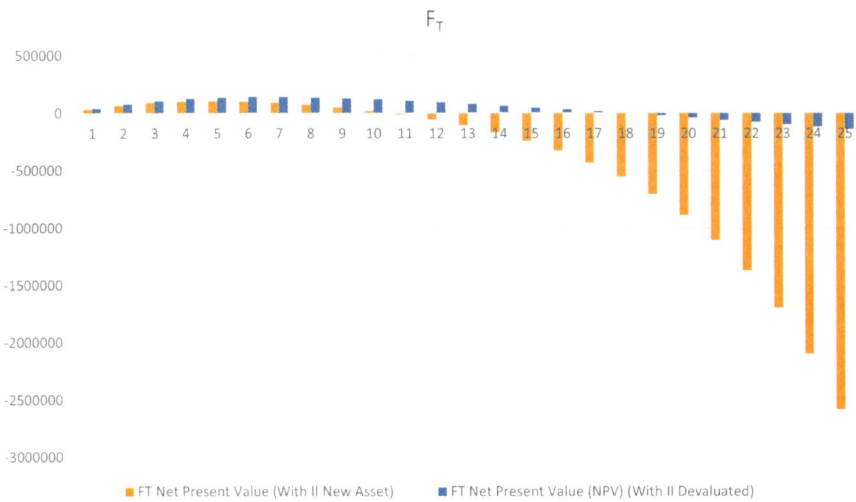

FIGURE 6.39 Results considering the acquisition of a new asset with devaluation and recovery.

REFERENCE

Farinha, J. T. (2018) *Asset Maintenance Engineering Methodologies* (1st Edition). Boca Raton, FL: CRC Press. https://doi.org/10.1201/9781315232867. https://www.taylorfrancis.com/books/mono/10.1201/9781315232867/asset-maintenance-engineering-methodologies-jos%C3%A9-manuel-torres-farinha

7 Norms of Physical Assets Management

7.1 INTRODUCTION

This chapter deals with the Physical Assets Management Standards, namely the ISO 55001, in the following perspectives:

- Main requisites of Physical Asset Norms;
- Diagnosis to implement Physical Asset Norms;
- Quantitative evaluation about the implementation of Physical Asset Norms based on the models previously discussed in Chapters 1–3:
  - Life Cycle Cost of Physical Assets;
  - Uniform Annual Income Method (UAI);
  - Minimization of Total Average Cost Method (MTAC);
  - MMTAC Reduced to Present Value (MTAC-RPV);
  - Life Cycle Investment of Physical Assets;
  - Life Cycle of Physical Assets with Recovery.

7.2 MAIN REQUISITES OF PHYSICAL ASSET NORMS

The main requisites of ISO 55001 are the following ones:

4. Context of the organization
   4.1 Understanding the organization and its context
   4.2 Understanding the needs and expectations of stakeholders
   4.3 Determining the scope of the asset management system
   4.4 Asset management system
5. Leadership
   5.1 Leadership and commitment
   5.2 Policy
   5.3 Organizational roles, responsibilities and authorities
6. Planning
   6.1 Actions to address risks and opportunities for the asset management system
   6.2 Asset management objectives and planning to achieve them
7. Support
   7.1 Resources
   7.2 Competence
   7.3 Awareness
   7.4 Communication

DOI: 10.1201/9781003395690-10

7.5 Information requirements
7.6 Documented information
8. Operation
8.1 Operational planning and control
8.2 Management of change
8.3 Outsourcing
9. Performance evaluation
9.1 Monitoring, measurement, analysis and evaluation
9.2 Internal audit
9.3 Management review
10. Improvement
10.1 Nonconformity and corrective action
10.2 Preventive action
10.3 Continual improvement

7.3 DIAGNOSIS TO IMPLEMENT PHYSICAL ASSET NORMS

Aiming to implement the ISO 55001 it is desirable to start the process making a diagnosis, following the same methodology presented in Chapter 6. Figure 7.1 shows an example of the requisite 5 (Leadership), namely the sub-requisite leadership and commitment.

7.4 QUANTITATIVE EVALUATION TO IMPLEMENT PHYSICAL ASSET NORMS

When an organization implements a new solution, it is important to evaluate quantitatively the investment and the return of the investment. It is by this reason because the next subsections evaluate, by simulations, the investments and the respective results in the Life Cycle of Assets, through the three Life Cycle models presented in this book.

ISO 5500X family of standards and, in particular, ISO 55001 that corresponds to the requisites is, probably, the most important standard that an organization can implement to create a new culture of Asset Management.

In this type of standard, it is not easy to demonstrate immediately the relation between the value of investment to implement the Norm in the Organization and the results, because these are diverse, like the followings:

- Helping to optimize the Life Cycle of Assets;
- Helping to support decisions about the best option to acquire an Asset;
- Helping to support decisions about the best option to withdrawal an Asset;
- Helping to evaluate the best option about the maintenance policy;
- Helping the Organization to reach the best image in the market.

It is based on the preceding that the simulations in the next three sub-sections are made.

Diagnostic survey 5

Asset Management

E. Leadership and commitment

	Questions	Always	Mostly	Generally	Hardly	Never
101	There is a Strategic Asset Management Plan (SAMP)					
102	There is an Asset Management policy well defined in SAMP					
103	There are Asset Management objectives well defined in SAMP					
104	There is an integration of Asset Management System (AMS) with the business plan					
105	There are internal resources to implement the AMS					
106	There are financial resources to implement the AMS					
107	The organization has an Asset Management department					
108	Does the Asset Managment departmant has its own budjet					
109	The SAMP has well defined objectives and goals					
110	The Organisation communicates about the AMS requirements and its importance					
111	There is an internal system to assure that the AMS achieves its intended outcome(s)					
112	There are internal audits and other cross functional evaluation					
113	There are KPI					
114	Are KPI reviewed periodically in a perspective of a continuous improvement					
115	Are other relevant management roles supporting the leadership related to AMS					
116	The Asset Management risk is aligned with organization's managing risk					
117	Is the Asset Management risk alined with ISO 31000 series					
118						
119						
	score					

Rating

Score	Answers			Mark
		to		
		to		
		to		
		to		
Total Mark (M)				

Category					
1		<	P	≤	
2		<	P	≤	
3		<	P	≤	
4		<	P	≤	
5		<	P	≤	

Explanatory sheet

Questions to review					

Company

Name	Email	Phone
Name of the person who answered	**Company function**	**Date**

FIGURE 7.1 Diagnosis sheet referring to requisite leadership and commitment.

7.4.1 Life Cycle Cost of Physical Assets

This section presents the use of Life Cycle models described in Chapter 1 taking into account the implication of implementing Physical Asset Norms, aiming to support quantitatively the decision of the responsible.

However, the simulations presented in the next sub-sections are the traditional, called Life Cycle Cost (LCC), they were added the profits that results from the use of Physical Assets, which increases a lot the evaluation of situations that can happen in day-to-day companies.

The simulations ought to follow always the same approach, that is, to compare values aiming to demonstrate how the methods presented in this book, namely in Chapter 1, can help the manager to decide if the implementation of the standards discussed in this book add value to the organization or not.

7.4.1.1 Uniform Annual Income Method (UAI)

The first simulation uses the following values:

- Initial Investment (II) –12000
- Residual value of equipment (RV) 10
- Functioning (*F*) –4500
- Energy –200
- Maintenance (*M*) –300
- Non Production (NP) 0
- Benefits (*B*) **50000**
- Internal Rate of Return (*r*) 0.12
- Risk (*R*) 0.11
- Apparent Rate 0.2432
- Maintenance Increase Rate 0.01
- Availability 0.8929
- MTBF 500
- MTTR 50
- MWT 10

The remaining values are presented in Table 7.1, and the graphical results are shown in Figures 7.2–7.5.

As referred in the preceding introduction on this section, the evaluation of introduction of Physical Asset Norms, the simulation of their implementation is made through the increase of Benefits. They may be used other indicators, but may be not so intuitive.

TABLE 7.1
Values Used for Simulation based on UAI Method Related to Physical Assets Standard – First Simulation

Time	0	1	2	3	4	5	6
Initial Investment (II)	−12000						
Initial Investment (II)	−12000						
II Present Value (IIPV) - Return	−12000	−9652.51	−7764.25	−6245.37	−5023.63	−4040.88	−3250.39
Life Time simulated	25						
Residual value of equipment (RV)	10						
Linear Depreciation Method (Vn)	−12000	−11519.60	−11039.20	−10558.80	−10078.40	−9598.00	−9117.60
LDM Present Value (Vn)		−9266.09	−7142.59	−5495.30	−4219.18	−3232.03	−2469.64
Sum of Digits Method (Vn)	−12000	−11076.15	−10189.26	−9339.32	−8526.34	−7750.31	−7011.23
SDM Present Value (Vn)		−8909.39	−6592.66	−4860.63	−3569.43	−2609.84	−1899.10
Exponential Method (Vn)	−12000	−9036.79	−6805.29	−5124.83	−3859.33	−2906.33	−2188.66
EM Present Value (Pv)		−7268.97	−4403.16	−2667.21	−1615.65	−978.68	−592.83
Functioning (*F*)	−4500	−4500.00	−4545.00	−4590.45	−4636.35	−4682.72	−4729.55
Present Value Accumulated (NPV)		−3619.69	−6560.40	−8949.49	−10890.43	−12467.29	−13748.36
Present Value (NPV)		−3619.69	−2940.71	−2389.09	−1940.94	−1576.86	−1281.07
Energy	−200	−200.00	−400.00	−600.00	−800.00	−1000.00	−1200.00
Energy (NPV)		−160.88	−258.81	−312.27	−334.91	−336.74	−325.04
Accumulated (FPV)		−15780.57	−22599.77	−31861.53	−43086.87	−55890.90	−69964.30
Maintenance (*M*)	−300	−300.00	−303.00	−306.03	−309.09	−312.18	−315.30
Present Value (MPV)		−241.31	−196.05	−159.27	−129.40	−105.12	−85.40
Accumulated (MPV)		−241.31	−437.36	−596.63	−726.03	−831.15	−916.56
Non Production (NP)	0	0.00	0.00	0.00	0.00	0.00	0.00
Present Value (NPPV)		0.00	0.00	0.00	0.00	0.00	0.00
Accumulated NPPV		0.00	0.00	0.00	0.00	0.00	0.00
Accumulated Total Costs (APV)		−16021.88	−23037.13	−32458.16	−43812.90	−56722.05	−70880.85
Accumulated Total Costs		−5000.00	−10248.00	−15744.48	−21489.92	−27484.82	−33729.67
Benefit (*B*)	50000	44642.86	44642.86	44642.86	44642.86	44642.86	44642.86
Present Value (BPV)		35909.63	28884.84	23234.27	18689.08	15033.05	12092.22
Accumulated Benefits BPV		35909.63	64794.47	88028.74	106717.82	121750.87	133843.09
BPV + FPV + MPV + NPPV + IIPV		7887.76	29757.34	43570.58	50904.93	53028.82	50962.23
Uniform Annual Income UAIn (LDM)		−11519.60	−4921.18	−2786.88	−1764.99	−1185.10	−823.74
Uniform Annual Income UAIn (SDM)		−11076.15	−4542.29	−2465.02	−1493.18	−956.96	−633.44
Uniform Annual Income UAIn (EM)		−9036.79	−3033.74	−1352.65	−675.87	−358.86	−197.74

(Continued)

TABLE 7.1 (*Continued*)
Values Used for Simulation based on UAI Method Related to Physical Assets Standard – First Simulation

Time	7	8	9	10	11	12
Initial Investment (II)						
Initial Investment (II)						
II Present Value (IIPV) - Return	−2614.53	−2103.07	−1691.66	−1360.73	−1094.54	−880.42
Life Time simulated						
Residual value of equipment (RV)						
Linear Depreciation Method (Vn)	−8637.20	−8156.80	−7676.40	−7196.00	−6715.60	−6235.20
LDM Present Value (Vn)	−1881.85	−1429.53	−1082.15	−815.98	−612.54	−457.47
Sum of Digits Method (Vn)	−6309.11	−5643.94	−5015.72	−4424.46	−3870.15	−3352.80
SDM Present Value (Vn)	−1374.61	−989.13	−707.07	−501.71	−353.00	−245.99
Exponential Method (Vn)	−1648.20	−1241.20	−934.71	−703.90	−530.08	−399.19
EM Present Value (Pv)	−359.11	−217.53	−131.77	−79.82	−48.35	−29.29
Functioning (*F*)	−4776.84	−4824.61	−4872.86	−4921.58	−4970.80	−5020.51
Present Value Accumulated (NPV)	−14789.13	−15634.67	−16321.60	−16879.68	−17333.07	−17701.42
Present Value (NPV)	−1040.77	−845.54	−686.93	−558.08	−453.39	−368.35
Energy	−1400.00	−1600.00	−1800.00	−2000.00	−2200.00	−2400.00
Energy (NPV)	−305.03	−280.41	−253.75	−226.79	−200.67	−176.08
Accumulated (FPV)	−85058.45	−100973.53	−117548.88	−134655.35	−152189.09	−170066.59
Maintenance (*M*)	−318.46	−321.64	−324.86	−328.11	−331.39	−334.70
Present Value (MPV)	−69.38	−56.37	−45.80	−37.21	−30.23	−24.56
Accumulated (MPV)	−985.94	−1042.31	−1088.11	−1125.31	−1155.54	−1180.09
Non Production (NP)	0.00	0.00	0.00	0.00	0.00	0.00
Present Value (NPPV)	0.00	0.00	0.00	0.00	0.00	0.00
Accumulated NPPV	0.00	0.00	0.00	0.00	0.00	0.00
Accumulated Total Costs (APV)	−86044.40	−102015.84	−118636.99	−135780.66	−153344.63	−171246.68
Accumulated Total Costs	−40224.97	−46971.22	−53968.93	−61218.62	−68720.81	−76476.01
Benefit (*B*)	44642.86	44642.86	44642.86	44642.86	44642.86	44642.86
Present Value (BPV)	9726.69	7823.91	6293.37	5062.23	4071.94	3275.37
Accumulated Benefits BPV	143569.78	151393.69	157687.05	162749.28	166821.22	170096.59
BPV+FPV+MPV+NPPV+IIPV	45525.38	37377.84	27050.06	14968.62	1476.59	−13150.10
Uniform Annual Income UAIn (LDM)	−585.16	−421.54	−306.37	−223.83	−163.92	−120.06
Uniform Annual Income UAIn (SDM)	−427.43	−291.67	−200.18	−137.62	−94.47	−64.56
Uniform Annual Income UAIn (EM)	−111.66	−64.14	−37.30	−21.89	−12.94	−7.69

(*Continued*)

TABLE 7.1 (*Continued*)
Values Used for Simulation based on UAI Method Related to Physical Assets Standard – First Simulation

Time	13	14	15	16	17	18
Initial Investment (II)						
Initial Investment (II)						
II Present Value (IIPV) - Return	−708.19	−569.65	−458.21	−368.57	−296.47	−238.48
Life Time simulated						
Residual value of equipment (RV)						
Linear Depreciation Method (Vn)	−5754.80	−5274.40	−4794.00	−4313.60	−3833.20	−3352.80
LDM Present Value (Vn)	−339.62	−250.38	−183.06	−132.49	−94.70	−66.63
Sum of Digits Method (Vn)	−2872.40	−2428.95	−2022.46	−1652.92	−1320.34	−1024.71
SDM Present Value (Vn)	−169.52	−115.30	−77.23	−50.77	−32.62	−20.36
Exponential Method (Vn)	−300.61	−226.38	−170.48	−128.38	−96.68	−72.81
EM Present Value (Pv)	−17.74	−10.75	−6.51	−3.94	−2.39	−1.45
Functioning (*F*)	−5070.71	−5121.42	−5172.63	−5224.36	−5276.60	−5329.37
Present Value Accumulated (NPV)	−18000.67	−18243.79	−18441.30	−18601.76	−18732.13	−18838.04
Present Value (NPV)	−299.25	−243.12	−197.51	−160.46	−130.36	−105.91
Energy	−2600.00	−2800.00	−3000.00	−3200.00	−3400.00	−3600.00
Energy (NPV)	−153.44	−132.92	−114.55	−98.29	−84.00	−71.54
Accumulated (FPV)	−188220.70	−206597.41	−225153.26	−243853.31	−262669.44	−281579.02
Maintenance (*M*)	−338.05	−341.43	−344.84	−348.29	−351.77	−355.29
Present Value (MPV)	−19.95	−16.21	−13.17	−10.70	−8.69	−7.06
Accumulated (MPV)	−1200.04	−1216.25	−1229.42	−1240.12	−1248.81	−1255.87
Non Production (NP)	0.00	0.00	0.00	0.00	0.00	0.00
Present Value (NPPV)	0.00	0.00	0.00	0.00	0.00	0.00
Accumulated NPPV	0.00	0.00	0.00	0.00	0.00	0.00
Accumulated Total Costs (APV)	−189420.75	−207813.66	−226382.68	−245093.43	−263918.25	−282834.89
Accumulated Total Costs	−84484.77	−92747.62	−101265.10	−110037.75	−119066.13	−128350.79
Benefit (*B*)	44642.86	44642.86	44642.86	44642.86	44642.86	44642.86
Present Value (BPV)	2634.63	2119.23	1704.66	1371.18	1102.95	887.18
Accumulated Benefits BPV	172731.21	174850.44	176555.10	177926.28	179029.23	179916.42
BPV+FPV+MPV+NPPV+IIPV	−28689.53	−44963.22	−61827.58	−79167.14	−96889.02	−114918.47
Uniform Annual Income UAIn (LDM)	−87.78	−63.93	−46.29	−33.24	−23.62	−16.53
Uniform Annual Income UAIn (SDM)	−43.81	−29.44	−19.53	−12.74	−8.13	−5.05
Uniform Annual Income UAIn (EM)	−4.59	−2.74	−1.65	−0.99	−0.60	−0.36

(*Continued*)

TABLE 7.1 (*Continued*)
Values Used for Simulation based on UAI Method Related to Physical Assets Standard – First Simulation

Time	19	20	21	22	23	24	25
Initial Investment (II)							
Initial Investment (II)							
II Present Value (IIPV) - Return	−191.82	−154.30	−124.11	−99.83	−80.30	−64.59	−51.96
Life Time simulated							
Residual value of equipment (RV)							
Linear Depreciation Method (Vn)	−2872.40	−2392.00	−1911.60	−1431.20	−950.80	−470.40	10.00
LDM Present Value (Vn)	−45.92	−30.76	−19.77	−11.91	−6.36	−2.53	0.04
Sum of Digits Method (Vn)	−766.03	−544.31	−359.54	−211.72	−100.86	−26.95	10.00
SDM Present Value (Vn)	−12.25	−7.00	−3.72	−1.76	−0.67	−0.15	0.04
Exponential Method (Vn)	−54.83	−41.29	−31.09	−23.42	−17.63	−13.28	−10.00
EM Present Value (Pv)	−0.88	−0.53	−0.32	−0.19	−0.12	−0.07	−0.04
Functioning (*F*)	−5382.66	−5436.49	−5490.86	−5545.76	−5601.22	−5657.23	−5713.81
Present Value Accumulated (NPV)	−18924.08	−18993.99	−19050.78	−19096.91	−19134.40	−19164.85	−19189.59
Present Value (NPV)	−86.04	−69.90	−56.79	−46.14	−37.48	−30.45	−24.74
Energy	−3800.00	−4000.00	−4200.00	−4400.00	−4600.00	−4800.00	−5000.00
Energy (NPV)	−60.74	−51.43	−43.44	−36.61	−30.78	−25.84	−21.65
Accumulated (FPV)	−300563.85	−319609.27	−338703.48	−357837.01	−377002.19	−396192.88	−415404.12
Maintenance (*M*)	−358.84	−362.43	−366.06	−369.72	−373.41	−377.15	−380.92
Present Value (MPV)	−5.74	−4.66	−3.79	−3.08	−2.50	−2.03	−1.65
Accumulated (MPV)	−1261.61	−1266.27	−1270.05	−1273.13	−1275.63	−1277.66	−1279.31
Non Production (NP)	0.00	0.00	0.00	0.00	0.00	0.00	0.00
Present Value (NPPV)	0.00	0.00	0.00	0.00	0.00	0.00	0.00
Accumulated NPPV	0.00	0.00	0.00	0.00	0.00	0.00	0.00
Accumulated Total Costs (APV)	−301825.45	−320875.53	−339973.54	−359110.13	−378277.81	−397470.53	−416683.42
Accumulated Total Costs	−137892.30	−147691.22	−157748.13	−168063.61	−178638.25	−189472.63	−200567.36
Benefit (*B*)	44642.86	44642.86	44642.86	44642.86	44642.86	44642.86	44642.86
Present Value (BPV)	713.63	574.03	461.73	371.41	298.75	240.31	193.30
Accumulated Benefits BPV	180630.05	181204.07	181665.81	182037.21	182335.96	182576.27	182769.57
BPV + FPV + MPV + NPPV + IIPV	−133195.41	−151671.46	−170307.73	−189072.92	−207941.85	−226894.26	−245913.85
Uniform Annual Income UAIn (LDM)	−11.35	−7.58	−4.86	−2.92	−1.56	−0.62	0.01
Uniform Annual Income UAIn (SDM)	−3.03	−1.72	−0.91	−0.43	−0.17	−0.04	0.01
Uniform Annual Income UAIn (EM)	−0.22	−0.13	−0.08	−0.05	−0.03	−0.02	−0.01

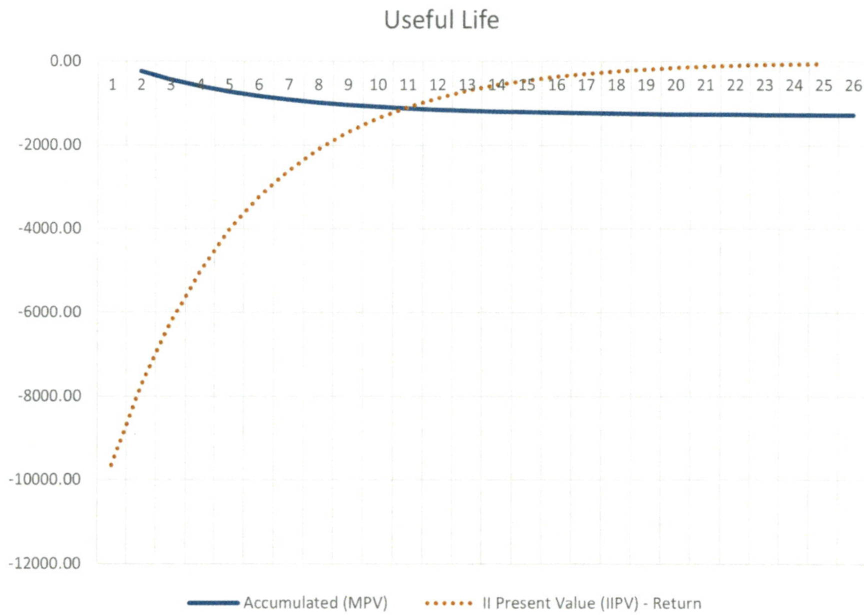

FIGURE 7.2 Physical Asset useful life.

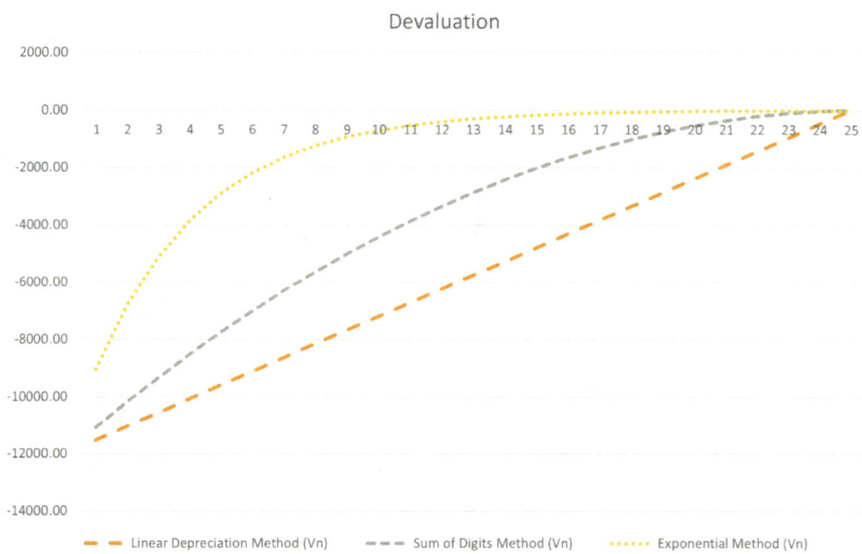

FIGURE 7.3 The three devaluation methods.

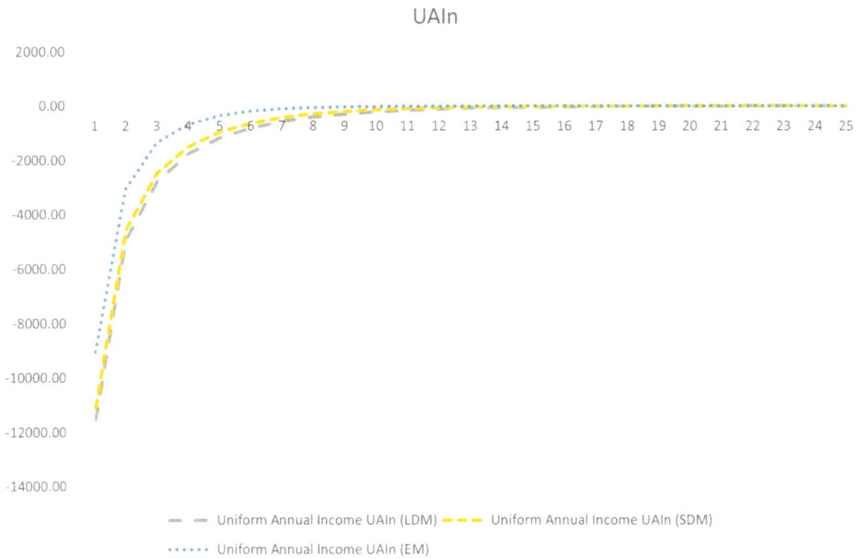

FIGURE 7.4 The UAI results from the three devaluation methods.

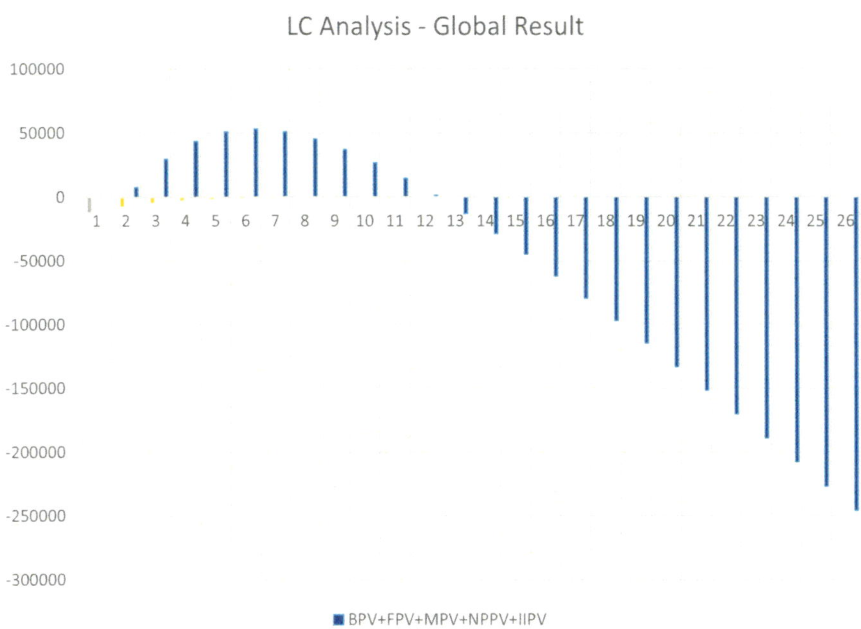

FIGURE 7.5 Life cycle analysis global result including profits.

The second simulation uses the following values (Benefits are increased):

- Initial Investment (II) –12000
- Residual value of equipment (RV) 10
- Functioning (*F*) –4500
- Energy –200
- Maintenance (*M*) –300
- Non Production (NP) 0
- Benefits (*B*) **70000**
- Internal Rate of Return (*r*) 0.12
- Risk (*R*) 0.11
- Apparent Rate 0.2432
- Maintenance Increase Rate 0.01
- Availability 0.8929
- MTBF 1000
- MTTR 25
- MWT 10

The remaining values are presented in Table 7.2, and the graphical results are shown in Figures 7.6–7.9.

As can be seen, namely comparing Figures 7.5 and 7.9, with the implementation of Physical Assets Standards, it is get four more positive years, and the positive incomes increases a lot.

7.4.1.2 Minimization of Total Average Cost Method (MMTAC)

The first simulation uses the following values:

- Initial Investment (II) –12000
- Life Time simulated 25
- Residual value of equipment (RV) 10
- Functioning (*F*) –4500
- Energy –200
- Maintenance (*M*) –300
- Non Production (NP) 0
- Benefits (*B*) **50000**
- Internal Rate of Return (*r*) 0.0
- Risk (*R*) 0.0
- Apparent Rate 0.0
- Maintenance Increase Rate 0.01
- Availability 0.8929
- MTBF 500
- MTTR 50
- MWT 10

TABLE 7.2
Values Used for Simulation based on UAI Method Related to Physical Assets Standard – Second Simulation

Time	0	1	2	3	4
Initial Investment (II)	−12000				
Initial Investment (II)	−12000				
II Present Value (IIPV) - Return	−12000	−9652.51	−7764.25	−6245.37	−5023.63
Life Time simulated	25				
Residual value of equipment (RV)	10				
Linear Depreciation Method (Vn)	−12000	−11519.60	−11039.20	−10558.80	−10078.40
LDM Present Value (Vn)		−9266.09	−7142.59	−5495.30	−4219.18
Sum of Digits Method (Vn)	−12000	−11076.15	−10189.26	−9339.32	−8526.34
SDM Present Value (Vn)		−8909.39	−6592.66	−4860.63	−3569.43
Exponential Method (Vn)	−12000	−9036.79	−6805.29	−5124.83	−3859.33
EM Present Value (Pv)		−7268.97	−4403.16	−2667.21	−1615.65
Functioning (*F*)	−4500	−4500.00	−4545.00	−4590.45	−4636.35
Present Value Accumulated (NPV)		−3619.69	−6560.40	−8949.49	−10890.43
Present Value (NPV)		−3619.69	−2940.71	−2389.09	−1940.94
Energy	−200	−200.00	−400.00	−600.00	−800.00
Energy (NPV)		−160.88	−258.81	−312.27	−334.91
Accumulated (FPV)		−15780.57	−22599.77	−31861.53	−43086.87
Maintenance (*M*)	−300	−300.00	−303.00	−306.03	−309.09
Present Value (MPV)		−241.31	−196.05	−159.27	−129.40
Accumulated (MPV)		−241.31	−437.36	−596.63	−726.03
Non Production (NP)	0	0.00	0.00	0.00	0.00
Present Value (NPPV)		0.00	0.00	0.00	0.00
Accumulated NPPV		0.00	0.00	0.00	0.00
Accumulated Total Costs (APV)		−16021.88	−23037.13	−32458.16	−43812.90
Accumulated Total Costs		−5000.00	−10248.00	−15744.48	−21489.92
Benefit (*B*)	70000	62500.00	62500.00	62500.00	62500.00
Present Value (BPV)		50273.49	40438.78	32527.97	26164.72
Accumulated Benefits BPV		50273.49	90712.26	123240.24	149404.95
BPV + FPV + MPV + NPPV + IIPV		22251.61	55675.13	78782.08	93592.06
Uniform Annual Income UAIn (LDM)		−11519.60	−4921.18	−2786.88	−1764.99
Uniform Annual Income UAIn (SDM)		−11076.15	−4542.29	−2465.02	−1493.18
Uniform Annual Income UAIn (EM)		−9036.79	−3033.74	−1352.65	−675.87

(Continued)

TABLE 7.2 (*Continued*)
Values Used for Simulation based on UAI Method Related to Physical Assets Standard – Second Simulation

Time	5	6	7	8	9
Initial Investment (II)					
Initial Investment (II)					
II Present Value (IIPV) - Return	−4040.88	−3250.39	−2614.53	−2103.07	−1691.66
Life Time simulated					
Residual value of equipment (RV)					
Linear Depreciation Method (Vn)	−9598.00	−9117.60	−8637.20	−8156.80	−7676.40
LDM Present Value (Vn)	−3232.03	−2469.64	−1881.85	−1429.53	−1082.15
Sum of Digits Method (Vn)	−7750.31	−7011.23	−6309.11	−5643.94	−5015.72
SDM Present Value (Vn)	−2609.84	−1899.10	−1374.61	−989.13	−707.07
Exponential Method (Vn)	−2906.33	−2188.66	−1648.20	−1241.20	−934.71
EM Present Value (Pv)	−978.68	−592.83	−359.11	−217.53	−131.77
Functioning (*F*)	−4682.72	−4729.55	−4776.84	−4824.61	−4872.86
Present Value Accumulated (NPV)	−12467.29	−13748.36	−14789.13	−15634.67	−16321.60
Present Value (NPV)	−1576.86	−1281.07	−1040.77	−845.54	−686.93
Energy	−1000.00	−1200.00	−1400.00	−1600.00	−1800.00
Energy (NPV)	−336.74	−325.04	−305.03	−280.41	−253.75
Accumulated (FPV)	−55890.90	−69964.30	−85058.45	−100973.53	−117548.88
Maintenance (*M*)	−312.18	−315.30	−318.46	−321.64	−324.86
Present Value (MPV)	−105.12	−85.40	−69.38	−56.37	−45.80
Accumulated (MPV)	−831.15	−916.56	−985.94	−1042.31	−1088.11
Non Production (NP)	0.00	0.00	0.00	0.00	0.00
Present Value (NPPV)	0.00	0.00	0.00	0.00	0.00
Accumulated NPPV	0.00	0.00	0.00	0.00	0.00
Accumulated Total Costs (APV)	−56722.05	−70880.85	−86044.40	−102015.84	−118636.99
Accumulated Total Costs	−27484.82	−33729.67	−40224.97	−46971.22	−53968.93
Benefit (*B*)	62500.00	62500.00	62500.00	62500.00	62500.00
Present Value (BPV)	21046.26	16929.11	13617.36	10953.48	8810.71
Accumulated Benefits BPV	170451.22	187380.32	200997.69	211951.16	220761.88
BPV+FPV+MPV+NPPV+IIPV	101729.17	104499.47	102953.29	97935.32	90124.89
Uniform Annual Income UAIn (LDM)	−1185.10	−823.74	−585.16	−421.54	−306.37
Uniform Annual Income UAIn (SDM)	−956.96	−633.44	−427.43	−291.67	−200.18
Uniform Annual Income UAIn (EM)	−358.86	−197.74	−111.66	−64.14	−37.30

(*Continued*)

TABLE 7.2 (*Continued*)
Values Used for Simulation based on UAI Method Related to Physical Assets Standard – Second Simulation

Time	10	11	12	13	14
Initial Investment (II)					
Initial Investment (II)					
II Present Value (IIPV) - Return	−1360.73	−1094.54	−880.42	−708.19	−569.65
Life Time simulated					
Residual value of equipment (RV)					
Linear Depreciation Method (Vn)	−7196.00	−6715.60	−6235.20	−5754.80	−5274.40
LDM Present Value (Vn)	−815.98	−612.54	−457.47	−339.62	−250.38
Sum of Digits Method (Vn)	−4424.46	−3870.15	−3352.80	−2872.40	−2428.95
SDM Present Value (Vn)	−501.71	−353.00	−245.99	−169.52	−115.30
Exponential Method (Vn)	−703.90	−530.08	−399.19	−300.61	−226.38
EM Present Value (Pv)	−79.82	−48.35	−29.29	−17.74	−10.75
Functioning (*F*)	−4921.58	−4970.80	−5020.51	−5070.71	−5121.42
Present Value Accumulated (NPV)	−16879.68	−17333.07	−17701.42	−18000.67	−18243.79
Present Value (NPV)	−558.08	−453.39	−368.35	−299.25	−243.12
Energy	−2000.00	−2200.00	−2400.00	−2600.00	−2800.00
Energy (NPV)	−226.79	−200.67	−176.08	−153.44	−132.92
Accumulated (FPV)	−134655.35	−152189.09	−170066.59	−188220.70	−206597.41
Maintenance (*M*)	−328.11	−331.39	−334.70	−338.05	−341.43
Present Value (MPV)	−37.21	−30.23	−24.56	−19.95	−16.21
Accumulated (MPV)	−1125.31	−1155.54	−1180.09	−1200.04	−1216.25
Non Production (NP)	0.00	0.00	0.00	0.00	0.00
Present Value (NPPV)	0.00	0.00	0.00	0.00	0.00
Accumulated NPPV	0.00	0.00	0.00	0.00	0.00
Accumulated Total Costs (APV)	−135780.66	−153344.63	−171246.68	−189420.75	−207813.66
Accumulated Total Costs	−61218.62	−68720.81	−76476.01	−84484.77	−92747.62
Benefit (*B*)	62500.00	62500.00	62500.00	62500.00	62500.00
Present Value (BPV)	7087.12	5700.71	4585.51	3688.48	2966.92
Accumulated Benefits BPV	227849.00	233549.71	238135.22	241823.70	244790.62
BPV + FPV + MPV + NPPV + IIPV	80068.34	68205.08	54888.54	40402.95	24976.96
Uniform Annual Income UAIn (LDM)	−223.83	−163.92	−120.06	−87.78	−63.93
Uniform Annual Income UAIn (SDM)	−137.62	−94.47	−64.56	−43.81	−29.44
Uniform Annual Income UAIn (EM)	−21.89	−12.94	−7.69	−4.59	−2.74

(*Continued*)

TABLE 7.2 (*Continued*)
Values Used for Simulation based on UAI Method Related to Physical Assets Standard – Second Simulation

Time	15	16	17	18	19
Initial Investment (II)					
Initial Investment (II)					
II Present Value (IIPV) - Return	−458.21	−368.57	−296.47	−238.48	−191.82
Life Time simulated					
Residual value of equipment (RV)					
Linear Depreciation Method (Vn)	−4794.00	−4313.60	−3833.20	−3352.80	−2872.40
LDM Present Value (Vn)	−183.06	−132.49	−94.70	−66.63	−45.92
Sum of Digits Method (Vn)	−2022.46	−1652.92	−1320.34	−1024.71	−766.03
SDM Present Value (Vn)	−77.23	−50.77	−32.62	−20.36	−12.25
Exponential Method (Vn)	−170.48	−128.38	−96.68	−72.81	−54.83
EM Present Value (Pv)	−6.51	−3.94	−2.39	−1.45	−0.88
Functioning (*F*)	−5172.63	−5224.36	−5276.60	−5329.37	−5382.66
Present Value Accumulated (NPV)	−18441.30	−18601.76	−18732.13	−18838.04	−18924.08
Present Value (NPV)	−197.51	−160.46	−130.36	−105.91	−86.04
Energy	−3000.00	−3200.00	−3400.00	−3600.00	−3800.00
Energy (NPV)	−114.55	−98.29	−84.00	−71.54	−60.74
Accumulated (FPV)	−225153.26	−243853.31	−262669.44	−281579.02	−300563.85
Maintenance (*M*)	−344.84	−348.29	−351.77	−355.29	−358.84
Present Value (MPV)	−13.17	−10.70	−8.69	−7.06	−5.74
Accumulated (MPV)	−1229.42	−1240.12	−1248.81	−1255.87	−1261.61
Non Production (NP)	0.00	0.00	0.00	0.00	0.00
Present Value (NPPV)	0.00	0.00	0.00	0.00	0.00
Accumulated NPPV	0.00	0.00	0.00	0.00	0.00
Accumulated Total Costs (APV)	−226382.68	−245093.43	−263918.25	−282834.89	−301825.45
Accumulated Total Costs	−101265.10	−110037.75	−119066.13	−128350.79	−137892.30
Benefit (*B*)	62500.00	62500.00	62500.00	62500.00	62500.00
Present Value (BPV)	2386.52	1919.66	1544.13	1242.06	999.08
Accumulated Benefits BPV	247177.14	249096.80	250640.93	251882.98	252882.07
BPV + FPV + MPV + NPPV + IIPV	8794.46	−7996.63	−25277.32	−42951.91	−60943.39
Uniform Annual Income UAIn (LDM)	−46.29	−33.24	−23.62	−16.53	−11.35
Uniform Annual Income UAIn (SDM)	−19.53	−12.74	−8.13	−5.05	−3.03
Uniform Annual Income UAIn (EM)	−1.65	−0.99	−0.60	−0.36	−0.22

(*Continued*)

TABLE 7.2 (*Continued*)
Values Used for Simulation based on UAI Method Related to Physical Assets Standard – Second Simulation

Time	20	21	22	23	24	25
Initial Investment (II)						
Initial Investment (II)						
II Present Value (IIPV) - Return	−154.30	−124.11	−99.83	−80.30	−64.59	−51.96
Life Time simulated						
Residual value of equipment (RV)						
Linear Depreciation Method (Vn)	−2392.00	−1911.60	−1431.20	−950.80	−470.40	10.00
LDM Present Value (Vn)	−30.76	−19.77	−11.91	−6.36	−2.53	0.04
Sum of Digits Method (Vn)	−544.31	−359.54	−211.72	−100.86	−26.95	10.00
SDM Present Value (Vn)	−7.00	−3.72	−1.76	−0.67	−0.15	0.04
Exponential Method (Vn)	−41.29	−31.09	−23.42	−17.63	−13.28	−10.00
EM Present Value (Pv)	−0.53	−0.32	−0.19	−0.12	−0.07	−0.04
Functioning (*F*)	−5436.49	−5490.86	−5545.76	−5601.22	−5657.23	−5713.81
Present Value Accumulated (NPV)	−18993.99	−19050.78	−19096.91	−19134.40	−19164.85	−19189.59
Present Value (NPV)	−69.90	−56.79	−46.14	−37.48	−30.45	−24.74
Energy	−4000.00	−4200.00	−4400.00	−4600.00	−4800.00	−5000.00
Energy (NPV)	−51.43	−43.44	−36.61	−30.78	−25.84	−21.65
Accumulated (FPV)	−319609.27	−338703.48	−357837.01	−377002.19	−396192.88	−415404.12
Maintenance (*M*)	−362.43	−366.06	−369.72	−373.41	−377.15	−380.92
Present Value (MPV)	−4.66	−3.79	−3.08	−2.50	−2.03	−1.65
Accumulated (MPV)	−1266.27	−1270.05	−1273.13	−1275.63	−1277.66	−1279.31
Non Production (NP)	0.00	0.00	0.00	0.00	0.00	0.00
Present Value (NPPV)	0.00	0.00	0.00	0.00	0.00	0.00
Accumulated NPPV	0.00	0.00	0.00	0.00	0.00	0.00
Accumulated Total Costs (APV)	−320875.53	−339973.54	−359110.13	−378277.81	−397470.53	−416683.42
Accumulated Total Costs	−147691.22	−157748.13	−168063.61	−178638.25	−189472.63	−200567.36
Benefit (*B*)	62500.00	62500.00	62500.00	62500.00	62500.00	62500.00
Present Value (BPV)	803.64	646.43	519.97	418.25	336.43	270.62
Accumulated Benefits BPV	253685.70	254332.13	254852.10	255270.35	255606.78	255877.40
BPV + FPV + MPV + NPPV + IIPV	−79189.83	−97641.41	−116258.03	−135007.46	−153863.75	−172806.02
Uniform Annual Income UAIn (LDM)	−7.58	−4.86	−2.92	−1.56	−0.62	0.01
Uniform Annual Income UAIn (SDM)	−1.72	−0.91	−0.43	−0.17	−0.04	0.01
Uniform Annual Income UAIn (EM)	−0.13	−0.08	−0.05	−0.03	−0.02	−0.01

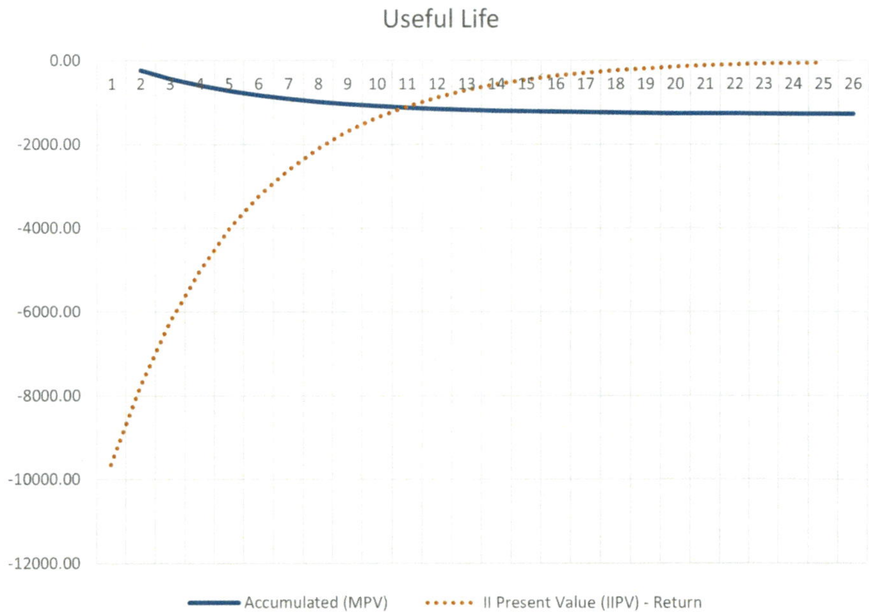

FIGURE 7.6 Physical Asset useful life.

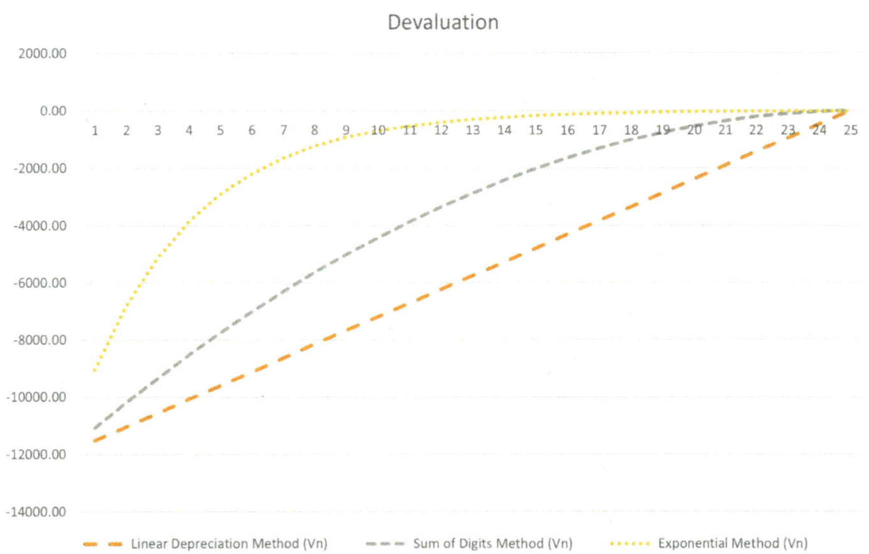

FIGURE 7.7 The three devaluation methods.

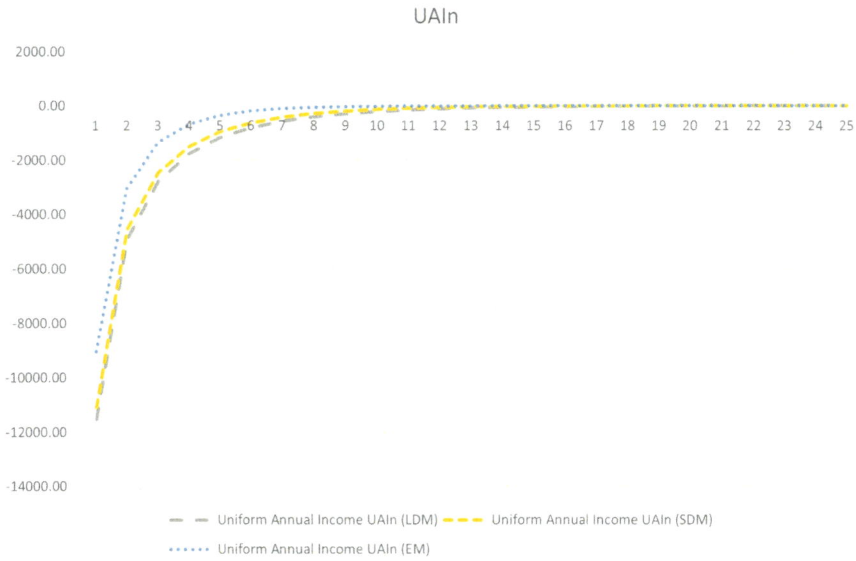

FIGURE 7.8 The UAI results from the three devaluation methods.

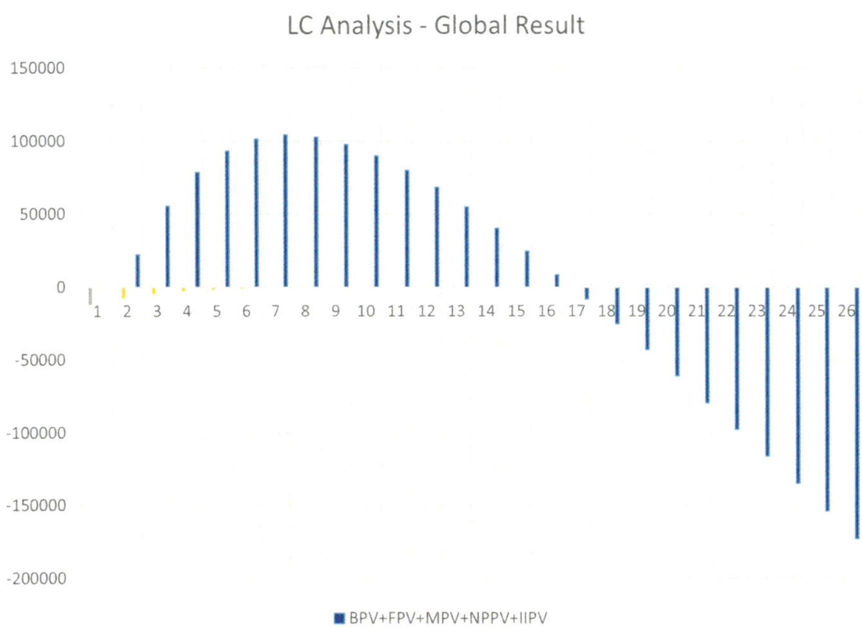

FIGURE 7.9 Life cycle analysis global result including profits.

The remaining values are presented in Table 7.3, and the graphical results are shown in Figures 7.10–7.13.

As referred in the preceding introduction on this section, the evaluation of introduction of Physical Asset Norms, the simulation of their implementation is made through the increase of Benefits. They may be used other indicators, but may be not so intuitive.

TABLE 7.3
Values Used for Simulation based on MMTAC Method Related to Physical Assets Standard – First Simulation

Time	0	1	2	3	4
Initial Investment (II)	−12000				
Initial Investment (II)	−12000				
II (IIPV) - Return	−12000	−12000.00	−12000.00	−12000.00	−12000.00
Useful Life	25				
Residual value of equipment (RV)	10				
Linear Depreciation Method (Vn)	−12000	−11519.60	−11039.20	−10558.80	−10078.40
LDM Present Value (PV)		−11519.60	−11039.20	−10558.80	−10078.40
Sum of Digits Method (Vn)	−12000	−11076.15	−10189.26	−9339.32	−8526.34
SDM Present Value (PV)		−11076.15	−10189.26	−9339.32	−8526.34
Exponential Method (Vn)	−12000	−9036.79	−6805.29	−5124.83	−3859.33
EM Present Value (Pv)		−9036.79	−6805.29	−5124.83	−3859.33
Functioning (*F*)	−4500	−4500.00	−4545.00	−4590.45	−4636.35
Functioning Accumulated (FA)		−4500.00	−9045.00	−13635.45	−18271.80
Present Value Accumulated (NPV)		−4500.00	−9045.00	−13635.45	−18271.80
Present Value (NPV)		−4500.00	−4545.00	−4590.45	−4636.35
Energy	−200	−200.00	−400.00	−600.00	−800.00
Energy Accumulated (EA)		−200.00	−600.00	−1200.00	−2000.00
Energy (NPV)		−200.00	−400.00	−600.00	−800.00
Accumulated Functioning (FPV)		−16700.00	−26145.00	−40380.45	−59452.25
Maintenance (*M*)	−300	−300.00	−303.00	−306.03	−309.09
Maintenane Accumulated (MA)		−300.00	−603.00	−909.03	−1218.12
Present Value (MPV)		−300.00	−303.00	−306.03	−309.09
Accumulated MPV		−300.00	−603.00	−909.03	−1218.12
Non Production (NP)	0	0.00	0.00	0.00	0.00
Non Production Accumulated (NPA)		0.00	0.00	0.00	0.00
Present Value (NPPV)		0.00	0.00	0.00	0.00
Accumulated NPPV		0.00	0.00	0.00	0.00
Accumulated Total Costs (APV)		−17000.00	−26748.00	−41289.48	−60670.37
Accumulated Total Costs		−5000.00	−10248.00	−15744.48	−21489.92
Benefit (B)	50000	44642.86	44642.86	44642.86	44642.86
Benefits Accumulated (BA)		44642.86	89285.71	133928.57	178571.43
Present Value (BPV)		44642.86	44642.86	44642.86	44642.86
Accumulated Benefits BPV		44642.86	89285.71	133928.57	178571.43
BPV+FPV+MPV+NPPV+IIPV		27642.86	67037.71	106184.09	145081.50
MTAC (LDM)		−5480.40	−11208.80	−17185.68	−23411.52
MTAC (SDM)		−5923.85	−12058.74	−18405.16	−24963.59
MTAC (EM)		−7963.21	−15442.71	−22619.65	−29630.59

(Continued)

TABLE 7.3 (*Continued*)
Values Used for Simulation based on MMTAC Method Related to Physical Assets Standard – First Simulation

Time	5	6	7	8	9
Initial Investment (II)					
Initial Investment (II)					
II (IIPV) - Return	−12000.00	−12000.00	−12000.00	−12000.00	−12000.00
Useful Life					
Residual value of equipment (RV)					
Linear Depreciation Method (Vn)	−9598.00	−9117.60	−8637.20	−8156.80	−7676.40
LDM Present Value (PV)	−9598.00	−9117.60	−8637.20	−8156.80	−7676.40
Sum of Digits Method (Vn)	−7750.31	−7011.23	−6309.11	−5643.94	−5015.72
SDM Present Value (PV)	−7750.31	−7011.23	−6309.11	−5643.94	−5015.72
Exponential Method (Vn)	−2906.33	−2188.66	−1648.20	−1241.20	−934.71
EM Present Value (Pv)	−2906.33	−2188.66	−1648.20	−1241.20	−934.71
Functioning (*F*)	−4682.72	−4729.55	−4776.84	−4824.61	−4872.86
Functioning Accumulated (FA)	−22954.52	−27684.07	−32460.91	−37285.52	−42158.37
Present Value Accumulated (NPV)	−22954.52	−27684.07	−32460.91	−37285.52	−42158.37
Present Value (NPV)	−4682.72	−4729.55	−4776.84	−4824.61	−4872.86
Energy	−1000.00	−1200.00	−1400.00	−1600.00	−1800.00
Energy Accumulated (EA)	−3000.00	−4200.00	−5600.00	−7200.00	−9000.00
Energy (NPV)	−1000.00	−1200.00	−1400.00	−1600.00	−1800.00
Accumulated Functioning (FPV)	−83406.78	−112290.84	−146151.75	−185037.27	−228995.64
Maintenance (*M*)	−312.18	−315.30	−318.46	−321.64	−324.86
Maintenane Accumulated (MA)	−1530.30	−1845.60	−2164.06	−2485.70	−2810.56
Present Value (MPV)	−312.18	−315.30	−318.46	−321.64	−324.86
Accumulated MPV	−1530.30	−1845.60	−2164.06	−2485.70	−2810.56
Non Production (NP)	0.00	0.00	0.00	0.00	0.00
Non Production Accumulated (NPA)	0.00	0.00	0.00	0.00	0.00
Present Value (NPPV)	0.00	0.00	0.00	0.00	0.00
Accumulated NPPV	0.00	0.00	0.00	0.00	0.00
Accumulated Total Costs (APV)	−84937.08	−114136.45	−148315.81	−187522.97	−231806.20
Accumulated Total Costs	−27484.82	−33729.67	−40224.97	−46971.22	−53968.93
Benefit (B)	44642.86	44642.86	44642.86	44642.86	44642.86
Benefits Accumulated (BA)	223214.29	267857.14	312500.00	357142.86	401785.71
Present Value (BPV)	44642.86	44642.86	44642.86	44642.86	44642.86
Accumulated Benefits BPV	223214.29	267857.14	312500.00	357142.86	401785.71
BPV + FPV + MPV + NPPV + IIPV	183729.46	222127.47	260275.03	298171.64	335816.78
MTAC (LDM)	−29886.82	−36612.07	−43587.77	−50814.42	−58292.53
MTAC (SDM)	−31734.52	−38718.44	−45915.86	−53327.28	−60953.21
MTAC (EM)	−36578.49	−43541.02	−50576.77	−57730.01	−65034.22

(*Continued*)

TABLE 7.3 (*Continued*)
Values Used for Simulation based on MMTAC Method Related to Physical Assets Standard – First Simulation

Time	10	11	12	13	14
Initial Investment (II)					
Initial Investment (II)					
II (IIPV) - Return	−12000.00	−12000.00	−12000.00	−12000.00	−12000.00
Useful Life					
Residual value of equipment (RV)					
Linear Depreciation Method (Vn)	−7196.00	−6715.60	−6235.20	−5754.80	−5274.40
LDM Present Value (PV)	−7196.00	−6715.60	−6235.20	−5754.80	−5274.40
Sum of Digits Method (Vn)	−4424.46	−3870.15	−3352.80	−2872.40	−2428.95
SDM Present Value (PV)	−4424.46	−3870.15	−3352.80	−2872.40	−2428.95
Exponential Method (Vn)	−703.90	−530.08	−399.19	−300.61	−226.38
EM Present Value (Pv)	−703.90	−530.08	−399.19	−300.61	−226.38
Functioning (*F*)	−4921.58	−4970.80	−5020.51	−5070.71	−5121.42
Functioning Accumulated (FA)	−47079.96	−52050.76	−57071.26	−62141.98	−67263.40
Present Value Accumulated (NPV)	−47079.96	−52050.76	−57071.26	−62141.98	−67263.40
Present Value (NPV)	−4921.58	−4970.80	−5020.51	−5070.71	−5121.42
Energy	−2000.00	−2200.00	−2400.00	−2600.00	−2800.00
Energy Accumulated (EA)	−11000.00	−13200.00	−15600.00	−18200.00	−21000.00
Energy (NPV)	−2000.00	−2200.00	−2400.00	−2600.00	−2800.00
Accumulated Functioning (FPV)	−278075.60	−332326.36	−391797.62	−456539.60	−526602.99
Maintenance (*M*)	−328.11	−331.39	−334.70	−338.05	−341.43
Maintenane Accumulated (MA)	−3138.66	−3470.05	−3804.75	−4142.80	−4484.23
Present Value (MPV)	−328.11	−331.39	−334.70	−338.05	−341.43
Accumulated MPV	−3138.66	−3470.05	−3804.75	−4142.80	−4484.23
Non Production (NP)	0.00	0.00	0.00	0.00	0.00
Non Production Accumulated (NPA)	0.00	0.00	0.00	0.00	0.00
Present Value (NPPV)	0.00	0.00	0.00	0.00	0.00
Accumulated NPPV	0.00	0.00	0.00	0.00	0.00
Accumulated Total Costs (APV)	−281214.26	−335796.41	−395602.37	−460682.39	−531087.22
Accumulated Total Costs	−61218.62	−68720.81	−76476.01	−84484.77	−92747.62
Benefit (B)	44642.86	44642.86	44642.86	44642.86	44642.86
Benefits Accumulated (BA)	446428.57	491071.43	535714.29	580357.14	625000.00
Present Value (BPV)	44642.86	44642.86	44642.86	44642.86	44642.86
Accumulated Benefits BPV	446428.57	491071.43	535714.29	580357.14	625000.00
BPV + FPV + MPV + NPPV + IIPV	373209.95	410350.62	447238.27	483872.37	520252.38
MTAC (LDM)	−66022.62	−74005.21	−82240.81	−90729.97	−99473.22
MTAC (SDM)	−68794.16	−76850.65	−85123.21	−93612.37	−102318.67
MTAC (EM)	−72514.72	−80190.73	−88076.83	−96184.16	−104521.24

(Continued)

TABLE 7.3 (*Continued*)
Values Used for Simulation based on MMTAC Method Related to Physical Assets Standard – First Simulation

Time	15	16	17	18	19
Initial Investment (II)					
Initial Investment (II)					
II (IIPV) - Return	−12000.00	−12000.00	−12000.00	−12000.00	−12000.00
Useful Life					
Residual value of equipment (RV)					
Linear Depreciation Method (Vn)	−4794.00	−4313.60	−3833.20	−3352.80	−2872.40
LDM Present Value (PV)	−4794.00	−4313.60	−3833.20	−3352.80	−2872.40
Sum of Digits Method (Vn)	−2022.46	−1652.92	−1320.34	−1024.71	−766.03
SDM Present Value (PV)	−2022.46	−1652.92	−1320.34	−1024.71	−766.03
Exponential Method (Vn)	−170.48	−128.38	−96.68	−72.81	−54.83
EM Present Value (Pv)	−170.48	−128.38	−96.68	−72.81	−54.83
Functioning (*F*)	−5172.63	−5224.36	−5276.60	−5329.37	−5382.66
Functioning Accumulated (FA)	−72436.03	−77660.39	−82936.99	−88266.36	−93649.03
Present Value Accumulated (NPV)	−72436.03	−77660.39	−82936.99	−88266.36	−93649.03
Present Value (NPV)	−5172.63	−5224.36	−5276.60	−5329.37	−5382.66
Energy	−3000.00	−3200.00	−3400.00	−3600.00	−3800.00
Energy Accumulated (EA)	−24000.00	−27200.00	−30600.00	−34200.00	−38000.00
Energy (NPV)	−3000.00	−3200.00	−3400.00	−3600.00	−3800.00
Accumulated Functioning (FPV)	−602039.02	−682899.41	−769236.41	−861102.77	−958551.80
Maintenance (*M*)	−344.84	−348.29	−351.77	−355.29	−358.84
Maintenane Accumulated (MA)	−4829.07	−5177.36	−5529.13	−5884.42	−6243.27
Present Value (MPV)	−344.84	−348.29	−351.77	−355.29	−358.84
Accumulated MPV	−4829.07	−5177.36	−5529.13	−5884.42	−6243.27
Non Production (NP)	0.00	0.00	0.00	0.00	0.00
Non Production Accumulated (NPA)	0.00	0.00	0.00	0.00	0.00
Present Value (NPPV)	0.00	0.00	0.00	0.00	0.00
Accumulated NPPV	0.00	0.00	0.00	0.00	0.00
Accumulated Total Costs (APV)	−606868.09	−688076.77	−774765.54	−866987.19	−964795.07
Accumulated Total Costs	−101265.10	−110037.75	−119066.13	−128350.79	−137892.30
Benefit (B)	44642.86	44642.86	44642.86	44642.86	44642.86
Benefits Accumulated (BA)	669642.86	714285.71	758928.57	803571.43	848214.29
Present Value (BPV)	44642.86	44642.86	44642.86	44642.86	44642.86
Accumulated Benefits BPV	669642.86	714285.71	758928.57	803571.43	848214.29
BPV + FPV + MPV + NPPV + IIPV	556377.76	592247.96	627862.44	663220.64	698321.99
MTAC (LDM)	−108471.10	−117724.15	−127232.93	−136997.99	−147019.90
MTAC (SDM)	−111242.64	−120384.83	−129745.79	−139326.08	−149126.27
MTAC (EM)	−113094.62	−121909.37	−130969.45	−140277.98	−149837.47

(*Continued*)

TABLE 7.3 (*Continued*)
Values Used for Simulation based on MMTAC Method Related to Physical Assets Standard – First Simulation

Time	20	21	22	23	24	25
Initial Investment (II)						
Initial Investment (II)						
II (IIPV) - Return	−12000.00	−12000.00	−12000.00	−12000.00	−12000.00	−12000.00
Useful Life						
Residual value of equipment (RV)						
Linear Depreciation Method (Vn)	−2392.00	−1911.60	−1431.20	−950.80	−470.40	10.00
LDM Present Value (PV)	−2392.00	−1911.60	−1431.20	−950.80	−470.40	10.00
Sum of Digits Method (Vn)	−544.31	−359.54	−211.72	−100.86	−26.95	10.00
SDM Present Value (PV)	−544.31	−359.54	−211.72	−100.86	−26.95	10.00
Exponential Method (Vn)	−41.29	−31.09	−23.42	−17.63	−13.28	−10.00
EM Present Value (Pv)	−41.29	−31.09	−23.42	−17.63	−13.28	−10.00
Functioning (*F*)	−5436.49	−5490.86	−5545.76	−5601.22	−5657.23	−5713.81
Functioning Accumulated (FA)	−99085.52	−104576.37	−110122.14	−115723.36	−121380.59	−127094.40
Present Value Accumulated (NPV)	−99085.52	−104576.37	−110122.14	−115723.36	−121380.59	−127094.40
Present Value (NPV)	−5436.49	−5490.86	−5545.76	−5601.22	−5657.23	−5713.81
Energy	−4000.00	−4200.00	−4400.00	−4600.00	−4800.00	−5000.00
Energy Accumulated (EA)	−42000.00	−46200.00	−50600.00	−55200.00	−60000.00	−65000.00
Energy (NPV)	−4000.00	−4200.00	−4400.00	−4600.00	−4800.00	−5000.00
Accumulated Functioning (FPV)	−1061637.32	−1170413.69	−1284935.83	−1405259.18	−1531439.78	−1663534.17
Maintenance (*M*)	−362.43	−366.06	−369.72	−373.41	−377.15	−380.92
Maintenane Accumulated (MA)	−6605.70	−6971.76	−7341.48	−7714.89	−8092.04	−8472.96
Present Value (MPV)	−362.43	−366.06	−369.72	−373.41	−377.15	−380.92
Accumulated MPV	−6605.70	−6971.76	−7341.48	−7714.89	−8092.04	−8472.96
Non Production (NP)	0.00	0.00	0.00	0.00	0.00	0.00
Non Production Accumulated (NPA)	0.00	0.00	0.00	0.00	0.00	0.00
Present Value (NPPV)	0.00	0.00	0.00	0.00	0.00	0.00
Accumulated NPPV	0.00	0.00	0.00	0.00	0.00	0.00
Accumulated Total Costs (APV)	−1068243.02	−1177385.45	−1292277.30	−1412974.07	−1539531.82	−1672007.13
Accumulated Total Costs	−147691.22	−157748.13	−168063.61	−178638.25	−189472.63	−200567.36
Benefit (B)	44642.86	44642.86	44642.86	44642.86	44642.86	44642.86
Benefits Accumulated (BA)	892857.14	937500.00	982142.86	1026785.71	1071428.57	1116071.43
Present Value (BPV)	44642.86	44642.86	44642.86	44642.86	44642.86	44642.86
Accumulated Benefits BPV	892857.14	937500.00	982142.86	1026785.71	1071428.57	1116071.43
BPV + FPV + MPV + NPPV + IIPV	733165.92	767751.87	802079.24	836147.47	869955.94	903504.07
MTAC (LDM)	−157299.22	−167836.53	−178632.41	−189687.45	−201002.23	−212577.36
MTAC (SDM)	−159146.91	−169388.59	−179851.89	−190537.39	−201445.68	−212577.36
MTAC (EM)	−159649.93	−169717.04	−180040.20	−190620.62	−201459.35	−212557.36

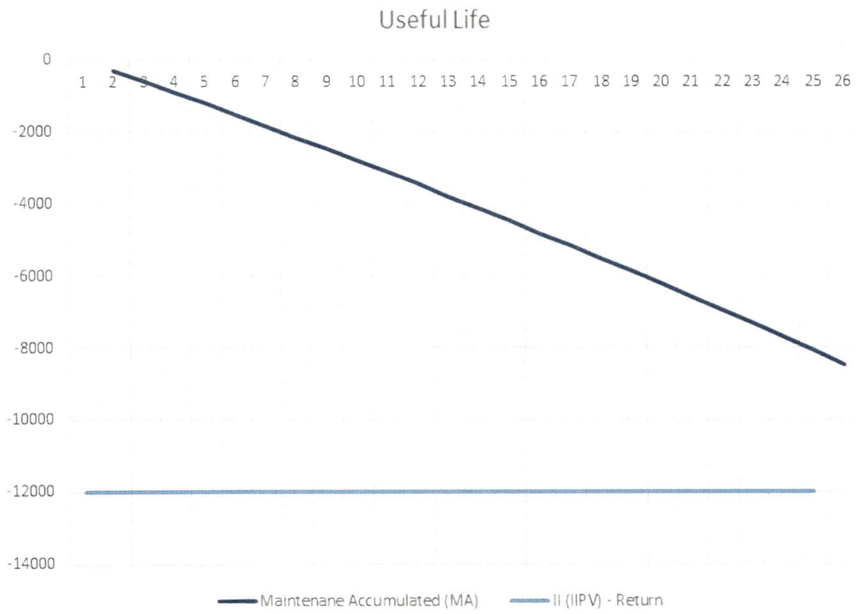

FIGURE 7.10 Physical Asset useful life.

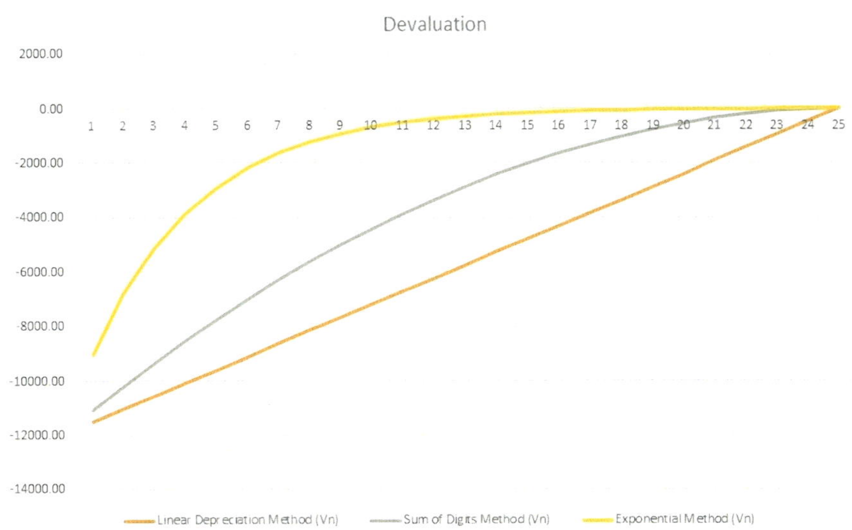

FIGURE 7.11 The three devaluation methods.

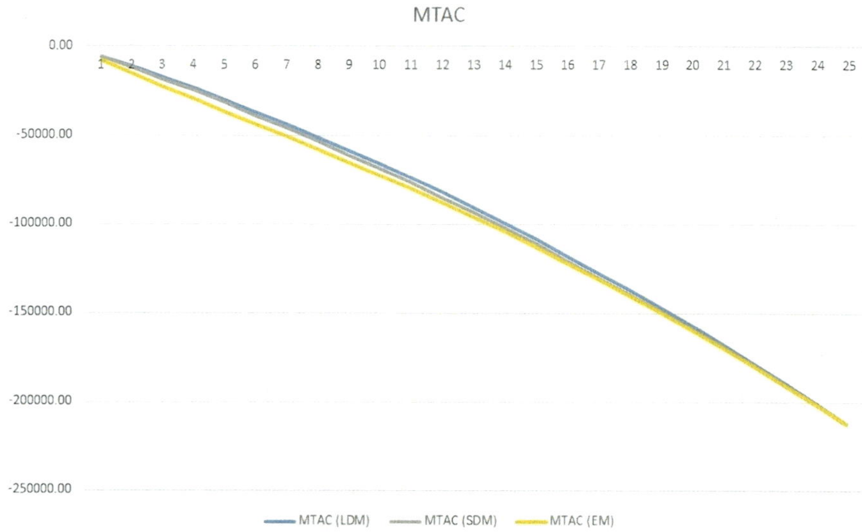

FIGURE 7.12 The MMTAC results from the three devaluation methods.

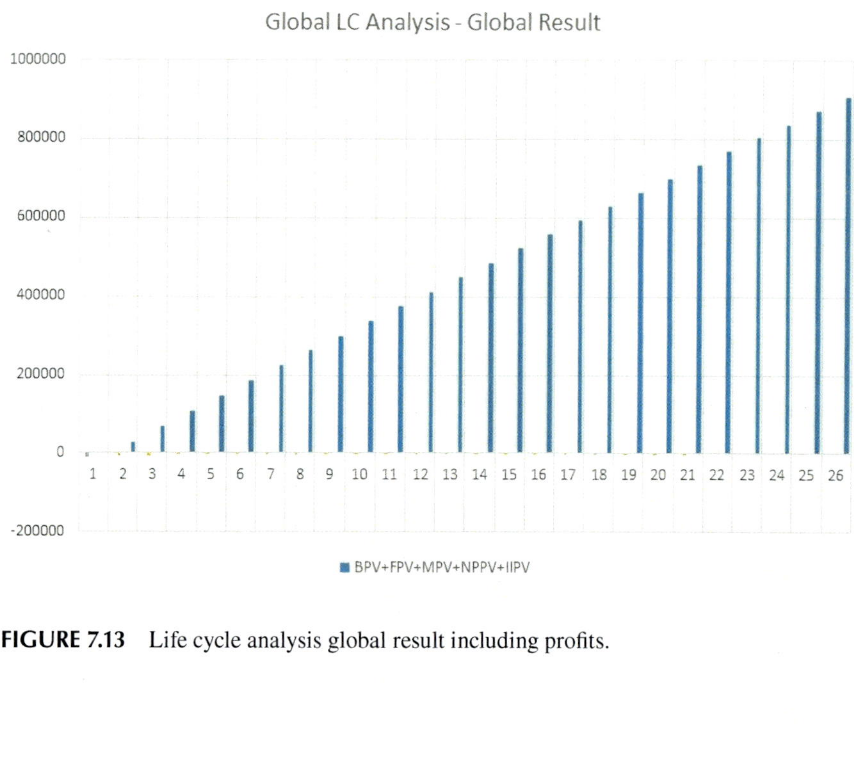

FIGURE 7.13 Life cycle analysis global result including profits.

The second simulation uses the following values (the MTBF is increased and MTTR decreases as a result of the Maintenance Standards):

- Initial Investment (II) –12000
- Life Time simulated 25
- Residual value of equipment (RV) 10
- Functioning (*F*) –4500
- Energy –200
- Maintenance (*M*) –300
- Non Production (NP) 0
- Benefit (*B*) **70000**
- Internal Rate of Return (*r*) 0.0
- Risk (*R*) 0.0
- Apparent Rate 0.0
- Maintenance Increase Rate 0.01
- Availability 0.9662
- MTBF 1000
- MTTR 25
- MWT 10

The remaining values are presented in Table 7.4, and the graphical results are shown in Figures 7.14–7.17.

As can be seen, namely comparing Figures 7.13 and 7.17, with the increasing of the maintenance management quality helped by the maintenance standards, it is get increasing positive results.

However, from both simulations, the results seem strange because the Apparent Rate is null, what does not happen in a real world. But, this type of analysis makes the contrast with the others analysis presented, emphasizing the importance of considering the rigorous values, both the Physical Assets and the market.

7.4.1.3 MMTAC Reduced to Present Value (MMTAC-RPV)

The first simulation uses the following values:

- Initial Investment (II) –12000
- Life Time simulated 25
- Residual value of equipment (RV) 10
- Functioning (*F*) –4500
- Energy (*E*) –200
- Maintenance (*M*) –300
- Non Production (NP) 0
- Benefits (*B*) **50000**
- Internal Rate of Return (*r*) 0.12
- Risk (R) 0.11
- Apparent Rate 0.2432
- Maintenance Increase Rate 0.01

- Availability 0.8929
- MTBF 500
- MTTR 50
- MWT 10

TABLE 7.4
Values Used for Simulation based on MTAC Method Related to Physical Assets Standard – Second Simulation

Time	0	1	2	3	4
Initial Investment (II)	−12000				
Initial Investment (II)	−12000				
II (IIPV) - Return	−12000	−12000.00	−12000.00	−12000.00	−12000.00
Useful Life	25				
Residual value of equipment (RV)	10				
Linear Depreciation Method (Vn)	−12000	−11519.60	−11039.20	−10558.80	−10078.40
LDM Present Value (PV)		−11519.60	−11039.20	−10558.80	−10078.40
Sum of Digits Method (Vn)	−12000	−11076.15	−10189.26	−9339.32	−8526.34
SDM Present Value (PV)		−11076.15	−10189.26	−9339.32	−8526.34
Exponential Method (Vn)	−12000	−9036.79	−6805.29	−5124.83	−3859.33
EM Present Value (Pv)		−9036.79	−6805.29	−5124.83	−3859.33
Functioning (*F*)	−4500	−4500.00	−4545.00	−4590.45	−4636.35
Functioning Accumulated (FA)		−4500.00	−9045.00	−13635.45	−18271.80
Present Value Accumulated (NPV)		−4500.00	−9045.00	−13635.45	−18271.80
Present Value (NPV)		−4500.00	−4545.00	−4590.45	−4636.35
Energy	−200	−200.00	−400.00	−600.00	−800.00
Energy Accumulated (EA)		−200.00	−600.00	−1200.00	−2000.00
Energy (NPV)		−200.00	−400.00	−600.00	−800.00
Accumulated Functioning (FPV)		−16700.00	−26145.00	−40380.45	−59452.25
Maintenance (*M*)	−300	−300.00	−303.00	−306.03	−309.09
Maintenane Accumulated (MA)		−300.00	−603.00	−909.03	−1218.12
Present Value (MPV)		−300.00	−303.00	−306.03	−309.09
Accumulated MPV		−300.00	−603.00	−909.03	−1218.12
Non Production (NP)	0	0.00	0.00	0.00	0.00
Non Production Accumulated (NPA)		0.00	0.00	0.00	0.00
Present Value (NPPV)		0.00	0.00	0.00	0.00
Accumulated NPPV		0.00	0.00	0.00	0.00
Accumulated Total Costs (APV)		−17000.00	−26748.00	−41289.48	−60670.37
Accumulated Total Costs		−5000.00	−10248.00	−15744.48	−21489.92
Benefit (*B*)	70000	67632.85	67632.85	67632.85	67632.85
Benefits Accumulated (BA)		67632.85	135265.70	202898.55	270531.40
Present Value (BPV)		67632.85	67632.85	67632.85	67632.85
Accumulated Benefits BPV		67632.85	135265.70	202898.55	270531.40
BPV + FPV + MPV + NPPV + IIPV		50632.85	113017.70	175154.07	237041.48
MTAC (LDM)		−5480.40	−11208.80	−17185.68	−23411.52
MTAC (SDM)		−5923.85	−12058.74	−18405.16	−24963.59
MTAC (EM)		−7963.21	−15442.71	−22619.65	−29630.59

(Continued)

TABLE 7.4 (*Continued*)
Values Used for Simulation based on MTAC Method Related to Physical Assets Standard – Second Simulation

Time	5	6	7	8	9
Initial Investment (II)					
Initial Investment (II)					
II (IIPV) - Return	−12000.00	−12000.00	−12000.00	−12000.00	−12000.00
Useful Life					
Residual value of equipment (RV)					
Linear Depreciation Method (Vn)	−9598.00	−9117.60	−8637.20	−8156.80	−7676.40
LDM Present Value (PV)	−9598.00	−9117.60	−8637.20	−8156.80	−7676.40
Sum of Digits Method (Vn)	−7750.31	−7011.23	−6309.11	−5643.94	−5015.72
SDM Present Value (PV)	−7750.31	−7011.23	−6309.11	−5643.94	−5015.72
Exponential Method (Vn)	−2906.33	−2188.66	−1648.20	−1241.20	−934.71
EM Present Value (Pv)	−2906.33	−2188.66	−1648.20	−1241.20	−934.71
Functioning (*F*)	−4682.72	−4729.55	−4776.84	−4824.61	−4872.86
Functioning Accumulated (FA)	−22954.52	−27684.07	−32460.91	−37285.52	−42158.37
Present Value Accumulated (NPV)	−22954.52	−27684.07	−32460.91	−37285.52	−42158.37
Present Value (NPV)	−4682.72	−4729.55	−4776.84	−4824.61	−4872.86
Energy	−1000.00	−1200.00	−1400.00	−1600.00	−1800.00
Energy Accumulated (EA)	−3000.00	−4200.00	−5600.00	−7200.00	−9000.00
Energy (NPV)	−1000.00	−1200.00	−1400.00	−1600.00	−1800.00
Accumulated Functioning (FPV)	−83406.78	−112290.84	−146151.75	−185037.27	−228995.64
Maintenance (*M*)	−312.18	−315.30	−318.46	−321.64	−324.86
Maintenane Accumulated (MA)	−1530.30	−1845.60	−2164.06	−2485.70	−2810.56
Present Value (MPV)	−312.18	−315.30	−318.46	−321.64	−324.86
Accumulated MPV	−1530.30	−1845.60	−2164.06	−2485.70	−2810.56
Non Production (NP)	0.00	0.00	0.00	0.00	0.00
Non Production Accumulated (NPA)	0.00	0.00	0.00	0.00	0.00
Present Value (NPPV)	0.00	0.00	0.00	0.00	0.00
Accumulated NPPV	0.00	0.00	0.00	0.00	0.00
Accumulated Total Costs (APV)	−84937.08	−114136.45	−148315.81	−187522.97	−231806.20
Accumulated Total Costs	−27484.82	−33729.67	−40224.97	−46971.22	−53968.93
Benefit (*B*)	67632.85	67632.85	67632.85	67632.85	67632.85
Benefits Accumulated (BA)	338164.25	405797.10	473429.95	541062.80	608695.65
Present Value (BPV)	67632.85	67632.85	67632.85	67632.85	67632.85
Accumulated Benefits BPV	338164.25	405797.10	473429.95	541062.80	608695.65
BPV+FPV+MPV+NPPV+IIPV	298679.43	360067.43	421204.98	482091.58	542726.72
MTAC (LDM)	−29886.82	−36612.07	−43587.77	−50814.42	−58292.53
MTAC (SDM)	−31734.52	−38718.44	−45915.86	−53327.28	−60953.21
MTAC (EM)	−36578.49	−43541.02	−50576.77	−57730.01	−65034.22

(*Continued*)

TABLE 7.4 (*Continued*)
Values Used for Simulation based on MTAC Method Related to Physical Assets Standard – Second Simulation

Time	10	11	12	13	14
Initial Investment (II)					
Initial Investment (II)					
II (IIPV) - Return	−12000.00	−12000.00	−12000.00	−12000.00	−12000.00
Useful Life					
Residual value of equipment (RV)					
Linear Depreciation Method (Vn)	−7196.00	−6715.60	−6235.20	−5754.80	−5274.40
LDM Present Value (PV)	−7196.00	−6715.60	−6235.20	−5754.80	−5274.40
Sum of Digits Method (Vn)	−4424.46	−3870.15	−3352.80	−2872.40	−2428.95
SDM Present Value (PV)	−4424.46	−3870.15	−3352.80	−2872.40	−2428.95
Exponential Method (Vn)	−703.90	−530.08	−399.19	−300.61	−226.38
EM Present Value (Pv)	−703.90	−530.08	−399.19	−300.61	−226.38
Functioning (*F*)	−4921.58	−4970.80	−5020.51	−5070.71	−5121.42
Functioning Accumulated (FA)	−47079.96	−52050.76	−57071.26	−62141.98	−67263.40
Present Value Accumulated (NPV)	−47079.96	−52050.76	−57071.26	−62141.98	−67263.40
Present Value (NPV)	−4921.58	−4970.80	−5020.51	−5070.71	−5121.42
Energy	−2000.00	−2200.00	−2400.00	−2600.00	−2800.00
Energy Accumulated (EA)	−11000.00	−13200.00	−15600.00	−18200.00	−21000.00
Energy (NPV)	−2000.00	−2200.00	−2400.00	−2600.00	−2800.00
Accumulated Functioning (FPV)	−278075.60	−332326.36	−391797.62	−456539.60	−526602.99
Maintenance (*M*)	−328.11	−331.39	−334.70	−338.05	−341.43
Maintenane Accumulated (MA)	−3138.66	−3470.05	−3804.75	−4142.80	−4484.23
Present Value (MPV)	−328.11	−331.39	−334.70	−338.05	−341.43
Accumulated MPV	−3138.66	−3470.05	−3804.75	−4142.80	−4484.23
Non Production (NP)	0.00	0.00	0.00	0.00	0.00
Non Production Accumulated (NPA)	0.00	0.00	0.00	0.00	0.00
Present Value (NPPV)	0.00	0.00	0.00	0.00	0.00
Accumulated NPPV	0.00	0.00	0.00	0.00	0.00
Accumulated Total Costs (APV)	−281214.26	−335796.41	−395602.37	−460682.39	−531087.22
Accumulated Total Costs	−61218.62	−68720.81	−76476.01	−84484.77	−92747.62
Benefit (*B*)	67632.85	67632.85	67632.85	67632.85	67632.85
Benefits Accumulated (BA)	676328.50	743961.35	811594.20	879227.05	946859.90
Present Value (BPV)	67632.85	67632.85	67632.85	67632.85	67632.85
Accumulated Benefits BPV	676328.50	743961.35	811594.20	879227.05	946859.90
BPV + FPV + MPV + NPPV + IIPV	603109.88	663240.55	723118.19	782742.28	842112.28
MTAC (LDM)	−66022.62	−74005.21	−82240.81	−90729.97	−99473.22
MTAC (SDM)	−68794.16	−76850.65	−85123.21	−93612.37	−102318.67
MTAC (EM)	−72514.72	−80190.73	−88076.83	−96184.16	−104521.24

(*Continued*)

TABLE 7.4 (*Continued*)
Values Used for Simulation based on MTAC Method Related to Physical Assets Standard – Second Simulation

Time	15	16	17	18	19
Initial Investment (II)					
Initial Investment (II)					
II (IIPV) - Return	−12000.00	−12000.00	−12000.00	−12000.00	−12000.00
Useful Life					
Residual value of equipment (RV)					
Linear Depreciation Method (Vn)	−4794.00	−4313.60	−3833.20	−3352.80	−2872.40
LDM Present Value (PV)	−4794.00	−4313.60	−3833.20	−3352.80	−2872.40
Sum of Digits Method (Vn)	−2022.46	−1652.92	−1320.34	−1024.71	−766.03
SDM Present Value (PV)	−2022.46	−1652.92	−1320.34	−1024.71	−766.03
Exponential Method (Vn)	−170.48	−128.38	−96.68	−72.81	−54.83
EM Present Value (Pv)	−170.48	−128.38	−96.68	−72.81	−54.83
Functioning (*F*)	−5172.63	−5224.36	−5276.60	−5329.37	−5382.66
Functioning Accumulated (FA)	−72436.03	−77660.39	−82936.99	−88266.36	−93649.03
Present Value Accumulated (NPV)	−72436.03	−77660.39	−82936.99	−88266.36	−93649.03
Present Value (NPV)	−5172.63	−5224.36	−5276.60	−5329.37	−5382.66
Energy	−3000.00	−3200.00	−3400.00	−3600.00	−3800.00
Energy Accumulated (EA)	−24000.00	−27200.00	−30600.00	−34200.00	−38000.00
Energy (NPV)	−3000.00	−3200.00	−3400.00	−3600.00	−3800.00
Accumulated Functioning (FPV)	−602039.02	−682899.41	−769236.41	−861102.77	−958551.80
Maintenance (*M*)	−344.84	−348.29	−351.77	−355.29	−358.84
Maintenane Accumulated (MA)	−4829.07	−5177.36	−5529.13	−5884.42	−6243.27
Present Value (MPV)	−344.84	−348.29	−351.77	−355.29	−358.84
Accumulated MPV	−4829.07	−5177.36	−5529.13	−5884.42	−6243.27
Non Production (NP)	0.00	0.00	0.00	0.00	0.00
Non Production Accumulated (NPA)	0.00	0.00	0.00	0.00	0.00
Present Value (NPPV)	0.00	0.00	0.00	0.00	0.00
Accumulated NPPV	0.00	0.00	0.00	0.00	0.00
Accumulated Total Costs (APV)	−606868.09	−688076.77	−774765.54	−866987.19	−964795.07
Accumulated Total Costs	−101265.10	−110037.75	−119066.13	−128350.79	−137892.30
Benefit (*B*)	67632.85	67632.85	67632.85	67632.85	67632.85
Benefits Accumulated (BA)	1014492.75	1082125.60	1149758.45	1217391.30	1285024.15
Present Value (BPV)	67632.85	67632.85	67632.85	67632.85	67632.85
Accumulated Benefits BPV	1014492.75	1082125.60	1149758.45	1217391.30	1285024.15
BPV + FPV + MPV + NPPV + IIPV	901227.66	960087.85	1018692.33	1077040.52	1135131.86
MTAC (LDM)	−108471.10	−117724.15	−127232.93	−136997.99	−147019.90
MTAC (SDM)	−111242.64	−120384.83	−129745.79	−139326.08	−149126.27
MTAC (EM)	−113094.62	−121909.37	−130969.45	−140277.98	−149837.47

(*Continued*)

TABLE 7.4 (*Continued*)
Values Used for Simulation based on MTAC Method Related to Physical Assets Standard – Second Simulation

Time	20	21	22	23	24	25
Initial Investment (II)						
Initial Investment (II)						
II (IIPV) - Return	−12000.00	−12000.00	−12000.00	−12000.00	−12000.00	−12000.00
Useful Life						
Residual value of equipment (RV)						
Linear Depreciation Method (Vn)	−2392.00	−1911.60	−1431.20	−950.80	−470.40	10.00
LDM Present Value (PV)	−2392.00	−1911.60	−1431.20	−950.80	−470.40	10.00
Sum of Digits Method (Vn)	−544.31	−359.54	−211.72	−100.86	−26.95	10.00
SDM Present Value (PV)	−544.31	−359.54	−211.72	−100.86	−26.95	10.00
Exponential Method (Vn)	−41.29	−31.09	−23.42	−17.63	−13.28	−10.00
EM Present Value (Pv)	−41.29	−31.09	−23.42	−17.63	−13.28	−10.00
Functioning (*F*)	−5436.49	−5490.86	−5545.76	−5601.22	−5657.23	−5713.81
Functioning Accumulated (FA)	−99085.52	−104576.37	−110122.14	−115723.36	−121380.59	−127094.40
Present Value Accumulated (NPV)	−99085.52	−104576.37	−110122.14	−115723.36	−121380.59	−127094.40
Present Value (NPV)	−5436.49	−5490.86	−5545.76	−5601.22	−5657.23	−5713.81
Energy	−4000.00	−4200.00	−4400.00	−4600.00	−4800.00	−5000.00
Energy Accumulated (EA)	−42000.00	−46200.00	−50600.00	−55200.00	−60000.00	−65000.00
Energy (NPV)	−4000.00	−4200.00	−4400.00	−4600.00	−4800.00	−5000.00
Accumulated Functioning (FPV)	−1061637.32	−1170413.69	−1284935.83	−1405259.18	−1531439.78	−1663534.17
Maintenance (*M*)	−362.43	−366.06	−369.72	−373.41	−377.15	−380.92
Maintenane Accumulated (MA)	−6605.70	−6971.76	−7341.48	−7714.89	−8092.04	−8472.96
Present Value (MPV)	−362.43	−366.06	−369.72	−373.41	−377.15	−380.92
Accumulated MPV	−6605.70	−6971.76	−7341.48	−7714.89	−8092.04	−8472.96
Non Production (NP)	0.00	0.00	0.00	0.00	0.00	0.00
Non Production Accumulated (NPA)	0.00	0.00	0.00	0.00	0.00	0.00
Present Value (NPPV)	0.00	0.00	0.00	0.00	0.00	0.00
Accumulated NPPV	0.00	0.00	0.00	0.00	0.00	0.00
Accumulated Total Costs (APV)	−1068243.02	−1177385.45	−1292277.30	−1412974.07	−1539531.82	−1672007.13
Accumulated Total Costs	−147691.22	−157748.13	−168063.61	−178638.25	−189472.63	−200567.36
Benefit (*B*)	67632.85	67632.85	67632.85	67632.85	67632.85	67632.85
Benefits Accumulated (BA)	1352657.00	1420289.86	1487922.71	1555555.56	1623188.41	1690821.26
Present Value (BPV)	67632.85	67632.85	67632.85	67632.85	67632.85	67632.85
Accumulated Benefits BPV	1352657.00	1420289.86	1487922.71	1555555.56	1623188.41	1690821.26
BPV + FPV + MPV + NPPV + IIPV	1192965.79	1250541.72	1307859.09	1364917.31	1421715.77	1478253.90
MTAC (LDM)	−157299.22	−167836.53	−178632.41	−189687.45	−201002.23	−212577.36
MTAC (SDM)	−159146.91	−169388.59	−179851.89	−190537.39	−201445.68	−212577.36
MTAC (EM)	−159649.93	−169717.04	−180040.20	−190620.62	−201459.35	−212557.36

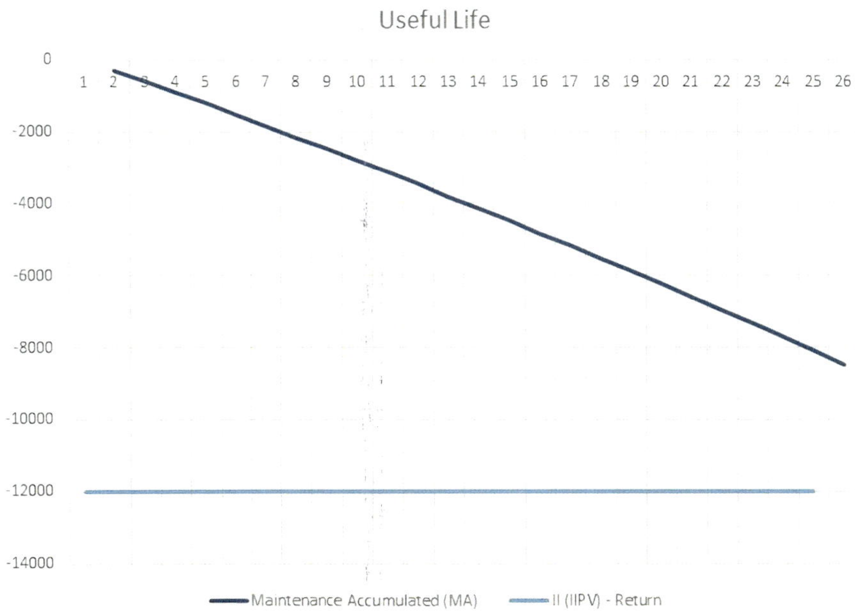

FIGURE 7.14 Physical Asset useful life.

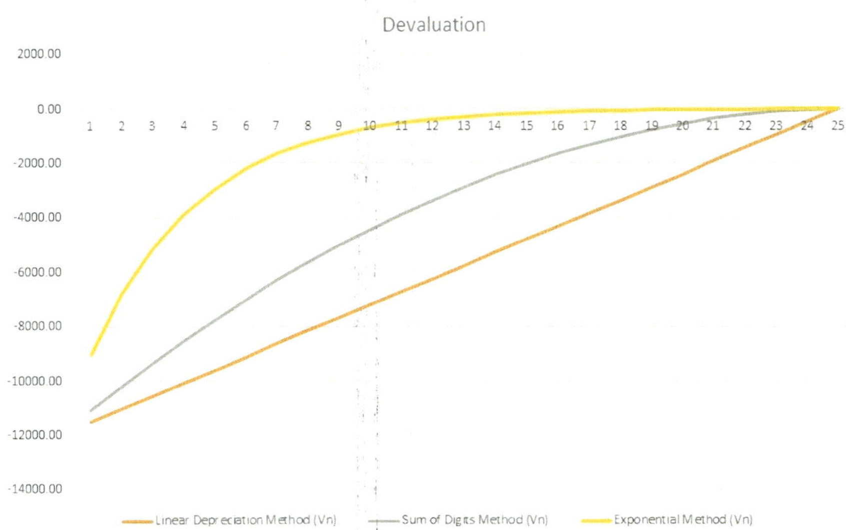

FIGURE 7.15 The three devaluation methods.

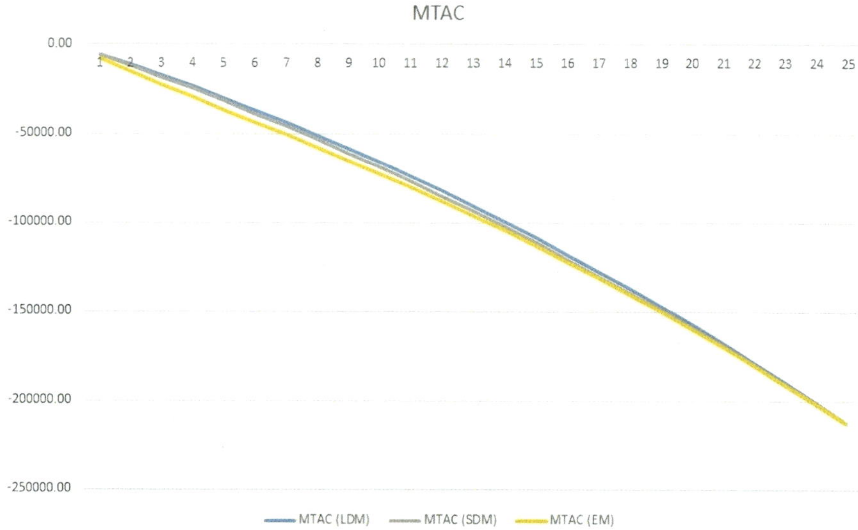

FIGURE 7.16 The MTAC results from the three devaluation methods.

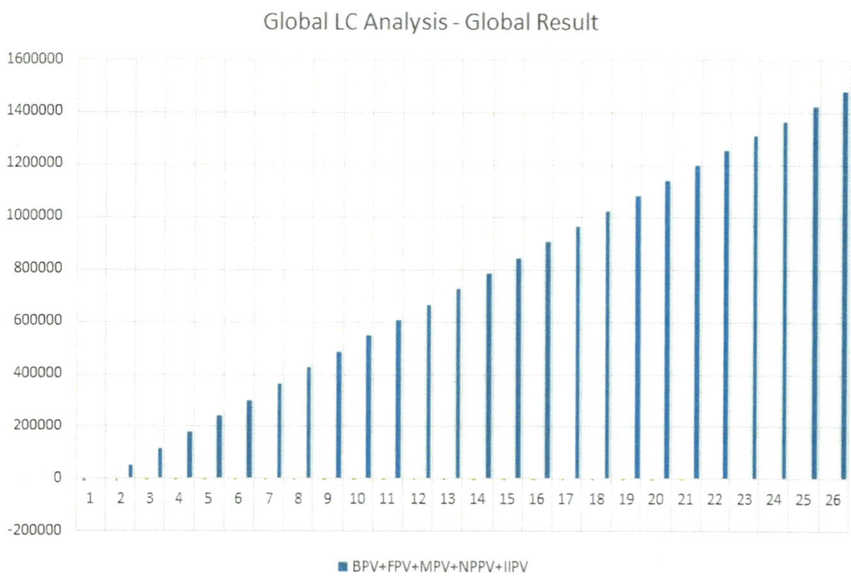

FIGURE 7.17 Life cycle analysis global result including profits.

The remaining values are presented in Table 7.5, and the graphical results are shown in Figures 7.18–7.21.

As referred in the preceding introduction on this section, the evaluation of introduction of Physical Asset Norms, the simulation of their implementation is made through the increase of Benefits. They may be used other indicators, but may be not so intuitive.

TABLE 7.5
Values Used for Simulation based on MMTAC-RPV Method Related to Physical Assets Standard – First Simulation

Time	0	1	2	3	4
Initial Investment (II)	−12000				
Initial Investment (II)	−12000				
II Present Value (IIPV) - Return	−12000	−9652.51	−7764.25	−6245.37	−5023.63
Life Time simulated	25				
Residual value of equipment (RV)	10				
Linear Depreciation Method (Vn)	−12000	−11519.60	−11039.20	−10558.80	−10078.40
LDM Present Value (Vn)		−9266.09	−7142.59	−5495.30	−4219.18
Sum of Digits Method (Vn)	−12000	−11076.15	−10189.26	−9339.32	−8526.34
SDM Present Value (Vn)		−8909.39	−6592.66	−4860.63	−3569.43
Exponential Method (Vn)	−12000	−9036.79	−6805.29	−5124.83	−3859.33
EM Present Value (Pv)		−7268.97	−4403.16	−2667.21	−1615.65
Functioning (*F*)	−4500	−4500.00	−4545.00	−4590.45	−4636.35
Present Value Acummulated (NPV)		−3619.69	−6560.40	−8949.49	−10890.43
Present Value (NPV)		−3619.69	−2940.71	−2389.09	−1940.94
Energy (*E*)	−200	−200.00	−400.00	−600.00	−800.00
Energy (NPV)		−160.88	−258.81	−312.27	−334.91
Accumulated FPV		−15780.57	−22599.77	−31861.53	−43086.87
Maintenance (*M*)	−300	−300.00	−303.00	−306.03	−309.09
Present Value (MPV)		−241.31	−196.05	−159.27	−129.40
Accumulated (MPV)		−241.31	−437.36	−596.63	−726.03
Non Production (NP)	0	0.00	0.00	0.00	0.00
Present Value (NPPV)		0.00	0.00	0.00	0.00
Accumulated NPPV		0.00	0.00	0.00	0.00
Accumulated Total Costs (APV)		−16021.88	−23037.13	−32458.16	−43812.90
Accumulated Total Costs		−5000.00	−10248.00	−15744.48	−21489.92
Benefit (*B*)	50000	44642.86	44642.86	44642.86	44642.86
Present Value (BPV)		35909.63	28884.84	23234.27	18689.08
Accumulated Benefits BPV		35909.63	64794.47	88028.74	106717.82
BPV + FPV + MPV + NPPV + IIPV		7887.76	29757.34	43570.58	50904.93
MTAC-PV (LDM)		−37287.97	−21089.86	−16651.15	−15008.02
MTAC-PV (SDM)		−36931.27	−20814.90	−16439.60	−14845.58
MTAC-PV (EM)		−35290.85	−19720.15	−15708.46	−14357.14

(Continued)

TABLE 7.5 (*Continued*)
Values Used for Simulation based on MMTAC-RPV Method Related to Physical Assets Standard – First Simulation

Time	5	6	7	8	9
Initial Investment (II)					
Initial Investment (II)					
II Present Value (IIPV) - Return	−4040.88	−3250.39	−2614.53	−2103.07	−1691.66
Life Time simulated					
Residual value of equipment (RV)					
Linear Depreciation Method (Vn)	−9598.00	−9117.60	−8637.20	−8156.80	−7676.40
LDM Present Value (Vn)	−3232.03	−2469.64	−1881.85	−1429.53	−1082.15
Sum of Digits Method (Vn)	−7750.31	−7011.23	−6309.11	−5643.94	−5015.72
SDM Present Value (Vn)	−2609.84	−1899.10	−1374.61	−989.13	−707.07
Exponential Method (Vn)	−2906.33	−2188.66	−1648.20	−1241.20	−934.71
EM Present Value (Pv)	−978.68	−592.83	−359.11	−217.53	−131.77
Functioning (*F*)	−4682.72	−4729.55	−4776.84	−4824.61	−4872.86
Present Value Acummulated (NPV)	−12467.29	−13748.36	−14789.13	−15634.67	−16321.60
Present Value (NPV)	−1576.86	−1281.07	−1040.77	−845.54	−686.93
Energy (*E*)	−1000.00	−1200.00	−1400.00	−1600.00	−1800.00
Energy (NPV)	−336.74	−325.04	−305.03	−280.41	−253.75
Accumulated FPV	−55890.90	−69964.30	−85058.45	−100973.53	−117548.88
Maintenance (*M*)	−312.18	−315.30	−318.46	−321.64	−324.86
Present Value (MPV)	−105.12	−85.40	−69.38	−56.37	−45.80
Accumulated (MPV)	−831.15	−916.56	−985.94	−1042.31	−1088.11
Non Production (NP)	0.00	0.00	0.00	0.00	0.00
Present Value (NPPV)	0.00	0.00	0.00	0.00	0.00
Accumulated NPPV	0.00	0.00	0.00	0.00	0.00
Accumulated Total Costs (APV)	−56722.05	−70880.85	−86044.40	−102015.84	−118636.99
Accumulated Total Costs	−27484.82	−33729.67	−40224.97	−46971.22	−53968.93
Benefit (*B*)	44642.86	44642.86	44642.86	44642.86	44642.86
Present Value (BPV)	15033.05	12092.22	9726.69	7823.91	6293.37
Accumulated Benefits BPV	121750.87	133843.09	143569.78	151393.69	157687.05
BPV+FPV+MPV+NPPV+IIPV	53028.82	50962.23	45525.38	37377.84	27050.06
MTAC-PV (LDM)	−14390.82	−14225.08	−14275.18	−14430.67	−14635.46
MTAC-PV (SDM)	−14266.38	−14129.99	−14202.72	−14375.62	−14593.78
MTAC-PV (EM)	−13940.15	−13912.28	−14057.64	−14279.17	−14529.86

(*Continued*)

TABLE 7.5 (*Continued*)
Values Used for Simulation based on MMTAC-RPV Method Related to Physical Assets Standard – First Simulation

Time	10	11	12	13	14
Initial Investment (II)					
Initial Investment (II)					
II Present Value (IIPV) - Return	−1360.73	−1094.54	−880.42	−708.19	−569.65
Life Time simulated					
Residual value of equipment (RV)					
Linear Depreciation Method (Vn)	−7196.00	−6715.60	−6235.20	−5754.80	−5274.40
LDM Present Value (Vn)	−815.98	−612.54	−457.47	−339.62	−250.38
Sum of Digits Method (Vn)	−4424.46	−3870.15	−3352.80	−2872.40	−2428.95
SDM Present Value (Vn)	−501.71	−353.00	−245.99	−169.52	−115.30
Exponential Method (Vn)	−703.90	−530.08	−399.19	−300.61	−226.38
EM Present Value (Pv)	−79.82	−48.35	−29.29	−17.74	−10.75
Functioning (*F*)	−4921.58	−4970.80	−5020.51	−5070.71	−5121.42
Present Value Acummulated (NPV)	−16879.68	−17333.07	−17701.42	−18000.67	−18243.79
Present Value (NPV)	−558.08	−453.39	−368.35	−299.25	−243.12
Energy (*E*)	−2000.00	−2200.00	−2400.00	−2600.00	−2800.00
Energy (NPV)	−226.79	−200.67	−176.08	−153.44	−132.92
Accumulated FPV	−134655.35	−152189.09	−170066.59	−188220.70	−206597.41
Maintenance (*M*)	−328.11	−331.39	−334.70	−338.05	−341.43
Present Value (MPV)	−37.21	−30.23	−24.56	−19.95	−16.21
Accumulated (MPV)	−1125.31	−1155.54	−1180.09	−1200.04	−1216.25
Non Production (NP)	0.00	0.00	0.00	0.00	0.00
Present Value (NPPV)	0.00	0.00	0.00	0.00	0.00
Accumulated NPPV	0.00	0.00	0.00	0.00	0.00
Accumulated Total Costs (APV)	−135780.66	−153344.63	−171246.68	−189420.75	−207813.66
Accumulated Total Costs	−61218.62	−68720.81	−76476.01	−84484.77	−92747.62
Benefit (*B*)	44642.86	44642.86	44642.86	44642.86	44642.86
Present Value (BPV)	5062.23	4071.94	3275.37	2634.63	2119.23
Accumulated Benefits BPV	162749.28	166821.22	170096.59	172731.21	174850.44
BPV + FPV + MPV + NPPV + IIPV	14968.62	1476.59	−13150.10	−28689.53	−44963.22
MTAC-PV (LDM)	−14859.66	−15087.02	−15308.68	−15520.03	−15718.86
MTAC-PV (SDM)	−14828.24	−15063.42	−15291.06	−15506.94	−15709.21
MTAC-PV (EM)	−14786.05	−15035.73	−15273.00	−15495.27	−15701.74

(*Continued*)

TABLE 7.5 (*Continued*)
Values Used for Simulation based on MMTAC-RPV Method Related to Physical Assets Standard – First Simulation

Time	15	16	17	18	19
Initial Investment (II)					
Initial Investment (II)					
II Present Value (IIPV) - Return	−458.21	−368.57	−296.47	−238.48	−191.82
Life Time simulated					
Residual value of equipment (RV)					
Linear Depreciation Method (Vn)	−4794.00	−4313.60	−3833.20	−3352.80	−2872.40
LDM Present Value (Vn)	−183.06	−132.49	−94.70	−66.63	−45.92
Sum of Digits Method (Vn)	−2022.46	−1652.92	−1320.34	−1024.71	−766.03
SDM Present Value (Vn)	−77.23	−50.77	−32.62	−20.36	−12.25
Exponential Method (Vn)	−170.48	−128.38	−96.68	−72.81	−54.83
EM Present Value (Pv)	−6.51	−3.94	−2.39	−1.45	−0.88
Functioning (*F*)	−5172.63	−5224.36	−5276.60	−5329.37	−5382.66
Present Value Acummulated (NPV)	−18441.30	−18601.76	−18732.13	−18838.04	−18924.08
Present Value (NPV)	−197.51	−160.46	−130.36	−105.91	−86.04
Energy (*E*)	−3000.00	−3200.00	−3400.00	−3600.00	−3800.00
Energy (NPV)	−114.55	−98.29	−84.00	−71.54	−60.74
Accumulated FPV	−225153.26	−243853.31	−262669.44	−281579.02	−300563.85
Maintenance (*M*)	−344.84	−348.29	−351.77	−355.29	−358.84
Present Value (MPV)	−13.17	−10.70	−8.69	−7.06	−5.74
Accumulated (MPV)	−1229.42	−1240.12	−1248.81	−1255.87	−1261.61
Non Production (NP)	0.00	0.00	0.00	0.00	0.00
Present Value (NPPV)	0.00	0.00	0.00	0.00	0.00
Accumulated NPPV	0.00	0.00	0.00	0.00	0.00
Accumulated Total Costs (APV)	−226382.68	−245093.43	−263918.25	−282834.89	−301825.45
Accumulated Total Costs	−101265.10	−110037.75	−119066.13	−128350.79	−137892.30
Benefit (*B*)	44642.86	44642.86	44642.86	44642.86	44642.86
Present Value (BPV)	1704.66	1371.18	1102.95	887.18	713.63
Accumulated Benefits BPV	176555.10	177926.28	179029.23	179916.42	180630.05
BPV+FPV+MPV+NPPV+IIPV	−61827.58	−79167.14	−96889.02	−114918.47	−133195.41
MTAC-PV (LDM)	−15904.38	−16076.62	−16236.06	−16383.42	−16519.55
MTAC-PV (SDM)	−15897.33	−16071.51	−16232.40	−16380.85	−16517.77
MTAC-PV (EM)	−15892.61	−16068.59	−16230.63	−16379.80	−16517.18

(*Continued*)

TABLE 7.5 (*Continued*)
Values Used for Simulation based on MMTAC-RPV Method Related to Physical Assets Standard – First Simulation

Time	20	21	22	23	24	25
Initial Investment (II)						
Initial Investment (II)						
II Present Value (IIPV) - Return	-154.30	-124.11	-99.83	-80.30	-64.59	-51.96
Life Time simulated						
Residual value of equipment (RV)						
Linear Depreciation Method (Vn)	-2392.00	-1911.60	-1431.20	-950.80	-470.40	10.00
LDM Present Value (Vn)	-30.76	-19.77	-11.91	-6.36	-2.53	0.04
Sum of Digits Method (Vn)	-544.31	-359.54	-211.72	-100.86	-26.95	10.00
SDM Present Value (Vn)	-7.00	-3.72	-1.76	-0.67	-0.15	0.04
Exponential Method (Vn)	-41.29	-31.09	-23.42	-17.63	-13.28	-10.00
EM Present Value (Pv)	-0.53	-0.32	-0.19	-0.12	-0.07	-0.04
Functioning (*F*)	-5436.49	-5490.86	-5545.76	-5601.22	-5657.23	-5713.81
Present Value Acummulated (NPV)	-18993.99	-19050.78	-19096.91	-19134.40	-19164.85	-19189.59
Present Value (NPV)	-69.90	-56.79	-46.14	-37.48	-30.45	-24.74
Energy (*E*)	-4000.00	-4200.00	-4400.00	-4600.00	-4800.00	-5000.00
Energy (NPV)	-51.43	-43.44	-36.61	-30.78	-25.84	-21.65
Accumulated FPV	-319609.27	-338703.48	-357837.01	-377002.19	-396192.88	-415404.12
Maintenance (*M*)	-362.43	-366.06	-369.72	-373.41	-377.15	-380.92
Present Value (MPV)	-4.66	-3.79	-3.08	-2.50	-2.03	-1.65
Accumulated (MPV)	-1266.27	-1270.05	-1273.13	-1275.63	-1277.66	-1279.31
Non Production (NP)	0.00	0.00	0.00	0.00	0.00	0.00
Present Value (NPPV)	0.00	0.00	0.00	0.00	0.00	0.00
Accumulated NPPV	0.00	0.00	0.00	0.00	0.00	0.00
Accumulated Total Costs (APV)	-320875.53	-339973.54	-359110.13	-378277.81	-397470.53	-416683.42
Accumulated Total Costs	-147691.22	-157748.13	-168063.61	-178638.25	-189472.63	-200567.36
Benefit (*B*)	44642.86	44642.86	44642.86	44642.86	44642.86	44642.86
Present Value (BPV)	574.03	461.73	371.41	298.75	240.31	193.30
Accumulated Benefits BPV	181204.07	181665.81	182037.21	182335.96	182576.27	182769.57
BPV+FPV+MPV+ NPPV+IIPV	-151671.46	-170307.73	-189072.92	-207941.85	-226894.26	-245913.85
MTAC-PV (LDM)	-16645.31	-16761.59	-16869.18	-16968.88	-17061.38	-17147.34
MTAC-PV (SDM)	-16644.13	-16760.82	-16868.72	-16968.63	-17061.28	-17147.34
MTAC-PV (EM)	-16643.80	-16760.66	-16868.65	-16968.61	-17061.28	-17147.34

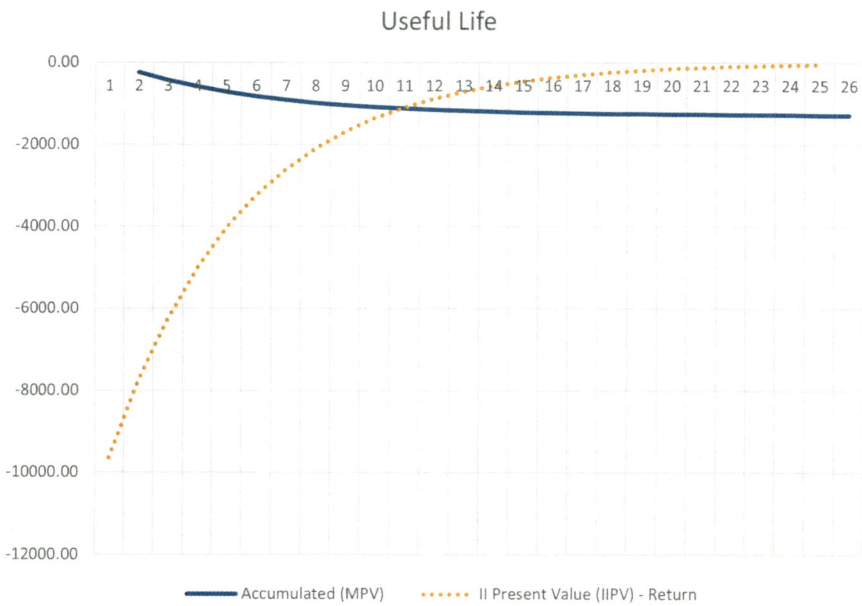

FIGURE 7.18 Physical Asset useful life.

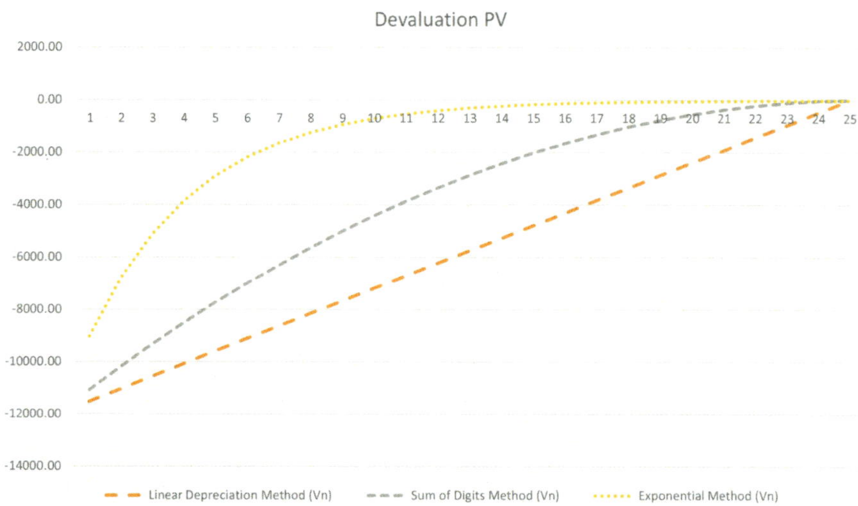

FIGURE 7.19 The three evaluation methods.

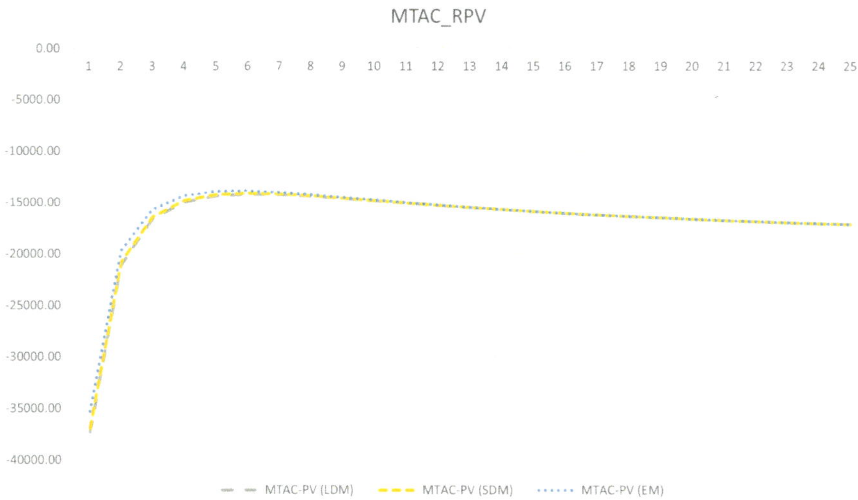

FIGURE 7.20 The MTAC-RPV results from the three devaluation methods.

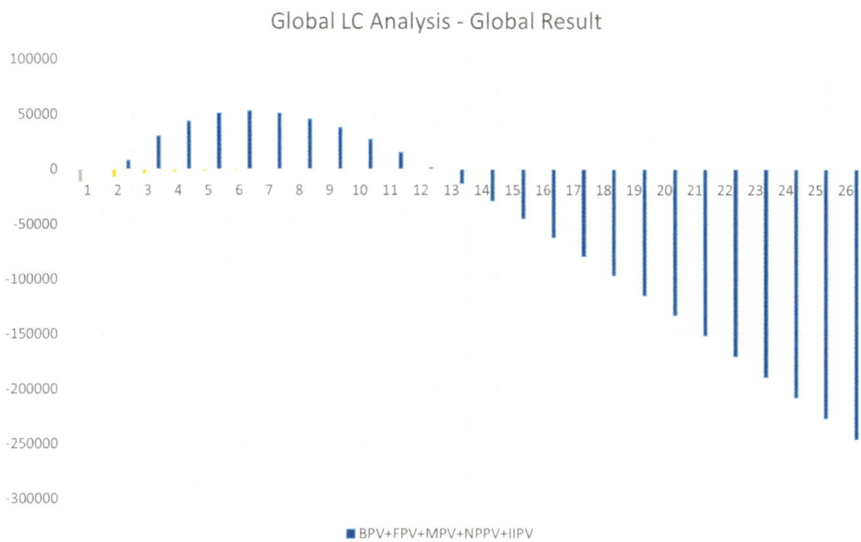

FIGURE 7.21 Life cycle analysis global result including profits.

The second simulation uses the following values (the MTBF is increased and MTTR decreases as a result of the Maintenance Standards):

- Initial Investment (II) –12000
- Life Time simulated 25
- Residual value of equipment (RV) 10
- Functioning (*F*) –4500
- Energy (*E*) –200
- Maintenance (*M*) –300
- Non Production (NP) 0
- Benefits (*B*) **70000**
- Internal Rate of Return (*r*) 0.12
- Risk (*R*) 0.11
- Apparent Rate 0.2432
- Maintenance Increase Rate 0.01
- Availability 0.8929
- MTBF 1000
- MTTR 25
- MWT 10

The remaining values are presented in Table 7.6, and the graphical results are shown in Figures 7.22–7.25.

As can be seen, namely comparing Figures 7.21 and 7.25, with the increasing of the maintenance management quality helped by the maintenance standards, it is get four more positive years, but the positive incomes increase a lot.

These results from the three methods are close, what reinforce a lot the robustness of the models under discussion.

7.4.2 Life Cycle Investment of Physical Assets

When a company decides to implement a new Standard, it is necessary to evaluate which is the expected result, quantitatively supported.

The way to demonstrate if the return of implementing the Standard under discussion is interesting or not, in this section it is done using the model of Life Cycle Investment, presented in Chapter 2.

TABLE 7.6
Values Used for Simulation based on MMTAC-RPV Method Related to Physical Assets Standard – Second Simulation

Time	0	1	2	3	4	5
Initial Investment (II)	−12000					
Initial Investment (II)	−12000					
II Present Value (IIPV) - Return	−12000	−9652.51	−7764.25	−6245.37	−5023.63	−4040.88
Life Time simulated	25					
Residual value of equipment (RV)	10					
Linear Depreciation Method (Vn)	−12000	−11519.60	−11039.20	−10558.80	−10078.40	−9598.00
LDM Present Value (Vn)		−9266.09	−7142.59	−5495.30	−4219.18	−3232.03
Sum of Digits Method (Vn)	−12000	−11076.15	−10189.26	−9339.32	−8526.34	−7750.31
SDM Present Value (Vn)		−8909.39	−6592.66	−4860.63	−3569.43	−2609.84
Exponential Method (Vn)	−12000	−9036.79	−6805.29	−5124.83	−3859.33	−2906.33
EM Present Value (Pv)		−7268.97	−4403.16	−2667.21	−1615.65	−978.68
Functioning (*F*)	−4500	−4500.00	−4545.00	−4590.45	−4636.35	−4682.72
Present Value Acummulated (NPV)		−3619.69	−6560.40	−8949.49	−10890.43	−12467.29
Present Value (NPV)		−3619.69	−2940.71	−2389.09	−1940.94	−1576.86
Energy (*E*)	−200	−200.00	−400.00	−600.00	−800.00	−1000.00
Energy (NPV)		−160.88	−258.81	−312.27	−334.91	−336.74
Accumulated FPV		−15780.57	−22599.77	−31861.53	−43086.87	−55890.90
Maintenance (*M*)	−300	−300.00	−303.00	−306.03	−309.09	−312.18
Present Value (MPV)		−241.31	−196.05	−159.27	−129.40	−105.12
Accumulated (MPV)		−241.31	−437.36	−596.63	−726.03	−831.15
Non Production (NP)	0	0.00	0.00	0.00	0.00	0.00
Present Value (NPPV)		0.00	0.00	0.00	0.00	0.00
Accumulated NPPV		0.00	0.00	0.00	0.00	0.00
Accumulated Total Costs (APV)		−16021.88	−23037.13	−32458.16	−43812.90	−56722.05
Accumulated Total Costs		−5000.00	−10248.00	−15744.48	−21489.92	−27484.82
Benefit (*B*)	70000	62500.00	62500.00	62500.00	62500.00	62500.00
Present Value (BPV)		50273.49	40438.78	32527.97	26164.72	21046.26
Accumulated Benefits BPV		50273.49	90712.26	123240.24	149404.95	170451.22
BPV+FPV+MPV+NPPV+IIPV		22251.61	55675.13	78782.08	93592.06	101729.17
MTAC-PV (LDM)		−37287.97	−21089.86	−16651.15	−15008.02	−14390.82
MTAC-PV (SDM)		−36931.27	−20814.90	−16439.60	−14845.58	−14266.38
MTAC-PV (EM)		−35290.85	−19720.15	−15708.46	−14357.14	−13940.15

(Continued)

TABLE 7.6 (*Continued*)
Values Used for Simulation based on MMTAC-RPV Method Related to Physical Assets Standard – Second Simulation

Time	6	7	8	9	10	11
Initial Investment (II)						
Initial Investment (II)						
II Present Value (IIPV) - Return	−3250.39	−2614.53	−2103.07	−1691.66	−1360.73	−1094.54
Life Time simulated						
Residual value of equipment (RV)						
Linear Depreciation Method (Vn)	−9117.60	−8637.20	−8156.80	−7676.40	−7196.00	−6715.60
LDM Present Value (Vn)	−2469.64	−1881.85	−1429.53	−1082.15	−815.98	−612.54
Sum of Digits Method (Vn)	−7011.23	−6309.11	−5643.94	−5015.72	−4424.46	−3870.15
SDM Present Value (Vn)	−1899.10	−1374.61	−989.13	−707.07	−501.71	−353.00
Exponential Method (Vn)	−2188.66	−1648.20	−1241.20	−934.71	−703.90	−530.08
EM Present Value (Pv)	−592.83	−359.11	−217.53	−131.77	−79.82	−48.35
Functioning (*F*)	−4729.55	−4776.84	−4824.61	−4872.86	−4921.58	−4970.80
Present Value Acummulated (NPV)	−13748.36	−14789.13	−15634.67	−16321.60	−16879.68	−17333.07
Present Value (NPV)	−1281.07	−1040.77	−845.54	−686.93	−558.08	−453.39
Energy (*E*)	−1200.00	−1400.00	−1600.00	−1800.00	−2000.00	−2200.00
Energy (NPV)	−325.04	−305.03	−280.41	−253.75	−226.79	−200.67
Accumulated FPV	−69964.30	−85058.45	-100973.53	−117548.88	-134655.35	−152189.09
Maintenance (*M*)	−315.30	−318.46	−321.64	−324.86	−328.11	−331.39
Present Value (MPV)	−85.40	−69.38	−56.37	−45.80	−37.21	−30.23
Accumulated (MPV)	−916.56	−985.94	−1042.31	−1088.11	−1125.31	−1155.54
Non Production (NP)	0.00	0.00	0.00	0.00	0.00	0.00
Present Value (NPPV)	0.00	0.00	0.00	0.00	0.00	0.00
Accumulated NPPV	0.00	0.00	0.00	0.00	0.00	0.00
Accumulated Total Costs (APV)	−70880.85	−86044.40	−102015.84	−118636.99	−135780.66	−153344.63
Accumulated Total Costs	−33729.67	−40224.97	−46971.22	−53968.93	−61218.62	−68720.81
Benefit (*B*)	62500.00	62500.00	62500.00	62500.00	62500.00	62500.00
Present Value (BPV)	16929.11	13617.36	10953.48	8810.71	7087.12	5700.71
Accumulated Benefits BPV	187380.32	200997.69	211951.16	220761.88	227849.00	233549.71
BPV + FPV + MPV + NPPV + IIPV	104499.47	102953.29	97935.32	90124.89	80068.34	68205.08
MTAC-PV (LDM)	−14225.08	−14275.18	−14430.67	−14635.46	−14859.66	−15087.02
MTAC-PV (SDM)	−14129.99	−14202.72	−14375.62	−14593.78	−14828.24	−15063.42
MTAC-PV (EM)	−13912.28	−14057.64	−14279.17	−14529.86	−14786.05	−15035.73

(*Continued*)

TABLE 7.6 (*Continued*)
Values Used for Simulation based on MMTAC-RPV Method Related to Physical Assets Standard – Second Simulation

Time	12	13	14	15	16
Initial Investment (II)					
Initial Investment (II)					
II Present Value (IIPV) - Return	−880.42	−708.19	−569.65	−458.21	−368.57
Life Time simulated					
Residual value of equipment (RV)					
Linear Depreciation Method (Vn)	−6235.20	−5754.80	−5274.40	−4794.00	−4313.60
LDM Present Value (Vn)	−457.47	−339.62	−250.38	−183.06	−132.49
Sum of Digits Method (Vn)	−3352.80	−2872.40	−2428.95	−2022.46	−1652.92
SDM Present Value (Vn)	−245.99	−169.52	−115.30	−77.23	−50.77
Exponential Method (Vn)	−399.19	−300.61	−226.38	−170.48	−128.38
EM Present Value (Pv)	−29.29	−17.74	−10.75	−6.51	−3.94
Functioning (*F*)	−5020.51	−5070.71	−5121.42	−5172.63	−5224.36
Present Value Acummulated (NPV)	−17701.42	−18000.67	−18243.79	−18441.30	−18601.76
Present Value (NPV)	−368.35	−299.25	−243.12	−197.51	−160.46
Energy (*E*)	−2400.00	−2600.00	−2800.00	−3000.00	−3200.00
Energy (NPV)	−176.08	−153.44	−132.92	−114.55	−98.29
Accumulated FPV	−170066.59	−188220.70	−206597.41	−225153.26	−243853.31
Maintenance (*M*)	−334.70	−338.05	−341.43	−344.84	−348.29
Present Value (MPV)	−24.56	−19.95	−16.21	−13.17	−10.70
Accumulated (MPV)	−1180.09	−1200.04	−1216.25	−1229.42	−1240.12
Non Production (NP)	0.00	0.00	0.00	0.00	0.00
Present Value (NPPV)	0.00	0.00	0.00	0.00	0.00
Accumulated NPPV	0.00	0.00	0.00	0.00	0.00
Accumulated Total Costs (APV)	−171246.68	−189420.75	−207813.66	−226382.68	−245093.43
Accumulated Total Costs	−76476.01	−84484.77	−92747.62	−101265.10	−110037.75
Benefit (*B*)	62500.00	62500.00	62500.00	62500.00	62500.00
Present Value (BPV)	4585.51	3688.48	2966.92	2386.52	1919.66
Accumulated Benefits BPV	238135.22	241823.70	244790.62	247177.14	249096.80
BPV + FPV + MPV + NPPV + IIPV	54888.54	40402.95	24976.96	8794.46	−7996.63
MTAC-PV (LDM)	−15308.68	−15520.03	−15718.86	−15904.38	−16076.62
MTAC-PV (SDM)	−15291.06	−15506.94	−15709.21	−15897.33	−16071.51
MTAC-PV (EM)	−15273.00	−15495.27	−15701.74	−15892.61	−16068.59

(*Continued*)

TABLE 7.6 (*Continued*)
Values Used for Simulation based on MMTAC-RPV Method Related to Physical Assets Standard – Second Simulation

Time	17	18	19	20	21
Initial Investment (II)					
Initial Investment (II)					
II Present Value (IIPV) - Return	−296.47	−238.48	−191.82	−154.30	−124.11
Life Time simulated					
Residual value of equipment (RV)					
Linear Depreciation Method (Vn)	−3833.20	−3352.80	−2872.40	−2392.00	−1911.60
LDM Present Value (Vn)	−94.70	−66.63	−45.92	−30.76	−19.77
Sum of Digits Method (Vn)	−1320.34	−1024.71	−766.03	−544.31	−359.54
SDM Present Value (Vn)	−32.62	−20.36	−12.25	−7.00	−3.72
Exponential Method (Vn)	−96.68	−72.81	−54.83	−41.29	−31.09
EM Present Value (Pv)	−2.39	−1.45	−0.88	−0.53	−0.32
Functioning (*F*)	−5276.60	−5329.37	−5382.66	−5436.49	−5490.86
Present Value Acummulated (NPV)	−18732.13	−18838.04	−18924.08	−18993.99	−19050.78
Present Value (NPV)	−130.36	−105.91	−86.04	−69.90	−56.79
Energy (*E*)	−3400.00	−3600.00	−3800.00	−4000.00	−4200.00
Energy (NPV)	−84.00	−71.54	−60.74	−51.43	−43.44
Accumulated FPV	−262669.44	−281579.02	−300563.85	−319609.27	−338703.48
Maintenance (*M*)	−351.77	−355.29	−358.84	−362.43	−366.06
Present Value (MPV)	−8.69	−7.06	−5.74	−4.66	−3.79
Accumulated (MPV)	−1248.81	−1255.87	−1261.61	−1266.27	−1270.05
Non Production (NP)	0.00	0.00	0.00	0.00	0.00
Present Value (NPPV)	0.00	0.00	0.00	0.00	0.00
Accumulated NPPV	0.00	0.00	0.00	0.00	0.00
Accumulated Total Costs (APV)	−263918.25	−282834.89	−301825.45	−320875.53	−339973.54
Accumulated Total Costs	−119066.13	−128350.79	−137892.30	−147691.22	−157748.13
Benefit (*B*)	62500.00	62500.00	62500.00	62500.00	62500.00
Present Value (BPV)	1544.13	1242.06	999.08	803.64	646.43
Accumulated Benefits BPV	250640.93	251882.98	252882.07	253685.70	254332.13
BPV + FPV + MPV + NPPV + IIPV	−25277.32	−42951.91	−60943.39	−79189.83	−97641.41
MTAC-PV (LDM)	−16236.06	−16383.42	−16519.55	−16645.31	−16761.59
MTAC-PV (SDM)	−16232.40	−16380.85	−16517.77	−16644.13	−16760.82
MTAC-PV (EM)	−16230.63	−16379.80	−16517.18	−16643.80	−16760.66

(*Continued*)

TABLE 7.6 (*Continued*)
Values Used for Simulation based on MMTAC-RPV Method Related to Physical Assets Standard – Second Simulation

Time	22	23	24	25
Initial Investment (II)				
Initial Investment (II)				
II Present Value (IIPV) - Return	−99.83	−80.30	−64.59	−51.96
Life Time simulated				
Residual value of equipment (RV)				
Linear Depreciation Method (Vn)	−1431.20	−950.80	−470.40	10.00
LDM Present Value (Vn)	−11.91	−6.36	−2.53	0.04
Sum of Digits Method (Vn)	−211.72	−100.86	−26.95	10.00
SDM Present Value (Vn)	−1.76	−0.67	−0.15	0.04
Exponential Method (Vn)	−23.42	−17.63	−13.28	−10.00
EM Present Value (Pv)	−0.19	−0.12	−0.07	−0.04
Functioning (*F*)	−5545.76	−5601.22	−5657.23	−5713.81
Present Value Acummulated (NPV)	−19096.91	−19134.40	−19164.85	−19189.59
Present Value (NPV)	−46.14	−37.48	−30.45	−24.74
Energy (*E*)	−4400.00	−4600.00	−4800.00	−5000.00
Energy (NPV)	−36.61	−30.78	−25.84	−21.65
Accumulated FPV	−357837.01	−377002.19	−396192.88	−415404.12
Maintenance (*M*)	−369.72	−373.41	−377.15	−380.92
Present Value (MPV)	−3.08	−2.50	−2.03	−1.65
Accumulated (MPV)	−1273.13	−1275.63	−1277.66	−1279.31
Non Production (NP)	0.00	0.00	0.00	0.00
Present Value (NPPV)	0.00	0.00	0.00	0.00
Accumulated NPPV	0.00	0.00	0.00	0.00
Accumulated Total Costs (APV)	−359110.13	−378277.81	−397470.53	−416683.42
Accumulated Total Costs	−168063.61	−178638.25	−189472.63	−200567.36
Benefit (*B*)	62500.00	62500.00	62500.00	62500.00
Present Value (BPV)	519.97	418.25	336.43	270.62
Accumulated Benefits BPV	254852.10	255270.35	255606.78	255877.40
BPV + FPV + MPV + NPPV + IIPV	−116258.03	−135007.46	−153863.75	−172806.02
MTAC-PV (LDM)	−16869.18	−16968.88	−17061.38	−17147.34
MTAC-PV (SDM)	−16868.72	−16968.63	−17061.28	−17147.34
MTAC-PV (EM)	−16868.65	−16968.61	−17061.28	−17147.34

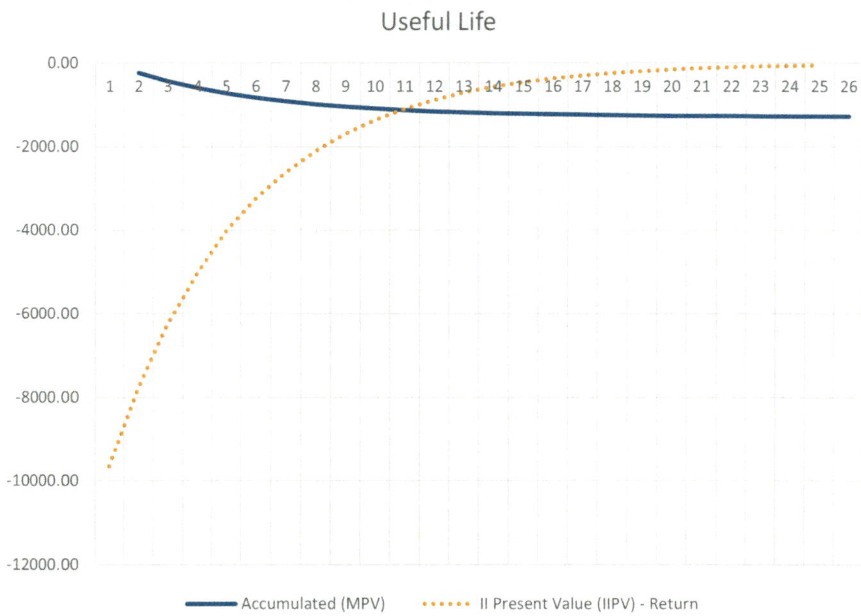

FIGURE 7.22 Physical Asset useful life.

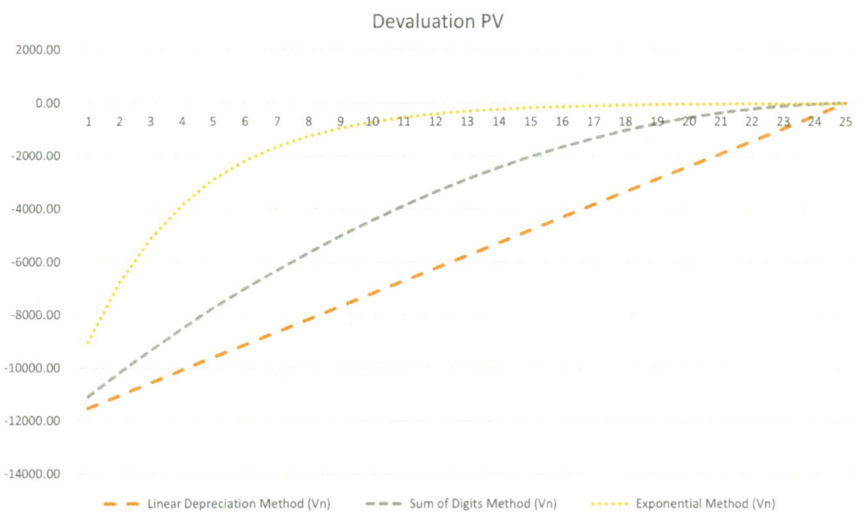

FIGURE 7.23 The three devaluation methods.

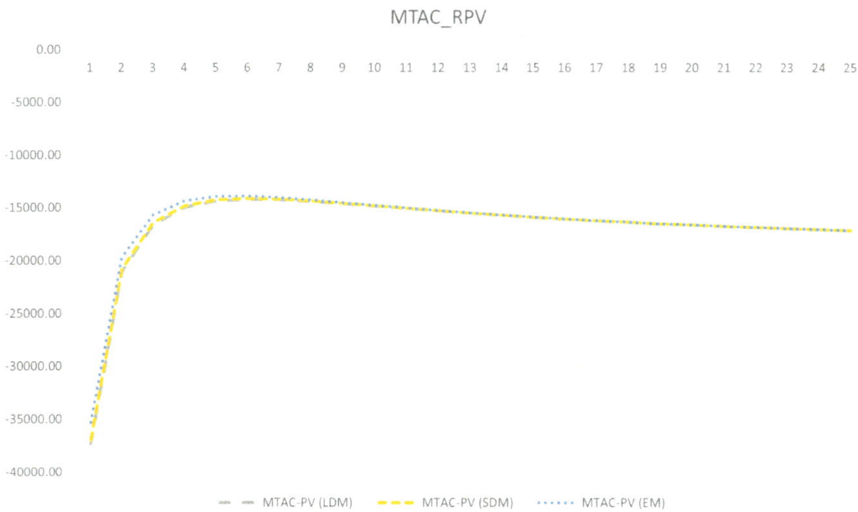

FIGURE 7.24 The MTAC results from the three devaluation methods.

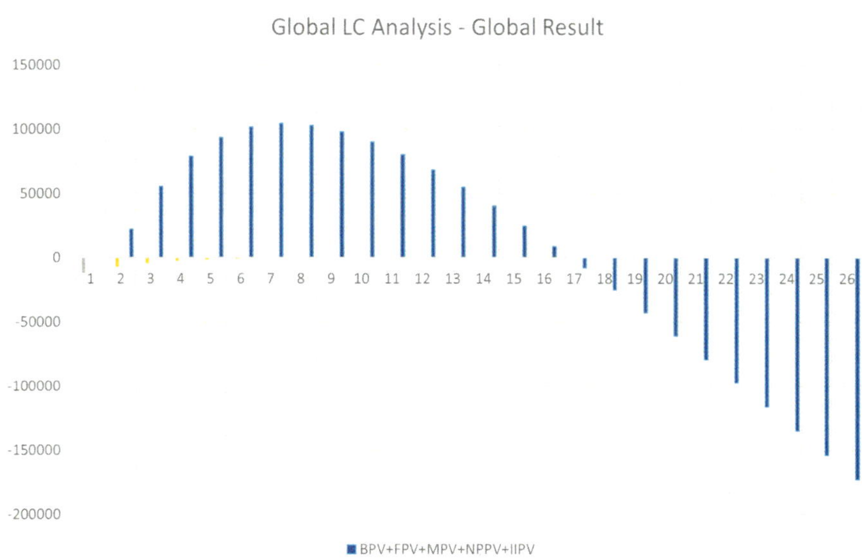

FIGURE 7.25 Life cycle analysis global result including profits.

The first simulation uses the following values:

- Initial Investment (II) –12000
- Life Time simulated 25
- Residual value of equipment (RV) 10
- Functioning (*F*) –4500
- Energy (*E*) –200
- Maintenance (*M*) –300
- Non Production (NP) 0
- Benefits (*B*) **50000**
- Internal Rate of Return (*r*) 0.12
- Risk (*B*) 0.11
- Apparent Rate 0.2432
- Maintenance Increase Rate 0.01
- Availability 0.8929
- MTBF 500
- MTTR 50
- MWT 10

The remaining values are presented in Table 7.7, and the graphical results are shown in Figures 7.26–7.28.

As referred in the preceding introduction to this section, the evaluation of introduction of Physical Asset Norms, the simulation of their implementation is made through the increase of benefits. They may be used other indicators, but may be not so intuitive.

The second simulation uses the following values (the MTBF is increased and MTTR decreases as a result of the Maintenance Standards implementation):

- Initial Investment (II) –12000
- Life Time simulated 25
- Residual value of equipment (RV) 10
- Functioning (*F*) –4500
- Energy (*E*) –200
- Maintenance (*M*) –300
- Non Production (NP) 0
- Benefit (*B*) **70000**
- Internal Rate of Return (*r*) 0.12
- Risk (*R*) 0.11
- Apparent Rate 0.2432
- Maintenance Increase Rate 0.01
- Availability 0.8929
- MTBF 1000
- MTTR 25
- MWT 10

TABLE 7.7
Values Used for Simulation based on Life Cycle Investment Related to Physical Assets Standard – First Simulation

Time	0	1	2	3	4	5	6	7	8
Initial Investment (II)	−12000								
II Present Value (IIPV) - Return	−12000	−9652.51	−7764.25	−6245.37	−5023.63	−4040.88	−3250.39	−2614.53	−2103.07
Functioning (*F*)	−4500	−4500.00	−4500.00	−4500.00	−4500.00	−4500.00	−4500.00	−4500.00	−4500.00
Present Value (PV) Accumulated		−3619.69	−6531.28	−8873.30	−10757.16	−12272.49	−13491.38	−14471.83	−15260.48
Energy (*E*)	−200	−200.00	−200.00	−200.00	−200.00	−200.00	−200.00	−200.00	−200.00
Energy (PV)		−160.88	−129.40	−104.09	−83.73	−67.35	−54.17	−43.58	−35.05
Accumulated (AFPV)		−15780.57	−6660.69	−15638.07	−26478.96	−38818.79	−52364.35	−66879.76	−82175.29
Maintenance (M)	−300	−300.00	−303.00	−306.03	−309.09	−312.18	−315.30	−318.46	−321.64
Present Value (PV)		−241.31	−194.11	−156.13	−125.59	−101.02	−81.26	−65.36	−52.58
Accumulated (AMPV)	−5000	−241.31	−435.42	−591.55	−717.14	−818.17	−899.43	−964.79	−1017.37
Non Production (NP)	0	0.00	0.00	0.00	0.00	0.00	0.00	0.00	0.00
Present Value (PV)		0.00	0.00	0.00	0.00	0.00	0.00	0.00	0.00
Accumulated (ANPPV)		0.00	0.00	0.00	0.00	0.00	0.00	0.00	0.00
Accumulated Total Costs (ATCPV)		−16021.88	−7096.11	−16229.63	−27196.10	−39636.96	−53263.78	−67844.55	−83192.66
Benefit (*B*) including IIPV	50000	34990.35	36878.61	38397.49	39619.23	40601.97	41392.47	42028.32	42539.79
Present Value (PV)		28145.39	23861.22	19983.88	16586.01	13672.32	11211.80	9157.04	7455.34
Accumulated Benefits (ABPV)		28145.39	52006.60	71990.48	88576.50	102248.82	113460.62	122617.65	130072.99
ATCPV+ABPV		12123.51	44910.50	55760.86	61380.40	62611.86	60196.84	54773.11	46880.33

(Continued)

TABLE 7.7 (*Continued*)

Values Used for Simulation based on Life Cycle Investment Related to Physical Assets Standard – First Simulation

Time	9	10	11	12	13	14	15	16	17
Initial Investment (II)									
II Present Value (IIPV) - Return	−1691.66	−1360.73	−1094.54	−880.42	−708.19	−569.65	−458.21	−368.57	−296.47
Functioning (*F*)	−4500.00	−4500.00	−4500.00	−4500.00	−4500.00	−4500.00	−4500.00	−4500.00	−4500.00
Present Value (PV) Accumulated	−15894.86	−16405.13	−16815.58	−17145.74	−17411.31	−17624.92	−17796.75	−17934.97	−18046.15
Energy (*E*)	−200.00	−200.00	−200.00	−200.00	−200.00	−200.00	−200.00	−200.00	−200.00
Energy (PV)	−28.19	−22.68	−18.24	−14.67	−11.80	−9.49	−7.64	−6.14	−4.94
Accumulated (AFPV)	−98098.34	−114526.15	−131359.97	−148520.38	−165943.49	−183577.91	−201382.30	−219323.41	−237374.50
Maintenance (M)	−324.86	−328.11	−331.39	−334.70	−338.05	−341.43	−344.84	−348.29	−351.77
Present Value (PV)	−42.29	−34.02	−27.36	−22.01	−17.70	−14.24	−11.46	−9.21	−7.41
Accumulated (AMPV)	−1059.66	−1093.68	−1121.04	−1143.05	−1160.75	−1174.99	−1186.45	−1195.66	−1203.08
Non Production (NP)	0.00	0.00	0.00	0.00	0.00	0.00	0.00	0.00	0.00
Present Value (PV)	0.00	0.00	0.00	0.00	0.00	0.00	0.00	0.00	0.00
Accumulated (ANPPV)	0.00	0.00	0.00	0.00	0.00	0.00	0.00	0.00	0.00
Accumulated Total Costs (ATCPV)	−99158.00	−115619.82	−132481.01	−149663.43	−167104.24	−184752.90	−202568.75	−220519.08	−238577.58
Benefit (*B*) including IIPV	42951.20	43282.13	43548.32	43762.44	43934.67	44073.21	44184.65	44274.28	44346.38
Present Value (PV)	6054.89	4907.93	3972.10	3210.77	2592.83	2092.19	1687.16	1359.86	1095.62
Accumulated Benefits (ABPV)	136127.88	141035.82	145007.92	148218.69	150811.52	152903.71	154590.87	155950.73	157046.36
ATCPV + ABPV	36969.88	25415.99	12526.91	−1444.74	−16292.72	−31849.19	−47977.88	−64568.34	−81531.22

(*Continued*)

TABLE 7.7 (*Continued*)
Values Used for Simulation based on Life Cycle Investment Related to Physical Assets Standard – First Simul

Time	18	19	20	21	22	23	24	25
Initial Investment (II)								
II Present Value (IIPV) - Return	−238.48	−191.82	−154.30	−124.11	−99.83	−80.30	−64.59	−51.96
Functioning (*F*)	−4500.00	−4500.00	−4500.00	−4500.00	−4500.00	−4500.00	−4500.00	−4500.00
Present Value (PV) Accumulated	−18135.57	−18207.51	−18265.37	−18311.91	−18349.35	−18379.47	−18403.69	−18423.17
Energy (*E*)	−200.00	−200.00	−200.00	−200.00	−200.00	−200.00	−200.00	−200.00
Energy (PV)	−3.97	−3.20	−2.57	−2.07	−1.66	−1.34	−1.08	−0.87
Accumulated (AFPV)	−255514.05	−273724.76	−291992.70	−310306.68	−328657.69	−347038.50	−365443.26	−383867.30
Maintenance (M)	−355.29	−358.84	−362.43	−366.06	−369.72	−373.41	−377.15	−380.92
Present Value (PV)	−5.96	−4.80	−3.86	−3.10	−2.50	−2.01	−1.61	−1.30
Accumulated (AMPV)	−1209.04	−1213.83	−1217.69	−1220.79	−1223.29	−1225.30	−1226.91	−1228.21
Non Production (NP)	0.00	0.00	0.00	0.00	0.00	0.00	0.00	0.00
Present Value (PV)	0.00	0.00	0.00	0.00	0.00	0.00	0.00	0.00
Accumulated (ANPPV)	0.00	0.00	0.00	0.00	0.00	0.00	0.00	0.00
Accumulated Total Costs (ATCPV)	−256723.09	−274938.59	−293210.39	−311527.47	−329880.98	−348263.80	−366670.18	−385095.51
Benefit (*B*) including IIPV	44404.38	44451.03	44488.56	44518.74	44543.02	44562.55	44578.26	44590.90
Present Value (PV)	882.45	710.56	572.04	460.45	370.58	298.21	239.96	193.07
Accumulated Benefits (ABPV)	157928.80	158639.37	159211.41	159671.86	160042.43	160340.65	160580.61	160773.68
ATCPV+ABPV	−98794.29	−116299.22	−133998.98	−151855.62	−169838.55	−187923.15	−206089.57	−224321.83

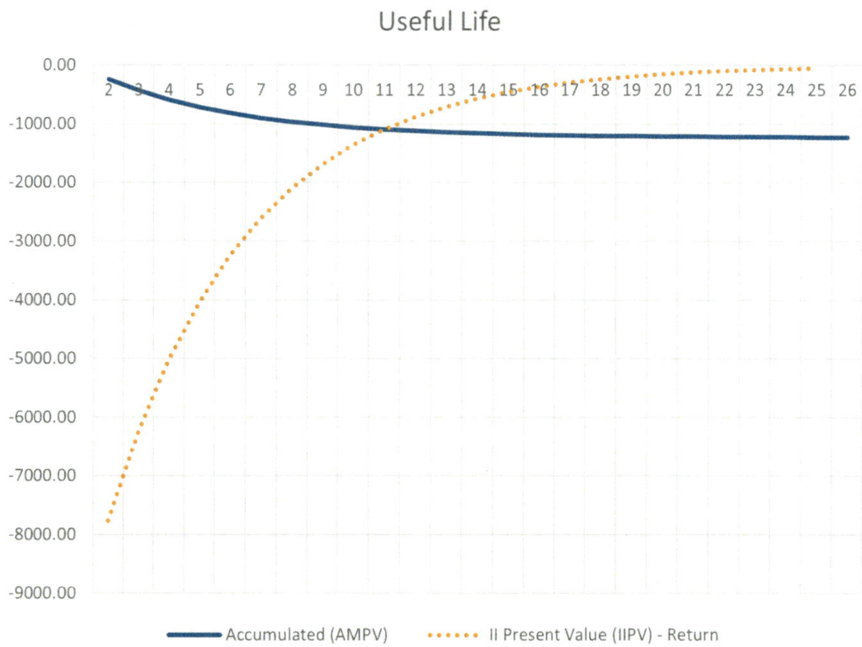

FIGURE 7.26 Physical Asset useful life.

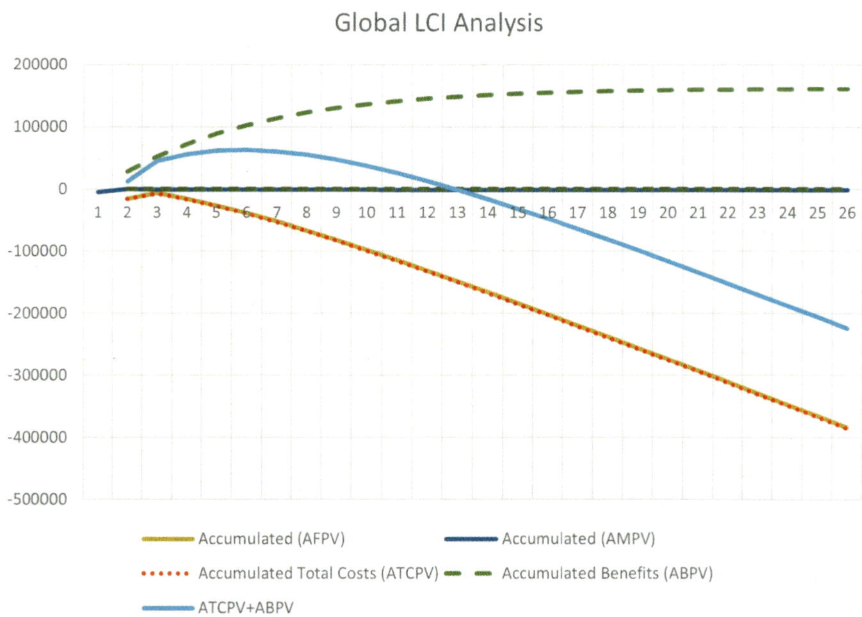

FIGURE 7.27 Global LCI analysis.

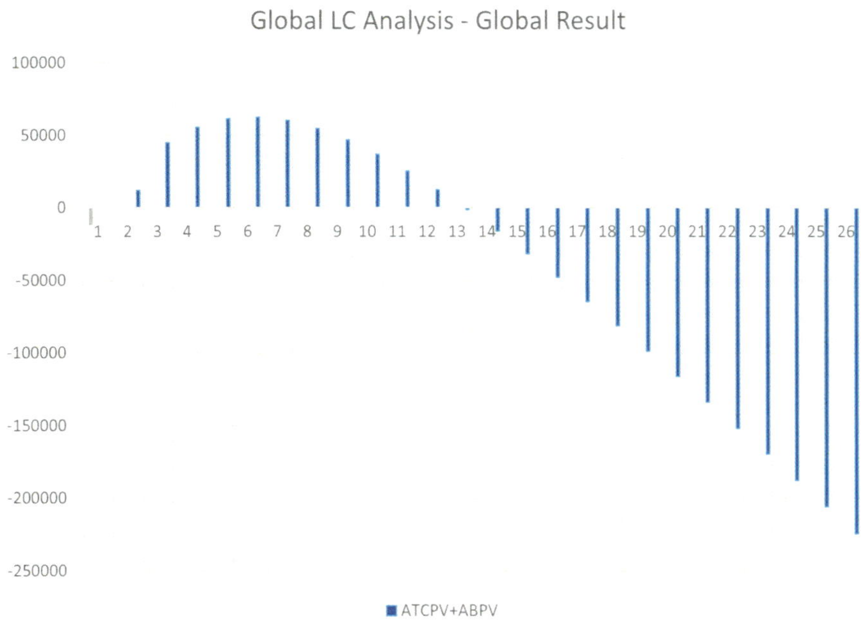

FIGURE 7.28 Life cycle analysis global result.

The remaining values are presented in Table 7.8, and the graphical results are shown in Figures 7.29–7.31.

As can be seen, when comparing Figures 7.28 and 7.31, with the increasing of the maintenance management quality helped by the maintenance standards, it gets five more positive years, and the positive incomes increase a lot.

7.4.3 Life Cycle of Physical Assets with Recovery

Considering the same arguments of the preceding sub-section and using the Life Cycle model of Physical Asset Life Cycle based with Recovery, the result of using this model, for a first scenario, is illustrated in Table 7.9.

Other relevant values considered are the following ones:

- Interest Rate (*r*) 0.05
- Inflation Rate (*h*) 0.02
- Profit Rate (*p*) 0.05
- Maintenance Increase Rate 0.01
- MTBF 500.0
- MTTR 60.0
- Risk (*R*) 0.10
- Time (*t*) 0
- Initial Investment (II) (C_0) –12000
- F_p Financial Movement (Production) **50000**
- Functioning Costs –4500
- Energy –200
- Maintenance Costs –300

TABLE 7.8
Values Used for Simulation based on Life Cycle Investment Related to Physical Assets Standard – Second Simulation

Time	0	1	2	3	4	5	6	7	8
Initial Investment (II)	−12000								
II Present Value (IIPV) – Return	−12000	−9652.51	−7764.25	−6245.37	−5023.63	−4040.88	−3250.39	−2614.53	−2103.07
Functioning (*F*)	−4500	−4500.00	−4500.00	−4500.00	−4500.00	−4500.00	−4500.00	−4500.00	−4500.00
Present Value (PV) Accumulated		−3619.69	−6531.28	−8873.30	−10757.16	−12272.49	−13491.38	−14471.83	−15260.48
Energy (*E*)	−200	−200.00	−200.00	−200.00	−200.00	−200.00	−200.00	−200.00	−200.00
Energy (PV)		−160.88	−129.40	−104.09	−83.73	−67.35	−54.17	−43.58	−35.05
Accumulated (AFPV)		−15780.57	−6660.69	−15638.07	−26478.96	−38818.79	−52364.35	−66879.76	−82175.29
Maintenance (M)	−300	−300.00	−303.00	−306.03	−309.09	−312.18	−315.30	−318.46	−321.64
Present Value (PV)		−241.31	−194.11	−156.13	−125.59	−101.02	−81.26	−65.36	−52.58
Accumulated (AMPV)	−5000	−241.31	−435.42	−591.55	−717.14	−818.17	−899.43	−964.79	−1017.37
Non Production (NP)	0	0.00	0.00	0.00	0.00	0.00	0.00	0.00	0.00
Present Value (PV)		0.00	0.00	0.00	0.00	0.00	0.00	0.00	0.00
Accumulated (ANPPV)		0.00	0.00	0.00	0.00	0.00	0.00	0.00	0.00
Accumulated Total Costs (ATCPV)		−16021.88	−7096.11	−16229.63	−27196.10	−39636.96	−53263.78	−67844.55	−83192.66
Benefit (*B*) including IIPV	70000	52847.49	54735.75	56254.63	57476.37	58459.12	59249.61	59885.47	60396.93
Present Value (PV)		42509.24	35415.15	29277.59	24061.65	19685.54	16048.69	13047.71	10584.90
Accumulated Benefits (ABPV)		42509.24	77924.39	107201.98	131263.63	150949.16	166997.85	180045.56	190630.47
ATCPV + ABPV		26487.36	70828.29	90972.35	104067.53	111312.20	113734.08	112201.02	107437.81

(*Continued*)

TABLE 7.8 (*Continued*)
Values Used for Simulation based on Life Cycle Investment Related to Physical Assets Standard – Second Simulation

Time	9	10	11	12	13	14	15	16
Initial Investment (II)								
II Present Value (IIPV) – Return	−1691.66	−1360.73	−1094.54	−880.42	−708.19	−569.65	−458.21	−368.57
Functioning (*F*)	−4500.00	−4500.00	−4500.00	−4500.00	−4500.00	−4500.00	−4500.00	−4500.00
Present Value (PV) Accumulated	−15894.86	−16405.13	−16815.58	−17145.74	−17411.31	−17624.92	−17796.75	−17934.97
Energy (*E*)	−200.00	−200.00	−200.00	−200.00	−200.00	−200.00	−200.00	−200.00
Energy (PV)	−28.19	−22.68	−18.24	−14.67	−11.80	−9.49	−7.64	−6.14
Accumulated (AFPV)	−98098.34	−114526.15	−131359.97	−148520.38	−165943.49	−183577.91	−201382.30	−219323.41
Maintenance (M)	−324.86	−328.11	−331.39	−334.70	−338.05	−341.43	−344.84	−348.29
Present Value (PV)	−42.29	−34.02	−27.36	−22.01	−17.70	−14.24	−11.46	−9.21
Accumulated (AMPV)	−1059.66	−1093.68	−1121.04	−1143.05	−1160.75	−1174.99	−1186.45	−1195.66
Non Production (NP)	0.00	0.00	0.00	0.00	0.00	0.00	0.00	0.00
Present Value (PV)	0.00	0.00	0.00	0.00	0.00	0.00	0.00	0.00
Accumulated (ANPPV)	0.00	0.00	0.00	0.00	0.00	0.00	0.00	0.00
Accumulated Total Costs (ATCPV)	−99158.00	−115619.82	−132481.01	−149663.43	−167104.24	−184752.90	−202568.75	−220519.08
Benefit (*B*) including IIPV	60808.34	61139.27	61405.46	61619.58	61791.81	61930.35	62041.79	62131.43
Present Value (PV)	8572.24	6932.83	5600.88	4520.92	3646.68	2939.88	2369.02	1908.34
Accumulated Benefits (ABPV)	199202.70	206135.53	211736.41	216257.32	219904.01	222843.89	225212.91	227121.25
ATCPV + ABPV	100044.70	90515.70	79255.40	66593.89	52799.76	38090.98	22644.16	6602.17

(*Continued*)

TABLE 7.8 (*Continued*)

Values Used for Simulation based on Life Cycle Investment Related to Physical Assets Standard – Second Simulation

Time	17	18	19	20	21	22	23	24	25
Initial Investment (II)									
II Present Value (IIPV) – Return	−296.47	−238.48	−191.82	−154.30	−124.11	−99.83	−80.30	−64.59	−51.96
Functioning (*F*)	−4500.00	−4500.00	−4500.00	−4500.00	−4500.00	−4500.00	−4500.00	−4500.00	−4500.00
Present Value (PV) Accumulated	−18046.15	−18135.57	−18207.51	−18265.37	−18311.91	−18349.35	−18379.47	−18403.69	−18423.17
Energy (*E*)	−200.00	−200.00	−200.00	−200.00	−200.00	−200.00	−200.00	−200.00	−200.00
Energy (PV)	−4.94	−3.97	−3.20	−2.57	−2.07	−1.66	−1.34	−1.08	−0.87
Accumulated (AFPV)	−237374.50	−255514.05	−273724.76	−291992.70	−310306.68	−328657.69	−347038.50	−365443.26	−383867.30
Maintenance (M)	−351.77	−355.29	−358.84	−362.43	−366.06	−369.72	−373.41	−377.15	−380.92
Present Value (PV)	−7.41	−5.96	−4.80	−3.86	−3.10	−2.50	−2.01	−1.61	−1.30
Accumulated (AMPV)	−1203.08	−1209.04	−1213.83	−1217.69	−1220.79	−1223.29	−1225.30	−1226.91	−1228.21
Non Production (NP)	0.00	0.00	0.00	0.00	0.00	0.00	0.00	0.00	0.00
Present Value (PV)	0.00	0.00	0.00	0.00	0.00	0.00	0.00	0.00	0.00
Accumulated (ANPPV)	0.00	0.00	0.00	0.00	0.00	0.00	0.00	0.00	0.00
Accumulated Total Costs (ATCPV)	−238577.58	−256723.09	−274938.59	−293210.39	−311527.47	−329880.98	−348263.80	−366670.18	−385095.51
Benefit (*B*) including IIPV	62203.53	62261.52	62308.18	62345.70	62375.89	62400.17	62419.70	62435.41	62448.04
Present Value (PV)	1536.80	1237.32	996.02	801.65	645.14	519.14	417.71	336.08	270.39
Accumulated Benefits (ABPV)	228658.05	229895.37	230891.38	231693.04	232338.18	232857.32	233275.03	233611.12	233881.51
ATCPV + ABPV	−9919.53	−26827.72	−44047.21	−61517.35	−79189.29	−97023.67	−114988.76	−133059.06	−151214.01

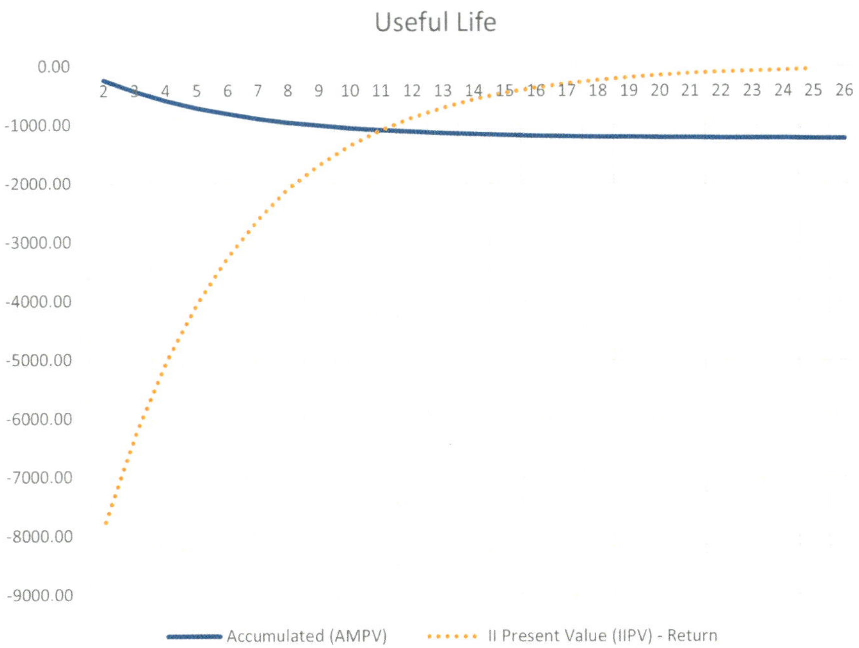

FIGURE 7.29 Physical Asset useful life.

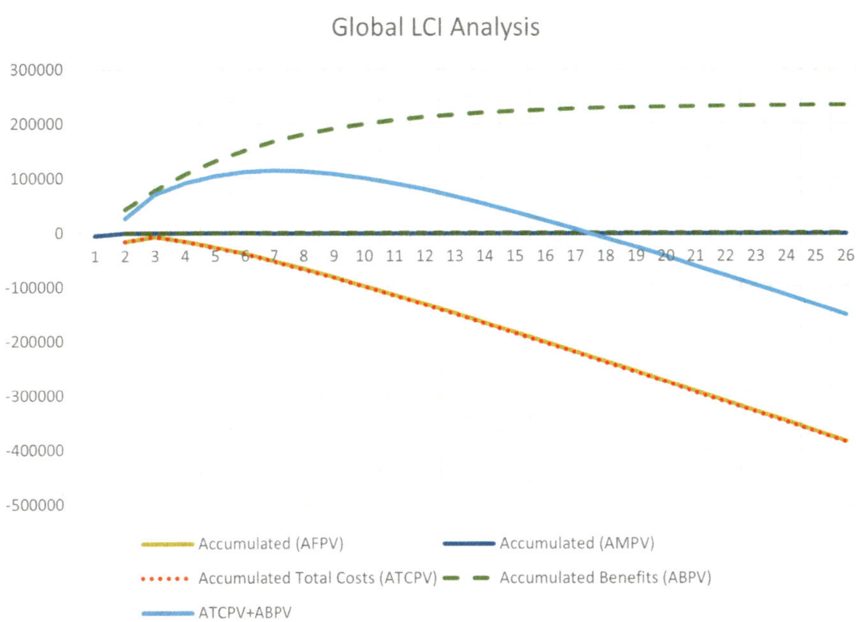

FIGURE 7.30 Global LCI analysis.

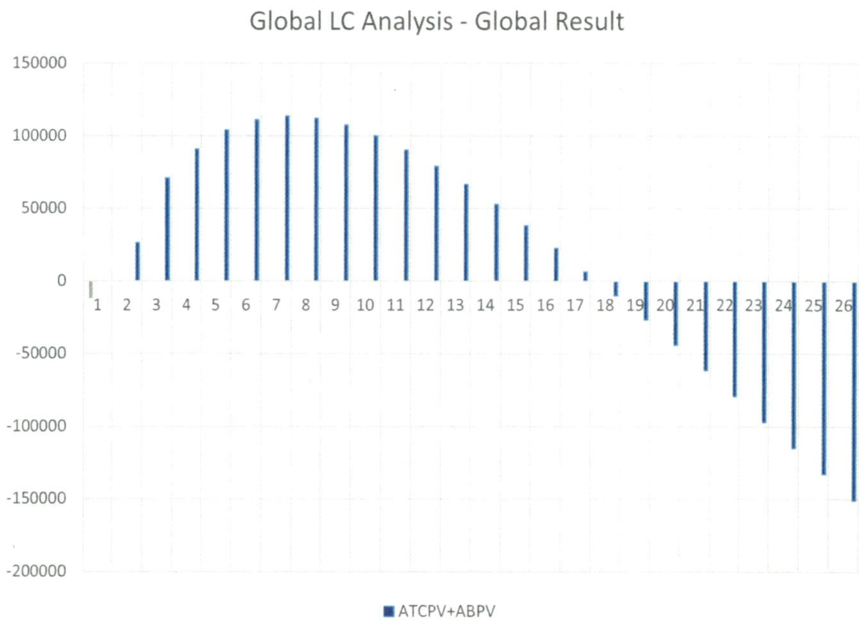

FIGURE 7.31 Life cycle analysis global result.

The graphical results are presented in Figures 7.32–7.34.

For the second scenario, other relevant values considered are the following ones:

- Interest Rate (r) 0.05
- Inflation Rate (h) 0.02
- Profit Rate (p) 0.05
- Maintenance Increase Rate 0.01
- MTBF 500.0
- MTTR 60.0
- Risk (R) 0.10
- Time (t) 0
- Initial Investment (II) (C_0) –12000
- F_p Financial Movement (Production) **70000**
- Functioning Costs –4500
- Energy –200
- Maintenance Costs –300

TABLE 7.9
Scenario for Life Cycle with Recovery Related to Physical Assets Standard – First Simulation

Time (t)	0	1	2	3	4	5	6	7	8
r – real interest rate $(1+r)$		1.05	1.05	1.05	1.05	1.05	1.05	1.05	1.05
h – inflation rate $(1+h)$		1.02	1.02	1.02	1.02	1.02	1.02	1.02	1.02
p – profit rate		1.05	1.05	1.05	1.05	1.05	1.05	1.05	1.05
i – money rate (nominal interest rate)		0.12	0.12	0.12	0.12	0.12	0.12	0.12	0.12
i – Apparent Rate with Risk		0.24	0.24	0.24	0.24	0.24	0.24	0.24	0.24
$1/((1+i)\ ^\wedge\ t))$		0.81	0.65	0.53	0.43	0.35	0.28	0.23	0.18
Initial Investment (II) (C_0)	–12000	–12000	–12000	–12000	–12000	–12000	–12000	–12000	–12000
Initial Investment (II) (C_0) with Devaluation (NPV)	–12000	–9701	–7842	–6340	–5125	–4143	–3349	–2708	–2189
Initial Investment (II) (C_0) considering NEW Asset (NPV)	–12000	–14844	–18362	–22714	–28097	–34757	–42994	–53184	–65789
Fp Financial Movement (Production)	50000	50000	50000	50000	50000	50000	50000	50000	50000
Fp Financial Movement (NPV)	50000	36746	29705	24014	19413	15694	12687	10256	8291
Fp Financial Movement (NPV) Accumulated	50000	36746	66451	90465	109878	125571	138258	148514	156805
Functioning Costs	–4500	–4500	–4500	–4500	–4500	–4500	–4500	–4500	–4500
Functioning Costs (NPV) Accumulated	–4500	–3638	–6579	–8956	–10878	–12432	–13688	–14703	–15524
Energy	–200	–200	–200	–200	–200	–200	–200	–200	–200
Energy Accumulated (NPV) Accumulated	–200	–162	–292	–398	–483	–553	–608	–653	–690
Maintenance Costs	–300	–300	–300	–300	–300	–300	–300	–300	–300
Maintenance Costs (NPV) Accumulated	–300	–245	–443	–603	–732	–837	–922	–990	–1045
Total Costs (NPV) Accumulated	–5000	–4083	–11304	–21063	–32874	–46342	–61152	–77044	–93813
FT Net Present Value (NPV) (With II Devaluated)		22962	47304	63062	71879	75086	73757	68762	60803
FT Net Present Value (With II New Asset)		17819	36784	46687	48907	44472	34112	18286	–2797

(*Continued*)

TABLE 7.9 (*Continued*)

Scenario for Life Cycle with Recovery Related to Physical Assets Standard – First Simulation

Time (t)	9	10	11	12	13	14	15	16
r – real interest rate $(1+r)$	1.05	1.05	1.05	1.05	1.05	1.05	1.05	1.05
h – inflation rate $(1+h)$	1.02	1.02	1.02	1.02	1.02	1.02	1.02	1.02
p – profit rate	1.05	1.05	1.05	1.05	1.05	1.05	1.05	1.05
i – money rate (nominal interest rate)	0.12	0.12	0.12	0.12	0.12	0.12	0.12	0.12
i – Apparent Rate with Risk	0.24	0.24	0.24	0.24	0.24	0.24	0.24	0.24
$1/((1+i)\ ^\wedge\ t))$	0.15	0.12	0.10	0.08	0.06	0.05	0.04	0.03
Initial Investment (II) (C_0)	−12000	−12000	−12000	−12000	−12000	−12000	−12000	−12000
Initial Investment (II) (C_0) with Devaluation (NPV)	−1769	−1430	−1156	−935	−756	−611	−494	−399
Initial Investment (II) (C_0) considering NEW Asset (NPV)	−81381	−100669	−124528	−154042	−190551	−235712	−291577	−360682
Fp Financial Movement (Production)	50000	50000	50000	50000	50000	50000	50000	50000
Fp Financial Movement (NPV)	6702	5418	4380	3541	2863	2314	1871	1512
Fp Financial Movement (NPV) Accumulated	163507	168926	173306	176847	179709	182023	183894	185406
Functioning Costs	−4500	−4500	−4500	−4500	−4500	−4500	−4500	−4500
Functioning Costs (NPV) Accumulated	−16187	−16724	−17157	−17508	−17791	−18020	−18206	−18355
Energy	−200	−200	−200	−200	−200	−200	−200	−200
Energy Accumulated (NPV) Accumulated	−719	−743	−763	−778	−791	−801	−809	−816
Maintenance Costs	−300	−300	−300	−300	−300	−300	−300	−300
Maintenance Costs (NPV) Accumulated	−1090	−1126	−1155	−1179	−1198	−1213	−1226	−1236
Total Costs (NPV) Accumulated	−111291	−129340	−147853	−166740	−185929	−205362	−224994	−244785
FT Net Present Value (NPV) (With II Devaluated)	50447	38155	24297	9173	−6975	−23950	−41593	−59778
FT Net Present Value (With II New Asset)	−29164	−61084	−99075	−143935	−196770	−259051	−332676	−420061

(*Continued*)

TABLE 7.9 (*Continued*)
Scenario for Life Cycle with Recovery Related to Physical Assets Standard – First Simulation

Time (t)	17	18	19	20	21	22	23	24	25
r – real interest rate $(1+r)$	1.05	1.05	1.05	1.05	1.05	1.05	1.05	1.05	1.05
h – inflation rate $(1+h)$	1.02	1.02	1.02	1.02	1.02	1.02	1.02	1.02	1.02
p – profit rate	1.05	1.05	1.05	1.05	1.05	1.05	1.05	1.05	1.05
i – money rate (nominal interest rate)	0.12	0.12	0.12	0.12	0.12	0.12	0.12	0.12	0.12
i – Apparent Rate with Risk	0.24	0.24	0.24	0.24	0.24	0.24	0.24	0.24	0.24
$1/((1+i)$ ^ $t))$	0.03	0.02	0.02	0.01	0.01	0.01	0.01	0.01	0.00
Initial Investment (II) (C_0)	–12000	–12000	–12000	–12000	–12000	–12000	–12000	–12000	–12000
Initial Investment (II) (C_0) with Devaluation (NPV)	–323	–261	–211	–171	–138	–111	–90	–73	–59
Initial Investment (II) (C_0) considering NEW Asset (NPV)	–446165	–551909	–682714	–844521	–1044676	–1292270	–1598544	–1977407	–2446063
Fp Financial Movement (Production)	50000	50000	50000	50000	50000	50000	50000	50000	50000
Fp Financial Movement (NPV)	1223	988	799	646	522	422	341	276	223
Fp Financial Movement (NPV) Accumulated	186629	187617	188416	189062	189584	190006	190348	190623	190846
Functioning Costs	–4500	–4500	–4500	–4500	–4500	–4500	–4500	–4500	–4500
Functioning Costs (NPV) Accumulated	–18476	–18574	–18653	–18717	–18769	–18811	–18844	–18872	–18894
Energy	–200	–200	–200	–200	–200	–200	–200	–200	–200
Energy Accumulated (NPV) Accumulated	–821	–826	–829	–832	–834	–836	–838	–839	–840
Maintenance Costs	–300	–300	–300	–300	–300	–300	–300	–300	–300
Maintenance Costs (NPV) Accumulated	–1244	–1251	–1256	–1260	–1264	–1267	–1269	–1271	–1272
Total Costs (NPV) Accumulated	–264705	–284730	–304839	–325017	–345249	–365527	–385840	–406182	–426548
FT Net Present Value (NPV) (With II Devaluated)	–78399	–97374	–116634	–136125	–155803	–175632	–195582	–215632	–235761
FT Net Present Value (With II New Asset)	–524242	–649022	–799137	–980475	–1200341	–1467790	–1794037	–2192966	–2681765

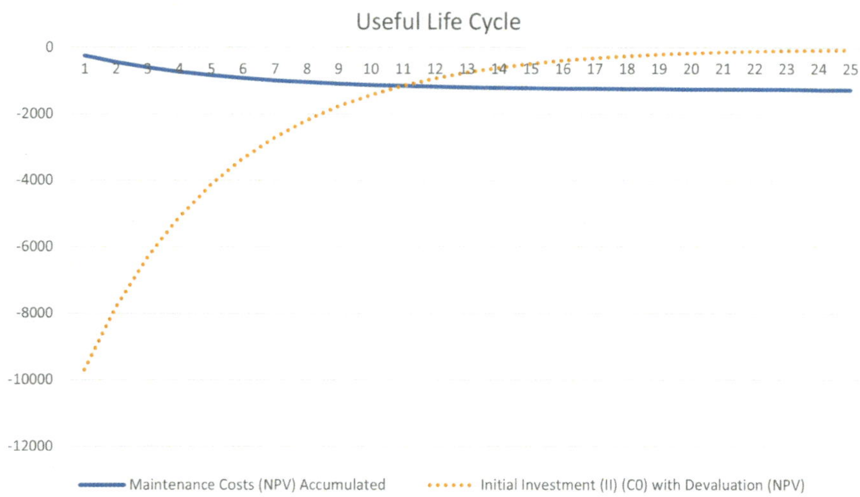

FIGURE 7.32 Useful life cycle.

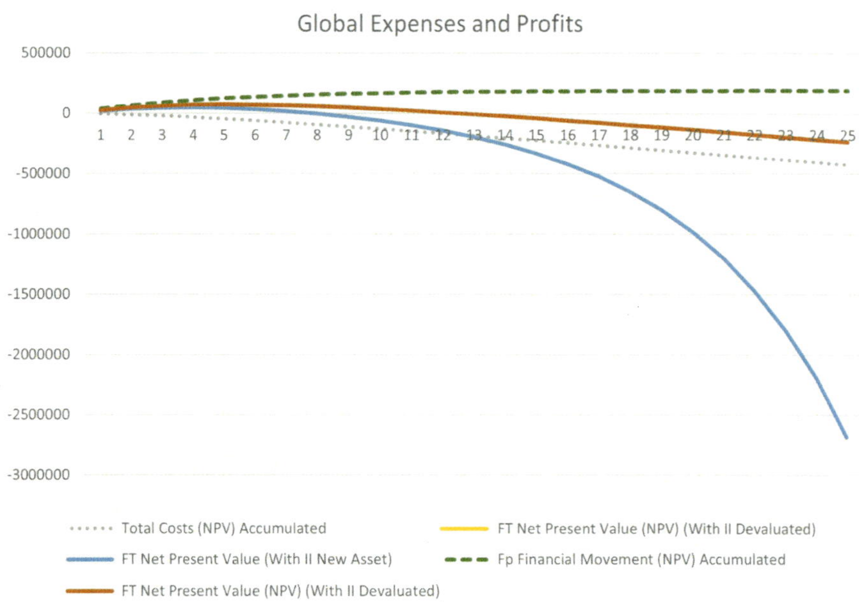

FIGURE 7.33 Global investments and profits.

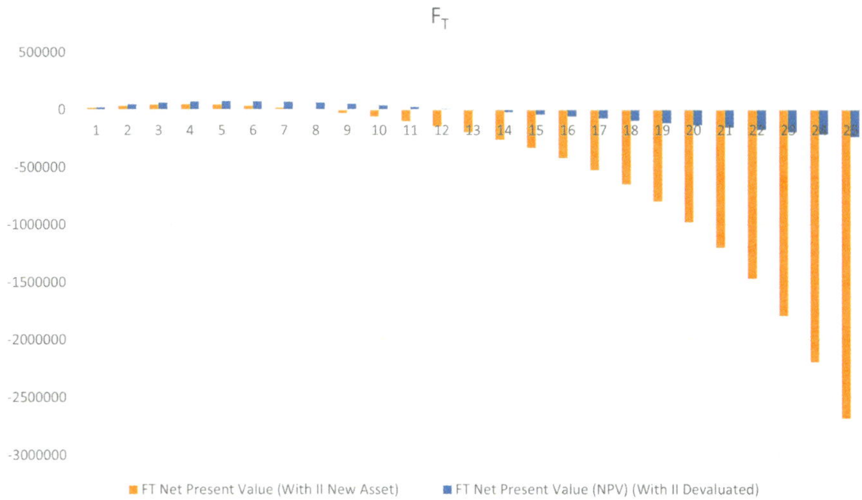

FIGURE 7.34 Results considering the acquisition of a new asset with devaluation and recovery.

Table 7.10 and Figures 7.35–7.37 show the results.

As can be observed in the Figure of Global Results, according to this Life Cycle model and second scenario, the implementation of this standard continues to be interesting, gaining three years in positive results and increasing the absolute values.

TABLE 7.10
Scenario for Life Cycle with Recovery Related to Physical Assets Standard – Second Simulation

Time (t)	0	1	2	3	4	5	6	7	8	9	10	11	12
r – real interest rate $(1+r)$		1.05	1.05	1.05	1.05	1.05	1.05	1.05	1.05	1.05	1.05	1.05	1.05
h – inflation rate $(1+h)$		1.02	1.02	1.02	1.02	1.02	1.02	1.02	1.02	1.02	1.02	1.02	1.02
p – profit rate		1.05	1.05	1.05	1.05	1.05	1.05	1.05	1.05	1.05	1.05	1.05	1.05
i – money rate (nominal interest rate)		0.12	0.12	0.12	0.12	0.12	0.12	0.12	0.12	0.12	0.12	0.12	0.12
i – Apparent Rate with Risk		0.24	0.24	0.24	0.24	0.24	0.24	0.24	0.24	0.24	0.24	0.24	0.24
$1/((1+i)\wedge t))$		0.81	0.65	0.53	0.43	0.35	0.28	0.23	0.18	0.15	0.12	0.10	0.08
Initial Investment (II) (C_0)	–12000	–12000	–12000	–12000	–12000	–12000	–12000	–12000	–12000	–12000	–12000	–12000	–12000
Initial Investment (II) (C_0) with Devaluation (NPV)	–12000	–9701	–7842	–6340	–5125	–4143	–3349	–2708	–2189	–1769	–1430	–1156	–935
Initial Investment (II) (C_0) considering NEW Asset (NPV)	–12000	–14844	–18362	–22714	–28097	–34757	–42994	–53184	–65789	–81381	–100669	–124528	–154042
Fp Financial Movement (Production)	70000	70000	70000	70000	70000	70000	70000	70000	70000	70000	70000	70000	70000
Fp Financial Movement (NPV)	70000	51444	41587	33619	27178	21971	17761	14358	11607	9383	7586	6132	4957
Fp Financial Movement (NPV) Accumulated	70000	51444	93031	126651	153829	175800	193561	207920	219527	228910	236496	242628	247586
Functioning Costs	–4500	–4500	–4500	–4500	–4500	–4500	–4500	–4500	–4500	–4500	–4500	–4500	–4500
Functioning Costs (NPV) Accumulated	–4500	–3638	–6579	–8956	–10878	–12432	–13688	–14703	–15524	–16187	–16724	–17157	–17508
Energy	–200	–200	–200	–200	–200	–200	–200	–200	–200	–200	–200	–200	–200
Energy Accumulated (NPV) Accumulated	–200	–162	–292	–398	–483	–553	–608	–653	–690	–719	–743	–763	–778
Maintenance Costs	–300	–300	–300	–300	–300	–300	–300	–300	–300	–300	–300	–300	–300
Maintenance Costs (NPV) Accumulated	–300	–245	–443	–603	–732	–837	–922	–990	–1045	–1090	–1126	–1155	–1179
Total Costs (NPV) Accumulated	–5000	–4083	–11304	–21063	–32874	–46342	–61152	–77044	–93813	–111291	–129340	–147853	–166740
FT Net Present Value (NPV) (With II Devaluated)		37660	73885	99248	115830	125314	129060	128168	123525	115850	105725	93619	79911
FT Net Present Value (With II New Asset)		32517	63365	82873	92858	94701	89415	77691	59925	36239	6487	–29753	–73196

(Continued)

TABLE 7.10 (*Continued*)
Scenario for Life Cycle with Recovery Related to Physical Assets Standard – Second Simulation

Time (t)	13	14	15	16	17	18	19	20	21	22	23	24	25
r – real interest rate $(1+r)$	1.05	1.05	1.05	1.05	1.05	1.05	1.05	1.05	1.05	1.05	1.05	1.05	1.05
h – inflation rate $(1+h)$	1.02	1.02	1.02	1.02	1.02	1.02	1.02	1.02	1.02	1.02	1.02	1.02	1.02
p – profit rate	1.05	1.05	1.05	1.05	1.05	1.05	1.05	1.05	1.05	1.05	1.05	1.05	1.05
i – money rate (nominal interest rate)	0.12	0.12	0.12	0.12	0.12	0.12	0.12	0.12	0.12	0.12	0.12	0.12	0.12
i – Apparent Rate with Risk	0.24	0.24	0.24	0.24	0.24	0.24	0.24	0.24	0.24	0.24	0.24	0.24	0.24
$1/((1+i)\wedge t))$	0.06	0.05	0.04	0.03	0.03	0.02	0.02	0.01	0.01	0.01	0.01	0.01	0.00
Initial Investment (II) (C_0)	−12000	−12000	−12000	−12000	−12000	−12000	−12000	−12000	−12000	−12000	−12000	−12000	−12000
Initial Investment (II) (C_0) with Devaluation (NPV)	−756	−611	−494	−399	−323	−261	−211	−171	−138	−111	−90	−73	−59
Initial Investment (II) (C_0) considering NEW Asset (NPV)	−190551	−235712	−291577	−360682	−446165	−551909	−682714	−844521	−1044676	−1292270	−1598544	−1977407	−2446063
Fp Financial Movement (Production)	70000	70000	70000	70000	70000	70000	70000	70000	70000	70000	70000	70000	70000
Fp Financial Movement (NPV)	4008	3240	2619	2117	1712	1384	1119	904	731	591	478	386	312
Fp Financial Movement (NPV) Accumulated	251593	254833	257452	259569	261281	262664	263783	264687	265418	266009	266487	266873	267185
Functioning Costs	−4500	−4500	−4500	−4500	−4500	−4500	−4500	−4500	−4500	−4500	−4500	−4500	−4500
Functioning Costs (NPV) Accumulated	−17791	−18020	−18206	−18355	−18476	−18574	−18653	−18717	−18769	−18811	−18844	−18872	−18894
Energy	−200	−200	−200	−200	−200	−200	−200	−200	−200	−200	−200	−200	−200
Energy Accumulated (NPV) Accumulated	−791	−801	−809	−816	−821	−826	−829	−832	−834	−836	−838	−839	−840
Maintenance Costs	−300	−300	−300	−300	−300	−300	−300	−300	−300	−300	−300	−300	−300
Maintenance Costs (NPV) Accumulated	−1198	−1213	−1226	−1236	−1244	−1251	−1256	−1260	−1264	−1267	−1269	−1271	−1272
Total Costs (NPV) Accumulated	−185929	−205362	−224994	−244785	−264705	−284730	−304839	−325017	−345249	−365527	−385840	−406182	−426548
FT Net Present Value (NPV) (With II Devaluated)	64909	48860	31964	14385	−3747	−22327	−41267	−60500	−79969	−99629	−119443	−139382	−159422
FT Net Present Value (With II New Asset)	−124886	−186242	−259119	−345898	−449590	−573975	−723771	−904850	−1124508	−1391788	−1717898	−2116717	−2605426

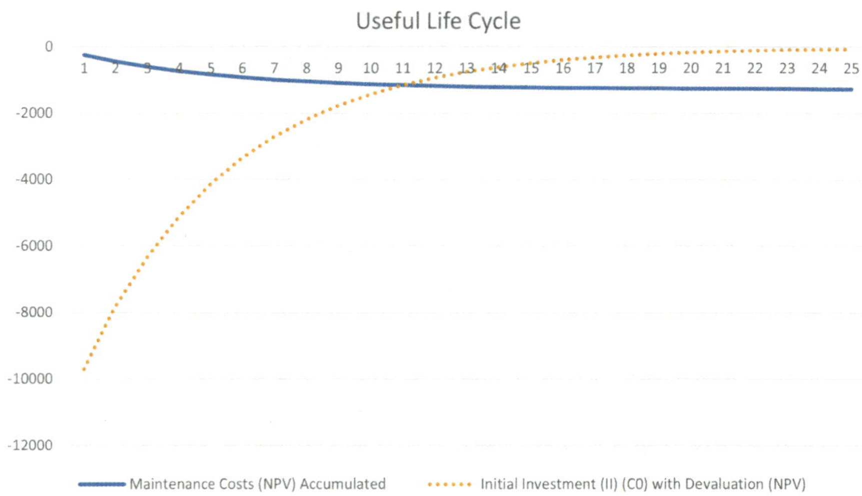

FIGURE 7.35 Useful life cycle.

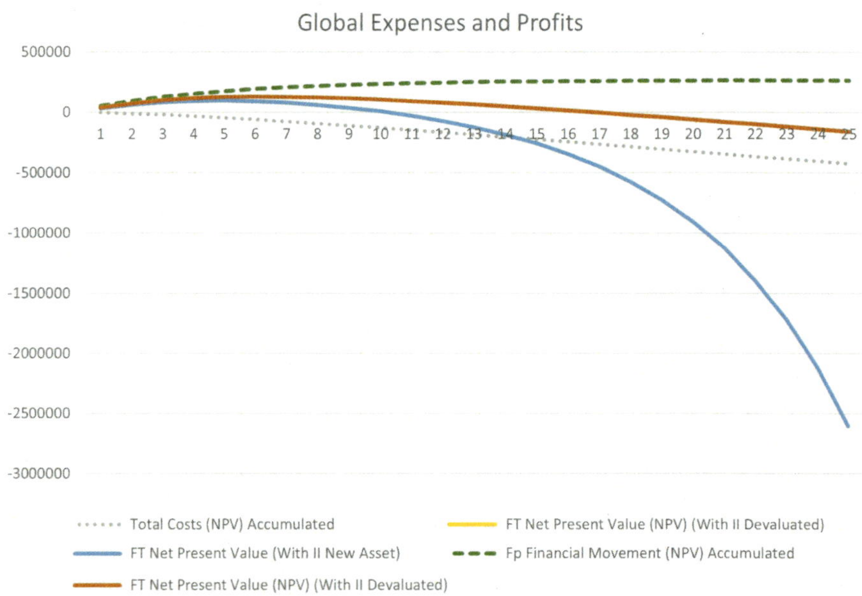

FIGURE 7.36 Global investments and profits.

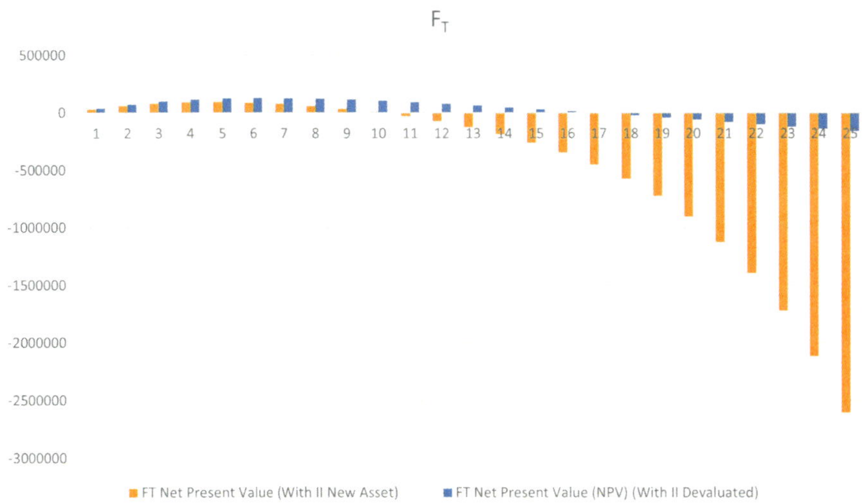

FIGURE 7.37 Results considering the acquisition of a new asset with devaluation and recovery.

8 Risk Norms

8.1 INTRODUCTION

This chapter deals with the implementation of Risk Standards, namely ISO 31000, being structured in the following sections:

- Characterization of Risk Norms;
- Diagnosis to implement Risk Norms;
- Quantitative evaluation about the implementation of Risk Norms.

8.2 CHARACTERIZATION OF RISK NORMS

The main relevant Risk Norms are the following (Woellner, Voorhees & Bell, 2020):

- ISO 31000- Risk management – Guidelines
  - Provides guidelines on managing risk faced by organizations. The application of these guidelines can be customized to any organization and its context;
  - Provides a common approach to managing any type of risk and is not industry or sector-specific;
  - Can be used throughout the life of the organization and can be applied to any activity, including decision-making at all levels.
- ISO 31010- Risk management - Risk assessment techniques
  - Provides guidance on the selection and application of techniques for assessing risk in a wide range of situations;
  - The techniques are used to assist in making decisions where there is uncertainty, to provide information about particular risks and as part of a process for managing risk;
  - Provides summaries of a range of techniques, with references to other documents where the techniques are described in more detail.
- ISO 31022- Risk management - Guidelines for the management of legal risk
  - Gives guidelines for managing the specific challenges of legal risk faced by organizations, as a complementary document to ISO 31000;
  - The guidelines can be customized to any organization and its context;
  - Provides a common approach to the management of legal risk and is not industry or sector-specific.
- ISO 31030- Managing travel risks - Guidance for organizations
  - Provides a structured approach to the development, implementation, evaluation, and review of:
    - policy;
    - programme development;

DOI: 10.1201/9781003395690-11

    - threat and hazard identification;
    - opportunities and strengths;
    - risk assessment;
    - prevention and mitigation strategies.
  - Is applicable to any type of organization, irrespective of sector or size, including but not limited to:
    - commercial organizations;
    - charitable and not-for-profit organizations;
    - governmental organizations;
    - non-governmental organizations;
    - educational organizations.
  - Does not apply to tourism and leisure-related travel, except in relation to travellers travelling on behalf of the organization.
- ISO 31050- Guidance for managing risks to enhance resilience
  - Provides a structured context (e.g. definitions, drivers, metrics, …) for dealing with emerging risks;
  - Provides an emerging risk management framework;
  - Process;
  - Guidance for common format(s) for interoperability, common/agreed indicators, considerations related to emerging risks in resilience assessment;
  - Emerging risk application examples.
- ISO Guide 73- Risk management – Vocabulary
  - Provides the definitions of generic terms related to risk management;
  - Aims to encourage a mutual and consistent understanding of, and a coherent approach to, the description of activities relating to the management of risk, and the use of uniform risk management terminology in processes and frameworks dealing with the management of risk.

From the preceding standards, it is emphasized ISO 31010 (Risk management - Risk assessment techniques) which presents several tools: Risk Identification; Risk Analysis; and Risk Evaluation. Some of these tools are the following:

- Brainstorming;
- Structured or semi-structured interviews;
- Check-lists;
- Primary hazard analysis;
- Hazard and Operability Studies (HAZOP);
- Hazard Analysis and Critical Control Points (HACCP);
- Root cause analysis;
- Failure Mode Effect Analysis (FMEA);
- Fault tree analysis;
- Decision tree;
- Bow tie analysis;
- Reliability Centred Maintenance (RCM);
- Markov analysis;

- Monte Carlo simulation;
- Cost/benefit analysis;
- Multi-criteria decision analysis.

8.3 DIAGNOSIS TO IMPLEMENT RISK NORMS

To implement Risk Norms, the proposed methodology is like the other standards (see Section 6.3), that is, it ought to perform the diagnosis of the organization aiming to prepare it to give the next step (Figure 8.1).

8.4 QUANTITATIVE EVALUATION ABOUT THE IMPLEMENTATION OF RISK NORMS

The evaluation about the implementation of Risk Evaluation norms is something different from the previously discussed norms and the norms discussed in the next chapters.

The proposal to evaluate the Risk is based on Equation 8.1 reproduced next.

$$(1+i_A)=(1+r)\times(1+h)\times(1+p) \tag{8.1}$$

where

i is the Apparent Rate
r is the Interest Rate
h is the Inflation Rate
p is the Profit Rate

The proposal is to add a new variable to the Apparent Rate Equation (8.1) and, as will be demonstrated, if behaves well translating the financial implications resulting from the Risk Rate in the Apparent Rate, according to Equation 8.2.

$$(1+i_A)=(1+r)\times(1+h)\times(1+p)\times(1+R) \tag{8.2}$$

where

R – is the Risk Rate

8.4.1 Life Cycle Cost of Physical Assets

This section presents the use of Life Cycle models described in Chapter 1 taking into account the implication of implementing Risk Norms, aiming to support quantitatively the decision of the responsible.

The simulations discussed in the following sub-sections adhere to the traditional Life Cycle Cost (LCC) approach, but they incorporate the profits derived from the use of Physical Assets, significantly enhancing the assessment of scenarios that may occur in the daily operations of companies.

ISO 31000

6.4 Risk assessment

	Questions	Always	Mostly	Generally	Hardly	Never
101	Is the Risk assessment conducted systematically, iteratively and collaboratively?					
102	The organization identify risks, whether or not their sources are under its control?					
103	The uncertainties, risk sources, consequences, likelihood, events, scenarios, controls and their effectiveness are considered					
104	Are the results of the risk analysis compared with the established risk criteria to determine where additional actions are required?					
105						
106						
107						
108						
109						
110						
111						
112						
113						
114						
115						
116						
117						
118						
119						
	score					

Rating

Score	Answers			Mark
		to		
		to		
		to		
		to		
Total Mark (M)				

Category					
1		<	P	≤	
2		<	P	≤	
3		<	P	≤	
4		<	P	≤	
5		<	P	≤	

Explanatory sheet

Questions to review					

Company

Name	Email	Phone

Name of the person who answered	Company function	Date

FIGURE 8.1 Diagnosis sheet referring to risk assessment.

The simulations ought to always follow the same approach, that is, to compare values aiming to demonstrate how the methods presented in this book, namely in Chapter 1, can help the manager to decide if the implementation of the standards discussed in this book add value to the organization or not.

8.4.1.1 Uniform Annual Income Method (UAI)

The first simulation uses the following values:

- Initial Investment (II) –12000
- Life Time simulated 25
- Residual value of equipment (RV) 10
- Linear Depreciation Method (Vn) –12000
- Functioning (*F*) –4500
- Energy –200
- Maintenance (*M*) –300
- Non-Production (NP) 0
- Benefit (B) 50000
- Internal Rate of Return (*r*) 0.12
- Risk (*R*) **0.30**
- Apparent Rate 0.456
- Maintenance Increase Rate 0.01
- Availability 0.8929
- MTBF 500
- MTTR 50
- MWT 10

The remaining values are presented in Table 8.1, and the graphical results are shown in Figures 8.2–8.5.

As referred in the preceding introduction of this section, the evaluation of introduction of Risk Norms, the simulation of their implementation is made through the variation of Risk tax. They may be used other indicators, but may be not so intuitive.

The second simulation uses the following values (Risk is decreased):

- Initial Investment (II) –12000
- Life Time simulated 25
- Residual value of equipment (RV) 10
- Linear Depreciation Method (Vn) –12000
- Functioning (*F*) –4500
- Energy –200
- Maintenance (*M*) –300
- Non-Production (NP) 0
- Benefit (*B*) 50000
- Internal Rate of Return (*r*) 0.12
- Risk (*R*) **0.01**
- Apparent Rate 0.456

TABLE 8.1
Values Used for Simulation based on UAI Method Related to Risk Standards. First Simulation

Time	0	1	2	3	4
Initial Investment (II)	−12000				
Initial Investment (II)	−12000				
II Present Value (IIPV) - Return	−12000	−8241.76	−5660.55	−3887.74	−2670.15
Life Time simulated	25				
Residual value of equipment (RV)	10				
Linear Depreciation Method (Vn)	−12000	−11519.60	−11039.20	−10558.80	−10078.40
LDM Present Value (Vn)		−7911.81	−5207.33	−3420.82	−2242.57
Sum of Digits Method (Vn)	−12000	−11076.15	−10189.26	−9339.32	−8526.34
SDM Present Value (Vn)		−7607.25	−4806.40	−3025.74	−1897.22
Exponential Method (Vn)	−12000	−9036.79	−6805.29	−5124.83	−3859.33
EM Present Value (Pv)		−6206.58	−3210.14	−1660.33	−858.75
Functioning (*F*)	−4500	−4500.00	−4545.00	−4590.45	−4636.35
Present Value Accumulated (NPV)		−3090.66	−5234.59	−6721.80	−7753.45
Present Value (NPV)		−3090.66	−2143.93	−1487.21	−1031.65
Energy	−200	−200.00	−400.00	−600.00	−800.00
Energy (NPV)		−137.36	−188.68	−194.39	−178.01
Accumulated (FPV)		−15228.02	−20651.30	−27567.48	−35498.94
Maintenance (*M*)	−300	−300.00	−303.00	−306.03	−309.09
Present Value (MPV)		−206.04	−142.93	−99.15	−68.78
Accumulated (MPV)		−206.04	−348.97	−448.12	−516.90
Non-Production (NP)	0	0.00	0.00	0.00	0.00
Present Value (NPPV)		0.00	0.00	0.00	0.00
Accumulated NPPV		0.00	0.00	0.00	0.00
Accumulated Total Costs (APV)		−15434.07	−21000.27	−28015.60	−36015.84
Accumulated Total Costs		−5000.00	−10248.00	−15744.48	−21489.92
Benefit (*B*)	50000	44642.86	44642.86	44642.86	44642.86
Present Value (BPV)		30661.30	21058.59	14463.32	9933.60
Accumulated Benefits BPV		30661.30	51719.89	66183.21	76116.80
BPV + FPV + MPV + NPPV + IIPV		3227.24	18719.62	26167.60	28100.97
Uniform Annual Income UAIn (LDM)		−11519.60	−4494.79	−2307.46	−1315.28
Uniform Annual Income UAIn (SDM)		−11076.15	−4148.72	−2040.96	−1112.73
Uniform Annual Income UAIn (EM)		−9036.79	−2770.88	−1119.95	−503.66

(*Continued*)

TABLE 8.1 (*Continued*)
Values Used for Simulation based on UAI Method Related to Risk Standards. First Simulation

Time	5	6	7	8	9
Initial Investment (II)					
Initial Investment (II)					
II Present Value (IIPV) - Return	−1833.89	−1259.54	−865.07	−594.14	−408.06
Life Time simulated					
Residual value of equipment (RV)					
Linear Depreciation Method (Vn)	−9598.00	−9117.60	−8637.20	−8156.80	−7676.40
LDM Present Value (Vn)	−1466.81	−957.00	−622.65	−403.86	−261.04
Sum of Digits Method (Vn)	−7750.31	−7011.23	−6309.11	−5643.94	−5015.72
SDM Present Value (Vn)	−1184.44	−735.91	−454.82	−279.44	−170.56
Exponential Method (Vn)	−2906.33	−2188.66	−1648.20	−1241.20	−934.71
EM Present Value (Pv)	−444.16	−229.73	−118.82	−61.45	−31.79
Functioning (*F*)	−4682.72	−4729.55	−4776.84	−4824.61	−4872.86
Present Value Accumulated (NPV)	−8469.08	−8965.50	−9309.86	−9548.74	−9714.44
Present Value (NPV)	−715.63	−496.42	−344.36	−238.88	−165.70
Energy	−1000.00	−1200.00	−1400.00	−1600.00	−1800.00
Energy (NPV)	−152.82	−125.95	−100.92	−79.22	−61.21
Accumulated (FPV)	−44120.84	−53212.30	−62623.08	−72251.04	−82026.69
Maintenance (*M*)	−312.18	−315.30	−318.46	−321.64	−324.86
Present Value (MPV)	−47.71	−33.09	−22.96	−15.93	−11.05
Accumulated (MPV)	−564.61	−597.70	−620.66	−636.58	−647.63
Non-Production (NP)	0.00	0.00	0.00	0.00	0.00
Present Value (NPPV)	0.00	0.00	0.00	0.00	0.00
Accumulated NPPV	0.00	0.00	0.00	0.00	0.00
Accumulated Total Costs (APV)	−44685.45	−53810.00	−63243.74	−72887.62	−82674.32
Accumulated Total Costs	−27484.82	−33729.67	−40224.97	−46971.22	−53968.93
Benefit (*B*)	44642.86	44642.86	44642.86	44642.86	44642.86
Present Value (BPV)	6822.52	4685.80	3218.27	2210.35	1518.10
Accumulated Benefits BPV	82939.33	87625.13	90843.39	93053.74	94571.84
BPV + FPV + MPV + NPPV + IIPV	26253.88	21815.13	15599.65	8166.12	−102.47
Uniform Annual Income UAIn (LDM)	−789.52	−487.57	−305.99	−193.75	−123.22
Uniform Annual Income UAIn (SDM)	−637.53	−374.93	−223.51	−134.06	−80.51
Uniform Annual Income UAIn (EM)	−239.07	−117.04	−58.39	−29.48	−15.00

(*Continued*)

TABLE 8.1 (*Continued*)
Values Used for Simulation based on UAI Method Related to Risk Standards. First Simulation

Time	10	11	12	13	14
Initial Investment (II)					
Initial Investment (II)					
II Present Value (IIPV) - Return	−280.26	−192.49	−132.20	−90.80	−62.36
Life Time simulated					
Residual value of equipment (RV)					
Linear Depreciation Method (Vn)	−7196.00	−6715.60	−6235.20	−5754.80	−5274.40
LDM Present Value (Vn)	−168.07	−107.72	−68.69	−43.54	−27.41
Sum of Digits Method (Vn)	−4424.46	−3870.15	−3352.80	−2872.40	−2428.95
SDM Present Value (Vn)	−103.33	−62.08	−36.94	−21.73	−12.62
Exponential Method (Vn)	−703.90	−530.08	−399.19	−300.61	−226.38
EM Present Value (Pv)	−16.44	−8.50	−4.40	−2.27	−1.18
Functioning (*F*)	−4921.58	−4970.80	−5020.51	−5070.71	−5121.42
Present Value Accumulated (NPV)	−9829.38	−9909.12	−9964.43	−10002.80	−10029.41
Present Value (NPV)	−114.95	−79.74	−55.31	−38.37	−26.62
Energy	−2000.00	−2200.00	−2400.00	−2600.00	−2800.00
Energy (NPV)	−46.71	−35.29	−26.44	−19.67	−14.55
Accumulated (FPV)	−91902.78	−101847.19	−111838.06	−121860.53	−131904.50
Maintenance (*M*)	−328.11	−331.39	−334.70	−338.05	−341.43
Present Value (MPV)	−7.66	−5.32	−3.69	−2.56	−1.77
Accumulated (MPV)	−655.29	−660.61	−664.30	−666.85	−668.63
Non-Production (NP)	0.00	0.00	0.00	0.00	0.00
Present Value (NPPV)	0.00	0.00	0.00	0.00	0.00
Accumulated NPPV	0.00	0.00	0.00	0.00	0.00
Accumulated Total Costs (APV)	−92558.07	−102507.80	−112502.36	−122527.39	−132573.13
Accumulated Total Costs	−61218.62	−68720.81	−76476.01	−84484.77	−92747.62
Benefit (*B*)	44642.86	44642.86	44642.86	44642.86	44642.86
Present Value (BPV)	1042.65	716.11	491.83	337.80	232.00
Accumulated Benefits BPV	95614.49	96330.60	96822.43	97160.22	97392.23
BPV + FPV + MPV + NPPV + IIPV	−8943.58	−18177.20	−27679.93	−37367.16	−47180.90
Uniform Annual Income UAIn (LDM)	−78.47	−49.92	−31.67	−20.01	−12.56
Uniform Annual Income UAIn (SDM)	−48.25	−28.77	−17.03	−9.99	−5.79
Uniform Annual Income UAIn (EM)	−7.68	−3.94	−2.03	−1.05	−0.54

(*Continued*)

TABLE 8.1 (*Continued*)
Values Used for Simulation based on UAI Method Related to Risk Standards. First Simulation

Time	15	16	17	18	19
Initial Investment (II)					
Initial Investment (II)					
II Present Value (IIPV) - Return	−42.83	−29.42	−20.20	−13.88	−9.53
Life Time simulated					
Residual value of equipment (RV)					
Linear Depreciation Method (Vn)	−4794.00	−4313.60	−3833.20	−3352.80	−2872.40
LDM Present Value (Vn)	−17.11	−10.57	−6.45	−3.88	−2.28
Sum of Digits Method (Vn)	−2022.46	−1652.92	−1320.34	−1024.71	−766.03
SDM Present Value (Vn)	−7.22	−4.05	−2.22	−1.18	−0.61
Exponential Method (Vn)	−170.48	−128.38	−96.68	−72.81	−54.83
EM Present Value (Pv)	−0.61	−0.31	−0.16	−0.08	−0.04
Functioning (*F*)	−5172.63	−5224.36	−5276.60	−5329.37	−5382.66
Present Value Accumulated (NPV)	−10047.88	−10060.68	−10069.57	−10075.73	−10080.01
Present Value (NPV)	−18.46	−12.81	−8.88	−6.16	−4.27
Energy	−3000.00	−3200.00	−3400.00	−3600.00	−3800.00
Energy (NPV)	−10.71	−7.84	−5.72	−4.16	−3.02
Accumulated (FPV)	−141963.08	−152031.61	−162106.90	−172186.80	−182269.82
Maintenance (*M*)	−344.84	−348.29	−351.77	−355.29	−358.84
Present Value (MPV)	−1.23	−0.85	−0.59	−0.41	−0.28
Accumulated (MPV)	−669.86	−670.71	−671.30	−671.72	−672.00
Non-Production (NP)	0.00	0.00	0.00	0.00	0.00
Present Value (NPPV)	0.00	0.00	0.00	0.00	0.00
Accumulated NPPV	0.00	0.00	0.00	0.00	0.00
Accumulated Total Costs (APV)	−142632.94	−152702.32	−162778.21	−172858.51	−182941.82
Accumulated Total Costs	−101265.10	−110037.75	−119066.13	−128350.79	−137892.30
Benefit (*B*)	44642.86	44642.86	44642.86	44642.86	44642.86
Present Value (BPV)	159.34	109.44	75.16	51.62	35.46
Accumulated Benefits BPV	97551.57	97661.01	97736.17	97787.79	97823.25
BPV + FPV + MPV + NPPV + IIPV	−57081.37	−67041.32	−77042.04	−87070.72	−97118.57
Uniform Annual Income UAIn (LDM)	−7.83	−4.83	−2.95	−1.77	−1.04
Uniform Annual Income UAIn (SDM)	−3.30	−1.85	−1.02	−0.54	−0.28
Uniform Annual Income UAIn (EM)	−0.28	−0.14	−0.07	−0.04	−0.02

(*Continued*)

TABLE 8.1 (*Continued*)
Values Used for Simulation based on UAI Method Related to Risk Standards. First Simulation

Time	20	21	22	23	24	25
Initial Investment (II)						
Initial Investment (II)						
II Present Value (IIPV) - Return	−6.55	−4.50	−3.09	−2.12	−1.46	−1.00
Life Time simulated						
Residual value of equipment (RV)						
Linear Depreciation Method (Vn)	−2392.00	−1911.60	−1431.20	−950.80	−470.40	10.00
LDM Present Value (Vn)	−1.30	−0.72	−0.37	−0.17	−0.06	0.00
Sum of Digits Method (Vn)	−544.31	−359.54	−211.72	−100.86	−26.95	10.00
SDM Present Value (Vn)	−0.30	−0.13	−0.05	−0.02	0.00	0.00
Exponential Method (Vn)	−41.29	−31.09	−23.42	−17.63	−13.28	−10.00
EM Present Value (Pv)	−0.02	−0.01	−0.01	0.00	0.00	0.00
Functioning (*F*)	−5436.49	−5490.86	−5545.76	−5601.22	−5657.23	−5713.81
Present Value Accumulated (NPV)	−10082.97	−10085.03	−10086.45	−10087.44	−10088.13	−10088.61
Present Value (NPV)	−2.97	−2.06	−1.43	−0.99	−0.69	−0.48
Energy	−4000.00	−4200.00	−4400.00	−4600.00	−4800.00	−5000.00
Energy (NPV)	−2.18	−1.57	−1.13	−0.81	−0.58	−0.42
Accumulated (FPV)	−192354.97	−202441.57	−212529.16	−222617.42	−232706.13	−242795.15
Maintenance (*M*)	−362.43	−366.06	−369.72	−373.41	−377.15	−380.92
Present Value (MPV)	−0.20	−0.14	−0.10	−0.07	−0.05	−0.03
Accumulated (MPV)	−672.20	−672.34	−672.43	−672.50	−672.54	−672.57
Non-Production (NP)	0.00	0.00	0.00	0.00	0.00	0.00
Present Value (NPPV)	0.00	0.00	0.00	0.00	0.00	0.00
Accumulated NPPV	0.00	0.00	0.00	0.00	0.00	0.00
Accumulated Total Costs (APV)	−193027.17	−203113.91	−213201.59	−223289.91	−233378.67	−243467.73
Accumulated Total Costs	−147691.22	−157748.13	−168063.61	−178638.25	−189472.63	−200567.36
Benefit (*B*)	44642.86	44642.86	44642.86	44642.86	44642.86	44642.86
Present Value (BPV)	24.35	16.72	11.49	7.89	5.42	3.72
Accumulated Benefits BPV	97847.60	97864.33	97875.81	97883.70	97889.12	97892.84
BPV + FPV + MPV + NPPV + IIPV	−107179.57	−117249.58	−127325.78	−137406.21	−147489.55	−157574.89
Uniform Annual Income UAIn (LDM)	−0.60	−0.33	−0.17	−0.08	−0.03	0.00
Uniform Annual Income UAIn (SDM)	−0.14	−0.06	−0.02	−0.01	0.00	0.00
Uniform Annual Income UAIn (EM)	−0.01	−0.01	0.00	0.00	0.00	0.00

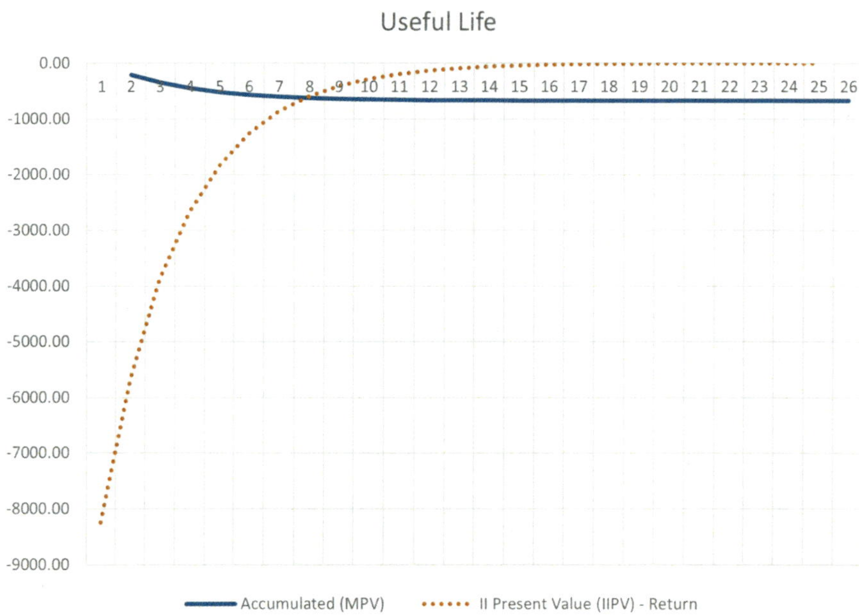

FIGURE 8.2 Physical Asset useful life.

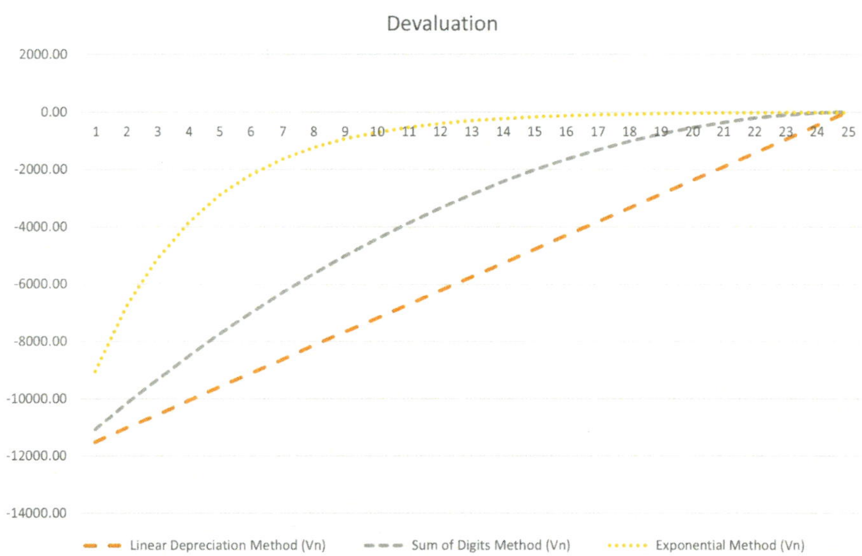

FIGURE 8.3 The three devaluation methods.

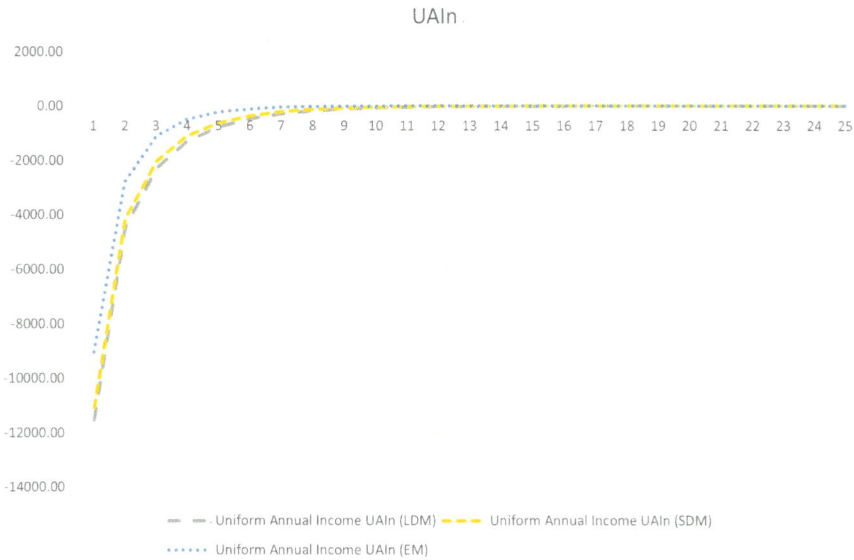

FIGURE 8.4 The UAI results from the three devaluation methods.

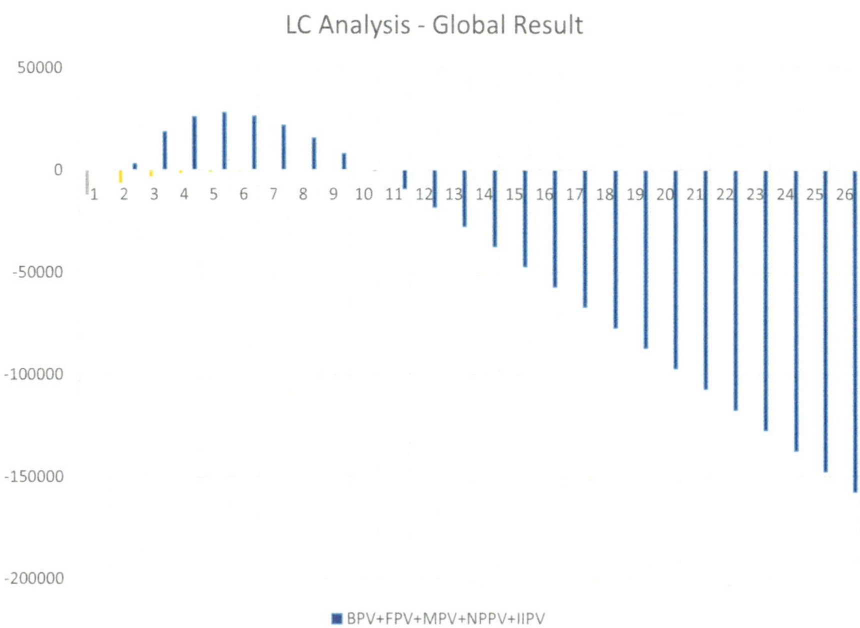

FIGURE 8.5 Life cycle analysis global result including profits.

- Maintenance Increase Rate 0.01
- Availability 0.8929
- MTBF 500
- MTTR 50
- MWT 10

The remaining values are presented in Table 8.2, and the graphical results are shown in Figures 8.6–8.9.

As can be seen, namely comparing Figures 8.5 and 8.9, with the implementation of Physical Assets Standards, it gets five more positive years, and the positive incomes increase a lot.

8.4.1.2 Minimization of Total Average Cost Method (MMTAC)

The first simulation uses the following values:

- Initial Investment (II) –12000
- Life Time simulated 25
- Residual value of equipment (RV) 10
- Functioning (*F*) –4500
- Benefit (B) –200
- Maintenance (*M*) –300
- Non-Production (NP) 0
- Benefit (*B*) 70000
- Internal Rate of Return (*r*) 0.12
- Risk (*R*) **0.30**
- Apparent Rate 0.456
- Maintenance Increase Rate 0.01
- Availability 0.9662
- MTBF 1000
- MTTR 25
- MWT 10

The remaining values are presented in Table 8.3, and the graphical results are shown in Figures 8.10–8.13.

As referred in the preceding introduction of this section, the evaluation of introduction of Risk Norms, the simulation of their implementation is made through the increase of Risk tax. They may be used other indicators, but may be not so intuitive.

The second simulation uses the following values (the value of Risk decreases a lot as a result of the Maintenance Standards):

- Initial Investment (II) –12000
- Life Time simulated 25
- Residual value of equipment (RV) 10
- Functioning (*F*) –4500
- Benefit (B) –200
- Maintenance (*M*) –300

TABLE 8.2
Values Used for Simulation based on UAI Method Related to Risk Standards. Second Simulation

Time	0	1	2	3	4	5
Initial Investment (II)	−12000					
Initial Investment (II)	−12000					
II Present Value (IIPV) - Return	−12000	−10608.20	−9377.83	−8290.16	−7328.64	−6478.65
Life Time simulated	25					
Residual value of equipment (RV)	10					
Linear Depreciation Method (Vn)	−12000	−11519.60	−11039.20	−10558.80	−10078.40	−9598.00
LDM Present Value (Vn)		−10183.52	−8626.98	−7294.51	−6155.08	−5181.84
Sum of Digits Method (Vn)	−12000	−11076.15	−10189.26	−9339.32	−8526.34	−7750.31
SDM Present Value (Vn)		−9791.51	−7962.77	−6452.04	−5207.21	−4184.29
Exponential Method (Vn)	−12000	−9036.79	−6805.29	−5124.83	−3859.33	−2906.33
EM Present Value (Pv)		−7988.67	−5318.24	−3540.47	−2356.97	−1569.09
Functioning (*F*)	−4500	−4500.00	−4545.00	−4590.45	−4636.35	−4682.72
Present Value Accumulated (NPV)		−3978.08	−7529.93	−10701.23	−13532.74	−16060.88
Present Value (NPV)		−3978.08	−3551.85	−3171.30	−2831.52	−2528.14
Energy	−200	−200.00	−400.00	−600.00	−800.00	−1000.00
Energy (NPV)		−176.80	−312.59	−414.51	−488.58	−539.89
Accumulated (FPV)		−16154.88	−23997.40	−35113.14	−49134.46	−65735.23
Maintenance (*M*)	−300	−300.00	−303.00	−306.03	−309.09	−312.18
Present Value (MPV)		−265.21	−236.79	−211.42	−188.77	−168.54
Accumulated (MPV)		−265.21	−502.00	−713.42	−902.18	−1070.73
Non-Production (NP)	0	0.00	0.00	0.00	0.00	0.00
Present Value (NPPV)		0.00	0.00	0.00	0.00	0.00
Accumulated NPPV		0.00	0.00	0.00	0.00	0.00
Accumulated Total Costs (APV)		−16420.08	−24499.40	−35826.56	−50036.64	−66805.96
Accumulated Total Costs		−5000.00	−10248.00	−15744.48	−21489.92	−27484.82
Benefit (*B*)	50000	44642.86	44642.86	44642.86	44642.86	44642.86
Present Value (BPV)		39465.04	34887.77	30841.38	27264.30	24102.11
Accumulated Benefits BPV		39465.04	74352.81	105194.19	132458.49	156560.60
BPV + FPV + MPV + NPPV + IIPV		11044.96	37853.41	57367.63	70421.85	77754.64
Uniform Annual Income UAIn (LDM)		−11519.60	−5179.80	−3095.68	−2074.47	−1477.59
Uniform Annual Income UAIn (SDM)		−11076.15	−4781.00	−2738.15	−1755.00	−1193.14
Uniform Annual Income UAIn (EM)		−9036.79	−3193.17	−1502.52	−794.38	−447.42

(*Continued*)

TABLE 8.2 (*Continued*)
Values Used for Simulation based on UAI Method Related to Risk Standards. Second Simulation

Time	6	7	8	9	10
Initial Investment (II)					
Initial Investment (II)					
II Present Value (IIPV) - Return	−5727.23	−5062.97	−4475.75	−3956.64	−3497.74
Life Time simulated					
Residual value of equipment (RV)					
Linear Depreciation Method (Vn)	−9117.60	−8637.20	−8156.80	−7676.40	−7196.00
LDM Present Value (Vn)	−4351.55	−3644.16	−3042.32	−2531.06	−2097.48
Sum of Digits Method (Vn)	−7011.23	−6309.11	−5643.94	−5015.72	−4424.46
SDM Present Value (Vn)	−3346.25	−2661.90	−2105.07	−1653.78	−1289.63
Exponential Method (Vn)	−2188.66	−1648.20	−1241.20	−934.71	−703.90
EM Present Value (Pv)	−1044.58	−695.40	−462.94	−308.19	−205.17
Functioning (*F*)	−4729.55	−4776.84	−4824.61	−4872.86	−4921.58
Present Value Accumulated (NPV)	−18318.15	−20333.57	−22133.05	−23739.73	−25174.26
Present Value (NPV)	−2257.27	−2015.42	−1799.48	−1606.68	−1434.53
Energy	−1200.00	−1400.00	−1600.00	−1800.00	−2000.00
Energy (NPV)	−572.72	−590.68	−596.77	−593.50	−582.96
Accumulated (FPV)	−84626.11	−105550.36	−128280.17	−152613.40	−178370.61
Maintenance (*M*)	−315.30	−318.46	−321.64	−324.86	−328.11
Present Value (MPV)	−150.48	−134.36	−119.97	−107.11	−95.64
Accumulated (MPV)	−1221.21	−1355.57	−1475.54	−1582.65	−1678.28
Non-Production (NP)	0.00	0.00	0.00	0.00	0.00
Present Value (NPPV)	0.00	0.00	0.00	0.00	0.00
Accumulated NPPV	0.00	0.00	0.00	0.00	0.00
Accumulated Total Costs (APV)	−85847.32	−106905.93	−129755.71	−154196.04	−180048.90
Accumulated Total Costs	−33729.67	−40224.97	−46971.22	−53968.93	−61218.62
Benefit (*B*)	44642.86	44642.86	44642.86	44642.86	44642.86
Present Value (BPV)	21306.67	18835.46	16650.87	14719.65	13012.42
Accumulated Benefits BPV	177867.27	196702.73	213353.60	228073.24	241085.66
BPV + FPV + MPV + NPPV + IIPV	80019.95	77796.80	71597.89	61877.20	49036.77
Uniform Annual Income UAIn (LDM)	−1092.19	−827.06	−636.59	−495.43	−388.40
Uniform Annual Income UAIn (SDM)	−839.87	−604.13	−440.47	−323.71	−238.81
Uniform Annual Income UAIn (EM)	−262.18	−157.83	−96.87	−60.33	−37.99

(*Continued*)

TABLE 8.2 (*Continued*)
Values Used for Simulation based on UAI Method Related to Risk Standards. Second Simulation

Time	11	12	13	14	15
Initial Investment (II)					
Initial Investment (II)					
II Present Value (IIPV) - Return	−3092.06	−2733.43	−2416.40	−2136.14	−1888.38
Life Time simulated					
Residual value of equipment (RV)					
Linear Depreciation Method (Vn)	−6715.60	−6235.20	−5754.80	−5274.40	−4794.00
LDM Present Value (Vn)	−1730.42	−1420.29	−1158.83	−938.90	−754.41
Sum of Digits Method (Vn)	−3870.15	−3352.80	−2872.40	−2428.95	−2022.46
SDM Present Value (Vn)	−997.23	−763.72	−578.41	−432.38	−318.27
Exponential Method (Vn)	−530.08	−399.19	−300.61	−226.38	−170.48
EM Present Value (Pv)	−136.59	−90.93	−60.53	−40.30	−26.83
Functioning (*F*)	−4970.80	−5020.51	−5070.71	−5121.42	−5172.63
Present Value Accumulated (NPV)	−26455.10	−27598.70	−28619.77	−29531.44	−30345.44
Present Value (NPV)	−1280.83	−1143.60	−1021.07	−911.67	−813.99
Energy	−2200.00	−2400.00	−2600.00	−2800.00	−3000.00
Energy (NPV)	−566.88	−546.69	−523.55	−498.43	−472.10
Accumulated (FPV)	−205392.59	−233537.97	−262681.29	−292711.17	−323528.70
Maintenance (*M*)	−331.39	−334.70	−338.05	−341.43	−344.84
Present Value (MPV)	−85.39	−76.24	−68.07	−60.78	−54.27
Accumulated (MPV)	−1763.67	−1839.91	−1907.98	−1968.76	−2023.03
Non-Production (NP)	0.00	0.00	0.00	0.00	0.00
Present Value (NPPV)	0.00	0.00	0.00	0.00	0.00
Accumulated NPPV	0.00	0.00	0.00	0.00	0.00
Accumulated Total Costs (APV)	−207156.26	−235377.88	−264589.28	−294679.93	−325551.73
Accumulated Total Costs	−68720.81	−76476.01	−84484.77	−92747.62	−101265.10
Benefit (*B*)	44642.86	44642.86	44642.86	44642.86	44642.86
Present Value (BPV)	11503.20	10169.02	8989.59	7946.95	7025.24
Accumulated Benefits BPV	252588.86	262757.88	271747.47	279694.42	286719.66
BPV + FPV + MPV + NPPV + IIPV	33432.60	15380.00	−4841.81	−26985.51	−50832.07
Uniform Annual Income UAIn (LDM)	−305.84	−241.31	−190.37	−149.86	−117.46
Uniform Annual Income UAIn (SDM)	−176.25	−129.76	−95.02	−69.01	−49.55
Uniform Annual Income UAIn (EM)	−24.14	−15.45	−9.94	−6.43	−4.18

(*Continued*)

TABLE 8.2 (*Continued*)
Values Used for Simulation based on UAI Method Related to Risk Standards. Second Simulation

Time	16	17	18	19	20
Initial Investment (II)					
Initial Investment (II)					
II Present Value (IIPV) - Return	−1669.36	−1475.75	−1304.58	−1153.27	−1019.51
Life Time simulated					
Residual value of equipment (RV)					
Linear Depreciation Method (Vn)	−4313.60	−3833.20	−3352.80	−2872.40	−2392.00
LDM Present Value (Vn)	−600.08	−471.40	−364.50	−276.06	−203.22
Sum of Digits Method (Vn)	−1652.92	−1320.34	−1024.71	−766.03	−544.31
SDM Present Value (Vn)	−229.94	−162.37	−111.40	−73.62	−46.24
Exponential Method (Vn)	−128.38	−96.68	−72.81	−54.83	−41.29
EM Present Value (Pv)	−17.86	−11.89	−7.92	−5.27	−3.51
Functioning (*F*)	−5224.36	−5276.60	−5329.37	−5382.66	−5436.49
Present Value Accumulated (NPV)	−31072.22	−31721.13	−32300.51	−32817.82	−33279.70
Present Value (NPV)	−726.78	−648.91	−579.38	−517.31	−461.88
Energy	−3200.00	−3400.00	−3600.00	−3800.00	−4000.00
Energy (NPV)	−445.16	−418.13	−391.38	−365.20	−339.84
Accumulated (FPV)	−355046.08	−387185.33	−419877.22	−453060.24	−486679.78
Maintenance (*M*)	−348.29	−351.77	−355.29	−358.84	−362.43
Present Value (MPV)	−48.45	−43.26	−38.63	−34.49	−30.79
Accumulated (MPV)	−2071.48	−2114.74	−2153.37	−2187.85	−2218.65
Non-Production (NP)	0.00	0.00	0.00	0.00	0.00
Present Value (NPPV)	0.00	0.00	0.00	0.00	0.00
Accumulated NPPV	0.00	0.00	0.00	0.00	0.00
Accumulated Total Costs (APV)	−357117.56	−389300.07	−422030.59	−455248.09	−488898.42
Accumulated Total Costs	−110037.75	−119066.13	−128350.79	−137892.30	−147691.22
Benefit (*B*)	44642.86	44642.86	44642.86	44642.86	44642.86
Present Value (BPV)	6210.43	5490.13	4853.36	4290.46	3792.84
Accumulated Benefits BPV	292930.09	298420.21	303273.58	307564.03	311356.87
BPV + FPV + MPV + NPPV + IIPV	−76187.47	−102879.86	−130757.01	−159684.06	−189541.55
Uniform Annual Income UAIn (LDM)	−91.45	−70.52	−53.66	−40.07	−29.14
Uniform Annual Income UAIn (SDM)	−35.04	−24.29	−16.40	−10.69	−6.63
Uniform Annual Income UAIn (EM)	−2.72	−1.78	−1.17	−0.76	−0.50

(*Continued*)

TABLE 8.2 (*Continued*)
Values Used for Simulation based on UAI Method Related to Risk Standards. Second Simulation

Time	21	22	23	24	25
Initial Investment (II)					
Initial Investment (II)					
II Present Value (IIPV) - Return	−901.27	−796.74	−704.33	−622.64	−550.42
Life Time simulated					
Residual value of equipment (RV)					
Linear Depreciation Method (Vn)	−1911.60	−1431.20	−950.80	−470.40	10.00
LDM Present Value (Vn)	−143.57	−95.02	−55.81	−24.41	0.46
Sum of Digits Method (Vn)	−359.54	−211.72	−100.86	−26.95	10.00
SDM Present Value (Vn)	−27.00	−14.06	−5.92	−1.40	0.46
Exponential Method (Vn)	−31.09	−23.42	−17.63	−13.28	−10.00
EM Present Value (Pv)	−2.34	−1.55	−1.03	−0.69	−0.46
Functioning (*F*)	−5490.86	−5545.76	−5601.22	−5657.23	−5713.81
Present Value Accumulated (NPV)	−33692.09	−34060.30	−34389.06	−34682.60	−34944.68
Present Value (NPV)	−412.39	−368.21	−328.76	−293.53	−262.08
Energy	−4200.00	−4400.00	−4600.00	−4800.00	−5000.00
Energy (NPV)	−315.44	−292.14	−269.99	−249.06	−229.34
Accumulated (FPV)	−520687.31	−555039.75	−589698.81	−624630.46	−659804.48
Maintenance (*M*)	−366.06	−369.72	−373.41	−377.15	−380.92
Present Value (MPV)	−27.49	−24.55	−21.92	−19.57	−17.47
Accumulated (MPV)	−2246.14	−2270.69	−2292.60	−2312.17	−2329.65
Non-Production (NP)	0.00	0.00	0.00	0.00	0.00
Present Value (NPPV)	0.00	0.00	0.00	0.00	0.00
Accumulated NPPV	0.00	0.00	0.00	0.00	0.00
Accumulated Total Costs (APV)	−522933.45	−557310.44	−591991.41	−626942.63	−662134.13
Accumulated Total Costs	−157748.13	−168063.61	−178638.25	−189472.63	−200567.36
Benefit (*B*)	44642.86	44642.86	44642.86	44642.86	44642.86
Present Value (BPV)	3352.93	2964.05	2620.27	2316.36	2047.70
Accumulated Benefits BPV	314709.80	317673.85	320294.12	322610.48	324658.18
BPV + FPV + MPV + NPPV + IIPV	−220223.65	−251636.59	−283697.29	−316332.15	−349475.94
Uniform Annual Income UAIn (LDM)	−20.37	−13.35	−7.78	−3.38	0.06
Uniform Annual Income UAIn (SDM)	−3.83	−1.98	−0.83	−0.19	0.06
Uniform Annual Income UAIn (EM)	−0.33	−0.22	−0.14	−0.10	−0.06

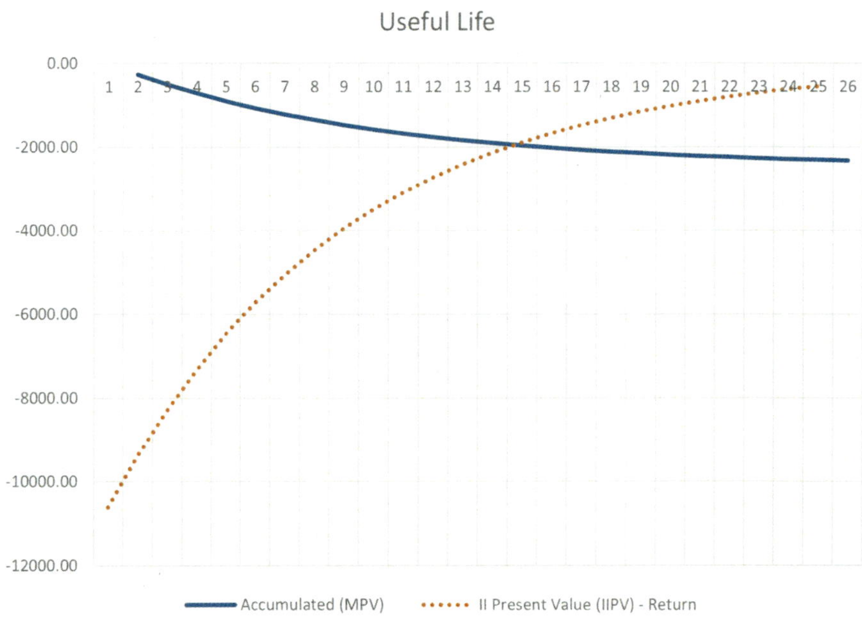

FIGURE 8.6 Physical Asset useful life.

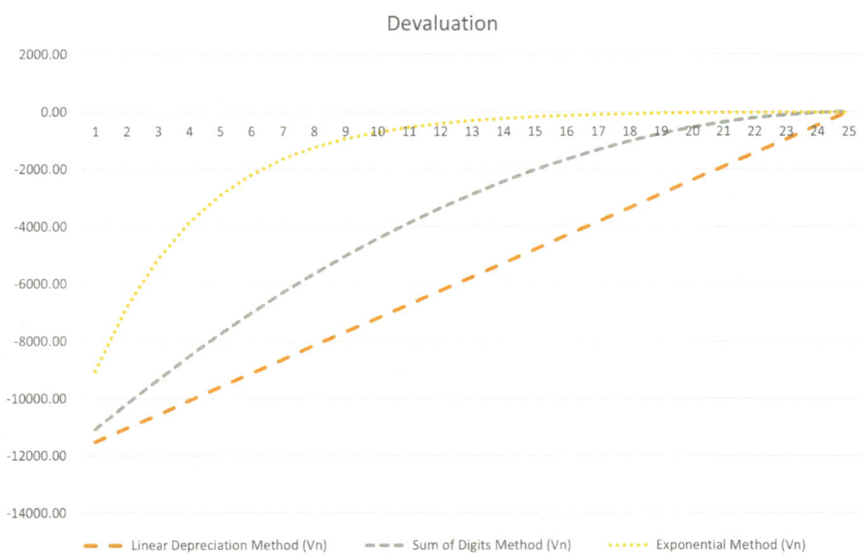

FIGURE 8.7 The three devaluation methods.

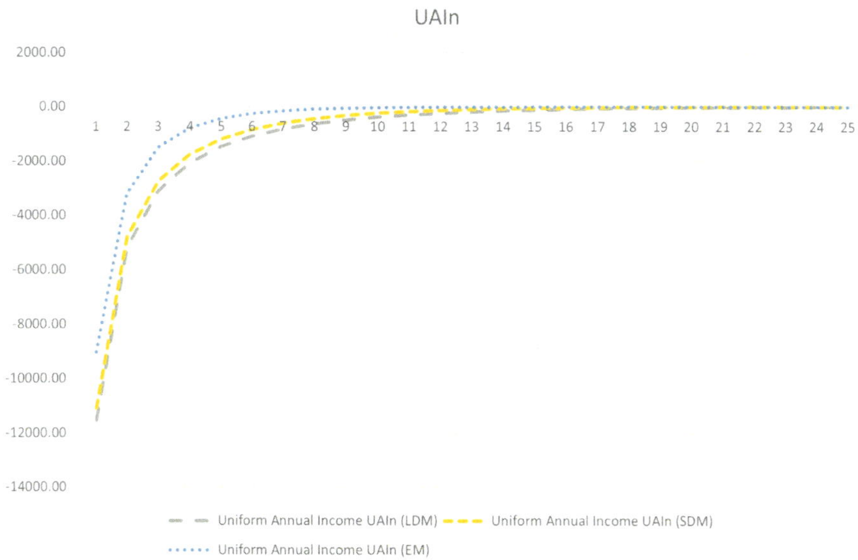

FIGURE 8.8 The UAI results from the three devaluation methods.

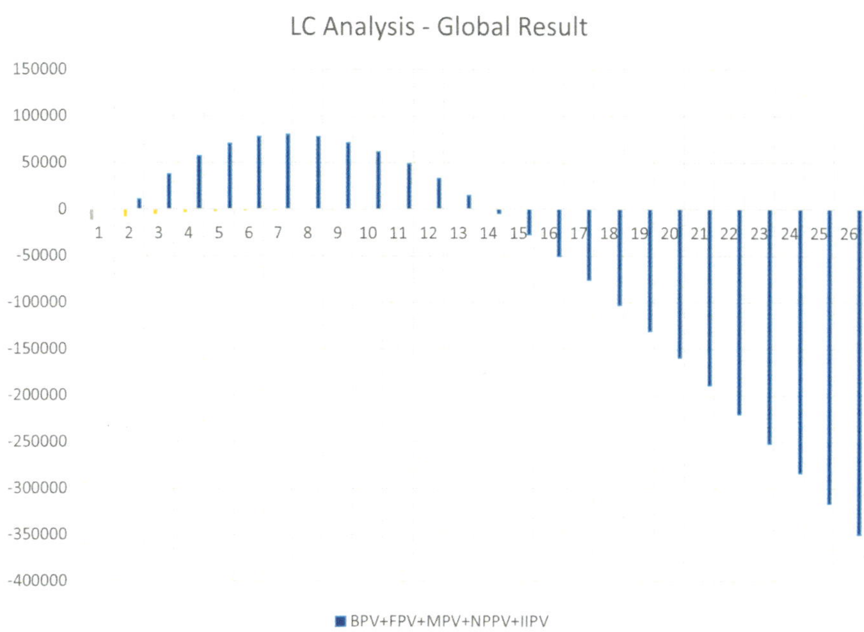

FIGURE 8.9 Life cycle analysis global result including profits.

TABLE 8.3

Values Used for Simulation based on MMTAC Method Related to Risk Standards. First Simulation

Time	0	1	2	3	4	5
Initial Investment (II)	−12000					
Initial Investment (II)	−12000					
II (IIPV) - Return	−12000	−12000.00	−12000.00	−12000.00	−12000.00	−12000.00
Useful Life	25					
Residual value of equipment (RV)	10					
Linear Depreciation Method (Vn)	−12000	−11519.60	−11039.20	−10558.80	−10078.40	−9598.00
LDM Present Value (PV)		−7911.81	−5207.33	−3420.82	−2242.57	−1466.81
Sum of Digits Method (Vn)	−12000	−11076.15	−10189.26	−9339.32	−8526.34	−7750.31
SDM Present Value (PV)		−7607.25	−4806.40	−3025.74	−1897.22	−1184.44
Exponential Method (Vn)	−12000	−9036.79	−6805.29	−5124.83	−3859.33	−2906.33
EM Present Value (Pv)		−6206.58	−3210.14	−1660.33	−858.75	−444.16
Functioning (*F*)	−4500	−4500.00	−4545.00	−4590.45	−4636.35	−4682.72
Functioning Accumulated (FA)		−4500.00	−9045.00	−13635.45	−18271.80	−22954.52
Present Value Accumulated (NPV)		−3090.66	−5234.59	−6721.80	−7753.45	−8469.08
Present Value (NPV)		−3090.66	−2143.93	−1487.21	−1031.65	−715.63
Energy	−200	−200.00	−400.00	−600.00	−800.00	−1000.00
Energy Accumulated (EA)		−200.00	−600.00	−1200.00	−2000.00	−3000.00
Energy (NPV)		−137.36	−188.68	−194.39	−178.01	−152.82
Accumulated Functioning (FPV)		−15228.02	−20651.30	−27567.48	−35498.94	−44120.84
Maintenance (*M*)	−300	−300.00	−303.00	−306.03	−309.09	−312.18
Maintenane Accumulated (MA)		−300.00	−603.00	−909.03	−1218.12	−1530.30
Present Value (MPV)		−206.04	−142.93	−99.15	−68.78	−47.71
Accumulated MPV		−206.04	−348.97	−448.12	−516.90	−564.61
Non-Production (NP)	0	0.00	0.00	0.00	0.00	0.00
Non-Production Accumulated (NPA)		0.00	0.00	0.00	0.00	0.00
Present Value (NPPV)		0.00	0.00	0.00	0.00	0.00
Accumulated NPPV		0.00	0.00	0.00	0.00	0.00
Accumulated Total Costs (APV)		−15434.07	−21000.27	−28015.60	−36015.84	−44685.45
Accumulated Total Costs		−5000.00	−10248.00	−15744.48	−21489.92	−27484.82
Benefit (*B*)	70000	67632.85	67632.85	67632.85	67632.85	67632.85
Benefits Accumulated (BA)		67632.85	135265.70	202898.55	270531.40	338164.25
Present Value (BPV)		46451.13	31903.25	21911.57	15049.16	10335.96
Accumulated Benefits BPV		46451.13	78354.38	100265.96	115315.12	125651.08
BPV + FPV + MPV + NPPV + IIPV		50632.85	113017.70	175154.07	237041.48	298679.43
MTAC (LDM)		−5480.40	−11208.80	−17185.68	−23411.52	−29886.82
MTAC (SDM)		−5923.85	−12058.74	−18405.16	−24963.59	−31734.52
MTAC (EM)		−7963.21	−15442.71	−22619.65	−29630.59	−36578.49

(*Continued*)

TABLE 8.3 (*Continued*)
Values Used for Simulation based on MMTAC Method Related to Risk Standards. First Simulation

Time	6	7	8	9	10
Initial Investment (II)					
Initial Investment (II)					
II (IIPV) - Return	−12000.00	−12000.00	−12000.00	−12000.00	−12000.00
Useful Life					
Residual value of equipment (RV)					
Linear Depreciation Method (Vn)	−9117.60	−8637.20	−8156.80	−7676.40	−7196.00
LDM Present Value (PV)	−957.00	−622.65	−403.86	−261.04	−168.07
Sum of Digits Method (Vn)	−7011.23	−6309.11	−5643.94	−5015.72	−4424.46
SDM Present Value (PV)	−735.91	−454.82	−279.44	−170.56	−103.33
Exponential Method (Vn)	−2188.66	−1648.20	−1241.20	−934.71	−703.90
EM Present Value (Pv)	−229.73	−118.82	−61.45	−31.79	−16.44
Functioning (*F*)	−4729.55	−4776.84	−4824.61	−4872.86	−4921.58
Functioning Accumulated (FA)	−27684.07	−32460.91	−37285.52	−42158.37	−47079.96
Present Value Accumulated (NPV)	−8965.50	−9309.86	−9548.74	−9714.44	−9829.38
Present Value (NPV)	−496.42	−344.36	−238.88	−165.70	−114.95
Energy	−1200.00	−1400.00	−1600.00	−1800.00	−2000.00
Energy Accumulated (EA)	−4200.00	−5600.00	−7200.00	−9000.00	−11000.00
Energy (NPV)	−125.95	−100.92	−79.22	−61.21	−46.71
Accumulated Functioning (FPV)	−53212.30	−62623.08	−72251.04	−82026.69	−91902.78
Maintenance (*M*)	−315.30	−318.46	−321.64	−324.86	−328.11
Maintenane Accumulated (MA)	−1845.60	−2164.06	−2485.70	−2810.56	−3138.66
Present Value (MPV)	−33.09	−22.96	−15.93	−11.05	−7.66
Accumulated MPV	−597.70	−620.66	−636.58	−647.63	−655.29
Non-Production (NP)	0.00	0.00	0.00	0.00	0.00
Non-Production Accumulated (NPA)	0.00	0.00	0.00	0.00	0.00
Present Value (NPPV)	0.00	0.00	0.00	0.00	0.00
Accumulated NPPV	0.00	0.00	0.00	0.00	0.00
Accumulated Total Costs (APV)	−53810.00	−63243.74	−72887.62	−82674.32	−92558.07
Accumulated Total Costs	−33729.67	−40224.97	−46971.22	−53968.93	−61218.62
Benefit (*B*)	67632.85	67632.85	67632.85	67632.85	67632.85
Benefits Accumulated (BA)	405797.10	473429.95	541062.80	608695.65	676328.50
Present Value (BPV)	7098.87	4875.60	3348.63	2299.88	1579.59
Accumulated Benefits BPV	132749.95	137625.55	140974.18	143274.06	144853.64
BPV + FPV + MPV + NPPV + IIPV	360067.43	421204.98	482091.58	542726.72	603109.88
MTAC (LDM)	−36612.07	−43587.77	−50814.42	−58292.53	−66022.62
MTAC (SDM)	−38718.44	−45915.86	−53327.28	−60953.21	−68794.16
MTAC (EM)	−43541.02	−50576.77	−57730.01	−65034.22	−72514.72

(*Continued*)

TABLE 8.3 (*Continued*)
Values Used for Simulation based on MMTAC Method Related to Risk Standards. First Simulation

Time	11	12	13	14	15
Initial Investment (II)					
Initial Investment (II)					
II (IIPV) - Return	−12000.00	−12000.00	−12000.00	−12000.00	−12000.00
Useful Life					
Residual value of equipment (RV)					
Linear Depreciation Method (Vn)	−6715.60	−6235.20	−5754.80	−5274.40	−4794.00
LDM Present Value (PV)	−107.72	−68.69	−43.54	−27.41	−17.11
Sum of Digits Method (Vn)	−3870.15	−3352.80	−2872.40	−2428.95	−2022.46
SDM Present Value (PV)	−62.08	−36.94	−21.73	−12.62	−7.22
Exponential Method (Vn)	−530.08	−399.19	−300.61	−226.38	−170.48
EM Present Value (Pv)	−8.50	−4.40	−2.27	−1.18	−0.61
Functioning (*F*)	−4970.80	−5020.51	−5070.71	−5121.42	−5172.63
Functioning Accumulated (FA)	−52050.76	−57071.26	−62141.98	−67263.40	−72436.03
Present Value Accumulated (NPV)	−9909.12	−9964.43	−10002.80	−10029.41	−10047.88
Present Value (NPV)	−79.74	−55.31	−38.37	−26.62	−18.46
Energy	−2200.00	−2400.00	−2600.00	−2800.00	−3000.00
Energy Accumulated (EA)	−13200.00	−15600.00	−18200.00	−21000.00	−24000.00
Energy (NPV)	−35.29	−26.44	−19.67	−14.55	−10.71
Accumulated Functioning (FPV)	−101847.19	−111838.06	−121860.53	−131904.50	−141963.08
Maintenance (*M*)	−331.39	−334.70	−338.05	−341.43	−344.84
Maintenane Accumulated (MA)	−3470.05	−3804.75	−4142.80	−4484.23	−4829.07
Present Value (MPV)	−5.32	−3.69	−2.56	−1.77	−1.23
Accumulated MPV	−660.61	−664.30	−666.85	−668.63	−669.86
Non-Production (NP)	0.00	0.00	0.00	0.00	0.00
Non-Production Accumulated (NPA)	0.00	0.00	0.00	0.00	0.00
Present Value (NPPV)	0.00	0.00	0.00	0.00	0.00
Accumulated NPPV	0.00	0.00	0.00	0.00	0.00
Accumulated Total Costs (APV)	−102507.80	−112502.36	−122527.39	−132573.13	−142632.94
Accumulated Total Costs	−68720.81	−76476.01	−84484.77	−92747.62	−101265.10
Benefit (*B*)	67632.85	67632.85	67632.85	67632.85	67632.85
Benefits Accumulated (BA)	743961.35	811594.20	879227.05	946859.90	1014492.75
Present Value (BPV)	1084.88	745.11	511.75	351.48	241.40
Accumulated Benefits BPV	145938.53	146683.64	147195.39	147546.87	147788.27
BPV + FPV + MPV + NPPV + IIPV	663240.55	723118.19	782742.28	842112.28	901227.66
MTAC (LDM)	−74005.21	−82240.81	−90729.97	−99473.22	−108471.10
MTAC (SDM)	−76850.65	−85123.21	−93612.37	−102318.67	−111242.64
MTAC (EM)	−80190.73	−88076.83	−96184.16	−104521.24	−113094.62

(*Continued*)

TABLE 8.3 (*Continued*)
Values Used for Simulation based on MMTAC Method Related to Risk Standards. First Simulation

Time	16	17	18	19	20
Initial Investment (II)					
Initial Investment (II)					
II (IIPV) - Return	−12000.00	−12000.00	−12000.00	−12000.00	−12000.00
Useful Life					
Residual value of equipment (RV)					
Linear Depreciation Method (Vn)	−4313.60	−3833.20	−3352.80	−2872.40	−2392.00
LDM Present Value (PV)	−10.57	−6.45	−3.88	−2.28	−1.30
Sum of Digits Method (Vn)	−1652.92	−1320.34	−1024.71	−766.03	−544.31
SDM Present Value (PV)	−4.05	−2.22	−1.18	−0.61	−0.30
Exponential Method (Vn)	−128.38	−96.68	−72.81	−54.83	−41.29
EM Present Value (Pv)	−0.31	−0.16	−0.08	−0.04	−0.02
Functioning (*F*)	−5224.36	−5276.60	−5329.37	−5382.66	−5436.49
Functioning Accumulated (FA)	−77660.39	−82936.99	−88266.36	−93649.03	−99085.52
Present Value Accumulated (NPV)	−10060.68	−10069.57	−10075.73	−10080.01	−10082.97
Present Value (NPV)	−12.81	−8.88	−6.16	−4.27	−2.97
Energy	−3200.00	−3400.00	−3600.00	−3800.00	−4000.00
Energy Accumulated (EA)	−27200.00	−30600.00	−34200.00	−38000.00	−42000.00
Energy (NPV)	−7.84	−5.72	−4.16	−3.02	−2.18
Accumulated Functioning (FPV)	−152031.61	−162106.90	−172186.80	−182269.82	−192354.97
Maintenance (*M*)	−348.29	−351.77	−355.29	−358.84	−362.43
Maintenane Accumulated (MA)	−5177.36	−5529.13	−5884.42	−6243.27	−6605.70
Present Value (MPV)	−0.85	−0.59	−0.41	−0.28	−0.20
Accumulated MPV	−670.71	−671.30	−671.72	−672.00	−672.20
Non-Production (NP)	0.00	0.00	0.00	0.00	0.00
Non-Production Accumulated (NPA)	0.00	0.00	0.00	0.00	0.00
Present Value (NPPV)	0.00	0.00	0.00	0.00	0.00
Accumulated NPPV	0.00	0.00	0.00	0.00	0.00
Accumulated Total Costs (APV)	−152702.32	−162778.21	−172858.51	−182941.82	−193027.17
Accumulated Total Costs	−110037.75	−119066.13	−128350.79	−137892.30	−147691.22
Benefit (*B*)	67632.85	67632.85	67632.85	67632.85	67632.85
Benefits Accumulated (BA)	1082125.60	1149758.45	1217391.30	1285024.15	1352657.00
Present Value (BPV)	165.80	113.87	78.21	53.71	36.89
Accumulated Benefits BPV	147954.06	148067.94	148146.14	148199.86	148236.75
BPV + FPV + MPV + NPPV + IIPV	960087.85	1018692.33	1077040.52	1135131.86	1192965.79
MTAC (LDM)	−117724.15	−127232.93	−136997.99	−147019.90	−157299.22
MTAC (SDM)	−120384.83	−129745.79	−139326.08	−149126.27	−159146.91
MTAC (EM)	−121909.37	−130969.45	−140277.98	−149837.47	−159649.93

(*Continued*)

TABLE 8.3 (*Continued*)
Values Used for Simulation based on MMTAC Method Related to Risk Standards. First Simulation

Time	21	22	23	24	25
Initial Investment (II)					
Initial Investment (II)					
II (IIPV) - Return	−12000.00	−12000.00	−12000.00	−12000.00	−12000.00
Useful Life					
Residual value of equipment (RV)					
Linear Depreciation Method (Vn)	−1911.60	−1431.20	−950.80	−470.40	10.00
LDM Present Value (PV)	−0.72	−0.37	−0.17	−0.06	0.00
Sum of Digits Method (Vn)	−359.54	−211.72	−100.86	−26.95	10.00
SDM Present Value (PV)	−0.13	−0.05	−0.02	0.00	0.00
Exponential Method (Vn)	−31.09	−23.42	−17.63	−13.28	−10.00
EM Present Value (Pv)	−0.01	−0.01	0.00	0.00	0.00
Functioning (*F*)	−5490.86	−5545.76	−5601.22	−5657.23	−5713.81
Functioning Accumulated (FA)	−104576.37	−110122.14	−115723.36	−121380.59	−127094.40
Present Value Accumulated (NPV)	−10085.03	−10086.45	−10087.44	−10088.13	−10088.61
Present Value (NPV)	−2.06	−1.43	−0.99	−0.69	−0.48
Energy	−4200.00	−4400.00	−4600.00	−4800.00	−5000.00
Energy Accumulated (EA)	−46200.00	−50600.00	−55200.00	−60000.00	−65000.00
Energy (NPV)	−1.57	−1.13	−0.81	−0.58	−0.42
Accumulated Functioning (FPV)	−202441.57	−212529.16	−222617.42	−232706.13	−242795.15
Maintenance (*M*)	−366.06	−369.72	−373.41	−377.15	−380.92
Maintenane Accumulated (MA)	−6971.76	−7341.48	−7714.89	−8092.04	−8472.96
Present Value (MPV)	−0.14	−0.10	−0.07	−0.05	−0.03
Accumulated MPV	−672.34	−672.43	−672.50	−672.54	−672.57
Non-Production (NP)	0.00	0.00	0.00	0.00	0.00
Non-Production Accumulated (NPA)	0.00	0.00	0.00	0.00	0.00
Present Value (NPPV)	0.00	0.00	0.00	0.00	0.00
Accumulated NPPV	0.00	0.00	0.00	0.00	0.00
Accumulated Total Costs (APV)	−203113.91	−213201.59	−223289.91	−233378.67	−243467.73
Accumulated Total Costs	−157748.13	−168063.61	−178638.25	−189472.63	−200567.36
Benefit (*B*)	67632.85	67632.85	67632.85	67632.85	67632.85
Benefits Accumulated (BA)	1420289.86	1487922.71	1555555.56	1623188.41	1690821.26
Present Value (BPV)	25.34	17.40	11.95	8.21	5.64
Accumulated Benefits BPV	148262.09	148279.49	148291.44	148299.65	148305.29
BPV + FPV + MPV + NPPV + IIPV	1250541.72	1307859.09	1364917.31	1421715.77	1478253.90
MTAC (LDM)	−167836.53	−178632.41	−189687.45	−201002.23	−212577.36
MTAC (SDM)	−169388.59	−179851.89	−190537.39	−201445.68	−212577.36
MTAC (EM)	−169717.04	−180040.20	−190620.62	−201459.35	−212557.36

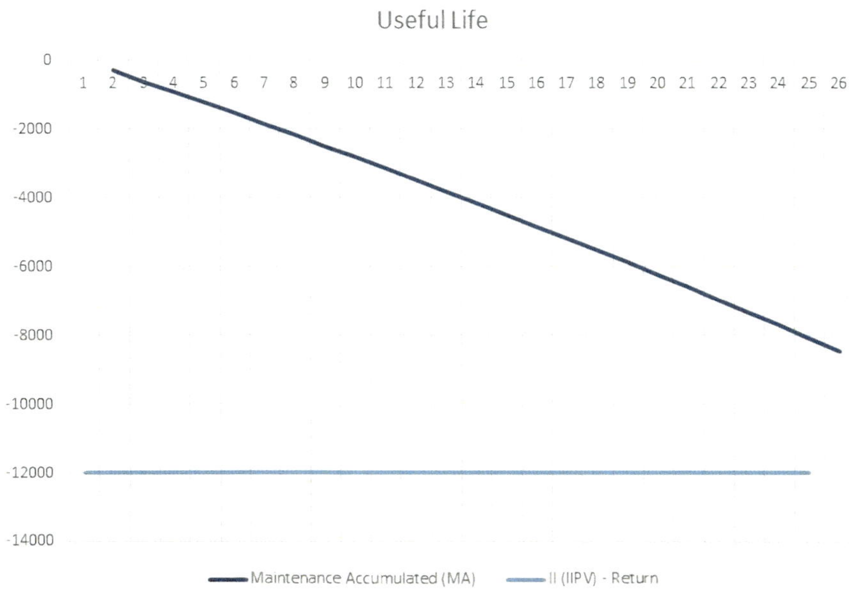

FIGURE 8.10 Physical Asset useful life.

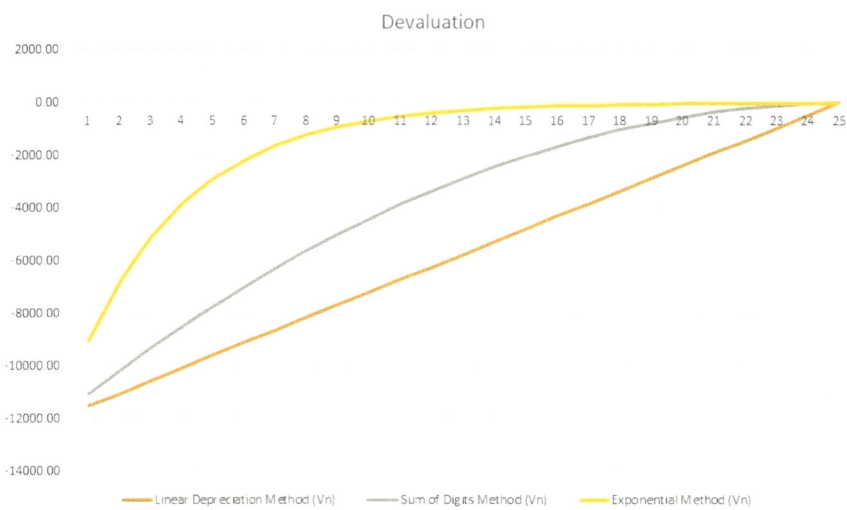

FIGURE 8.11 The three devaluation methods.

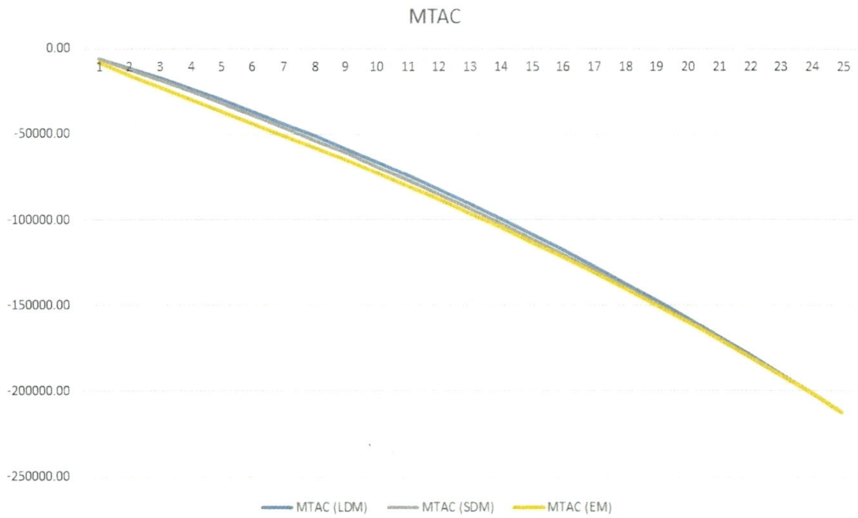

FIGURE 8.12 The MMTAC results from the three devaluation methods.

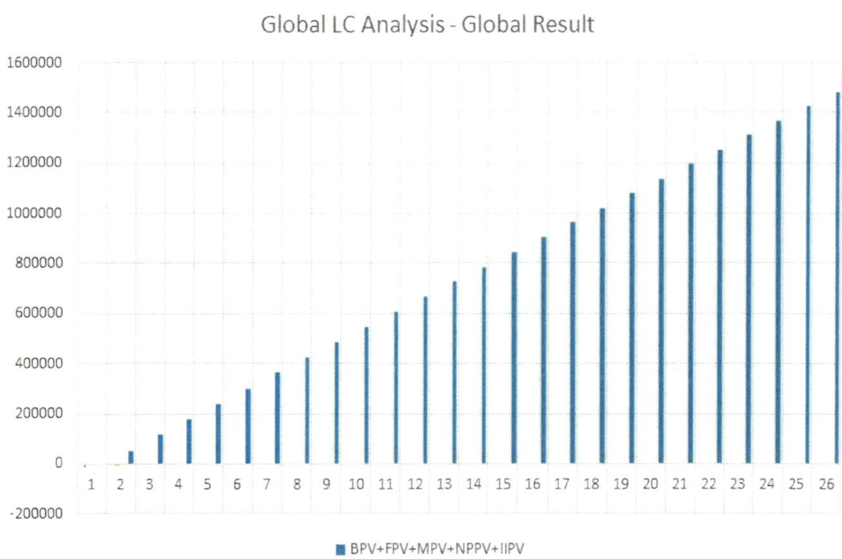

FIGURE 8.13 Life cycle analysis global result including profits.

- Non-Production (NP) 0
- Benefit (B) 70000
- Internal Rate of Return (r) 0.12
- Risk (R) **0.01**
- Apparent Rate 0.1312
- Maintenance Increase Rate 0.01
- Availability 0.9662
- MTBF 1000
- MTTR 25
- MWT 10

The remaining values are presented in Table 8.4, and the graphical results are shown in Figures 8.14–8.17.

As can be seen, namely comparing figures of the first and second simulation, the results are the same, despite the variation of Risk value, because there is no Reduction to Present Value (RPV). This is the main reason because this method may not be used for this situation, but the other is presented in this book.

8.4.1.3 MMTAC Reduced to Present Value (MMTAC-RPV)

The first simulation uses the following values:

- Initial Investment (II) –12000
- Life Time simulated 25
- Residual value of equipment (RV) 10
- Linear Depreciation Method (Vn) –12000
- Functioning (F) –4500
- Energy (E) –200
- Maintenance (M) –300
- Non-Production (NP) 0
- Benefit (B) 70000
- Internal Rate of Return (r) 0.12
- Risk (R) **0.30**
- Apparent Rate 0.2432
- Maintenance Increase Rate 0.01
- Availability 0.9662
- MTBF 1000
- MTTR 25
- MWT 10

The remaining values are presented in Table 8.5, and the graphical results are shown in Figures 8.18–8.21.

As referred in the preceding introduction of this section, the evaluation of introduction of Risk Norms, the simulation of their implementation is made through the increase of Risk tax. They may be used other indicators, but may be not so intuitive.

TABLE 8.4
Values Used for Simulation based on MMTAC Method Related to Risk Standards. Second Simulation

Time	0	1	2	3	4	5
Initial Investment (II)	−12000					
Initial Investment (II)	−12000					
II (IIPV) - Return	−12000	−12000.00	−12000.00	−12000.00	−12000.00	−12000.00
Useful Life	25					
Residual value of equipment (RV)	10					
Linear Depreciation Method (Vn)	−12000	−11519.60	−11039.20	−10558.80	−10078.40	−9598.00
LDM Present Value (PV)		−10183.52	−8626.98	−7294.51	−6155.08	−5181.84
Sum of Digits Method (Vn)	−12000	−11076.15	−10189.26	−9339.32	−8526.34	−7750.31
SDM Present Value (PV)		−9791.51	−7962.77	−6452.04	−5207.21	−4184.29
Exponential Method (Vn)	−12000	−9036.79	−6805.29	−5124.83	−3859.33	−2906.33
EM Present Value (Pv)		−7988.67	−5318.24	−3540.47	−2356.97	−1569.09
Functioning (*F*)	−4500	−4500.00	−4545.00	−4590.45	−4636.35	−4682.72
Functioning Accumulated (FA)		−4500.00	−9045.00	−13635.45	−18271.80	−22954.52
Present Value Accumulated (NPV)		−3978.08	−7529.93	−10701.23	−13532.74	−16060.88
Present Value (NPV)		−3978.08	−3551.85	−3171.30	−2831.52	−2528.14
Energy	−200	−200.00	−400.00	−600.00	−800.00	−1000.00
Energy Accumulated (EA)		−200.00	−600.00	−1200.00	−2000.00	−3000.00
Energy (NPV)		−176.80	−312.59	−414.51	−488.58	−539.89
Accumulated Functioning (FPV)		−16154.88	−23997.40	−35113.14	−49134.46	−65735.23
Maintenance (*M*)	−300	−300.00	−303.00	−306.03	−309.09	−312.18
Maintenane Accumulated (MA)		−300.00	−603.00	−909.03	−1218.12	−1530.30
Present Value (MPV)		−265.21	−236.79	−211.42	−188.77	−168.54
Accumulated MPV		−265.21	−502.00	−713.42	−902.18	−1070.73
Non-Production (NP)	0	0.00	0.00	0.00	0.00	0.00
Non-Production Accumulated (NPA)		0.00	0.00	0.00	0.00	0.00
Present Value (NPPV)		0.00	0.00	0.00	0.00	0.00
Accumulated NPPV		0.00	0.00	0.00	0.00	0.00
Accumulated Total Costs (APV)		−16420.08	−24499.40	−35826.56	−50036.64	−66805.96
Accumulated Total Costs		−5000.00	−10248.00	−15744.48	−21489.92	−27484.82
Benefit (*B*)	70000	67632.85	67632.85	67632.85	67632.85	67632.85
Benefits Accumulated (BA)		67632.85	135265.70	202898.55	270531.40	338164.25
Present Value (BPV)		59788.59	52854.13	46723.94	41304.76	36514.11
Accumulated Benefits BPV		59788.59	112642.71	159366.66	200671.42	237185.53
BPV + FPV + MPV + NPPV + IIPV		50632.85	113017.70	175154.07	237041.48	298679.43
MTAC (LDM)		−5480.40	−11208.80	−17185.68	−23411.52	−29886.82
MTAC (SDM)		−5923.85	−12058.74	−18405.16	−24963.59	−31734.52
MTAC (EM)		−7963.21	−15442.71	−22619.65	−29630.59	−36578.49

(*Continued*)

TABLE 8.4 (*Continued*)
Values Used for Simulation based on MMTAC Method Related to Risk Standards. Second Simulation

Time	6	7	8	9	10
Initial Investment (II)					
Initial Investment (II)					
II (IIPV) - Return	−12000.00	−12000.00	−12000.00	−12000.00	−12000.00
Useful Life					
Residual value of equipment (RV)					
Linear Depreciation Method (Vn)	−9117.60	−8637.20	−8156.80	−7676.40	−7196.00
LDM Present Value (PV)	−4351.55	−3644.16	−3042.32	−2531.06	−2097.48
Sum of Digits Method (Vn)	−7011.23	−6309.11	−5643.94	−5015.72	−4424.46
SDM Present Value (PV)	−3346.25	−2661.90	−2105.07	−1653.78	−1289.63
Exponential Method (Vn)	−2188.66	−1648.20	−1241.20	−934.71	−703.90
EM Present Value (Pv)	−1044.58	−695.40	−462.94	−308.19	−205.17
Functioning (*F*)	−4729.55	−4776.84	−4824.61	−4872.86	−4921.58
Functioning Accumulated (FA)	−27684.07	−32460.91	−37285.52	−42158.37	−47079.96
Present Value Accumulated (NPV)	−18318.15	−20333.57	−22133.05	−23739.73	−25174.26
Present Value (NPV)	−2257.27	−2015.42	−1799.48	−1606.68	−1434.53
Energy	−1200.00	−1400.00	−1600.00	−1800.00	−2000.00
Energy Accumulated (EA)	−4200.00	−5600.00	−7200.00	−9000.00	−11000.00
Energy (NPV)	−572.72	−590.68	−596.77	−593.50	−582.96
Accumulated Functioning (FPV)	−84626.11	−105550.36	−128280.17	−152613.40	−178370.61
Maintenance (*M*)	−315.30	−318.46	−321.64	−324.86	−328.11
Maintenane Accumulated (MA)	−1845.60	−2164.06	−2485.70	−2810.56	−3138.66
Present Value (MPV)	−150.48	−134.36	−119.97	−107.11	−95.64
Accumulated MPV	−1221.21	−1355.57	−1475.54	−1582.65	−1678.28
Non-Production (NP)	0.00	0.00	0.00	0.00	0.00
Non-Production Accumulated (NPA)	0.00	0.00	0.00	0.00	0.00
Present Value (NPPV)	0.00	0.00	0.00	0.00	0.00
Accumulated NPPV	0.00	0.00	0.00	0.00	0.00
Accumulated Total Costs (APV)	−85847.32	−106905.93	−129755.71	−154196.04	−180048.90
Accumulated Total Costs	−33729.67	−40224.97	−46971.22	−53968.93	−61218.62
Benefit (*B*)	67632.85	67632.85	67632.85	67632.85	67632.85
Benefits Accumulated (BA)	405797.10	473429.95	541062.80	608695.65	676328.50
Present Value (BPV)	32279.09	28535.27	25225.66	22299.91	19713.50
Accumulated Benefits BPV	269464.62	297999.89	323225.54	345525.45	365238.95
BPV + FPV + MPV + NPPV + IIPV	360067.43	421204.98	482091.58	542726.72	603109.88
MTAC (LDM)	−36612.07	−43587.77	−50814.42	−58292.53	−66022.62
MTAC (SDM)	−38718.44	−45915.86	−53327.28	−60953.21	−68794.16
MTAC (EM)	−43541.02	−50576.77	−57730.01	−65034.22	−72514.72

(*Continued*)

TABLE 8.4 (*Continued*)
Values Used for Simulation based on MMTAC Method Related to Risk Standards. Second Simulation

Time	11	12	13	14	15
Initial Investment (II)					
Initial Investment (II)					
II (IIPV) - Return	−12000.00	−12000.00	−12000.00	−12000.00	−12000.00
Useful Life					
Residual value of equipment (RV)					
Linear Depreciation Method (Vn)	−6715.60	−6235.20	−5754.80	−5274.40	−4794.00
LDM Present Value (PV)	−1730.42	−1420.29	−1158.83	−938.90	−754.41
Sum of Digits Method (Vn)	−3870.15	−3352.80	−2872.40	−2428.95	−2022.46
SDM Present Value (PV)	−997.23	−763.72	−578.41	−432.38	−318.27
Exponential Method (Vn)	−530.08	−399.19	−300.61	−226.38	−170.48
EM Present Value (Pv)	−136.59	−90.93	−60.53	−40.30	−26.83
Functioning (*F*)	−4970.80	−5020.51	−5070.71	−5121.42	−5172.63
Functioning Accumulated (FA)	−52050.76	−57071.26	−62141.98	−67263.40	−72436.03
Present Value Accumulated (NPV)	−26455.10	−27598.70	−28619.77	−29531.44	−30345.44
Present Value (NPV)	−1280.83	−1143.60	−1021.07	−911.67	−813.99
Energy	−2200.00	−2400.00	−2600.00	−2800.00	−3000.00
Energy Accumulated (EA)	−13200.00	−15600.00	−18200.00	−21000.00	−24000.00
Energy (NPV)	−566.88	−546.69	−523.55	−498.43	−472.10
Accumulated Functioning (FPV)	−205392.59	−233537.97	−262681.29	−292711.17	−323528.70
Maintenance (*M*)	−331.39	−334.70	−338.05	−341.43	−344.84
Maintenane Accumulated (MA)	−3470.05	−3804.75	−4142.80	−4484.23	−4829.07
Present Value (MPV)	−85.39	−76.24	−68.07	−60.78	−54.27
Accumulated MPV	−1763.67	−1839.91	−1907.98	−1968.76	−2023.03
Non-Production (NP)	0.00	0.00	0.00	0.00	0.00
Non-Production Accumulated (NPA)	0.00	0.00	0.00	0.00	0.00
Present Value (NPPV)	0.00	0.00	0.00	0.00	0.00
Accumulated NPPV	0.00	0.00	0.00	0.00	0.00
Accumulated Total Costs (APV)	−207156.26	−235377.88	−264589.28	−294679.93	−325551.73
Accumulated Total Costs	−68720.81	−76476.01	−84484.77	−92747.62	−101265.10
Benefit (*B*)	67632.85	67632.85	67632.85	67632.85	67632.85
Benefits Accumulated (BA)	743961.35	811594.20	879227.05	946859.90	1014492.75
Present Value (BPV)	17427.07	15405.82	13619.01	12039.44	10643.07
Accumulated Benefits BPV	382666.02	398071.85	411690.86	423730.29	434373.36
BPV + FPV + MPV + NPPV + IIPV	663240.55	723118.19	782742.28	842112.28	901227.66
MTAC (LDM)	−74005.21	−82240.81	−90729.97	−99473.22	−108471.10
MTAC (SDM)	−76850.65	−85123.21	−93612.37	−102318.67	−111242.64
MTAC (EM)	−80190.73	−88076.83	−96184.16	−104521.24	−113094.62

(Continued)

TABLE 8.4 (*Continued*)
Values Used for Simulation based on MMTAC Method Related to Risk Standards. Second Simulation

Time	16	17	18	19	20
Initial Investment (II)					
Initial Investment (II)					
II (IIPV) - Return	−12000.00	−12000.00	−12000.00	−12000.00	−12000.00
Useful Life					
Residual value of equipment (RV)					
Linear Depreciation Method (Vn)	−4313.60	−3833.20	−3352.80	−2872.40	−2392.00
LDM Present Value (PV)	−600.08	−471.40	−364.50	−276.06	−203.22
Sum of Digits Method (Vn)	−1652.92	−1320.34	−1024.71	−766.03	−544.31
SDM Present Value (PV)	−229.94	−162.37	−111.40	−73.62	−46.24
Exponential Method (Vn)	−128.38	−96.68	−72.81	−54.83	−41.29
EM Present Value (Pv)	−17.86	−11.89	−7.92	−5.27	−3.51
Functioning (*F*)	−5224.36	−5276.60	−5329.37	−5382.66	−5436.49
Functioning Accumulated (FA)	−77660.39	−82936.99	−88266.36	−93649.03	−99085.52
Present Value Accumulated (NPV)	−31072.22	−31721.13	−32300.51	−32817.82	−33279.70
Present Value (NPV)	−726.78	−648.91	−579.38	−517.31	−461.88
Energy	−3200.00	−3400.00	−3600.00	−3800.00	−4000.00
Energy Accumulated (EA)	−27200.00	−30600.00	−34200.00	−38000.00	−42000.00
Energy (NPV)	−445.16	−418.13	−391.38	−365.20	−339.84
Accumulated Functioning (FPV)	−355046.08	−387185.33	−419877.22	−453060.24	−486679.78
Maintenance (*M*)	−348.29	−351.77	−355.29	−358.84	−362.43
Maintenane Accumulated (MA)	−5177.36	−5529.13	−5884.42	−6243.27	−6605.70
Present Value (MPV)	−48.45	−43.26	−38.63	−34.49	−30.79
Accumulated MPV	−2071.48	−2114.74	−2153.37	−2187.85	−2218.65
Non-Production (NP)	0.00	0.00	0.00	0.00	0.00
Non-Production Accumulated (NPA)	0.00	0.00	0.00	0.00	0.00
Present Value (NPPV)	0.00	0.00	0.00	0.00	0.00
Accumulated NPPV	0.00	0.00	0.00	0.00	0.00
Accumulated Total Costs (APV)	−357117.56	−389300.07	−422030.59	−455248.09	−488898.42
Accumulated Total Costs	−110037.75	−119066.13	−128350.79	−137892.30	−147691.22
Benefit (*B*)	67632.85	67632.85	67632.85	67632.85	67632.85
Benefits Accumulated (BA)	1082125.60	1149758.45	1217391.30	1285024.15	1352657.00
Present Value (BPV)	9408.65	8317.41	7352.73	6499.94	5746.05
Accumulated Benefits BPV	443782.01	452099.42	459452.14	465952.08	471698.14
BPV + FPV + MPV + NPPV + IIPV	960087.85	1018692.33	1077040.52	1135131.86	1192965.79
MTAC (LDM)	−117724.15	−127232.93	−136997.99	−147019.90	−157299.22
MTAC (SDM)	−120384.83	−129745.79	−139326.08	−149126.27	−159146.91
MTAC (EM)	−121909.37	−130969.45	−140277.98	−149837.47	−159649.93

(*Continued*)

TABLE 8.4 (*Continued*)
Values Used for Simulation based on MMTAC Method Related to Risk Standards. Second Simulation

Time	21	22	23	24	25
Initial Investment (II)					
Initial Investment (II)					
II (IIPV) - Return	−12000.00	−12000.00	−12000.00	−12000.00	−12000.00
Useful Life					
Residual value of equipment (RV)					
Linear Depreciation Method (Vn)	−1911.60	−1431.20	−950.80	−470.40	10.00
LDM Present Value (PV)	−143.57	−95.02	−55.81	−24.41	0.46
Sum of Digits Method (Vn)	−359.54	−211.72	−100.86	−26.95	10.00
SDM Present Value (PV)	−27.00	−14.06	−5.92	−1.40	0.46
Exponential Method (Vn)	−31.09	−23.42	−17.63	−13.28	−10.00
EM Present Value (Pv)	−2.34	−1.55	−1.03	−0.69	−0.46
Functioning (*F*)	−5490.86	−5545.76	−5601.22	−5657.23	−5713.81
Functioning Accumulated (FA)	−104576.37	−110122.14	−115723.36	−121380.59	−127094.40
Present Value Accumulated (NPV)	−33692.09	−34060.30	−34389.06	−34682.60	−34944.68
Present Value (NPV)	−412.39	−368.21	−328.76	−293.53	−262.08
Energy	−4200.00	−4400.00	−4600.00	−4800.00	−5000.00
Energy Accumulated (EA)	−46200.00	−50600.00	−55200.00	−60000.00	−65000.00
Energy (NPV)	−315.44	−292.14	−269.99	−249.06	−229.34
Accumulated Functioning (FPV)	−520687.31	−555039.75	−589698.81	−624630.46	−659804.48
Maintenance (*M*)	−366.06	−369.72	−373.41	−377.15	−380.92
Maintenane Accumulated (MA)	−6971.76	−7341.48	−7714.89	−8092.04	−8472.96
Present Value (MPV)	−27.49	−24.55	−21.92	−19.57	−17.47
Accumulated MPV	−2246.14	−2270.69	−2292.60	−2312.17	−2329.65
Non-Production (NP)	0.00	0.00	0.00	0.00	0.00
Non-Production Accumulated (NPA)	0.00	0.00	0.00	0.00	0.00
Present Value (NPPV)	0.00	0.00	0.00	0.00	0.00
Accumulated NPPV	0.00	0.00	0.00	0.00	0.00
Accumulated Total Costs (APV)	−522933.45	−557310.44	−591991.41	−626942.63	−662134.13
Accumulated Total Costs	−157748.13	−168063.61	−178638.25	−189472.63	−200567.36
Benefit (*B*)	67632.85	67632.85	67632.85	67632.85	67632.85
Benefits Accumulated (BA)	1420289.86	1487922.71	1555555.56	1623188.41	1690821.26
Present Value (BPV)	5079.61	4490.46	3969.64	3509.23	3102.22
Accumulated Benefits BPV	476777.75	481268.21	485237.85	488747.08	491849.31
BPV + FPV + MPV + NPPV + IIPV	1250541.72	1307859.09	1364917.31	1421715.77	1478253.90
MTAC (LDM)	−167836.53	−178632.41	−189687.45	−201002.23	−212577.36
MTAC (SDM)	−169388.59	−179851.89	−190537.39	−201445.68	−212577.36
MTAC (EM)	−169717.04	−180040.20	−190620.62	−201459.35	−212557.36

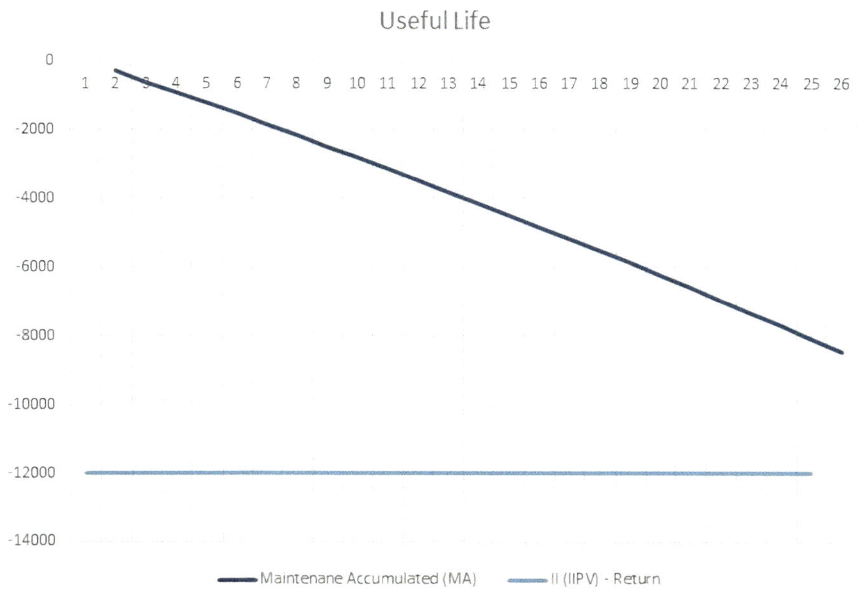

FIGURE 8.14 Physical Asset useful life.

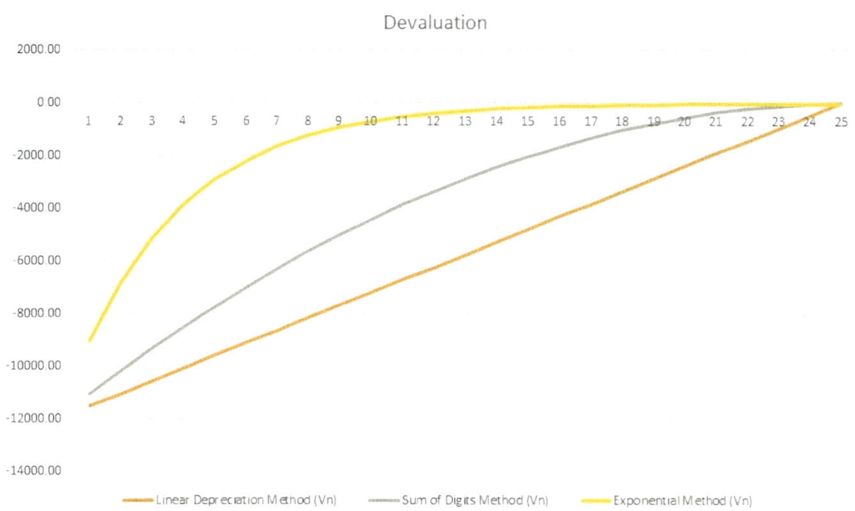

FIGURE 8.15 The three devaluation methods.

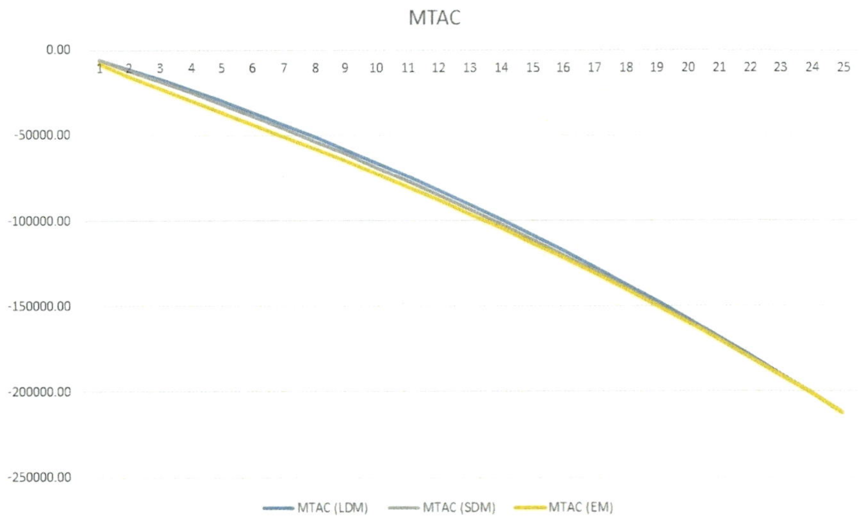

FIGURE 8.16 The MMTAC results from the three devaluation methods.

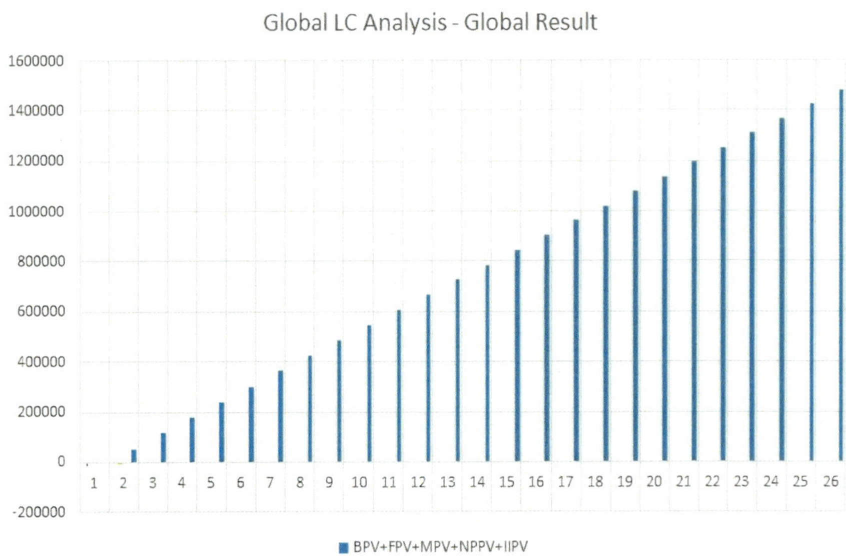

FIGURE 8.17 Life cycle analysis global result including profits.

The second simulation uses the following values:

- Initial Investment (II) –12000
- Life Time simulated 25
- Residual value of equipment (RV) 10
- Linear Depreciation Method (Vn) –12000
- Functioning (*F*) –4500
- Energy (*E*) –200
- Maintenance (*M*) –300
- Non-Production (NP) 0
- Benefit (*B*) 70000
- Internal Rate of Return (*r*) 0.12
- Risk (*R*) **0.01**
- Apparent Rate 0.2432
- Maintenance Increase Rate 0.01
- Availability 0.9662

The remaining values are presented in Table 8.6, and the graphical results are shown in Figures 8.22–8.25.

As can be seen, namely comparing Figures 8.21 and 8.25, with the increasing of the maintenance management quality helped by the maintenance standards, it gets two more positive years, but the positive incomes increase a lot.

These results from the three methods are close, which reinforce a lot the robustness of the models under discussion.

8.4.2 Life Cycle Investment of Physical Assets

As referred in the previous sections, ISO 31010 (Risk management - Risk assessment techniques) presents several tools to assess Risk. Additionally, other tools can be added, like Fuzzy Logic, among others.

The final result will be a Risk Rate that is inserted in Equation 8.1

The way to demonstrate whether the return of implementing the standard under discussion is interesting or not, in this section, is done using the model of Life Cycle Investment, presented in Chapter 2.

The first simulation uses the following values:

- Initial Investment (II) –12000
- Functioning (*F*) –4500
- Energy (*E*) –200
- Maintenance (*M*) –300
- Accumulated (AMPV) –5000
- Non-Production (NP) 0
- Benefit (*B*) including IIPV 70000
- Internal Rate of Return (*r*) 0.12
- Risk (*R*) **0.30**
- Apparent Rate 0.456

TABLE 8.5
Values Used for Simulation based on MMTAC-RPV Method Related to Risk Standards. First Simulation

Time	0	1	2	3	4	5
Initial Investment (II)	−12000					
Initial Investment (II)	−12000					
II Present Value (IIPV) - Return	−12000	−8241.76	−5660.55	−3887.74	−2670.15	−1833.89
Life Time simulated	25					
Residual value of equipment (RV)	10					
Linear Depreciation Method (Vn)	−12000	−11519.60	−11039.20	−10558.80	−10078.40	−9598.00
LDM Present Value (Vn)		−7911.81	−5207.33	−3420.82	−2242.57	−1466.81
Sum of Digits Method (Vn)	−12000	−11076.15	−10189.26	−9339.32	−8526.34	−7750.31
SDM Present Value (Vn)		−7607.25	−4806.40	−3025.74	−1897.22	−1184.44
Exponential Method (Vn)	−12000	−9036.79	−6805.29	−5124.83	−3859.33	−2906.33
EM Present Value (Pv)		−6206.58	−3210.14	−1660.33	−858.75	−444.16
Functioning (F)	−4500	−4500.00	−4545.00	−4590.45	−4636.35	−4682.72
Present Value Acummulated (NPV)		−3090.66	−5234.59	−6721.80	−7753.45	−8469.08
Present Value (NPV)		−3090.66	−2143.93	−1487.21	−1031.65	−715.63
Energy (*E*)	−200	−200.00	−400.00	−600.00	−800.00	−1000.00
Energy (NPV)		−137.36	−188.68	−194.39	−178.01	−152.82
Accumulated FPV		−15228.02	−20651.30	−27567.48	−35498.94	−44120.84
Maintenance (*M*)	−300	−300.00	−303.00	−306.03	−309.09	−312.18
Present Value (MPV)		−206.04	−142.93	−99.15	−68.78	−47.71
Accumulated (MPV)		−206.04	−348.97	−448.12	−516.90	−564.61
Non-Production (NP)	0	0.00	0.00	0.00	0.00	0.00
Present Value (NPPV)		0.00	0.00	0.00	0.00	0.00
Accumulated NPPV		0.00	0.00	0.00	0.00	0.00
Accumulated Total Costs (APV)		−15434.07	−21000.27	−28015.60	−36015.84	−44685.45
Accumulated Total Costs		−5000.00	−10248.00	−15744.48	−21489.92	−27484.82
Benefit (*B*)	70000	62500.00	62500.00	62500.00	62500.00	62500.00
Present Value (BPV)		42925.82	29482.02	20248.64	13907.03	9551.53
Accumulated Benefits BPV		42925.82	72407.85	92656.49	106563.52	116115.06
BPV + FPV + MPV + NPPV + IIPV		15491.76	39407.57	52640.88	58547.69	59429.61
MTAC-PV (LDM)		−35345.88	−19103.80	−14478.81	−12564.60	−11630.45
MTAC-PV (SDM)		−35041.31	−18903.34	−14347.11	−12478.26	−11573.98
MTAC-PV (EM)		−33640.65	−18105.21	−13891.98	−12218.65	−11425.92

(Continued)

TABLE 8.5 (*Continued*)
Values Used for Simulation based on MMTAC-RPV Method Related to Risk Standards. First Simulation

Time	6	7	8	9	10
Initial Investment (II)					
Initial Investment (II)					
II Present Value (IIPV) - Return	−1259.54	−865.07	−594.14	−408.06	−280.26
Life Time simulated					
Residual value of equipment (RV)					
Linear Depreciation Method (Vn)	−9117.60	−8637.20	−8156.80	−7676.40	−7196.00
LDM Present Value (Vn)	−957.00	−622.65	−403.86	−261.04	−168.07
Sum of Digits Method (Vn)	−7011.23	−6309.11	−5643.94	−5015.72	−4424.46
SDM Present Value (Vn)	−735.91	−454.82	−279.44	−170.56	−103.33
Exponential Method (Vn)	−2188.66	−1648.20	−1241.20	−934.71	−703.90
EM Present Value (Pv)	−229.73	−118.82	−61.45	−31.79	−16.44
Functioning (F)	−4729.55	−4776.84	−4824.61	−4872.86	−4921.58
Present Value Acummulated (NPV)	−8965.50	−9309.86	−9548.74	−9714.44	−9829.38
Present Value (NPV)	−496.42	−344.36	−238.88	−165.70	−114.95
Energy (*E*)	−1200.00	−1400.00	−1600.00	−1800.00	−2000.00
Energy (NPV)	−125.95	−100.92	−79.22	−61.21	−46.71
Accumulated FPV	−53212.30	−62623.08	−72251.04	−82026.69	−91902.78
Maintenance (*M*)	−315.30	−318.46	−321.64	−324.86	−328.11
Present Value (MPV)	−33.09	−22.96	−15.93	−11.05	−7.66
Accumulated (MPV)	−597.70	−620.66	−636.58	−647.63	−655.29
Non-Production (NP)	0.00	0.00	0.00	0.00	0.00
Present Value (NPPV)	0.00	0.00	0.00	0.00	0.00
Accumulated NPPV	0.00	0.00	0.00	0.00	0.00
Accumulated Total Costs (APV)	−53810.00	−63243.74	−72887.62	−82674.32	−92558.07
Accumulated Total Costs	−33729.67	−40224.97	−46971.22	−53968.93	−61218.62
Benefit (*B*)	62500.00	62500.00	62500.00	62500.00	62500.00
Present Value (BPV)	6560.12	4505.58	3094.49	2125.34	1459.71
Accumulated Benefits BPV	122675.18	127180.75	130275.24	132400.58	133860.29
BPV + FPV + MPV + NPPV + IIPV	56865.18	51937.01	45387.62	37726.26	29302.21
MTAC-PV (LDM)	−11127.83	−10838.06	−10661.43	−10548.37	−10472.61
MTAC-PV (SDM)	−11090.99	−10814.08	−10645.88	−10538.32	−10466.14
MTAC-PV (EM)	−11006.62	−10766.08	−10618.63	−10522.90	−10457.45

(*Continued*)

TABLE 8.5 (*Continued*)
Values Used for Simulation based on MMTAC-RPV Method Related to Risk Standards. First Simulation

Time	11	12	13	14	15
Initial Investment (II)					
Initial Investment (II)					
II Present Value (IIPV) - Return	−192.49	−132.20	−90.80	−62.36	−42.83
Life Time simulated					
Residual value of equipment (RV)					
Linear Depreciation Method (Vn)	−6715.60	−6235.20	−5754.80	−5274.40	−4794.00
LDM Present Value (Vn)	−107.72	−68.69	−43.54	−27.41	−17.11
Sum of Digits Method (Vn)	−3870.15	−3352.80	−2872.40	−2428.95	−2022.46
SDM Present Value (Vn)	−62.08	−36.94	−21.73	−12.62	−7.22
Exponential Method (Vn)	−530.08	−399.19	−300.61	−226.38	−170.48
EM Present Value (Pv)	−8.50	−4.40	−2.27	−1.18	−0.61
Functioning (F)	−4970.80	−5020.51	−5070.71	−5121.42	−5172.63
Present Value Acummulated (NPV)	−9909.12	−9964.43	−10002.80	−10029.41	−10047.88
Present Value (NPV)	−79.74	−55.31	−38.37	−26.62	−18.46
Energy (*E*)	−2200.00	−2400.00	−2600.00	−2800.00	−3000.00
Energy (NPV)	−35.29	−26.44	−19.67	−14.55	−10.71
Accumulated FPV	−101847.19	−111838.06	−121860.53	−131904.50	−141963.08
Maintenance (*M*)	−331.39	−334.70	−338.05	−341.43	−344.84
Present Value (MPV)	−5.32	−3.69	−2.56	−1.77	−1.23
Accumulated (MPV)	−660.61	−664.30	−666.85	−668.63	−669.86
Non-Production (NP)	0.00	0.00	0.00	0.00	0.00
Present Value (NPPV)	0.00	0.00	0.00	0.00	0.00
Accumulated NPPV	0.00	0.00	0.00	0.00	0.00
Accumulated Total Costs (APV)	−102507.80	−112502.36	−122527.39	−132573.13	−142632.94
Accumulated Total Costs	−68720.81	−76476.01	−84484.77	−92747.62	−101265.10
Benefit (*B*)	62500.00	62500.00	62500.00	62500.00	62500.00
Present Value (BPV)	1002.55	688.56	472.91	324.80	223.08
Accumulated Benefits BPV	134862.83	135551.40	136024.31	136349.12	136572.19
BPV + FPV + MPV + NPPV + IIPV	20355.04	11049.04	1496.93	−8224.01	−18060.75
MTAC-PV (LDM)	−10419.59	−10380.92	−10351.61	−10328.61	−10310.00
MTAC-PV (SDM)	−10415.44	−10378.27	−10349.93	−10327.55	−10309.34
MTAC-PV (EM)	−10410.57	−10375.56	−10348.44	−10326.74	−10308.90

(*Continued*)

TABLE 8.5 (*Continued*)
Values Used for Simulation based on MMTAC-RPV Method Related to Risk Standards. First Simulation

Time	16	17	18	19	20
Initial Investment (II)					
Initial Investment (II)					
II Present Value (IIPV) - Return	−29.42	−20.20	−13.88	−9.53	−6.55
Life Time simulated					
Residual value of equipment (RV)					
Linear Depreciation Method (Vn)	−4313.60	−3833.20	−3352.80	−2872.40	−2392.00
LDM Present Value (Vn)	−10.57	−6.45	−3.88	−2.28	−1.30
Sum of Digits Method (Vn)	−1652.92	−1320.34	−1024.71	−766.03	−544.31
SDM Present Value (Vn)	−4.05	−2.22	−1.18	−0.61	−0.30
Exponential Method (Vn)	−128.38	−96.68	−72.81	−54.83	−41.29
EM Present Value (Pv)	−0.31	−0.16	−0.08	−0.04	−0.02
Functioning (F)	−5224.36	−5276.60	−5329.37	−5382.66	−5436.49
Present Value Acummulated (NPV)	−10060.68	−10069.57	−10075.73	−10080.01	−10082.97
Present Value (NPV)	−12.81	−8.88	−6.16	−4.27	−2.97
Energy (*E*)	−3200.00	−3400.00	−3600.00	−3800.00	−4000.00
Energy (NPV)	−7.84	−5.72	−4.16	−3.02	−2.18
Accumulated FPV	−152031.61	−162106.90	−172186.80	−182269.82	−192354.97
Maintenance (*M*)	−348.29	−351.77	−355.29	−358.84	−362.43
Present Value (MPV)	−0.85	−0.59	−0.41	−0.28	−0.20
Accumulated (MPV)	−670.71	−671.30	−671.72	−672.00	−672.20
Non-Production (NP)	0.00	0.00	0.00	0.00	0.00
Present Value (NPPV)	0.00	0.00	0.00	0.00	0.00
Accumulated NPPV	0.00	0.00	0.00	0.00	0.00
Accumulated Total Costs (APV)	−152702.32	−162778.21	−172858.51	−182941.82	−193027.17
Accumulated Total Costs	−110037.75	−119066.13	−128350.79	−137892.30	−147691.22
Benefit (*B*)	62500.00	62500.00	62500.00	62500.00	62500.00
Present Value (BPV)	153.21	105.23	72.27	49.64	34.09
Accumulated Benefits BPV	136725.41	136830.64	136902.91	136952.55	136986.64
BPV + FPV + MPV + NPPV + IIPV	−27976.91	−37947.57	−47955.60	−57989.27	−68040.53
MTAC-PV (LDM)	−10294.56	−10281.45	−10270.13	−10260.22	−10251.42
MTAC-PV (SDM)	−10294.15	−10281.20	−10269.98	−10260.13	−10251.37
MTAC-PV (EM)	−10293.91	−10281.08	−10269.92	−10260.10	−10251.36

(*Continued*)

TABLE 8.5 (*Continued*)
Values Used for Simulation based on MMTAC-RPV Method Related to Risk Standards. First Simulation

Time	21	22	23	24	25
Initial Investment (II)					
Initial Investment (II)					
II Present Value (IIPV) - Return	−4.50	−3.09	−2.12	−1.46	−1.00
Life Time simulated					
Residual value of equipment (RV)					
Linear Depreciation Method (Vn)	−1911.60	−1431.20	−950.80	−470.40	10.00
LDM Present Value (Vn)	−0.72	−0.37	−0.17	−0.06	0.00
Sum of Digits Method (Vn)	−359.54	−211.72	−100.86	−26.95	10.00
SDM Present Value (Vn)	−0.13	−0.05	−0.02	0.00	0.00
Exponential Method (Vn)	−31.09	−23.42	−17.63	−13.28	−10.00
EM Present Value (Pv)	−0.01	−0.01	0.00	0.00	0.00
Functioning (F)	−5490.86	−5545.76	−5601.22	−5657.23	−5713.81
Present Value Acummulated (NPV)	−10085.03	−10086.45	−10087.44	−10088.13	−10088.61
Present Value (NPV)	−2.06	−1.43	−0.99	−0.69	−0.48
Energy (*E*)	−4200.00	−4400.00	−4600.00	−4800.00	−5000.00
Energy (NPV)	−1.57	−1.13	−0.81	−0.58	−0.42
Accumulated FPV	−202441.57	−212529.16	−222617.42	−232706.13	−242795.15
Maintenance (*M*)	−366.06	−369.72	−373.41	−377.15	−380.92
Present Value (MPV)	−0.14	−0.10	−0.07	−0.05	−0.03
Accumulated (MPV)	−672.34	−672.43	−672.50	−672.54	−672.57
Non-Production (NP)	0.00	0.00	0.00	0.00	0.00
Present Value (NPPV)	0.00	0.00	0.00	0.00	0.00
Accumulated NPPV	0.00	0.00	0.00	0.00	0.00
Accumulated Total Costs (APV)	−203113.91	−213201.59	−223289.91	−233378.67	−243467.73
Accumulated Total Costs	−157748.13	−168063.61	−178638.25	−189472.63	−200567.36
Benefit (*B*)	62500.00	62500.00	62500.00	62500.00	62500.00
Present Value (BPV)	23.41	16.08	11.05	7.59	5.21
Accumulated Benefits BPV	137010.06	137026.14	137037.18	137044.77	137049.98
BPV + FPV + MPV + NPPV + IIPV	−78103.85	−88175.45	−98252.73	−108333.90	−118417.75
MTAC-PV (LDM)	−10243.55	−10236.45	−10230.00	−10224.11	−10218.71
MTAC-PV (SDM)	−10243.53	−10236.44	−10230.00	−10224.11	−10218.71
MTAC-PV (EM)	−10243.52	−10236.44	−10230.00	−10224.11	−10218.71

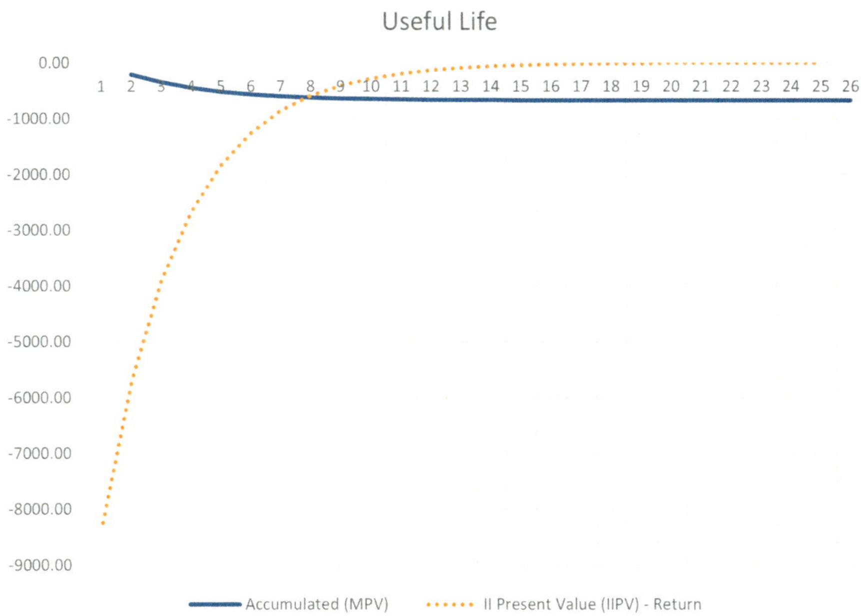

FIGURE 8.18 Physical Asset useful life.

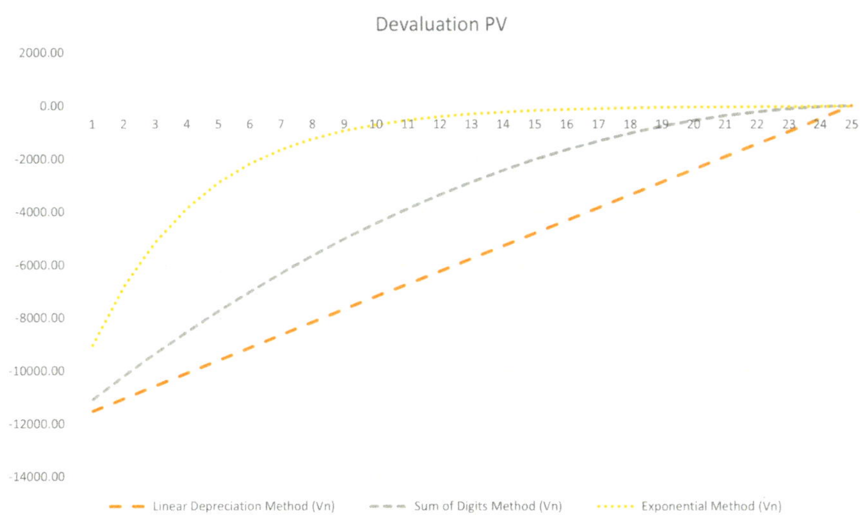

FIGURE 8.19 The three devaluation methods.

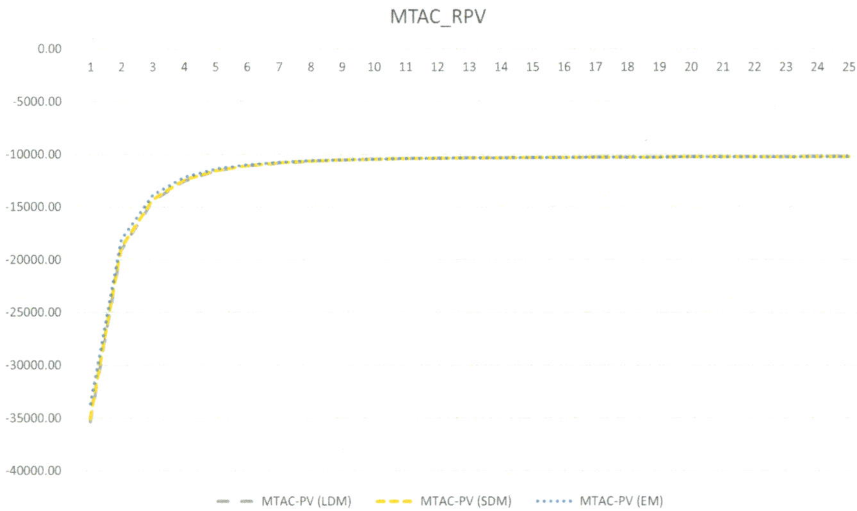

FIGURE 8.20 The MMTAC-RPV results from the three devaluation methods.

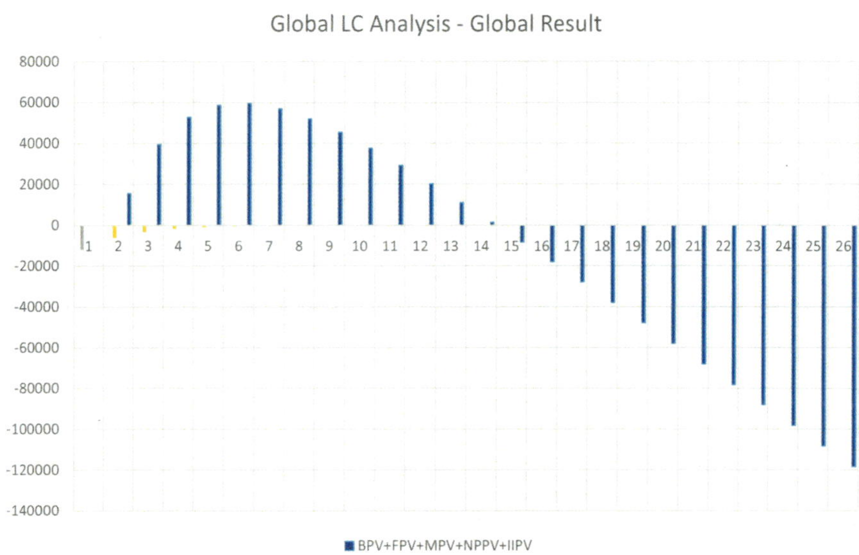

FIGURE 8.21 Life cycle analysis global result including profits.

TABLE 8.6
Values Used for Simulation based on MMTAC-RPV Method Related to Risk Standards. Second Simulation

Time	0	1	2	3	4	5
Initial Investment (II)	−12000					
Initial Investment (II)	−12000					
II Present Value (IIPV) - Return	−12000	−10608.20	−9377.83	−8290.16	−7328.64	−6478.65
Life Time simulated	25					
Residual value of equipment (RV)	10					
Linear Depreciation Method (Vn)	−12000	−11519.60	−11039.20	−10558.80	−10078.40	−9598.00
LDM Present Value (Vn)		−10183.52	−8626.98	−7294.51	−6155.08	−5181.84
Sum of Digits Method (Vn)	−12000	−11076.15	−10189.26	−9339.32	−8526.34	−7750.31
SDM Present Value (Vn)		−9791.51	−7962.77	−6452.04	−5207.21	−4184.29
Exponential Method (Vn)	−12000	−9036.79	−6805.29	−5124.83	−3859.33	−2906.33
EM Present Value (Pv)		−7988.67	−5318.24	−3540.47	−2356.97	−1569.09
Functioning (*F*)	−4500	−4500.00	−4545.00	−4590.45	−4636.35	−4682.72
Present Value Acummulated (NPV)		−3978.08	−7529.93	−10701.23	−13532.74	−16060.88
Present Value (NPV)		−3978.08	−3551.85	−3171.30	−2831.52	−2528.14
Energy (*E*)	−200	−200.00	−400.00	−600.00	−800.00	−1000.00
Energy (NPV)		−176.80	−312.59	−414.51	−488.58	−539.89
Accumulated FPV		−16154.88	−23997.40	−35113.14	−49134.46	−65735.23
Maintenance (*M*)	−300	−300.00	−303.00	−306.03	−309.09	−312.18
Present Value (MPV)		−265.21	−236.79	−211.42	−188.77	−168.54
Accumulated (MPV)		−265.21	−502.00	−713.42	−902.18	−1070.73
Non-Production (NP)	0	0.00	0.00	0.00	0.00	0.00
Present Value (NPPV)		0.00	0.00	0.00	0.00	0.00
Accumulated NPPV		0.00	0.00	0.00	0.00	0.00
Accumulated Total Costs (APV)		−16420.08	−24499.40	−35826.56	−50036.64	−66805.96
Accumulated Total Costs		−5000.00	−10248.00	−15744.48	−21489.92	−27484.82
Benefit (*B*)	70000	62500.00	62500.00	62500.00	62500.00	62500.00
Present Value (BPV)		55251.06	48842.88	43177.93	38170.02	33742.95
Accumulated Benefits BPV		55251.06	104093.94	147271.87	185441.89	219184.84
BPV + FPV + MPV + NPPV + IIPV		26830.98	67594.54	99445.31	123405.25	140378.88
MTAC-PV (LDM)		−38603.61	−22563.19	−18373.69	−17047.93	−16797.56
MTAC-PV (SDM)		−38211.59	−22231.08	−18092.87	−16810.96	−16598.05
MTAC-PV (EM)		−36408.76	−20908.82	−17122.34	−16098.40	−16075.01

(Continued)

TABLE 8.6 (*Continued*)
Values Used for Simulation based on MMTAC-RPV Method Related to Risk Standards. Second Simulation

Time	6	7	8	9	10
Initial Investment (II)					
Initial Investment (II)					
II Present Value (IIPV) - Return	−5727.23	−5062.97	−4475.75	−3956.64	−3497.74
Life Time simulated					
Residual value of equipment (RV)					
Linear Depreciation Method (Vn)	−9117.60	−8637.20	−8156.80	−7676.40	−7196.00
LDM Present Value (Vn)	−4351.55	−3644.16	−3042.32	−2531.06	−2097.48
Sum of Digits Method (Vn)	−7011.23	−6309.11	−5643.94	−5015.72	−4424.46
SDM Present Value (Vn)	−3346.25	−2661.90	−2105.07	−1653.78	−1289.63
Exponential Method (Vn)	−2188.66	−1648.20	−1241.20	−934.71	−703.90
EM Present Value (Pv)	−1044.58	−695.40	−462.94	−308.19	−205.17
Functioning (*F*)	−4729.55	−4776.84	−4824.61	−4872.86	−4921.58
Present Value Acummulated (NPV)	−18318.15	−20333.57	−22133.05	−23739.73	−25174.26
Present Value (NPV)	−2257.27	−2015.42	−1799.48	−1606.68	−1434.53
Energy (*E*)	−1200.00	−1400.00	−1600.00	−1800.00	−2000.00
Energy (NPV)	−572.72	−590.68	−596.77	−593.50	−582.96
Accumulated FPV	−84626.11	−105550.36	−128280.17	−152613.40	−178370.61
Maintenance (*M*)	−315.30	−318.46	−321.64	−324.86	−328.11
Present Value (MPV)	−150.48	−134.36	−119.97	−107.11	−95.64
Accumulated (MPV)	−1221.21	−1355.57	−1475.54	−1582.65	−1678.28
Non-Production (NP)	0.00	0.00	0.00	0.00	0.00
Present Value (NPPV)	0.00	0.00	0.00	0.00	0.00
Accumulated NPPV	0.00	0.00	0.00	0.00	0.00
Accumulated Total Costs (APV)	−85847.32	−106905.93	−129755.71	−154196.04	−180048.90
Accumulated Total Costs	−33729.67	−40224.97	−46971.22	−53968.93	−61218.62
Benefit (*B*)	62500.00	62500.00	62500.00	62500.00	62500.00
Present Value (BPV)	29829.34	26369.64	23311.21	20607.51	18217.39
Accumulated Benefits BPV	249014.18	275383.82	298695.03	319302.54	337519.93
BPV + FPV + MPV + NPPV + IIPV	151166.86	156477.89	156939.33	153106.50	145471.03
MTAC-PV (LDM)	−17033.14	−17507.16	−18099.75	−18747.46	−19414.64
MTAC-PV (SDM)	−16865.59	−17366.83	−17982.60	−18649.98	−19333.85
MTAC-PV (EM)	−16481.98	−17085.90	−17777.33	−18500.47	−19225.41

(*Continued*)

TABLE 8.6 (*Continued*)
Values Used for Simulation based on MMTAC-RPV Method Related to Risk Standards. Second Simulation

Time	11	12	13	14	15
Initial Investment (II)					
Initial Investment (II)					
II Present Value (IIPV) - Return	−3092.06	−2733.43	−2416.40	−2136.14	−1888.38
Life Time simulated					
Residual value of equipment (RV)					
Linear Depreciation Method (Vn)	−6715.60	−6235.20	−5754.80	−5274.40	−4794.00
LDM Present Value (Vn)	−1730.42	−1420.29	−1158.83	−938.90	−754.41
Sum of Digits Method (Vn)	−3870.15	−3352.80	−2872.40	−2428.95	−2022.46
SDM Present Value (Vn)	−997.23	−763.72	−578.41	−432.38	−318.27
Exponential Method (Vn)	−530.08	−399.19	−300.61	−226.38	−170.48
EM Present Value (Pv)	−136.59	−90.93	−60.53	−40.30	−26.83
Functioning (*F*)	−4970.80	−5020.51	−5070.71	−5121.42	−5172.63
Present Value Acummulated (NPV)	−26455.10	−27598.70	−28619.77	−29531.44	−30345.44
Present Value (NPV)	−1280.83	−1143.60	−1021.07	−911.67	−813.99
Energy (*E*)	−2200.00	−2400.00	−2600.00	−2800.00	−3000.00
Energy (NPV)	−566.88	−546.69	−523.55	−498.43	−472.10
Accumulated FPV	−205392.59	−233537.97	−262681.29	−292711.17	−323528.70
Maintenance (*M*)	−331.39	−334.70	−338.05	−341.43	−344.84
Present Value (MPV)	−85.39	−76.24	−68.07	−60.78	−54.27
Accumulated (MPV)	−1763.67	−1839.91	−1907.98	−1968.76	−2023.03
Non-Production (NP)	0.00	0.00	0.00	0.00	0.00
Present Value (NPPV)	0.00	0.00	0.00	0.00	0.00
Accumulated NPPV	0.00	0.00	0.00	0.00	0.00
Accumulated Total Costs (APV)	−207156.26	−235377.88	−264589.28	−294679.93	−325551.73
Accumulated Total Costs	−68720.81	−76476.01	−84484.77	−92747.62	−101265.10
Benefit (*B*)	62500.00	62500.00	62500.00	62500.00	62500.00
Present Value (BPV)	16104.48	14236.63	12585.42	11125.73	9835.33
Accumulated Benefits BPV	353624.40	367861.04	380446.46	391572.19	401407.52
BPV + FPV + MPV + NPPV + IIPV	134468.15	120483.15	103857.18	84892.26	63855.79
MTAC-PV (LDM)	−20080.61	−20733.18	−21365.24	−21972.77	−22553.74
MTAC-PV (SDM)	−20013.95	−20678.47	−21320.59	−21936.59	−22524.67
MTAC-PV (EM)	−19935.71	−20622.40	−21280.75	−21908.59	−22505.24

(*Continued*)

TABLE 8.6 (*Continued*)
Values Used for Simulation based on MMTAC-RPV Method Related to Risk Standards. Second Simulation

Time	16	17	18	19	20
Initial Investment (II)					
Initial Investment (II)					
II Present Value (IIPV) - Return	−1669.36	−1475.75	−1304.58	−1153.27	−1019.51
Life Time simulated					
Residual value of equipment (RV)					
Linear Depreciation Method (Vn)	−4313.60	−3833.20	−3352.80	−2872.40	−2392.00
LDM Present Value (Vn)	−600.08	−471.40	−364.50	−276.06	−203.22
Sum of Digits Method (Vn)	−1652.92	−1320.34	−1024.71	−766.03	−544.31
SDM Present Value (Vn)	−229.94	−162.37	−111.40	−73.62	−46.24
Exponential Method (Vn)	−128.38	−96.68	−72.81	−54.83	−41.29
EM Present Value (Pv)	−17.86	−11.89	−7.92	−5.27	−3.51
Functioning (*F*)	−5224.36	−5276.60	−5329.37	−5382.66	−5436.49
Present Value Acummulated (NPV)	−31072.22	−31721.13	−32300.51	−32817.82	−33279.70
Present Value (NPV)	−726.78	−648.91	−579.38	−517.31	−461.88
Energy (*E*)	−3200.00	−3400.00	−3600.00	−3800.00	−4000.00
Energy (NPV)	−445.16	−418.13	−391.38	−365.20	−339.84
Accumulated FPV	−355046.08	−387185.33	−419877.22	−453060.24	−486679.78
Maintenance (*M*)	−348.29	−351.77	−355.29	−358.84	−362.43
Present Value (MPV)	−48.45	−43.26	−38.63	−34.49	−30.79
Accumulated (MPV)	−2071.48	−2114.74	−2153.37	−2187.85	−2218.65
Non-Production (NP)	0.00	0.00	0.00	0.00	0.00
Present Value (NPPV)	0.00	0.00	0.00	0.00	0.00
Accumulated NPPV	0.00	0.00	0.00	0.00	0.00
Accumulated Total Costs (APV)	−357117.56	−389300.07	−422030.59	−455248.09	−488898.42
Accumulated Total Costs	−110037.75	−119066.13	−128350.79	−137892.30	−147691.22
Benefit (*B*)	62500.00	62500.00	62500.00	62500.00	62500.00
Present Value (BPV)	8694.60	7686.18	6794.71	6006.64	5309.97
Accumulated Benefits BPV	410102.12	417788.30	424583.01	430589.65	435899.62
BPV + FPV + MPV + NPPV + IIPV	40984.56	16488.22	−9447.58	−36658.45	−64998.81
MTAC-PV (LDM)	−23107.35	−23633.62	−24133.06	−24606.53	−25055.08
MTAC-PV (SDM)	−23084.22	−23615.44	−24119.00	−24595.88	−25047.23
MTAC-PV (EM)	−23070.96	−23606.59	−24113.25	−24592.28	−25045.10

(*Continued*)

TABLE 8.6 (*Continued*)

Values Used for Simulation based on MMTAC-RPV Method Related to Risk Standards. Second Simulation

Time	21	22	23	24	25
Initial Investment (II)					
Initial Investment (II)					
II Present Value (IIPV) - Return	−901.27	−796.74	−704.33	−622.64	−550.42
Life Time simulated					
Residual value of equipment (RV)					
Linear Depreciation Method (Vn)	−1911.60	−1431.20	−950.80	−470.40	10.00
LDM Present Value (Vn)	−143.57	−95.02	−55.81	−24.41	0.46
Sum of Digits Method (Vn)	−359.54	−211.72	−100.86	−26.95	10.00
SDM Present Value (Vn)	−27.00	−14.06	−5.92	−1.40	0.46
Exponential Method (Vn)	−31.09	−23.42	−17.63	−13.28	−10.00
EM Present Value (Pv)	−2.34	−1.55	−1.03	−0.69	−0.46
Functioning (*F*)	−5490.86	−5545.76	−5601.22	−5657.23	−5713.81
Present Value Acummulated (NPV)	−33692.09	−34060.30	−34389.06	−34682.60	−34944.68
Present Value (NPV)	−412.39	−368.21	−328.76	−293.53	−262.08
Energy (*E*)	−4200.00	−4400.00	−4600.00	−4800.00	−5000.00
Energy (NPV)	−315.44	−292.14	−269.99	−249.06	−229.34
Accumulated FPV	−520687.31	−555039.75	−589698.81	−624630.46	−659804.48
Maintenance (*M*)	−366.06	−369.72	−373.41	−377.15	−380.92
Present Value (MPV)	−27.49	−24.55	−21.92	−19.57	−17.47
Accumulated (MPV)	−2246.14	−2270.69	−2292.60	−2312.17	−2329.65
Non-Production (NP)	0.00	0.00	0.00	0.00	0.00
Present Value (NPPV)	0.00	0.00	0.00	0.00	0.00
Accumulated NPPV	0.00	0.00	0.00	0.00	0.00
Accumulated Total Costs (APV)	−522933.45	−557310.44	−591991.41	−626942.63	−662134.13
Accumulated Total Costs	−157748.13	−168063.61	−178638.25	−189472.63	−200567.36
Benefit (*B*)	62500.00	62500.00	62500.00	62500.00	62500.00
Present Value (BPV)	4694.10	4149.67	3668.38	3242.91	2866.78
Accumulated Benefits BPV	440593.72	444743.39	448411.76	451654.67	454521.46
BPV + FPV + MPV + NPPV + IIPV	−94339.73	−124567.05	−155579.65	−187287.96	−219612.67
MTAC-PV (LDM)	−25479.86	−25882.07	−26262.92	−26623.63	−26965.35
MTAC-PV (SDM)	−25474.31	−25878.39	−26260.75	−26622.67	−26965.35
MTAC-PV (EM)	−25473.13	−25877.82	−26260.54	−26622.64	−26965.38

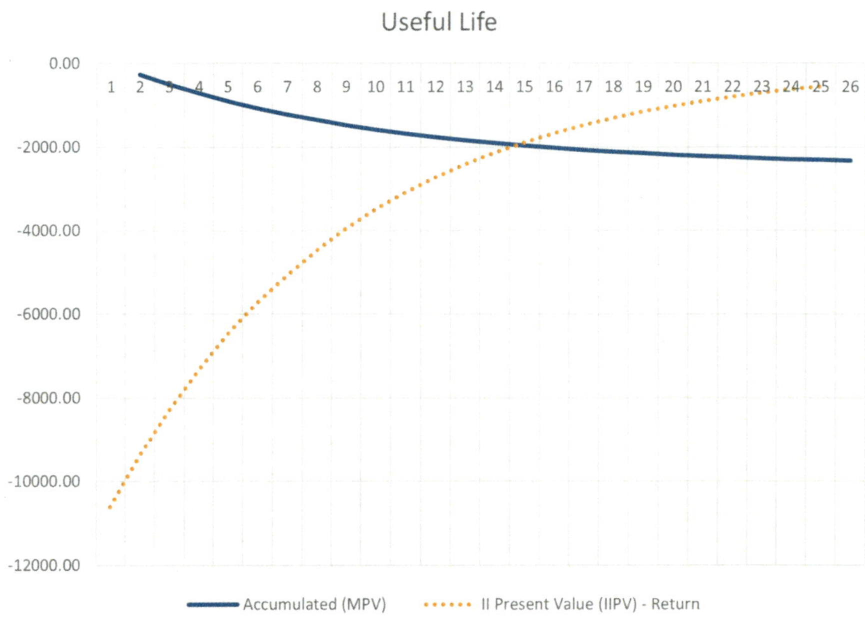

FIGURE 8.22 Physical Asset useful life.

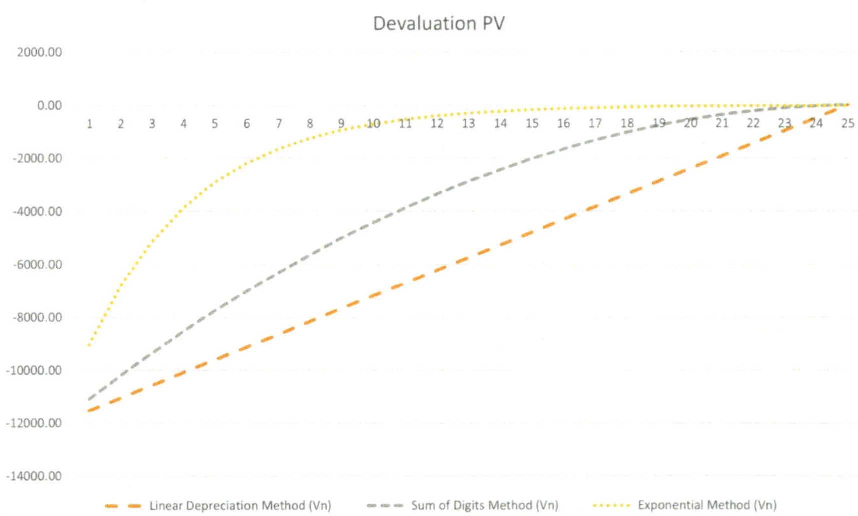

FIGURE 8.23 The three devaluation methods.

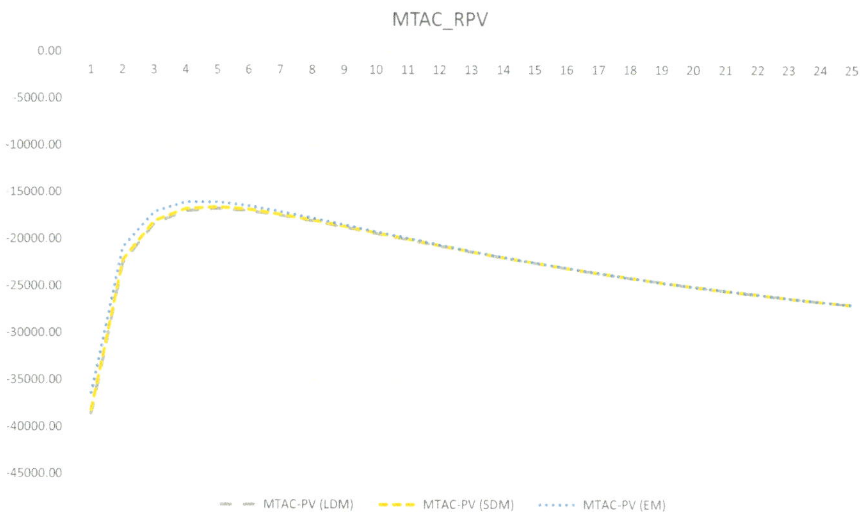

FIGURE 8.24 The MMTAC results from the three devaluation methods.

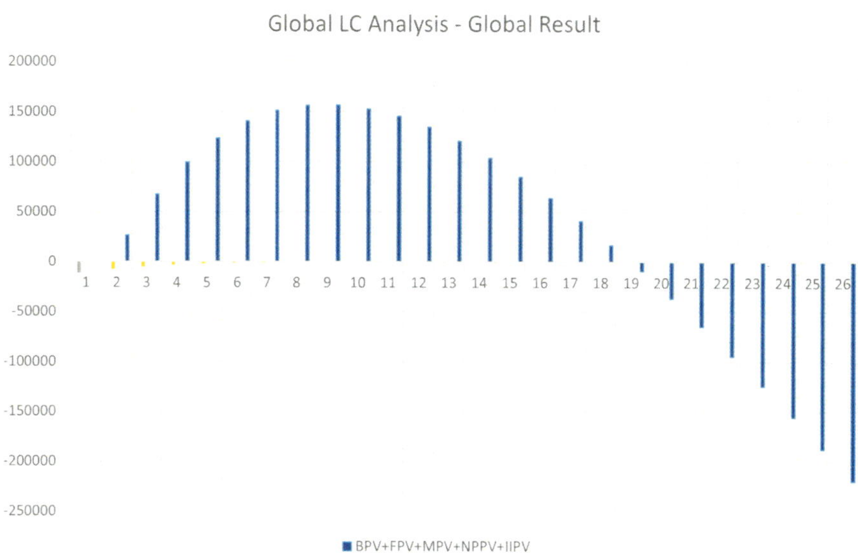

FIGURE 8.25 Life cycle analysis global result including profits.

- Maintenance Increase Rate 0.01
- Availability 0.9662
- MTBF 1000
- MTTR 25
- MWT 10

As referred in the preceding introduction of this section, the evaluation of the introduction of Risk Norms, the simulation of their implementation is made through the variation of the Risk tax. They may be used other indicators, but may be not so intuitive.

In the first scenario, a Risk Rate of 0.30 is used; the result is shown in Table 8.7 and Figures 8.26–8.28.

The second simulation uses the following values:

- Initial Investment (II) –12000
- Functioning (*F*) –4500
- Energy (*E*) –200
- Maintenance (*M*) –300
- Accumulated (AMPV) –5000
- Non-Production (NP) 0
- Benefit (*B*) including IIPV 70000
- Internal Rate of Return (*r*) 0.12
- Risk (*R*) **0.01**
- Apparent Rate 0.456
- Maintenance Increase Rate 0.01
- Availability 0.9662
- MTBF 1000
- MTTR 25
- MWT 10

The remaining values are presented in Table 8.8, and the graphical results are shown in Figures 8.29–8.31.

As can be seen, when comparing Figures 8.28 and 8.31, with the increasing of the maintenance management quality helped by the maintenance standards, it gets two more positive years, and the positive incomes increase a lot.

TABLE 8.7
Values Used for Simulation based on Life Cycle Investment Related to Risk Standards. First Simulation

Time	0	1	2	3	4	5	6	7	8
Initial Investment (II)	−12000								
II Present Value (IIPV) - Return	−12000	−8241.76	−5660.55	−3887.74	−2670.15	−1833.89	−1259.54	−865.07	−594.14
Functioning (*F*)	−4500	−4500.00	−4500.00	−4500.00	−4500.00	−4500.00	−4500.00	−4500.00	−4500.00
Present Value (PV) Accumulated		−3090.66	−5213.36	−6671.27	−7672.57	−8360.28	−8832.61	−9157.01	−9379.82
Energy (E)	−200	−200.00	−200.00	−200.00	−200.00	−200.00	−200.00	−200.00	−200.00
Energy (PV)		−137.36	−94.34	−64.80	−44.50	−30.56	−20.99	−14.42	−9.90
Accumulated (AFPV)		−15228.02	−5307.71	−12043.77	−19760.85	−28151.70	−37005.30	−46176.73	−55566.45
Maintenance (*M*)	−300	−300.00	−303.00	−306.03	−309.09	−312.18	−315.30	−318.46	−321.64
Present Value (PV)		−206.04	−141.51	−97.19	−66.75	−45.85	−31.49	−21.63	−14.85
Accumulated (AMPV)	−5000	−206.04	−347.56	−444.75	−511.50	−557.35	−588.84	−610.47	−625.32
Non-Production (NP)	0	0.00	0.00	0.00	0.00	0.00	0.00	0.00	0.00
Present Value (PV)		0.00	0.00	0.00	0.00	0.00	0.00	0.00	0.00
Accumulated (ANPPV)		0.00	0.00	0.00	0.00	0.00	0.00	0.00	0.00
Accumulated Total Costs (ATCPV)		−15434.07	−5655.27	−12488.52	−20272.35	−28709.05	−37594.14	−46787.20	−56191.77
Benefit (*B*) including IIPV	70000	54258.24	56839.45	58612.26	59829.85	60666.11	61240.46	61634.93	61905.86
Present Value (PV)		37265.28	26811.87	18989.10	13312.89	9271.27	6427.92	4443.21	3065.07
Accumulated Benefits (ABPV)		37265.28	64077.15	83066.25	96379.14	105650.41	112078.32	116521.54	119586.61
ATCPV + ABPV		21831.21	58421.88	70577.72	76106.79	76941.36	74484.18	69734.34	63394.84

(Continued)

TABLE 8.7 (*Continued*)
Values Used for Simulation based on Life Cycle Investment Related to Risk Standards. First Simulation

Time	9	10	11	12	13	14	15	16	17
Initial Investment (II)									
II Present Value (IIPV) - Return	−408.06	−280.26	−192.49	−132.20	−90.80	−62.36	−42.83	−29.42	−20.20
Functioning (*F*)	−4500.00	−4500.00	−4500.00	−4500.00	−4500.00	−4500.00	−4500.00	−4500.00	−4500.00
Present Value (PV) Accumulated	−9532.84	−9637.94	−9710.12	−9759.70	−9793.75	−9817.14	−9833.20	−9844.23	−9851.81
Energy (E)	−200.00	−200.00	−200.00	−200.00	−200.00	−200.00	−200.00	−200.00	−200.00
Energy (PV)	−6.80	−4.67	−3.21	−2.20	−1.51	−1.04	−0.71	−0.49	−0.34
Accumulated (AFPV)	−65106.09	−74748.71	−84462.04	−94223.94	−104019.21	−113837.38	−123671.29	−133516.01	−143368.16
Maintenance (*M*)	−324.86	−328.11	−331.39	−334.70	−338.05	−341.43	−344.84	−348.29	−351.77
Present Value (PV)	−10.20	−7.01	−4.81	−3.31	−2.27	−1.56	−1.07	−0.74	−0.51
Accumulated (AMPV)	−635.52	−642.53	−647.34	−650.65	−652.92	−654.48	−655.55	−656.28	−656.79
Non-Production (NP)	0.00	0.00	0.00	0.00	0.00	0.00	0.00	0.00	0.00
Present Value (PV)	0.00	0.00	0.00	0.00	0.00	0.00	0.00	0.00	0.00
Accumulated (ANPPV)	0.00	0.00	0.00	0.00	0.00	0.00	0.00	0.00	0.00
Accumulated Total Costs (ATCPV)	−65741.62	−75391.24	−85109.38	−94874.59	−104672.12	−114491.86	−124326.84	−134172.30	−144024.94
Benefit (*B*) including IIPV	62091.94	62219.74	62307.51	62367.80	62409.20	62437.64	62457.17	62470.58	62479.80
Present Value (PV)	2111.46	1453.16	999.46	687.11	472.23	324.48	222.93	153.14	105.20
Accumulated Benefits (ABPV)	121698.07	123151.23	124150.69	124837.80	125310.03	125634.51	125857.43	126010.57	126115.77
ATCPV + ABPV	55956.45	47760.00	39041.31	29963.21	20637.90	11142.65	1530.59	−8161.72	−17909.17

(Continued)

TABLE 8.7 (*Continued*)
Values Used for Simulation based on Life Cycle Investment Related to Risk Standards. First Simulation

Time	18	19	20	21	22	23	24	25
Initial Investment (II)								
II Present Value (IIPV) - Return	−13.88	−9.53	−6.55	−4.50	−3.09	−2.12	−1.46	−1.00
Functioning (*F*)	−4500.00	−4500.00	−4500.00	−4500.00	−4500.00	−4500.00	−4500.00	−4500.00
Present Value (PV) Accumulated	−9857.01	−9860.58	−9863.04	−9864.72	−9865.88	−9866.68	−9867.22	−9867.60
Energy (E)	−200.00	−200.00	−200.00	−200.00	−200.00	−200.00	−200.00	−200.00
Energy (PV)	−0.23	−0.16	−0.11	−0.07	−0.05	−0.04	−0.02	−0.02
Accumulated (AFPV)	−153225.40	−163086.14	−172949.29	−182814.09	−192680.02	−202546.73	−212413.98	−222281.59
Maintenance (*M*)	−355.29	−358.84	−362.43	−366.06	−369.72	−373.41	−377.15	−380.92
Present Value (PV)	−0.35	−0.24	−0.16	−0.11	−0.08	−0.05	−0.04	−0.03
Accumulated (AMPV)	−657.13	−657.37	−657.54	−657.65	−657.73	−657.78	−657.81	−657.84
Non-Production (NP)	0.00	0.00	0.00	0.00	0.00	0.00	0.00	0.00
Present Value (PV)	0.00	0.00	0.00	0.00	0.00	0.00	0.00	0.00
Accumulated (ANPPV)	0.00	0.00	0.00	0.00	0.00	0.00	0.00	0.00
Accumulated Total Costs (ATCPV)	−153882.53	−163743.51	−173606.82	−183471.73	−193337.74	−203204.51	−213071.79	−222939.43
Benefit (*B*) including IIPV	62486.12	62490.47	62493.45	62495.50	62496.91	62497.88	62498.54	62499.00
Present Value (PV)	72.26	49.63	34.09	23.41	16.08	11.04	7.59	5.21
Accumulated Benefits (ABPV)	126188.03	126237.66	126271.75	126295.16	126311.24	126322.28	126329.87	126335.08
ATCPV + ABPV	−27694.50	−37505.85	−47335.08	−57176.58	−67026.50	−76882.23	−86741.92	−96604.35

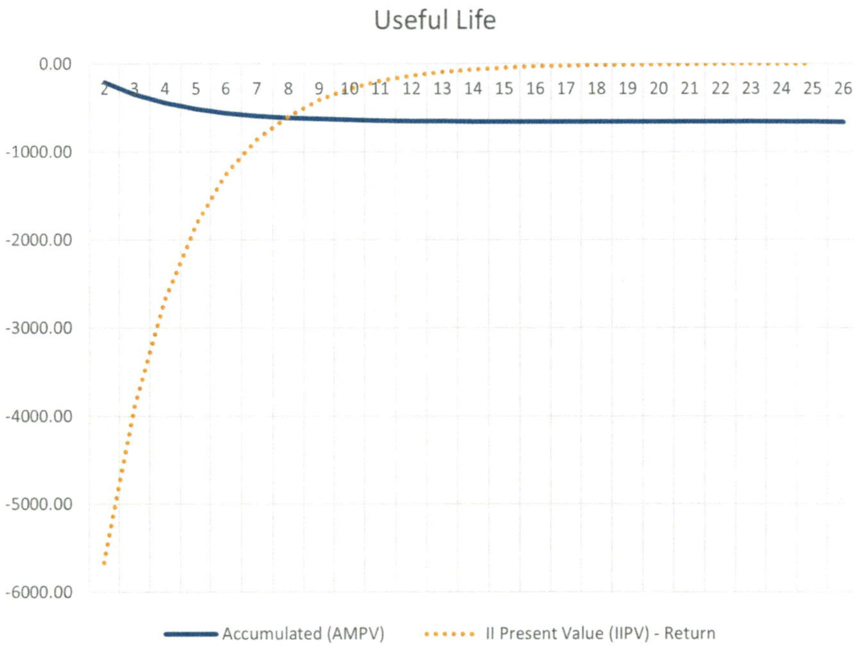

FIGURE 8.26 Physical Asset useful life.

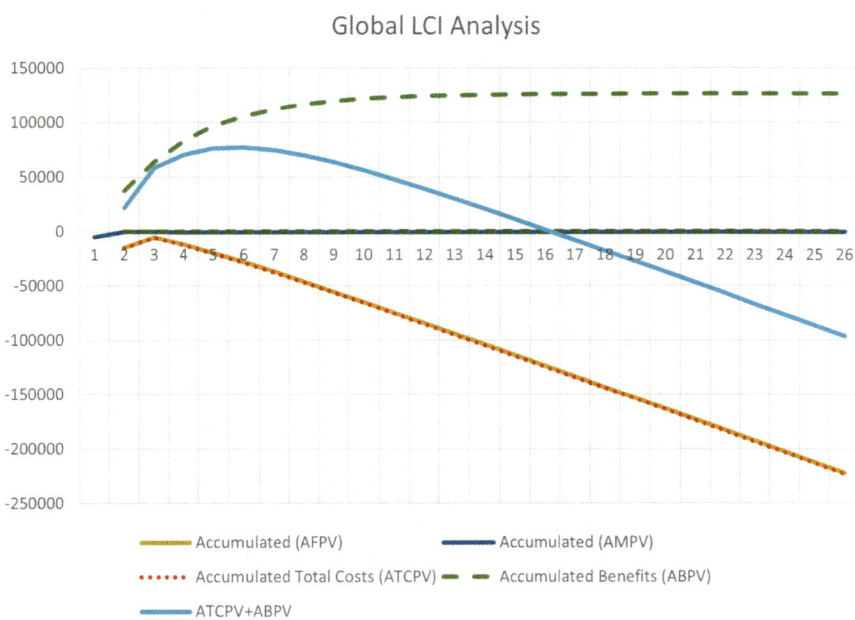

FIGURE 8.27 Global LCI analysis.

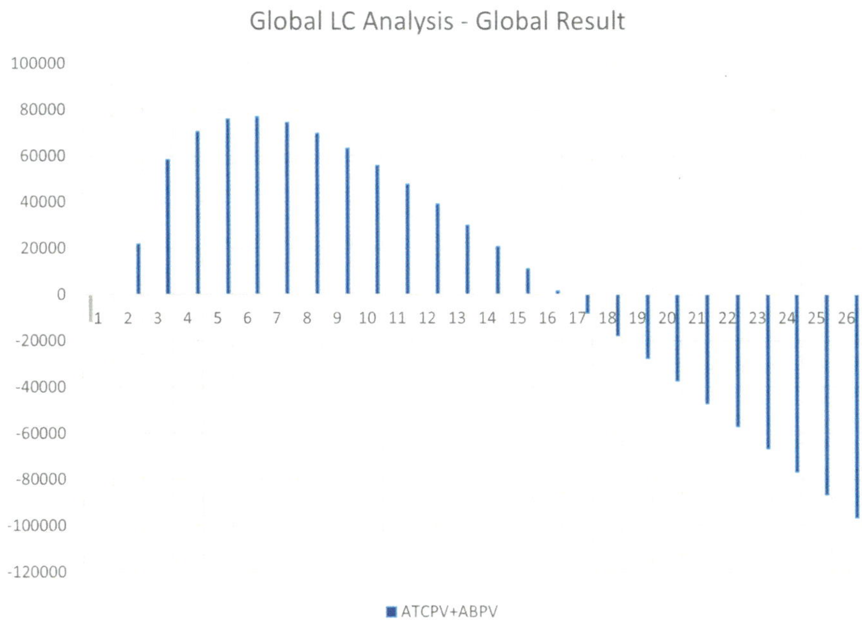

FIGURE 8.28 Life cycle analysis global result.

8.4.3 Life Cycle of Physical Assets with Recovery

This section follows the same approach like the preceding, being the first scenario designed with a Risk Rate of 0.3. The result is shown in Table 8.9 and Figures 8.32–8.34.

Other relevant values considered are the following ones:

- Interest Rate (r) 0.05
- Inflation Rate (h) 0.02
- Profit Rate (p) 0.05
- Maintenance Increase Rate 0.01
- MTBF 1000.0
- MTTR 30.0
- Risk (R) **0.30**
- Initial Investment (II) (C_0) –12000
- F_p Financial Movement (Production) 70000
- Functioning Costs –4500
- Energy –200
- Maintenance Costs –300

TABLE 8.8
Values Used for Simulation based on Life Cycle Investment Related to Risk Standards. Second Simulation

Time	0	1	2	3	4	5	6	7	8
Initial Investment (II)	−12000								
II Present Value (IIPV) - Return	−12000	−10608.20	−9377.83	−8290.16	−7328.64	−6478.65	−5727.23	−5062.97	−4475.75
Functioning (*F*)	−4500	−4500.00	−4500.00	−4500.00	−4500.00	−4500.00	−4500.00	−4500.00	−4500.00
Present Value (PV) Accumulated		−3978.08	−7494.76	−10603.57	−13351.82	−15781.31	−17929.02	−19827.64	−21506.04
Energy (*E*)	−200	−200.00	−200.00	−200.00	−200.00	−200.00	−200.00	−200.00	−200.00
Energy (PV)		−176.80	−156.30	−138.17	−122.14	−107.98	−95.45	−84.38	−74.60
Accumulated (AFPV)		−16154.88	−7651.06	−18392.80	−31866.76	−47756.05	−65780.53	−85692.54	−107273.18
Maintenance (*M*)	−300	−300.00	−303.00	−306.03	−309.09	−312.18	−315.30	−318.46	−321.64
Present Value (PV)		−265.21	−234.45	−207.25	−183.22	−161.97	−143.18	−126.57	−111.89
Accumulated (AMPV)	−5000	−265.21	−499.65	−706.90	−890.12	−1052.09	−1195.27	−1321.84	−1433.74
Non-Production (NP)	0	0.00	0.00	0.00	0.00	0.00	0.00	0.00	0.00
Present Value (PV)		0.00	0.00	0.00	0.00	0.00	0.00	0.00	0.00
Accumulated (ANPPV)		0.00	0.00	0.00	0.00	0.00	0.00	0.00	0.00
Accumulated Total Costs (ATCPV)		−16420.08	−8150.71	−19099.71	−32756.89	−48808.14	−66975.79	−87014.39	−108706.92
Benefit (*B*) including IIPV	70000	51891.80	53122.17	54209.84	55171.36	56021.35	56772.77	57437.03	58024.25
Present Value (PV)		45873.23	41514.23	37450.70	33694.27	30245.21	27095.91	24233.50	21641.85
Accumulated Benefits (ABPV)		45873.23	87387.46	124838.16	158532.43	188777.64	215873.55	240107.05	261748.90
ATCPV + ABPV		29453.14	79236.75	105738.45	125775.54	139969.50	148897.75	153092.66	153041.98

(Continued)

TABLE 8.8 (*Continued*)
Values Used for Simulation based on Life Cycle Investment Related to Risk Standards. Second Simulation

Time	9	10	11	12	13	14	15	16	17
Initial Investment (II)									
II Present Value (IIPV) - Return	−3956.64	−3497.74	−3092.06	−2733.43	−2416.40	−2136.14	−1888.38	−1669.36	−1475.75
Functioning (*F*)	−4500.00	−4500.00	−4500.00	−4500.00	−4500.00	−4500.00	−4500.00	−4500.00	−4500.00
Present Value (PV) Accumulated	−22989.78	−24301.43	−25460.96	−26485.99	−27392.15	−28193.20	−28901.34	−29527.35	−30080.76
Energy (*E*)	−200.00	−200.00	−200.00	−200.00	−200.00	−200.00	−200.00	−200.00	−200.00
Energy (PV)	−65.94	−58.30	−51.53	−45.56	−40.27	−35.60	−31.47	−27.82	−24.60
Accumulated (AFPV)	−130328.91	−154688.64	−180201.13	−206732.68	−234165.10	−262393.90	−291326.72	−320881.89	−350987.24
Maintenance (*M*)	−324.86	−328.11	−331.39	−334.70	−338.05	−341.43	−344.84	−348.29	−351.77
Present Value (PV)	−98.92	−87.44	−77.30	−68.34	−60.41	−53.40	−47.21	−41.73	−36.89
Accumulated (AMPV)	−1532.65	−1620.10	−1697.40	−1765.73	−1826.14	−1879.55	−1926.76	−1968.49	−2005.38
Non-Production (NP)	0.00	0.00	0.00	0.00	0.00	0.00	0.00	0.00	0.00
Present Value (PV)	0.00	0.00	0.00	0.00	0.00	0.00	0.00	0.00	0.00
Accumulated (ANPPV)	0.00	0.00	0.00	0.00	0.00	0.00	0.00	0.00	0.00
Accumulated Total Costs (ATCPV)	−131861.56	−156308.73	−181898.53	−208498.42	−235991.24	−264273.45	−293253.47	−322850.38	−352992.63
Benefit (*B*) including IIPV	58543.36	59002.26	59407.94	59766.57	60083.60	60363.86	60611.62	60830.64	61024.25
Present Value (PV)	19302.92	17197.87	15307.74	13613.99	12098.84	10745.47	9538.17	8462.37	7504.69
Accumulated Benefits (ABPV)	281051.82	298249.69	313557.43	327171.43	339270.27	350015.74	359553.91	368016.28	375520.97
ATCPV + ABPV	149190.26	141940.96	131658.90	118673.01	103279.02	85742.29	66300.43	45165.89	22528.34

(*Continued*)

TABLE 8.8 (*Continued*)
Values Used for Simulation based on Life Cycle Investment Related to Risk Standards. Second Simulation

Time	18	19	20	21	22	23	24	25
Initial Investment (II)								
II Present Value (IIPV) - Return	−1304.58	−1153.27	−1019.51	−901.27	−796.74	−704.33	−622.64	−550.42
Functioning (*F*)	−4500.00	−4500.00	−4500.00	−4500.00	−4500.00	−4500.00	−4500.00	−4500.00
Present Value (PV) Accumulated	−30569.98	−31002.45	−31384.77	−31722.75	−32021.52	−32285.65	−32519.14	−32725.54
Energy (*E*)	−200.00	−200.00	−200.00	−200.00	−200.00	−200.00	−200.00	−200.00
Energy (PV)	−21.74	−19.22	−16.99	−15.02	−13.28	−11.74	−10.38	−9.17
Accumulated (AFPV)	−381578.96	−412600.64	−444002.40	−475740.17	−507774.98	−540072.36	−572601.88	−605336.59
Maintenance (*M*)	−355.29	−358.84	−362.43	−366.06	−369.72	−373.41	−377.15	−380.92
Present Value (PV)	−32.61	−28.83	−25.49	−22.53	−19.92	−17.61	−15.57	−13.76
Accumulated (AMPV)	−2038.00	−2066.83	−2092.32	−2114.85	−2134.77	−2152.38	−2167.94	−2181.70
Non-Production (NP)	0.00	0.00	0.00	0.00	0.00	0.00	0.00	0.00
Present Value (PV)	0.00	0.00	0.00	0.00	0.00	0.00	0.00	0.00
Accumulated (ANPPV)	0.00	0.00	0.00	0.00	0.00	0.00	0.00	0.00
Accumulated Total Costs (ATCPV)	−383616.96	−414667.47	−446094.72	−477855.02	−509909.74	−542224.74	−574769.82	−607518.30
Benefit (*B*) including IIPV	61195.42	61346.73	61480.49	61598.73	61703.26	61795.67	61877.36	61949.58
Present Value (PV)	6652.88	5895.80	5223.35	4626.41	4096.77	3627.04	3210.60	2841.54
Accumulated Benefits (ABPV)	382173.85	388069.65	393293.00	397919.41	402016.18	405643.22	408853.82	411695.36
ATCPV + ABPV	−1443.12	−26597.82	−52801.72	−79935.61	−107893.56	−136581.52	−165916.00	−195822.94

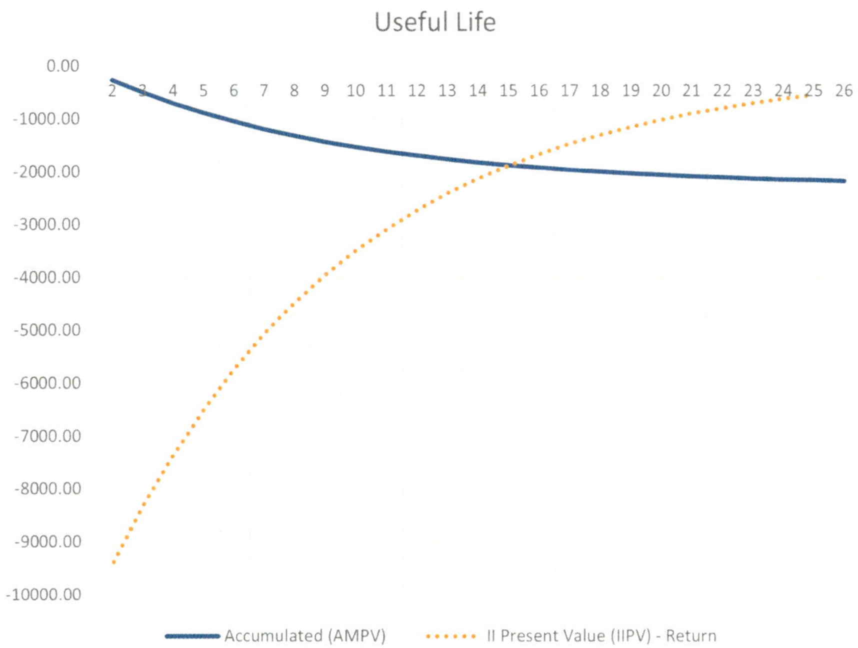

FIGURE 8.29 Physical Asset useful life.

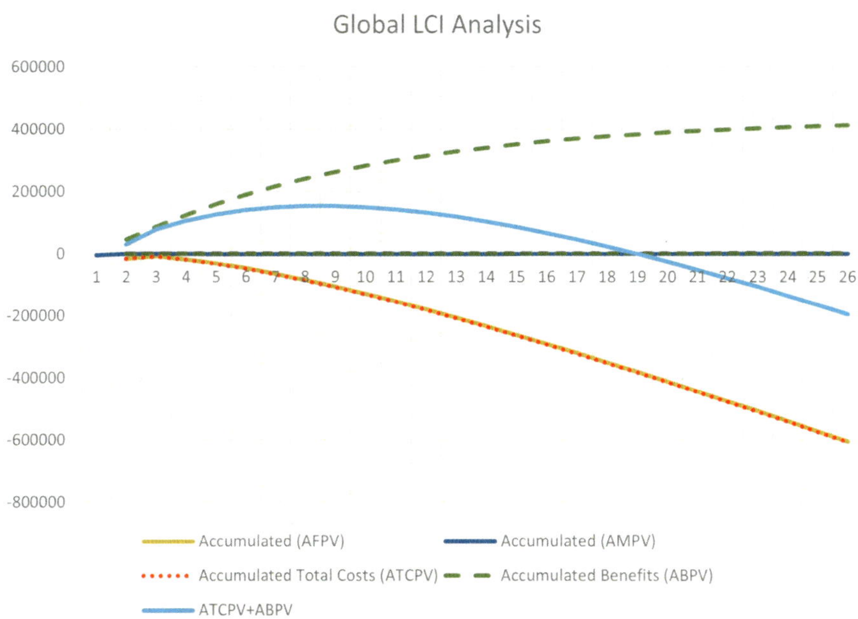

FIGURE 8.30 Global LCI analysis.

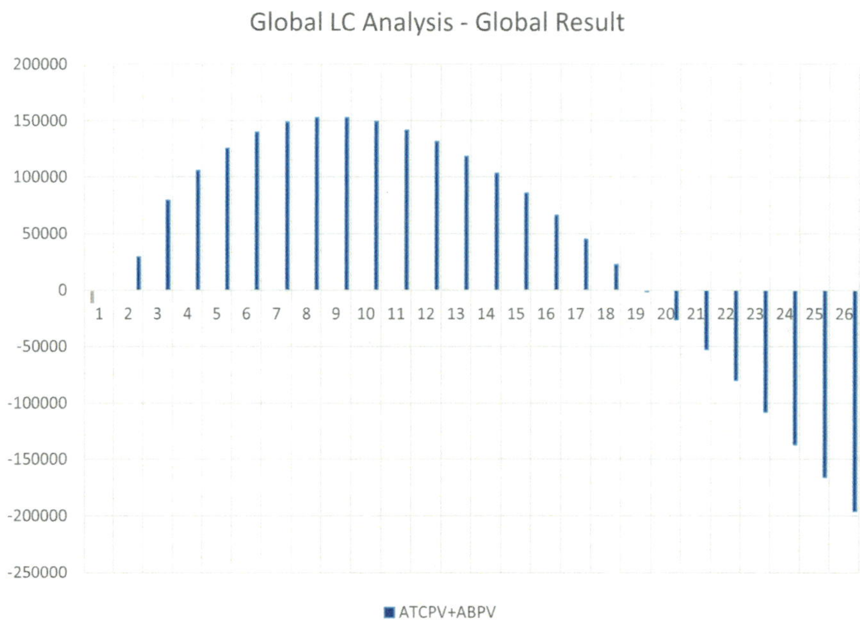

FIGURE 8.31 Life cycle analysis global result.

For the second scenario, other relevant values considered are the following ones:

- Interest Rate (*r*) 0.05
- Inflation Rate (*h*) 0.02
- Profit Rate (*p*) 0.05
- Maintenance Increase Rate 0.01
- MTBF 1000.0
- MTTR 30.0
- Risk (R) **0.01**
- Initial Investment (II) (C_0) –12000
- F_p Financial Movement (Production) 70000
- Functioning Costs –4500
- Energy –200
- Maintenance Costs –300

The results and graphical results are presented in Table 8.10 and Figures 8.35–8.37.

Comparing Figures 8.34 and 8.37, it can be seen that if the risk decreases, the positive results increase massively.

TABLE 8.9
Values Used for Simulation based on Life Cycle with Recovery Related to Risk Standards. First Simulation

Time (t)	**0**	**1**	**2**	**3**	**4**	**5**	**6**	**7**	**8**
r - real interest rate $(1+r)$		1.05	1.05	1.05	1.05	1.05	1.05	1.05	1.05
h - inflation rate $(1+h)$		1.02	1.02	1.02	1.02	1.02	1.02	1.02	1.02
p - profit rate		1.05	1.05	1.05	1.05	1.05	1.05	1.05	1.05
i - money rate (nominal interest rate)		0.12	0.12	0.12	0.12	0.12	0.12	0.12	0.12
i - Apparent Rate with Risk		0.46	0.46	0.46	0.46	0.46	0.46	0.46	0.46
$1/((1+i)\ ^\wedge\ t))$		0.68	0.47	0.32	0.22	0.15	0.10	0.07	0.05
Initial Investment (II) (C_0)	−12000	−12000	−12000	−12000	−12000	−12000	−12000	−12000	−12000
Initial Investment (II) (C_0) with Devaluation (NPV)	−12000	−8208	−5615	−3841	−2627	−1797	−1229	−841	−575
Initial Investment (II) (C_0) considering NEW Asset (NPV)	−12000	−17543	−25646	−37493	−54811	−80129	−117142	−171252	−250356
Fp Financial Movement (Production)	70000	70000	70000	70000	70000	70000	70000	70000	70000
Fp Financial Movement (NPV)	70000	43529	29776	20368	13932	9530	6519	4459	3050
Fp Financial Movement (NPV) Accumulated	70000	43529	73305	93673	107605	117135	123654	128113	131163
Functioning Costs	−4500	−4500	−4500	−4500	−4500	−4500	−4500	−4500	−4500
Functioning Costs (NPV) Accumulated	−4500	−3078	−5184	−6624	−7609	−8283	−8744	−9059	−9275
Energy	−200	−200	−200	−200	−200	−200	−200	−200	−200
Energy Accumulated (NPV) Accumulated	−200	−137	−230	−294	−338	−368	−389	−403	−412
Maintenance Costs	−300	−300	−300	−300	−300	−300	−300	−300	−300
Maintenance Costs (NPV) Accumulated	−300	−207	−349	−446	−512	−558	−589	−610	−625
Total Costs (NPV) Accumulated	−5000	−3485	−9218	−16488	−24810	−33851	−43383	−53253	−63352
FT Net Present Value (NPV) (With II Devaluated)		31836	58472	73344	80168	81487	79041	74019	67235
FT Net Present Value (With II New Asset)		22501	38441	39692	27984	3155	−36872	−96392	−182546

(Continued)

TABLE 8.9 (*Continued*)

Values Used for Simulation based on Life Cycle with Recovery Related to Risk Standards. First Simulation

Time (t)	9	10	11	12	13	14	15	16	17
r - real interest rate $(1+r)$	1.05	1.05	1.05	1.05	1.05	1.05	1.05	1.05	1.05
h - inflation rate $(1+h)$	1.02	1.02	1.02	1.02	1.02	1.02	1.02	1.02	1.02
p - profit rate	1.05	1.05	1.05	1.05	1.05	1.05	1.05	1.05	1.05
i - money rate (nominal interest rate)	0.12	0.12	0.12	0.12	0.12	0.12	0.12	0.12	0.12
i - Apparent Rate with Risk	0.46	0.46	0.46	0.46	0.46	0.46	0.46	0.46	0.46
$1/((1+i)\ ^\wedge\ t))$	0.03	0.02	0.02	0.01	0.01	0.00	0.00	0.00	0.00
Initial Investment (II) (C_0)	−12000	−12000	−12000	−12000	−12000	−12000	−12000	−12000	−12000
Initial Investment (II) (C_0) with Devaluation (NPV)	−393	−269	−184	−126	−86	−59	−40	−28	−19
Initial Investment (II) (C_0) considering NEW Asset (NPV)	−365999	−535060	−782212	−1143528	−1671740	−2443942	−3572836	−5223183	−7635849
Fp Financial Movement (Production)	70000	70000	70000	70000	70000	70000	70000	70000	70000
Fp Financial Movement (NPV)	2086	1427	976	668	457	312	214	146	100
Fp Financial Movement (NPV) Accumulated	133249	134677	135653	136321	136777	137090	137304	137450	137550
Functioning Costs	−4500	−4500	−4500	−4500	−4500	−4500	−4500	−4500	−4500
Functioning Costs (NPV) Accumulated	−9423	−9524	−9593	−9640	−9672	−9694	−9709	−9720	−9727
Energy	−200	−200	−200	−200	−200	−200	−200	−200	−200
Energy Accumulated (NPV) Accumulated	−419	−423	−426	−428	−430	−431	−432	−432	−432
Maintenance Costs	−300	−300	−300	−300	−300	−300	−300	−300	−300
Maintenance Costs (NPV) Accumulated	−634	−641	−646	−649	−651	−653	−654	−654	−655
Total Costs (NPV) Accumulated	−73610	−83974	−94413	−104902	−115425	−125972	−136535	−147109	−157691
FT Net Present Value (NPV) (With II Devaluated)	59246	50433	41056	31293	21266	11059	728	−9687	−20160
FT Net Present Value (With II New Asset)	−306359	−484358	−740972	−1112109	−1650388	−2432824	−3572067	−5232842	−7655990

(*Continued*)

TABLE 8.9 (*Continued*)
Values Used for Simulation based on Life Cycle with Recovery Related to Risk Standards. First Simulation

Time (t)	18	19	20	21	22	23	24	25
r - real interest rate $(1+r)$	1.05	1.05	1.05	1.05	1.05	1.05	1.05	1.05
h - inflation rate $(1+h)$	1.02	1.02	1.02	1.02	1.02	1.02	1.02	1.02
p - profit rate	1.05	1.05	1.05	1.05	1.05	1.05	1.05	1.05
i - money rate (nominal interest rate)	0.12	0.12	0.12	0.12	0.12	0.12	0.12	0.12
i - Apparent Rate with Risk	0.46	0.46	0.46	0.46	0.46	0.46	0.46	0.46
$1/((1+i)\wedge t))$	0.00	0.00	0.00	0.00	0.00	0.00	0.00	0.00
Initial Investment (II) (C_0)	−12000	−12000	−12000	−12000	−12000	−12000	−12000	−12000
Initial Investment (II) (C_0) with Devaluation (NPV)	−13	−9	−6	−4	−3	−2	−1	−1
Initial Investment (II) (C_0) considering NEW Asset (NPV)	−11162962	−16319302	−23857432	−34877537	−50987995	−74540115	−108971312	−159306795
Fp Financial Movement (Production)	70000	70000	70000	70000	70000	70000	70000	70000
Fp Financial Movement (NPV)	68	47	32	22	15	10	7	5
Fp Financial Movement (NPV) Accumulated	137618	137665	137697	137719	137734	137744	137751	137756
Functioning Costs	−4500	−4500	−4500	−4500	−4500	−4500	−4500	−4500
Functioning Costs (NPV) Accumulated	−9732	−9735	−9737	−9739	−9740	−9740	−9741	−9741
Energy	−200	−200	−200	−200	−200	−200	−200	−200
Energy Accumulated (NPV) Accumulated	−433	−433	−433	−433	−433	−433	−433	−433
Maintenance Costs	−300	−300	−300	−300	−300	−300	−300	−300
Maintenance Costs (NPV) Accumulated	−655	−655	−656	−656	−656	−656	−656	−656
Total Costs (NPV) Accumulated	−168278	−178868	−189461	−200055	−210651	−221247	−231844	−242441
FT Net Present Value (NPV) (With II Devaluated)	−30672	−41212	−51770	−62341	−72920	−83505	−94094	−104686
FT Net Present Value (With II New Asset)	−11193622	−16360505	−23909196	−34939874	−51060912	−74623618	−109065405	−159411481

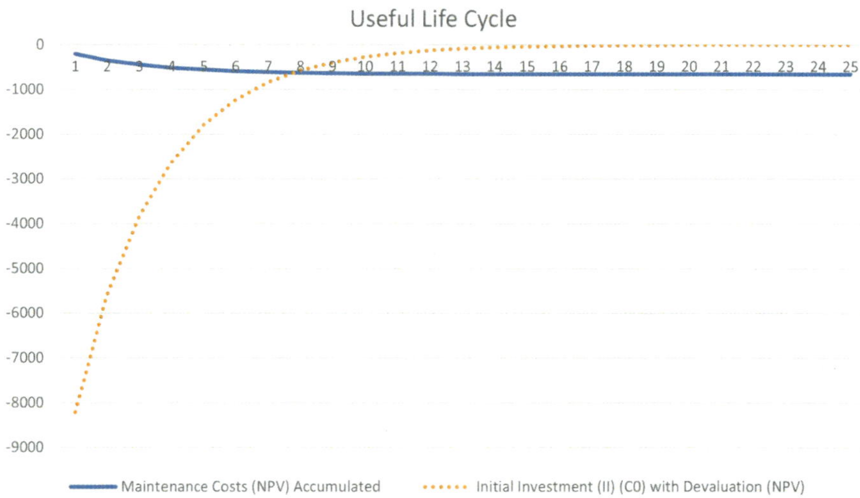

FIGURE 8.32 Useful life cycle.

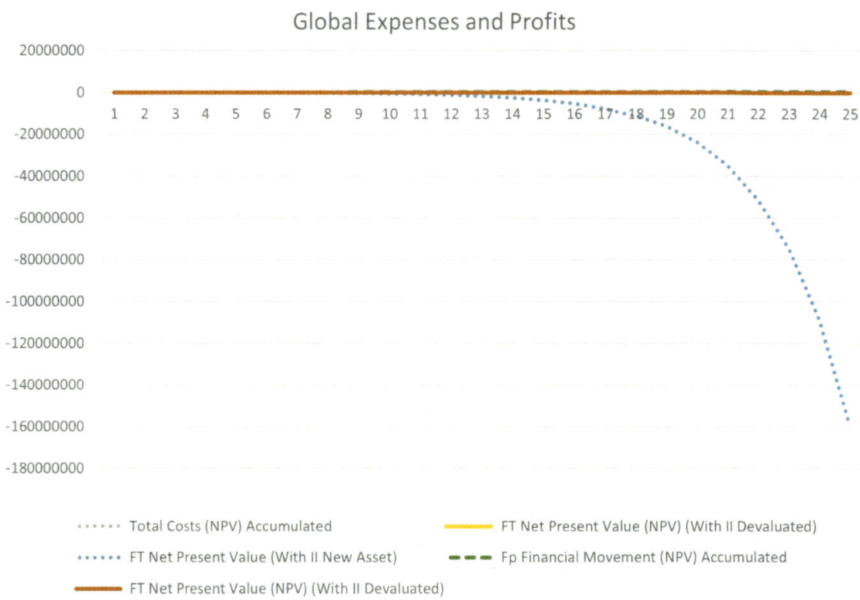

FIGURE 8.33 Global investments and profits.

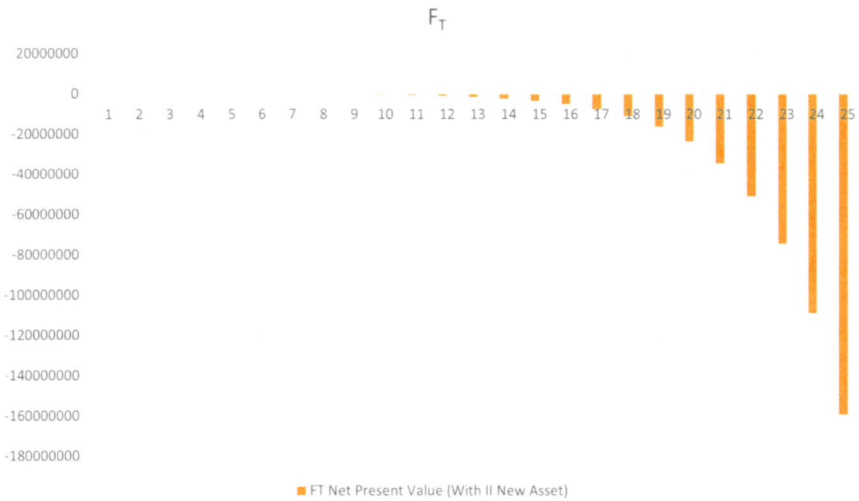

FIGURE 8.34 Results considering the acquisition of a new asset with devaluation and recovery.

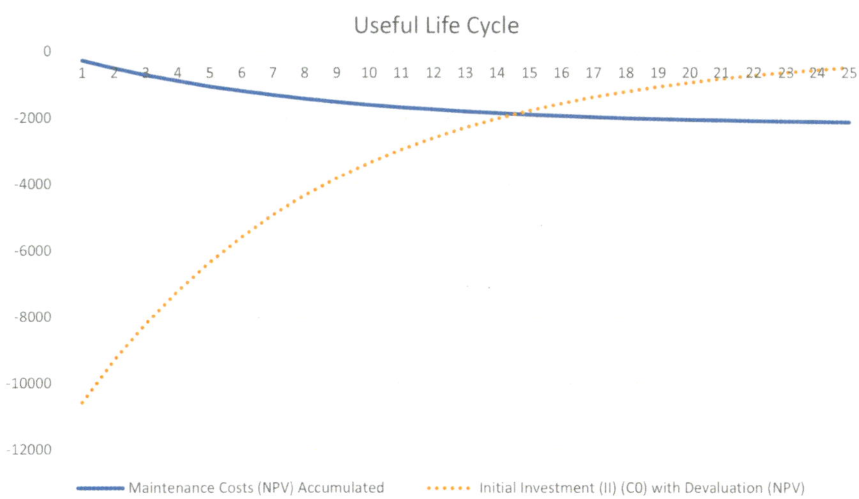

FIGURE 8.35 Useful life cycle.

TABLE 8.10
Values Used for Simulation based on Life Cycle with Recovery Related to Risk Standards. Second Simulation

Time (t)	**0**	**1**	**2**	**3**	**4**	**5**	**6**	**7**	**8**
r - real interest rate $(1+r)$		1.05	1.05	1.05	1.05	1.05	1.05	1.05	1.05
h - inflation rate $(1+h)$		1.02	1.02	1.02	1.02	1.02	1.02	1.02	1.02
p - profit rate		1.05	1.05	1.05	1.05	1.05	1.05	1.05	1.05
i - money rate (nominal interest rate)		0.12	0.12	0.12	0.12	0.12	0.12	0.12	0.12
i - Apparent Rate with Risk		0.14	0.14	0.14	0.14	0.14	0.14	0.14	0.14
$1/((1+i)\ ^\wedge\ t))$		0.88	0.78	0.68	0.60	0.53	0.47	0.41	0.36
Initial Investment (II) (C0)	−12000	−12000	−12000	−12000	−12000	−12000	−12000	−12000	−12000
Initial Investment (II) (C_0) with Devaluation (NPV)	−12000	−10565	−9302	−8190	−7211	−6349	−5590	−4921	−4333
Initial Investment (II) (C_0) considering NEW Asset (NPV)	−12000	−13630	−15480	−17583	−19970	−22682	−25762	−29261	−33234
Fp Financial Movement (Production)	70000	70000	70000	70000	70000	70000	70000	70000	70000
Fp Financial Movement (NPV)	70000	56028	49329	43432	38239	33667	29642	26098	22978
Fp Financial Movement (NPV) Accumulated	70000	56028	105357	148789	187028	220695	250336	276434	299412
Functioning Costs	−4500	−4500	−4500	−4500	−4500	−4500	−4500	−4500	−4500
Functioning Costs (NPV) Accumulated	−4500	−3962	−7450	−10521	−13226	−15606	−17702	−19548	−21173
Energy	−200	−200	−200	−200	−200	−200	−200	−200	−200
Energy Accumulated (NPV) Accumulated	−200	−176	−331	−468	−588	−694	−787	−869	−941
Maintenance Costs	−300	−300	−300	−300	−300	−300	−300	−300	−300
Maintenance Costs (NPV) Accumulated	−300	−267	−502	−708	−891	−1051	−1192	−1316	−1426
Total Costs (NPV) Accumulated	−5000	−4429	−12581	−24011	−38327	−55184	−74278	−95342	−118140
FT Net Present Value (NPV) (With II Devaluated)		41034	83475	116588	141490	159162	170469	176171	176938
FT Net Present Value (With II New Asset)		37970	77296	107196	128731	142829	150296	151832	148037

(Continued)

TABLE 8.10 (*Continued*)
Values Used for Simulation based on Life Cycle with Recovery Related to Risk Standards. Second Simulation

Time (t)	9	10	11	12	13	14	15	16	17
r - real interest rate $(1+r)$	1.05	1.05	1.05	1.05	1.05	1.05	1.05	1.05	1.05
h - inflation rate $(1+h)$	1.02	1.02	1.02	1.02	1.02	1.02	1.02	1.02	1.02
p - profit rate	1.05	1.05	1.05	1.05	1.05	1.05	1.05	1.05	1.05
i - money rate (nominal interest rate)	0.12	0.12	0.12	0.12	0.12	0.12	0.12	0.12	0.12
i - Apparent Rate with Risk	0.14	0.14	0.14	0.14	0.14	0.14	0.14	0.14	0.14
$1/((1+i)\wedge t))$	0.32	0.28	0.25	0.22	0.19	0.17	0.15	0.13	0.11
Initial Investment (II) (C0)	−12000	−12000	−12000	−12000	−12000	−12000	−12000	−12000	−12000
Initial Investment (II) (C_0) with Devaluation (NPV)	−3815	−3359	−2957	−2604	−2292	−2018	−1777	−1565	−1377
Initial Investment (II) (C_0) considering NEW Asset (NPV)	−37747	−42873	−48695	−55307	−62818	−71348	−81037	−92041	−104540
Fp Financial Movement (Production)	70000	70000	70000	70000	70000	70000	70000	70000	70000
Fp Financial Movement (NPV)	20230	17812	15682	13807	12156	10703	9423	8297	7305
Fp Financial Movement (NPV) Accumulated	319642	337454	353136	366943	379099	389802	399226	407522	414827
Functioning Costs	−4500	−4500	−4500	−4500	−4500	−4500	−4500	−4500	−4500
Functioning Costs (NPV) Accumulated	−22603	−23863	−24972	−25948	−26808	−27565	−28231	−28818	−29334
Energy	−200	−200	−200	−200	−200	−200	−200	−200	−200
Energy Accumulated (NPV) Accumulated	−1005	−1061	−1110	−1153	−1191	−1225	−1255	−1281	−1304
Maintenance Costs	−300	−300	−300	−300	−300	−300	−300	−300	−300
Maintenance Costs (NPV) Accumulated	−1522	−1607	−1681	−1747	−1805	−1856	−1901	−1940	−1975
Total Costs (NPV) Accumulated	−142466	−168135	−194988	−222884	−251697	−281317	−311649	−342607	−374116
FT Net Present Value (NPV) (With II Devaluated)	173362	165960	155190	141456	125111	106467	85800	63351	39333
FT Net Present Value (With II New Asset)	139430	126446	109453	88752	64585	37137	6540	−27126	−63830

(*Continued*)

TABLE 8.10 (*Continued*)
Values Used for Simulation based on Life Cycle with Recovery Related to Risk Standards. Second Simulation

Time (t)	18	19	20	21	22	23	24	25
r - real interest rate $(1+r)$	1.05	1.05	1.05	1.05	1.05	1.05	1.05	1.05
h - inflation rate $(1+h)$	1.02	1.02	1.02	1.02	1.02	1.02	1.02	1.02
p - profit rate	1.05	1.05	1.05	1.05	1.05	1.05	1.05	1.05
i - money rate (nominal interest rate)	0.12	0.12	0.12	0.12	0.12	0.12	0.12	0.12
i - Apparent Rate with Risk	0.14	0.14	0.14	0.14	0.14	0.14	0.14	0.14
$1/((1+i)\ ^\wedge\ t))$	0.10	0.09	0.08	0.07	0.06	0.05	0.05	0.04
Initial Investment (II) (C0)	−12000	−12000	−12000	−12000	−12000	−12000	−12000	−12000
Initial Investment (II) (C_0) with Devaluation (NPV)	−1213	−1068	−940	−828	−729	−642	−565	−497
Initial Investment (II) (C_0) considering NEW Asset (NPV)	−118736	−134860	−153174	−173974	−197599	−224432	−254909	−289524
Fp Financial Movement (Production)	70000	70000	70000	70000	70000	70000	70000	70000
Fp Financial Movement (NPV)	6431	5662	4985	4389	3865	3403	2996	2638
Fp Financial Movement (NPV) Accumulated	421258	426921	431906	436296	440160	443563	446559	449196
Functioning Costs	−4500	−4500	−4500	−4500	−4500	−4500	−4500	−4500
Functioning Costs (NPV) Accumulated	−29789	−30189	−30542	−30852	−31126	−31366	−31578	−31765
Energy	−200	−200	−200	−200	−200	−200	−200	−200
Energy Accumulated (NPV) Accumulated	−1324	−1342	−1357	−1371	−1383	−1394	−1403	−1412
Maintenance Costs	−300	−300	−300	−300	−300	−300	−300	−300
Maintenance Costs (NPV) Accumulated	−2006	−2033	−2056	−2077	−2096	−2112	−2126	−2139
Total Costs (NPV) Accumulated	−406111	−438533	−471332	−504462	−537883	−571561	−605466	−639569
FT Net Present Value (NPV) (With II Devaluated)	13934	−12680	−40366	−68994	−98451	−128640	−159472	−190870
FT Net Present Value (With II New Asset)	−103589	−146473	−192599	−242140	−295322	−352430	−413816	−479897

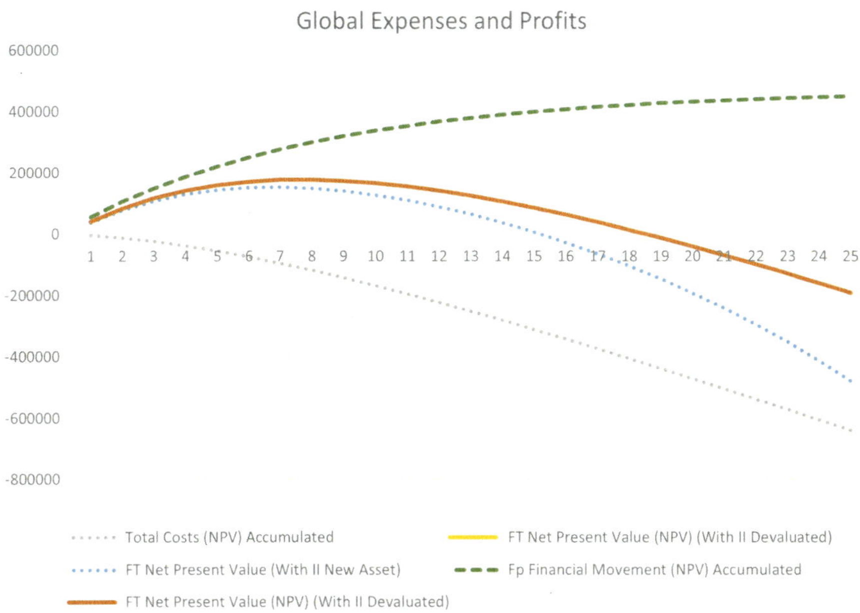

FIGURE 8.36 Global investments and profits.

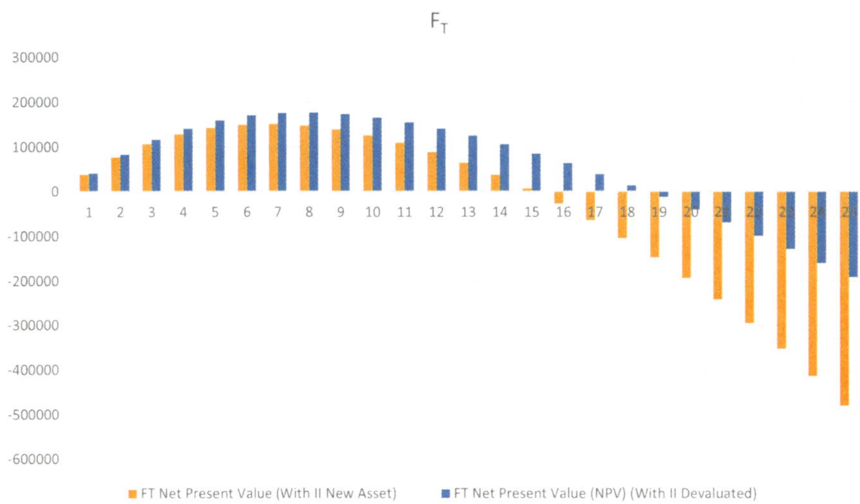

FIGURE 8.37 Results considering the acquisition of a new asset with devaluation and recovery.

REFERENCE

Woellner, R. A., Voorhees, J., & Bell, C. L. (2020). *International Environmental Risk Management* (2nd Edition). London: CRC Press. ISBN: 978-1-138-05452-3

9 Energy Norms

9.1 INTRODUCTION

This chapter deals with the implementation of Energy Standards, namely the ISO 50001, being structured in the following sections:

- Characterization of Energy Norms;
- Diagnosis to Implement Energy Norms;
- Quantitative Evaluation about the Implementation of Energy Norms.

9.2 CHARACTERIZATION OF ENERGY NORMS

The ISO 50000 family includes:

- ISO 50001, Energy management systems – Requirements and guidelines for its use;
- ISO 50002, Energy audits – Requirements with guidance for use;
- ISO 50003, Energy management systems – Requirements for bodies providing audit and certification of energy management systems;
- ISO 50004, Energy management systems – Guidance for the implementation, maintenance, and improvement of an energy management system;
- ISO 50006, Energy management systems – Measuring energy performance using Energy Baselines (EnB) and Energy Performance Indicators (EnPI) – General principles and guidance;
- ISO 50007/1, Energy services – Guidelines for the assessment and improvement of the energy service to users;
- ISO 50015, Energy management systems – Measurement and verification of energy performance of organizations – General principles and guidance;
- ISO 50047, Energy savings – Determination of energy savings in organizations;
- ISO 17741, General technical rules for measurement, calculation, and verification of energy savings of projects;
- ISO 17742, Energy efficiency and savings calculation for countries, regions, and cities;
- ISO 17743, Energy savings – Definition of a methodological framework applicable to calculation and reporting on energy savings;
- ISO/IEC 13273-1, Energy efficiency and renewable energy sources – Common international terminology – Part 1: Energy efficiency;
- ISO/IEC 13273-2, Energy efficiency and renewable energy sources – Common international terminology – Part 2: Renewable energy sources.

DOI: 10.1201/9781003395690-12

9.3 DIAGNOSIS TO IMPLEMENT ENERGY NORMS

To implement Risk Norms, the proposed methodology is like the other standards (see Section 6.3), that is to start with the diagnosis before to give the next step (Figure 9.1).

9.4 QUANTITATIVE EVALUATION ABOUT THE IMPLEMENTATION OF ENERGY NORMS

9.4.1 Life Cycle Cost of Physical Assets

This Chapter and sections follow the same approach as the preceding homologous about LCC. The reference value for Energy was 200 Monetary Units.

9.4.1.1 Uniform Annual Income Method (UAI)

The first simulation uses the following values:

- Initial Investment (II) –12000
- Life Time simulated 25
- Functioning (*F*) –4500
- Energy **–1000**
- Maintenance (*M*) –300
- Non Production (NP) 0
- Benefit (*B*) 50000
- Internal Rate of Return (*r*) 0.12
- Risk (*R*) 0.01
- Apparent Rate 0.1312
- Maintenance Increase Rate 0.01
- Availability 0.8929
- MTBF 500
- MTTR 50
- MWT 10

The remaining values are presented in Table 9.1, and the graphical results are shown in Figures 9.2–9.5.

As referred in the preceding introduction of this section, the evaluation of the introduction of Energy Norms, the simulation of their implementation is made through the variation of value of Energy. Other indicators may be used, but may not be so intuitive.

The second simulation uses the following values (the value of energy is decreased):

- Initial Investment (II) –12000
- Life Time simulated 25
- Functioning (*F*) –4500
- Energy **–200**
- Maintenance (*M*) –300
- Non Production (NP) 0

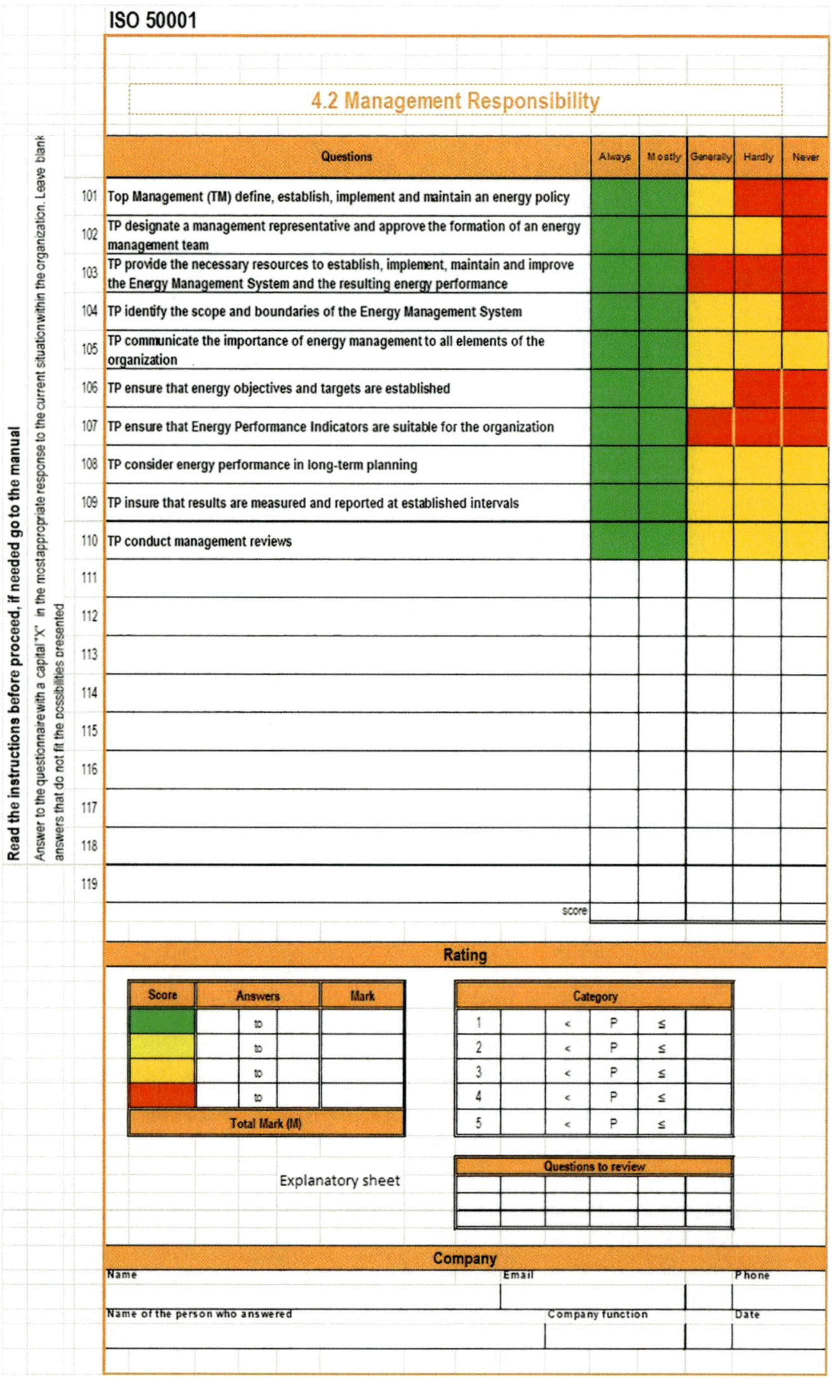

Read the instructions before proceed, if needed go to the manual

Answer to the questionnaire with a capital "X" in the mostappropriate response to the current situation within the organization. Leave blank answers that do not fit the possibilities presented

ISO 50001

4.2 Management Responsibility

	Questions	Always	Mostly	Generally	Hardly	Never
101	Top Management (TM) define, establish, implement and maintain an energy policy					
102	TP designate a management representative and approve the formation of an energy management team					
103	TP provide the necessary resources to establish, implement, maintain and improve the Energy Management System and the resulting energy performance					
104	TP identify the scope and boundaries of the Energy Management System					
105	TP communicate the importance of energy management to all elements of the organization					
106	TP ensure that energy objectives and targets are established					
107	TP ensure that Energy Performance Indicators are suitable for the organization					
108	TP consider energy performance in long-term planning					
109	TP insure that results are measured and reported at established intervals					
110	TP conduct management reviews					
111						
112						
113						
114						
115						
116						
117						
118						
119						
	score					

Rating

Score	Answers			Mark
		to		
		to		
		to		
		to		
Total Mark (M)				

Category					
1		<	P	≤	
2		<	P	≤	
3		<	P	≤	
4		<	P	≤	
5		<	P	≤	

Explanatory sheet

Questions to review					

Company

Name	Email	Phone

Name of the person who answered	Company function	Date

FIGURE 9.1 Diagnosis form.

TABLE 9.1
Values Used for Simulation based on UAI Method Related to Energy Standards. First Simulation

Time	0	1	2	3	4	5
Initial Investment (II)	−12000					
Initial Investment (II)	−12000					
II Present Value (IIPV) - Return	−12000	−10608.20	−9377.83	−8290.16	−7328.64	−6478.65
Life Time simulated	25					
Residual value of equipment (RV)	10					
Linear Depreciation Method (Vn)	−12000	−11519.60	−11039.20	−10558.80	−10078.40	−9598.00
LDM Present Value (Vn)		−10183.52	−8626.98	−7294.51	−6155.08	−5181.84
Sum of Digits Method (Vn)	−12000	−11076.15	−10189.26	−9339.32	−8526.34	−7750.31
SDM Present Value (Vn)		−9791.51	−7962.77	−6452.04	−5207.21	−4184.29
Exponential Method (Vn)	−12000	−9036.79	−6805.29	−5124.83	−3859.33	−2906.33
EM Present Value (Pv)		−7988.67	−5318.24	−3540.47	−2356.97	−1569.09
Functioning (*F*)	−4500	−4500.00	−4545.00	−4590.45	−4636.35	−4682.72
Present Value Accumulated (NPV)		−3978.08	−7529.93	−10701.23	−13532.74	−16060.88
Present Value (NPV)		−3978.08	−3551.85	−3171.30	−2831.52	−2528.14
Energy	−1000	−1000.00	−2000.00	−3000.00	−4000.00	−5000.00
Energy (NPV)		−884.02	−1562.97	−2072.54	−2442.88	−2699.44
Accumulated (FPV)		−16862.09	−25955.00	−38728.76	−54704.39	−73464.71
Maintenance (*M*)	−300	−300.00	−303.00	−306.03	−309.09	−312.18
Present Value (MPV)		−265.21	−236.79	−211.42	−188.77	−168.54
Accumulated (MPV)		−265.21	−502.00	−713.42	−902.18	−1070.73
Non Production (NP)	0	0.00	0.00	0.00	0.00	0.00
Present Value (NPPV)		0.00	0.00	0.00	0.00	0.00
Accumulated NPPV		0.00	0.00	0.00	0.00	0.00
Accumulated Total Costs (APV)		−17127.30	−26456.99	−39442.18	−55606.57	−74535.44
Accumulated Total Costs		−5800.00	−12648.00	−20544.48	−29489.92	−39484.82
Benefit (*B*)	50000	44642.86	44642.86	44642.86	44642.86	44642.86
Present Value (BPV)		39465.04	34887.77	30841.38	27264.30	24102.11
Accumulated Benefits BPV		39465.04	74352.81	105194.19	132458.49	156560.60
BPV + FPV + MPV + NPPV + IIPV		10337.74	35895.82	53752.01	64851.92	70025.16
Uniform Annual Income UAIn (LDM)		−11519.60	−5179.80	−3095.68	−2074.47	−1477.59
Uniform Annual Income UAIn (SDM)		−11076.15	−4781.00	−2738.15	−1755.00	−1193.14
Uniform Annual Income UAIn (EM)		−9036.79	−3193.17	−1502.52	−794.38	−447.42

(Continued)

TABLE 9.1 (*Continued*)
Values Used for Simulation based on UAI Method Related to Energy Standards. First Simulation

Time	6	7	8	9	10
Initial Investment (II)					
Initial Investment (II)					
II Present Value (IIPV) - Return	−5727.23	−5062.97	−4475.75	−3956.64	−3497.74
Life Time simulated					
Residual value of equipment (RV)					
Linear Depreciation Method (Vn)	−9117.60	−8637.20	−8156.80	−7676.40	−7196.00
LDM Present Value (Vn)	−4351.55	−3644.16	−3042.32	−2531.06	−2097.48
Sum of Digits Method (Vn)	−7011.23	−6309.11	−5643.94	−5015.72	−4424.46
SDM Present Value (Vn)	−3346.25	−2661.90	−2105.07	−1653.78	−1289.63
Exponential Method (Vn)	−2188.66	−1648.20	−1241.20	−934.71	−703.90
EM Present Value (Pv)	−1044.58	−695.40	−462.94	−308.19	−205.17
Functioning (*F*)	−4729.55	−4776.84	−4824.61	−4872.86	−4921.58
Present Value Accumulated (NPV)	−18318.15	−20333.57	−22133.05	−23739.73	−25174.26
Present Value (NPV)	−2257.27	−2015.42	−1799.48	−1606.68	−1434.53
Energy	−6000.00	−7000.00	−8000.00	−9000.00	−10000.00
Energy (NPV)	−2863.62	−2953.40	−2983.84	−2967.48	−2914.78
Accumulated (FPV)	−94646.48	−117933.45	−143050.33	−169757.54	−197846.58
Maintenance (*M*)	−315.30	−318.46	−321.64	−324.86	−328.11
Present Value (MPV)	−150.48	−134.36	−119.97	−107.11	−95.64
Accumulated (MPV)	−1221.21	−1355.57	−1475.54	−1582.65	−1678.28
Non Production (NP)	0.00	0.00	0.00	0.00	0.00
Present Value (NPPV)	0.00	0.00	0.00	0.00	0.00
Accumulated NPPV	0.00	0.00	0.00	0.00	0.00
Accumulated Total Costs (APV)	−95867.69	−119289.02	−144525.87	−171340.19	−199524.87
Accumulated Total Costs	−50529.67	−62624.97	−75771.22	−89968.93	−105218.62
Benefit (*B*)	44642.86	44642.86	44642.86	44642.86	44642.86
Present Value (BPV)	21306.67	18835.46	16650.87	14719.65	13012.42
Accumulated Benefits BPV	177867.27	196702.73	213353.60	228073.24	241085.66
BPV + FPV + MPV + NPPV + IIPV	69999.58	65413.71	56827.73	44733.06	29560.80
Uniform Annual Income UAIn (LDM)	−1092.19	−827.06	−636.59	−495.43	−388.40
Uniform Annual Income UAIn (SDM)	−839.87	−604.13	−440.47	−323.71	−238.81
Uniform Annual Income UAIn (EM)	−262.18	−157.83	−96.87	−60.33	−37.99

(*Continued*)

TABLE 9.1 (*Continued*)
Values Used for Simulation based on UAI Method Related to Energy Standards. First Simulation

Time	11	12	13	14	15
Initial Investment (II)					
Initial Investment (II)					
II Present Value (IIPV) - Return	−3092.06	−2733.43	−2416.40	−2136.14	−1888.38
Life Time simulated					
Residual value of equipment (RV)					
Linear Depreciation Method (Vn)	−6715.60	−6235.20	−5754.80	−5274.40	−4794.00
LDM Present Value (Vn)	−1730.42	−1420.29	−1158.83	−938.90	−754.41
Sum of Digits Method (Vn)	−3870.15	−3352.80	−2872.40	−2428.95	−2022.46
SDM Present Value (Vn)	−997.23	−763.72	−578.41	−432.38	−318.27
Exponential Method (Vn)	−530.08	−399.19	−300.61	−226.38	−170.48
EM Present Value (Pv)	−136.59	−90.93	−60.53	−40.30	−26.83
Functioning (*F*)	−4970.80	−5020.51	−5070.71	−5121.42	−5172.63
Present Value Accumulated (NPV)	−26455.10	−27598.70	−28619.77	−29531.44	−30345.44
Present Value (NPV)	−1280.83	−1143.60	−1021.07	−911.67	−813.99
Energy	−11000.00	−12000.00	−13000.00	−14000.00	−15000.00
Energy (NPV)	−2834.39	−2733.43	−2617.77	−2492.16	−2360.48
Accumulated (FPV)	−227136.06	−257468.20	−288705.73	−320729.34	−353435.26
Maintenance (*M*)	−331.39	−334.70	−338.05	−341.43	−344.84
Present Value (MPV)	−85.39	−76.24	−68.07	−60.78	−54.27
Accumulated (MPV)	−1763.67	−1839.91	−1907.98	−1968.76	−2023.03
Non Production (NP)	0.00	0.00	0.00	0.00	0.00
Present Value (NPPV)	0.00	0.00	0.00	0.00	0.00
Accumulated NPPV	0.00	0.00	0.00	0.00	0.00
Accumulated Total Costs (APV)	−228899.74	−259308.11	−290613.72	−322698.10	−355458.28
Accumulated Total Costs	−121520.81	−138876.01	−157284.77	−176747.62	−197265.10
Benefit (*B*)	44642.86	44642.86	44642.86	44642.86	44642.86
Present Value (BPV)	11503.20	10169.02	8989.59	7946.95	7025.24
Accumulated Benefits BPV	252588.86	262757.88	271747.47	279694.42	286719.66
BPV + FPV + MPV + NPPV + IIPV	11689.12	−8550.23	−30866.25	−55003.68	−80738.63
Uniform Annual Income UAIn (LDM)	−305.84	−241.31	−190.37	−149.86	−117.46
Uniform Annual Income UAIn (SDM)	−176.25	−129.76	−95.02	−69.01	−49.55
Uniform Annual Income UAIn (EM)	−24.14	−15.45	−9.94	−6.43	−4.18

(*Continued*)

TABLE 9.1 (*Continued*)
Values Used for Simulation based on UAI Method Related to Energy Standards. First Simulation

Time	16	17	18	19	20
Initial Investment (II)					
Initial Investment (II)					
II Present Value (IIPV) - Return	−1669.36	−1475.75	−1304.58	−1153.27	−1019.51
Life Time simulated					
Residual value of equipment (RV)					
Linear Depreciation Method (Vn)	−4313.60	−3833.20	−3352.80	−2872.40	−2392.00
LDM Present Value (Vn)	−600.08	−471.40	−364.50	−276.06	−203.22
Sum of Digits Method (Vn)	−1652.92	−1320.34	−1024.71	−766.03	−544.31
SDM Present Value (Vn)	−229.94	−162.37	−111.40	−73.62	−46.24
Exponential Method (Vn)	−128.38	−96.68	−72.81	−54.83	−41.29
EM Present Value (Pv)	−17.86	−11.89	−7.92	−5.27	−3.51
Functioning (*F*)	−5224.36	−5276.60	−5329.37	−5382.66	−5436.49
Present Value Accumulated (NPV)	−31072.22	−31721.13	−32300.51	−32817.82	−33279.70
Present Value (NPV)	−726.78	−648.91	−579.38	−517.31	−461.88
Energy	−16000.00	−17000.00	−18000.00	−19000.00	−20000.00
Energy (NPV)	−2225.82	−2090.64	−1956.88	−1826.02	−1699.19
Accumulated (FPV)	−386733.29	−420545.05	−454802.44	−489446.28	−524425.17
Maintenance (*M*)	−348.29	−351.77	−355.29	−358.84	−362.43
Present Value (MPV)	−48.45	−43.26	−38.63	−34.49	−30.79
Accumulated (MPV)	−2071.48	−2114.74	−2153.37	−2187.85	−2218.65
Non Production (NP)	0.00	0.00	0.00	0.00	0.00
Present Value (NPPV)	0.00	0.00	0.00	0.00	0.00
Accumulated NPPV	0.00	0.00	0.00	0.00	0.00
Accumulated Total Costs (APV)	−388804.77	−422659.80	−456955.81	−491634.13	−526643.81
Accumulated Total Costs	−218837.75	−241466.13	−265150.79	−289892.30	−315691.22
Benefit (*B*)	44642.86	44642.86	44642.86	44642.86	44642.86
Present Value (BPV)	6210.43	5490.13	4853.36	4290.46	3792.84
Accumulated Benefits BPV	292930.09	298420.21	303273.58	307564.03	311356.87
BPV + FPV + MPV + NPPV + IIPV	−107874.68	−136239.58	−165682.23	−196070.10	−227286.94
Uniform Annual Income UAIn (LDM)	−91.45	−70.52	−53.66	−40.07	−29.14
Uniform Annual Income UAIn (SDM)	−35.04	−24.29	−16.40	−10.69	−6.63
Uniform Annual Income UAIn (EM)	−2.72	−1.78	−1.17	−0.76	−0.50

(*Continued*)

TABLE 9.1 (*Continued*)
Values Used for Simulation based on UAI Method Related to Energy Standards. First Simulation

Time	21	22	23	24	25
Initial Investment (II)					
Initial Investment (II)					
II Present Value (IIPV) - Return	−901.27	−796.74	−704.33	−622.64	−550.42
Life Time simulated					
Residual value of equipment (RV)					
Linear Depreciation Method (Vn)	−1911.60	−1431.20	−950.80	−470.40	10.00
LDM Present Value (Vn)	−143.57	−95.02	−55.81	−24.41	0.46
Sum of Digits Method (Vn)	−359.54	−211.72	−100.86	−26.95	10.00
SDM Present Value (Vn)	−27.00	−14.06	−5.92	−1.40	0.46
Exponential Method (Vn)	−31.09	−23.42	−17.63	−13.28	−10.00
EM Present Value (Pv)	−2.34	−1.55	−1.03	−0.69	−0.46
Functioning (*F*)	−5490.86	−5545.76	−5601.22	−5657.23	−5713.81
Present Value Accumulated (NPV)	−33692.09	−34060.30	−34389.06	−34682.60	−34944.68
Present Value (NPV)	−412.39	−368.21	−328.76	−293.53	−262.08
Energy	−21000.00	−22000.00	−23000.00	−24000.00	−25000.00
Energy (NPV)	−1577.22	−1460.68	−1349.96	−1245.28	−1146.71
Accumulated (FPV)	−559694.48	−595215.46	−630954.49	−666882.36	−702973.75
Maintenance (*M*)	−366.06	−369.72	−373.41	−377.15	−380.92
Present Value (MPV)	−27.49	−24.55	−21.92	−19.57	−17.47
Accumulated (MPV)	−2246.14	−2270.69	−2292.60	−2312.17	−2329.65
Non Production (NP)	0.00	0.00	0.00	0.00	0.00
Present Value (NPPV)	0.00	0.00	0.00	0.00	0.00
Accumulated NPPV	0.00	0.00	0.00	0.00	0.00
Accumulated Total Costs (APV)	−561940.62	−597486.15	−633247.09	−669194.53	−705303.40
Accumulated Total Costs	−342548.13	−370463.61	−399438.25	−429472.63	−460567.36
Benefit (*B*)	44642.86	44642.86	44642.86	44642.86	44642.86
Present Value (BPV)	3352.93	2964.05	2620.27	2316.36	2047.70
Accumulated Benefits BPV	314709.80	317673.85	320294.12	322610.48	324658.18
BPV + FPV + MPV + NPPV + IIPV	−259230.82	−291812.30	−324952.97	−358584.05	−392645.22
Uniform Annual Income UAIn (LDM)	−20.37	−13.35	−7.78	−3.38	0.06
Uniform Annual Income UAIn (SDM)	−3.83	−1.98	−0.83	−0.19	0.06
Uniform Annual Income UAIn (EM)	−0.33	−0.22	−0.14	−0.10	−0.06

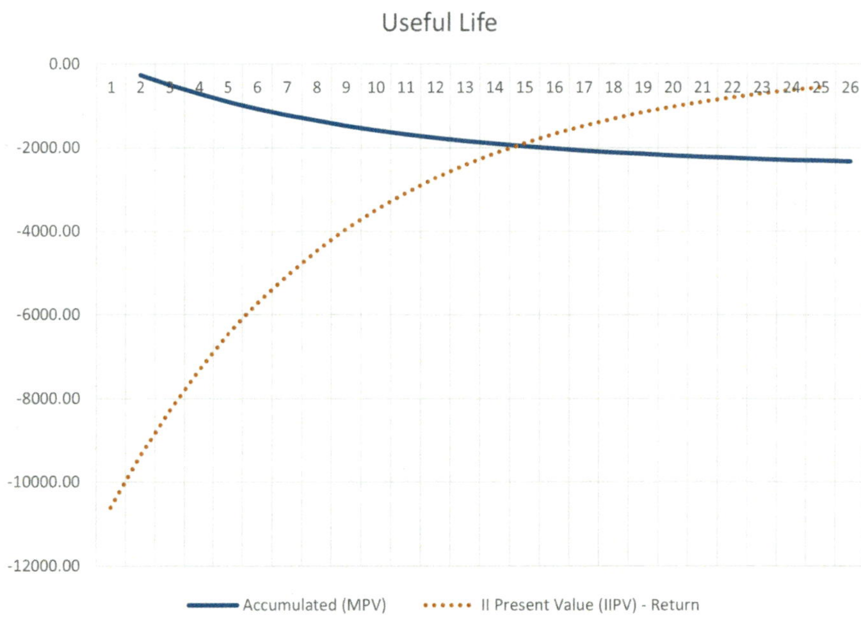

FIGURE 9.2 Physical Asset useful life.

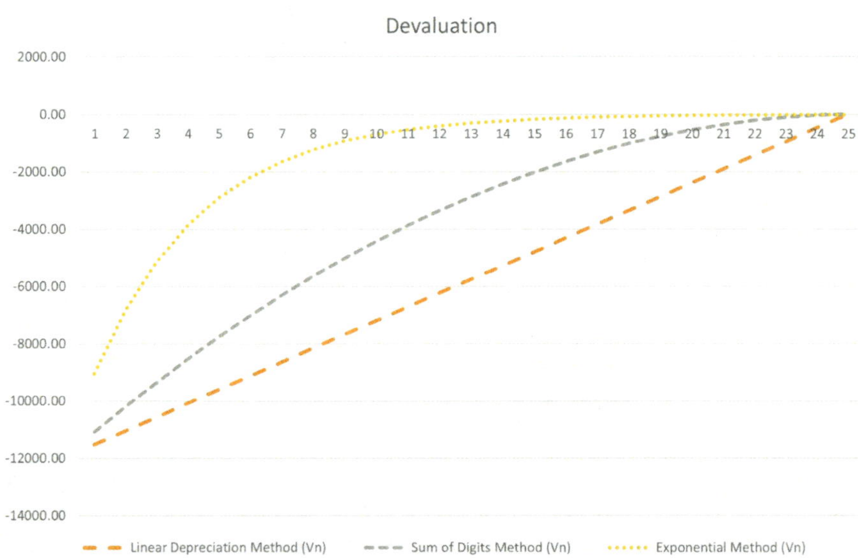

FIGURE 9.3 The three devaluation methods.

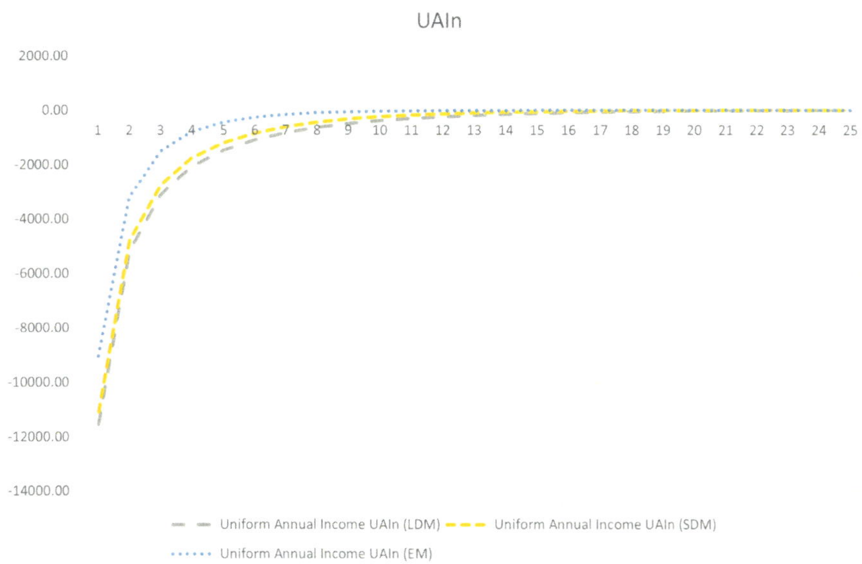

FIGURE 9.4 The UAI results from the three devaluation methods.

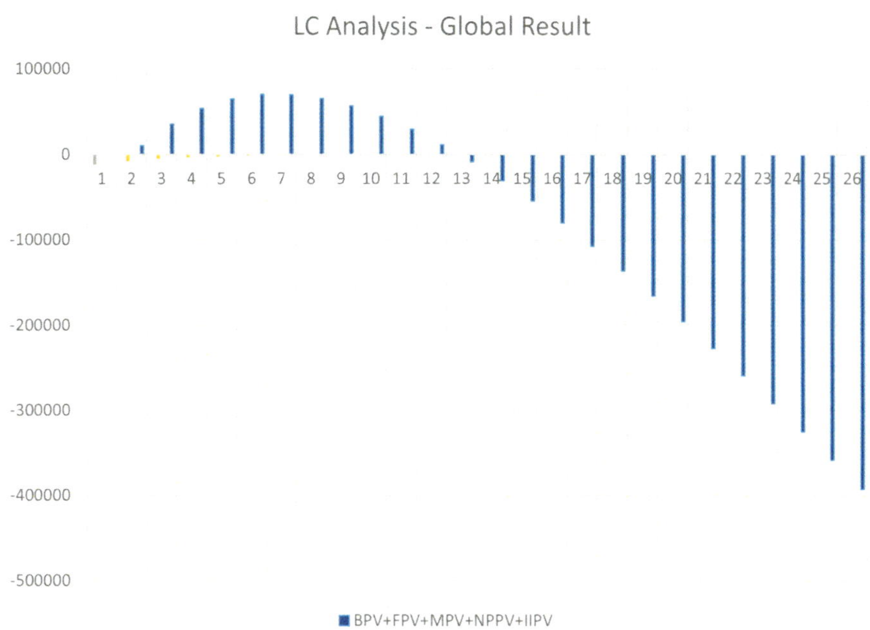

FIGURE 9.5 Life cycle analysis global result including profits.

- Benefit (B) 50000
- Internal Rate of Return (r) 0.12
- Risk (R) 0.01
- Apparent Rate 0.1312
- Maintenance Increase Rate 0.01
- Availability 0.8929
- MTBF 500
- MTTR 50
- MWT 10

The remaining values are presented in Table 9.2, and the results are shown in Figures 9.6–9.9.

As referred in the preceding introduction of this section, the evaluation of the introduction of Energy Norms, the simulation of their implementation is made through the variation of value of Energy. Other indicators may be used, but may not be so intuitive.

As can be seen, namely comparing Figures 9.5 and 9.9, with the implementation of Energy Standards, it gets one more positive year, and the positive incomes increase a lot.

9.4.1.2 Minimization of Total Average Cost Method (MMTAC)

The first simulation uses the following values:

- Initial Investment (II) −12000
- Life Time simulated 25
- Residual value of equipment (RV) 10
- Functioning (F) −4500
- Energy **−1000**
- Maintenance (M) −300
- Non Production (NP) 0
- Benefit (B) 70000
- Internal Rate of Return (r) 0.0
- Risk (R) 0.0
- Apparent Rate 0.0
- Maintenance Increase Rate 0.01
- Availability 0.9662
- MTBF 1000
- MTTR 25
- MWT 10

The remaining values are presented in Table 9.3, and the graphical results are shown in Figures 9.10–9.13.

As referred in the preceding introduction of this section, the evaluation of introduction of Energy Norms, the simulation of their implementation is made through the increase of variation of the value of Energy. Other indicators may be used, but may not be so intuitive.

TABLE 9.2
Values Used for Simulation based on UAI Method Related to Energy Standards. Second Simulation

Time	0	1	2	3	4	5
Initial Investment (II)	−12000					
Initial Investment (II)	−12000					
II Present Value (IIPV) - Return	−12000	−10608.20	−9377.83	−8290.16	−7328.64	−6478.65
Life Time simulated	25					
Residual value of equipment (RV)	10					
Linear Depreciation Method (Vn)	−12000	−11519.60	−11039.20	−10558.80	−10078.40	−9598.00
LDM Present Value (Vn)		−10183.52	−8626.98	−7294.51	−6155.08	−5181.84
Sum of Digits Method (Vn)	−12000	−11076.15	−10189.26	−9339.32	−8526.34	−7750.31
SDM Present Value (Vn)		−9791.51	−7962.77	−6452.04	−5207.21	−4184.29
Exponential Method (Vn)	−12000	−9036.79	−6805.29	−5124.83	−3859.33	−2906.33
EM Present Value (Pv)		−7988.67	−5318.24	−3540.47	−2356.97	−1569.09
Functioning (*F*)	−4500	−4500.00	−4545.00	−4590.45	−4636.35	−4682.72
Present Value Accumulated (NPV)		−3978.08	−7529.93	−10701.23	−13532.74	−16060.88
Present Value (NPV)		−3978.08	−3551.85	−3171.30	−2831.52	−2528.14
Energy	−200	−200.00	−400.00	−600.00	−800.00	−1000.00
Energy (NPV)		−176.80	−312.59	−414.51	−488.58	−539.89
Accumulated (FPV)		−16154.88	−23997.40	−35113.14	−49134.46	−65735.23
Maintenance (*M*)	−300	−300.00	−303.00	−306.03	−309.09	−312.18
Present Value (MPV)		−265.21	−236.79	−211.42	−188.77	−168.54
Accumulated (MPV)		−265.21	−502.00	−713.42	−902.18	−1070.73
Non Production (NP)	0	0.00	0.00	0.00	0.00	0.00
Present Value (NPPV)		0.00	0.00	0.00	0.00	0.00
Accumulated NPPV		0.00	0.00	0.00	0.00	0.00
Accumulated Total Costs (APV)		−16420.08	−24499.40	−35826.56	−50036.64	−66805.96
Accumulated Total Costs		−5000.00	−10248.00	−15744.48	−21489.92	−27484.82
Benefit (*B*)	50000	44642.86	44642.86	44642.86	44642.86	44642.86
Present Value (BPV)		39465.04	34887.77	30841.38	27264.30	24102.11
Accumulated Benefits BPV		39465.04	74352.81	105194.19	132458.49	156560.60
BPV + FPV + MPV + NPPV + IIPV		11044.96	37853.41	57367.63	70421.85	77754.64
Uniform Annual Income UAIn (LDM)		−11519.60	−5179.80	−3095.68	−2074.47	−1477.59
Uniform Annual Income UAIn (SDM)		−11076.15	−4781.00	−2738.15	−1755.00	−1193.14
Uniform Annual Income UAIn (EM)		−9036.79	−3193.17	−1502.52	−794.38	−447.42

(Continued)

TABLE 9.2 (*Continued*)
Values Used for Simulation based on UAI Method Related to Energy Standards. Second Simulation

Time	6	7	8	9	10
Initial Investment (II)					
Initial Investment (II)					
II Present Value (IIPV) - Return	−5727.23	−5062.97	−4475.75	−3956.64	−3497.74
Life Time simulated					
Residual value of equipment (RV)					
Linear Depreciation Method (Vn)	−9117.60	−8637.20	−8156.80	−7676.40	−7196.00
LDM Present Value (Vn)	−4351.55	−3644.16	−3042.32	−2531.06	−2097.48
Sum of Digits Method (Vn)	−7011.23	−6309.11	−5643.94	−5015.72	−4424.46
SDM Present Value (Vn)	−3346.25	−2661.90	−2105.07	−1653.78	−1289.63
Exponential Method (Vn)	−2188.66	−1648.20	−1241.20	−934.71	−703.90
EM Present Value (Pv)	−1044.58	−695.40	−462.94	−308.19	−205.17
Functioning (*F*)	−4729.55	−4776.84	−4824.61	−4872.86	−4921.58
Present Value Accumulated (NPV)	−18318.15	−20333.57	−22133.05	−23739.73	−25174.26
Present Value (NPV)	−2257.27	−2015.42	−1799.48	−1606.68	−1434.53
Energy	−1200.00	−1400.00	−1600.00	−1800.00	−2000.00
Energy (NPV)	−572.72	−590.68	−596.77	−593.50	−582.96
Accumulated (FPV)	−84626.11	−105550.36	−128280.17	−152613.40	−178370.61
Maintenance (*M*)	−315.30	−318.46	−321.64	−324.86	−328.11
Present Value (MPV)	−150.48	−134.36	−119.97	−107.11	−95.64
Accumulated (MPV)	−1221.21	−1355.57	−1475.54	−1582.65	−1678.28
Non Production (NP)	0.00	0.00	0.00	0.00	0.00
Present Value (NPPV)	0.00	0.00	0.00	0.00	0.00
Accumulated NPPV	0.00	0.00	0.00	0.00	0.00
Accumulated Total Costs (APV)	−85847.32	−106905.93	−129755.71	−154196.04	−180048.90
Accumulated Total Costs	−33729.67	−40224.97	−46971.22	−53968.93	−61218.62
Benefit (*B*)	44642.86	44642.86	44642.86	44642.86	44642.86
Present Value (BPV)	21306.67	18835.46	16650.87	14719.65	13012.42
Accumulated Benefits BPV	177867.27	196702.73	213353.60	228073.24	241085.66
BPV + FPV + MPV + NPPV + IIPV	80019.95	77796.80	71597.89	61877.20	49036.77
Uniform Annual Income UAIn (LDM)	−1092.19	−827.06	−636.59	−495.43	−388.40
Uniform Annual Income UAIn (SDM)	−839.87	−604.13	−440.47	−323.71	−238.81
Uniform Annual Income UAIn (EM)	−262.18	−157.83	−96.87	−60.33	−37.99

(*Continued*)

TABLE 9.2 (*Continued*)
Values Used for Simulation based on UAI Method Related to Energy Standards. Second Simulation

Time	11	12	13	14	15
Initial Investment (II)					
Initial Investment (II)					
II Present Value (IIPV) - Return	−3092.06	−2733.43	−2416.40	−2136.14	−1888.38
Life Time simulated					
Residual value of equipment (RV)					
Linear Depreciation Method (Vn)	−6715.60	−6235.20	−5754.80	−5274.40	−4794.00
LDM Present Value (Vn)	−1730.42	−1420.29	−1158.83	−938.90	−754.41
Sum of Digits Method (Vn)	−3870.15	−3352.80	−2872.40	−2428.95	−2022.46
SDM Present Value (Vn)	−997.23	−763.72	−578.41	−432.38	−318.27
Exponential Method (Vn)	−530.08	−399.19	−300.61	−226.38	−170.48
EM Present Value (Pv)	−136.59	−90.93	−60.53	−40.30	−26.83
Functioning (*F*)	−4970.80	−5020.51	−5070.71	−5121.42	−5172.63
Present Value Accumulated (NPV)	−26455.10	−27598.70	−28619.77	−29531.44	−30345.44
Present Value (NPV)	−1280.83	−1143.60	−1021.07	−911.67	−813.99
Energy	−2200.00	−2400.00	−2600.00	−2800.00	−3000.00
Energy (NPV)	−566.88	−546.69	−523.55	−498.43	−472.10
Accumulated (FPV)	−205392.59	−233537.97	−262681.29	−292711.17	−323528.70
Maintenance (*M*)	−331.39	−334.70	−338.05	−341.43	−344.84
Present Value (MPV)	−85.39	−76.24	−68.07	−60.78	−54.27
Accumulated (MPV)	−1763.67	−1839.91	−1907.98	−1968.76	−2023.03
Non Production (NP)	0.00	0.00	0.00	0.00	0.00
Present Value (NPPV)	0.00	0.00	0.00	0.00	0.00
Accumulated NPPV	0.00	0.00	0.00	0.00	0.00
Accumulated Total Costs (APV)	−207156.26	−235377.88	−264589.28	−294679.93	−325551.73
Accumulated Total Costs	−68720.81	−76476.01	−84484.77	−92747.62	−101265.10
Benefit (*B*)	44642.86	44642.86	44642.86	44642.86	44642.86
Present Value (BPV)	11503.20	10169.02	8989.59	7946.95	7025.24
Accumulated Benefits BPV	252588.86	262757.88	271747.47	279694.42	286719.66
BPV + FPV + MPV + NPPV + IIPV	33432.60	15380.00	−4841.81	−26985.51	−50832.07
Uniform Annual Income UAIn (LDM)	−305.84	−241.31	−190.37	−149.86	−117.46
Uniform Annual Income UAIn (SDM)	−176.25	−129.76	−95.02	−69.01	−49.55
Uniform Annual Income UAIn (EM)	−24.14	−15.45	−9.94	−6.43	−4.18

(*Continued*)

TABLE 9.2 (*Continued*)
Values Used for Simulation based on UAI Method Related to Energy Standards. Second Simulation

Time	16	17	18	19	20
Initial Investment (II)					
Initial Investment (II)					
II Present Value (IIPV) - Return	−1669.36	−1475.75	−1304.58	−1153.27	−1019.51
Life Time simulated					
Residual value of equipment (RV)					
Linear Depreciation Method (Vn)	−4313.60	−3833.20	−3352.80	−2872.40	−2392.00
LDM Present Value (Vn)	−600.08	−471.40	−364.50	−276.06	−203.22
Sum of Digits Method (Vn)	−1652.92	−1320.34	−1024.71	−766.03	−544.31
SDM Present Value (Vn)	−229.94	−162.37	−111.40	−73.62	−46.24
Exponential Method (Vn)	−128.38	−96.68	−72.81	−54.83	−41.29
EM Present Value (Pv)	−17.86	−11.89	−7.92	−5.27	−3.51
Functioning (*F*)	−5224.36	−5276.60	−5329.37	−5382.66	−5436.49
Present Value Accumulated (NPV)	−31072.22	−31721.13	−32300.51	−32817.82	−33279.70
Present Value (NPV)	−726.78	−648.91	−579.38	−517.31	−461.88
Energy	−3200.00	−3400.00	−3600.00	−3800.00	−4000.00
Energy (NPV)	−445.16	−418.13	−391.38	−365.20	−339.84
Accumulated (FPV)	−355046.08	−387185.33	−419877.22	−453060.24	−486679.78
Maintenance (*M*)	−348.29	−351.77	−355.29	−358.84	−362.43
Present Value (MPV)	−48.45	−43.26	−38.63	−34.49	−30.79
Accumulated (MPV)	−2071.48	−2114.74	−2153.37	−2187.85	−2218.65
Non Production (NP)	0.00	0.00	0.00	0.00	0.00
Present Value (NPPV)	0.00	0.00	0.00	0.00	0.00
Accumulated NPPV	0.00	0.00	0.00	0.00	0.00
Accumulated Total Costs (APV)	−357117.56	−389300.07	−422030.59	−455248.09	−488898.42
Accumulated Total Costs	−110037.75	−119066.13	−128350.79	−137892.30	−147691.22
Benefit (*B*)	44642.86	44642.86	44642.86	44642.86	44642.86
Present Value (BPV)	6210.43	5490.13	4853.36	4290.46	3792.84
Accumulated Benefits BPV	292930.09	298420.21	303273.58	307564.03	311356.87
BPV + FPV + MPV + NPPV + IIPV	−76187.47	−102879.86	−130757.01	−159684.06	−189541.55
Uniform Annual Income UAIn (LDM)	−91.45	−70.52	−53.66	−40.07	−29.14
Uniform Annual Income UAIn (SDM)	−35.04	−24.29	−16.40	−10.69	−6.63
Uniform Annual Income UAIn (EM)	−2.72	−1.78	−1.17	−0.76	−0.50

(*Continued*)

TABLE 9.2 (*Continued*)
Values Used for Simulation based on UAI Method Related to Energy Standards. Second Simulation

Time	21	22	23	24	25
Initial Investment (II)					
Initial Investment (II)					
II Present Value (IIPV) - Return	−901.27	−796.74	−704.33	−622.64	−550.42
Life Time simulated					
Residual value of equipment (RV)					
Linear Depreciation Method (Vn)	−1911.60	−1431.20	−950.80	−470.40	10.00
LDM Present Value (Vn)	−143.57	−95.02	−55.81	−24.41	0.46
Sum of Digits Method (Vn)	−359.54	−211.72	−100.86	−26.95	10.00
SDM Present Value (Vn)	−27.00	−14.06	−5.92	−1.40	0.46
Exponential Method (Vn)	−31.09	−23.42	−17.63	−13.28	−10.00
EM Present Value (Pv)	−2.34	−1.55	−1.03	−0.69	−0.46
Functioning (*F*)	−5490.86	−5545.76	−5601.22	−5657.23	−5713.81
Present Value Accumulated (NPV)	−33692.09	−34060.30	−34389.06	−34682.60	−34944.68
Present Value (NPV)	−412.39	−368.21	−328.76	−293.53	−262.08
Energy	−4200.00	−4400.00	−4600.00	−4800.00	−5000.00
Energy (NPV)	−315.44	−292.14	−269.99	−249.06	−229.34
Accumulated (FPV)	−520687.31	−555039.75	−589698.81	−624630.46	−659804.48
Maintenance (*M*)	−366.06	−369.72	−373.41	−377.15	−380.92
Present Value (MPV)	−27.49	−24.55	−21.92	−19.57	−17.47
Accumulated (MPV)	−2246.14	−2270.69	−2292.60	−2312.17	−2329.65
Non Production (NP)	0.00	0.00	0.00	0.00	0.00
Present Value (NPPV)	0.00	0.00	0.00	0.00	0.00
Accumulated NPPV	0.00	0.00	0.00	0.00	0.00
Accumulated Total Costs (APV)	−522933.45	−557310.44	−591991.41	−626942.63	−662134.13
Accumulated Total Costs	−157748.13	−168063.61	−178638.25	−189472.63	−200567.36
Benefit (*B*)	44642.86	44642.86	44642.86	44642.86	44642.86
Present Value (BPV)	3352.93	2964.05	2620.27	2316.36	2047.70
Accumulated Benefits BPV	314709.80	317673.85	320294.12	322610.48	324658.18
BPV + FPV + MPV + NPPV + IIPV	−220223.65	−251636.59	−283697.29	−316332.15	−349475.94
Uniform Annual Income UAIn (LDM)	−20.37	−13.35	−7.78	−3.38	0.06
Uniform Annual Income UAIn (SDM)	−3.83	−1.98	−0.83	−0.19	0.06
Uniform Annual Income UAIn (EM)	−0.33	−0.22	−0.14	−0.10	−0.06

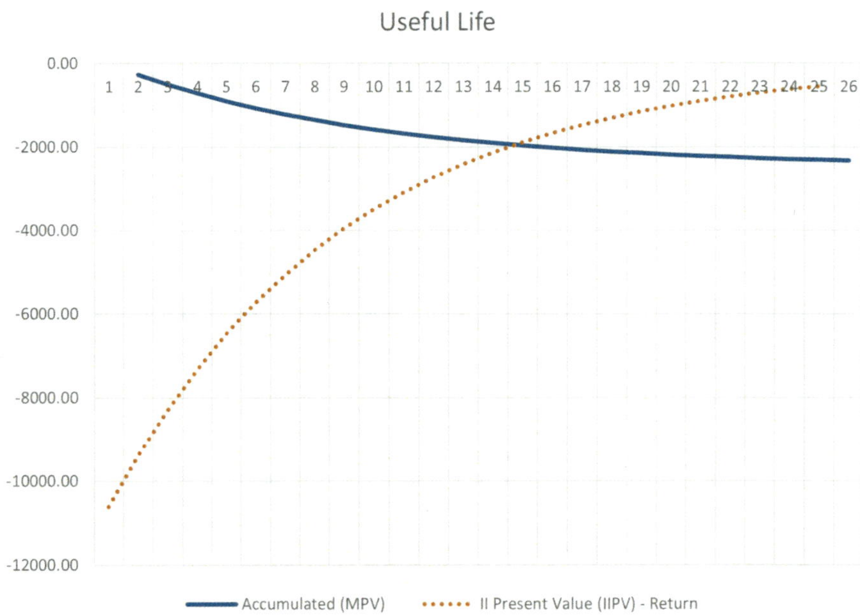

FIGURE 9.6 Physical Asset useful life.

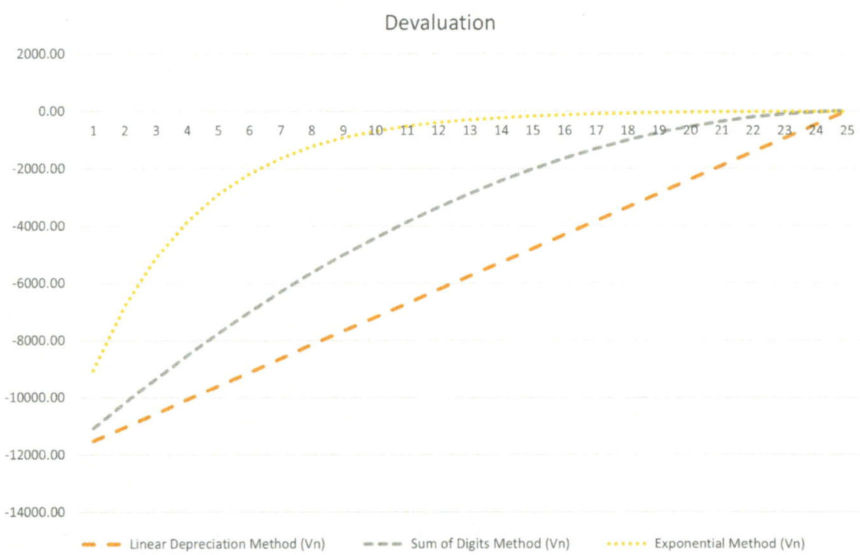

FIGURE 9.7 The three devaluation methods.

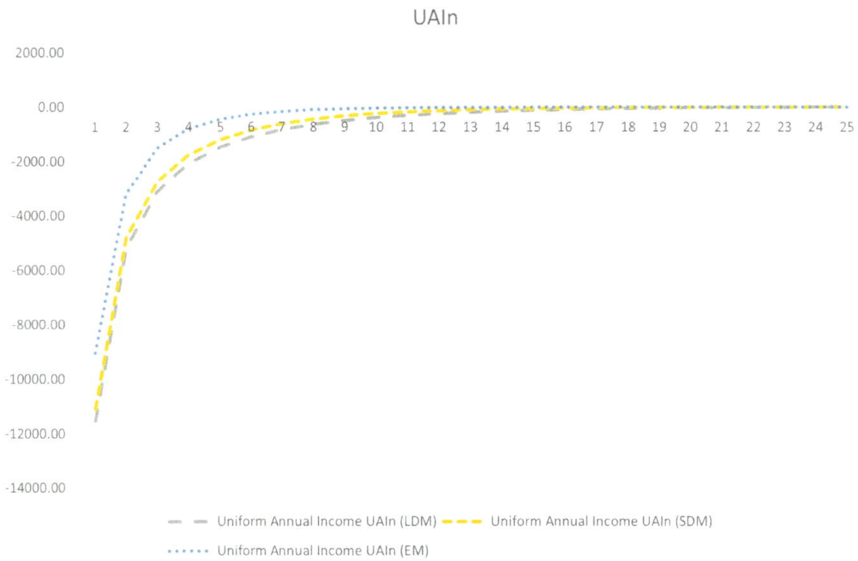

FIGURE 9.8 The UAI results from the three devaluation methods.

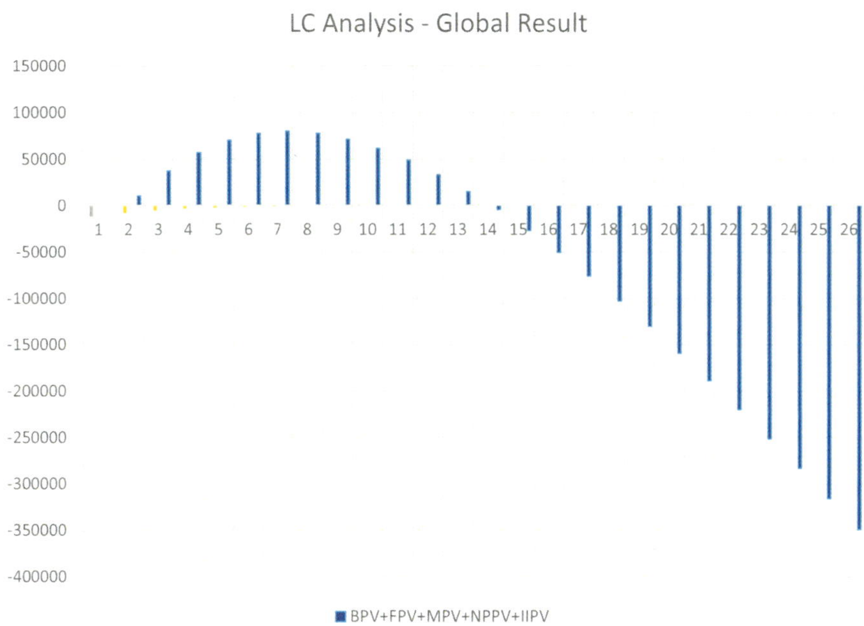

FIGURE 9.9 Life cycle analysis global result including profits.

TABLE 9.3
Values Used for Simulation based on MMTAC Method Related to Energy Standards. First Simulation

Time	0	1	2	3	4	5
Initial Investment (II)	−12000					
Initial Investment (II)	−12000					
II (IIPV) - Return	−12000	−12000.00	−12000.00	−12000.00	−12000.00	−12000.00
Useful Life	25					
Residual value of equipment (RV)	10					
Linear Depreciation Method (Vn)	−12000	−11519.60	−11039.20	−10558.80	−10078.40	−9598.00
LDM Present Value (PV)		−11519.60	−11039.20	−10558.80	−10078.40	−9598.00
Sum of Digits Method (Vn)	−12000	−11076.15	−10189.26	−9339.32	−8526.34	−7750.31
SDM Present Value (PV)		−11076.15	−10189.26	−9339.32	−8526.34	−7750.31
Exponential Method (Vn)	−12000	−9036.79	−6805.29	−5124.83	−3859.33	−2906.33
EM Present Value (Pv)		−9036.79	−6805.29	−5124.83	−3859.33	−2906.33
Functioning (*F*)	−4500	−4500.00	−4545.00	−4590.45	−4636.35	−4682.72
Functioning Accumulated (FA)		−4500.00	−9045.00	−13635.45	−18271.80	−22954.52
Present Value Accumulated (NPV)		−4500.00	−9045.00	−13635.45	−18271.80	−22954.52
Present Value (NPV)		−4500.00	−4545.00	−4590.45	−4636.35	−4682.72
Energy	−1000	−1000.00	−2000.00	−3000.00	−4000.00	−5000.00
Energy Accumulated (EA)		−1000.00	−3000.00	−6000.00	−10000.00	−15000.00
Energy (NPV)		−1000.00	−2000.00	−3000.00	−4000.00	−5000.00
Accumulated Functioning (FPV)		−17500.00	−28545.00	−45180.45	−67452.25	−95406.78
Maintenance (*M*)	−300	−300.00	−303.00	−306.03	−309.09	−312.18
Maintenane Accumulated (MA)		−300.00	−603.00	−909.03	−1218.12	−1530.30
Present Value (MPV)		−300.00	−303.00	−306.03	−309.09	−312.18
Accumulated MPV		−300.00	−603.00	−909.03	−1218.12	−1530.30
Non Production (NP)	0	0.00	0.00	0.00	0.00	0.00
Non Production Accumulated (NPA)		0.00	0.00	0.00	0.00	0.00
Present Value (NPPV)		0.00	0.00	0.00	0.00	0.00
Accumulated NPPV		0.00	0.00	0.00	0.00	0.00
Accumulated Total Costs (APV)		−17800.00	−29148.00	−46089.48	−68670.37	−96937.08
Accumulated Total Costs		−5800.00	−12648.00	−20544.48	−29489.92	−39484.82
Benefit (*B*)	70000	67632.85	67632.85	67632.85	67632.85	67632.85
Benefits Accumulated (BA)		67632.85	135265.70	202898.55	270531.40	338164.25
Present Value (BPV)		67632.85	67632.85	67632.85	67632.85	67632.85
Accumulated Benefits BPV		67632.85	135265.70	202898.55	270531.40	338164.25
BPV + FPV + MPV + NPPV + IIPV		49832.85	110617.70	170354.07	229041.48	286679.43
MTAC (LDM)		−6280.40	−13608.80	−21985.68	−31411.52	−41886.82
MTAC (SDM)		−6723.85	−14458.74	−23205.16	−32963.59	−43734.52
MTAC (EM)		−8763.21	−17842.71	−27419.65	−37630.59	−48578.49

(*Continued*)

TABLE 9.3 (*Continued*)
Values Used for Simulation based on MMTAC Method Related to Energy Standards. First Simulation

Time	6	7	8	9	10
Initial Investment (II)					
Initial Investment (II)					
II (IIPV) - Return	−12000.00	−12000.00	−12000.00	−12000.00	−12000.00
Useful Life					
Residual value of equipment (RV)					
Linear Depreciation Method (Vn)	−9117.60	−8637.20	−8156.80	−7676.40	−7196.00
LDM Present Value (PV)	−9117.60	−8637.20	−8156.80	−7676.40	−7196.00
Sum of Digits Method (Vn)	−7011.23	−6309.11	−5643.94	−5015.72	−4424.46
SDM Present Value (PV)	−7011.23	−6309.11	−5643.94	−5015.72	−4424.46
Exponential Method (Vn)	−2188.66	−1648.20	−1241.20	−934.71	−703.90
EM Present Value (Pv)	−2188.66	−1648.20	−1241.20	−934.71	−703.90
Functioning (*F*)	−4729.55	−4776.84	−4824.61	−4872.86	−4921.58
Functioning Accumulated (FA)	−27684.07	−32460.91	−37285.52	−42158.37	−47079.96
Present Value Accumulated (NPV)	−27684.07	−32460.91	−37285.52	−42158.37	−47079.96
Present Value (NPV)	−4729.55	−4776.84	−4824.61	−4872.86	−4921.58
Energy	−6000.00	−7000.00	−8000.00	−9000.00	−10000.00
Energy Accumulated (EA)	−21000.00	−28000.00	−36000.00	−45000.00	−55000.00
Energy (NPV)	−6000.00	−7000.00	−8000.00	−9000.00	−10000.00
Accumulated Functioning (FPV)	−129090.84	−168551.75	−213837.27	−264995.64	−322075.60
Maintenance (*M*)	−315.30	−318.46	−321.64	−324.86	−328.11
Maintenane Accumulated (MA)	−1845.60	−2164.06	−2485.70	−2810.56	−3138.66
Present Value (MPV)	−315.30	−318.46	−321.64	−324.86	−328.11
Accumulated MPV	−1845.60	−2164.06	−2485.70	−2810.56	−3138.66
Non Production (NP)	0.00	0.00	0.00	0.00	0.00
Non Production Accumulated (NPA)	0.00	0.00	0.00	0.00	0.00
Present Value (NPPV)	0.00	0.00	0.00	0.00	0.00
Accumulated NPPV	0.00	0.00	0.00	0.00	0.00
Accumulated Total Costs (APV)	−130936.45	−170715.81	−216322.97	−267806.20	−325214.26
Accumulated Total Costs	−50529.67	−62624.97	−75771.22	−89968.93	−105218.62
Benefit (*B*)	67632.85	67632.85	67632.85	67632.85	67632.85
Benefits Accumulated (BA)	405797.10	473429.95	541062.80	608695.65	676328.50
Present Value (BPV)	67632.85	67632.85	67632.85	67632.85	67632.85
Accumulated Benefits BPV	405797.10	473429.95	541062.80	608695.65	676328.50
BPV + FPV + MPV + NPPV + IIPV	343267.43	398804.98	453291.58	506726.72	559109.88
MTAC (LDM)	−53412.07	−65987.77	−79614.42	−94292.53	−110022.62
MTAC (SDM)	−55518.44	−68315.86	−82127.28	−96953.21	−112794.16
MTAC (EM)	−60341.02	−72976.77	−86530.01	−101034.22	−116514.72

(*Continued*)

TABLE 9.3 (*Continued*)
Values Used for Simulation based on MMTAC Method Related to Energy Standards. First Simulation

Time	11	12	13	14	15
Initial Investment (II)					
Initial Investment (II)					
II (IIPV) - Return	−12000.00	−12000.00	−12000.00	−12000.00	−12000.00
Useful Life					
Residual value of equipment (RV)					
Linear Depreciation Method (Vn)	−6715.60	−6235.20	−5754.80	−5274.40	−4794.00
LDM Present Value (PV)	−6715.60	−6235.20	−5754.80	−5274.40	−4794.00
Sum of Digits Method (Vn)	−3870.15	−3352.80	−2872.40	−2428.95	−2022.46
SDM Present Value (PV)	−3870.15	−3352.80	−2872.40	−2428.95	−2022.46
Exponential Method (Vn)	−530.08	−399.19	−300.61	−226.38	−170.48
EM Present Value (Pv)	−530.08	−399.19	−300.61	−226.38	−170.48
Functioning (*F*)	−4970.80	−5020.51	−5070.71	−5121.42	−5172.63
Functioning Accumulated (FA)	−52050.76	−57071.26	−62141.98	−67263.40	−72436.03
Present Value Accumulated (NPV)	−52050.76	−57071.26	−62141.98	−67263.40	−72436.03
Present Value (NPV)	−4970.80	−5020.51	−5070.71	−5121.42	−5172.63
Energy	−11000.00	−12000.00	−13000.00	−14000.00	−15000.00
Energy Accumulated (EA)	−66000.00	−78000.00	−91000.00	−105000.00	−120000.00
Energy (NPV)	−11000.00	−12000.00	−13000.00	−14000.00	−15000.00
Accumulated Functioning (FPV)	−385126.36	−454197.62	−529339.60	−610602.99	−698039.02
Maintenance (*M*)	−331.39	−334.70	−338.05	−341.43	−344.84
Maintenane Accumulated (MA)	−3470.05	−3804.75	−4142.80	−4484.23	−4829.07
Present Value (MPV)	−331.39	−334.70	−338.05	−341.43	−344.84
Accumulated MPV	−3470.05	−3804.75	−4142.80	−4484.23	−4829.07
Non Production (NP)	0.00	0.00	0.00	0.00	0.00
Non Production Accumulated (NPA)	0.00	0.00	0.00	0.00	0.00
Present Value (NPPV)	0.00	0.00	0.00	0.00	0.00
Accumulated NPPV	0.00	0.00	0.00	0.00	0.00
Accumulated Total Costs (APV)	−388596.41	−458002.37	−533482.39	−615087.22	−702868.09
Accumulated Total Costs	−121520.81	−138876.01	−157284.77	−176747.62	−197265.10
Benefit (*B*)	67632.85	67632.85	67632.85	67632.85	67632.85
Benefits Accumulated (BA)	743961.35	811594.20	879227.05	946859.90	1014492.75
Present Value (BPV)	67632.85	67632.85	67632.85	67632.85	67632.85
Accumulated Benefits BPV	743961.35	811594.20	879227.05	946859.90	1014492.75
BPV + FPV + MPV + NPPV + IIPV	610440.55	660718.19	709942.28	758112.28	805227.66
MTAC (LDM)	−126805.21	−144640.81	−163529.97	−183473.22	−204471.10
MTAC (SDM)	−129650.65	−147523.21	−166412.37	−186318.67	−207242.64
MTAC (EM)	−132990.73	−150476.83	−168984.16	−188521.24	−209094.62

(Continued)

TABLE 9.3 (*Continued*)
Values Used for Simulation based on MMTAC Method Related to Energy Standards. First Simulation

Time	16	17	18	19	20
Initial Investment (II)					
Initial Investment (II)					
II (IIPV) - Return	−12000.00	−12000.00	−12000.00	−12000.00	−12000.00
Useful Life					
Residual value of equipment (RV)					
Linear Depreciation Method (Vn)	−4313.60	−3833.20	−3352.80	−2872.40	−2392.00
LDM Present Value (PV)	−4313.60	−3833.20	−3352.80	−2872.40	−2392.00
Sum of Digits Method (Vn)	−1652.92	−1320.34	−1024.71	−766.03	−544.31
SDM Present Value (PV)	−1652.92	−1320.34	−1024.71	−766.03	−544.31
Exponential Method (Vn)	−128.38	−96.68	−72.81	−54.83	−41.29
EM Present Value (Pv)	−128.38	−96.68	−72.81	−54.83	−41.29
Functioning (*F*)	−5224.36	−5276.60	−5329.37	−5382.66	−5436.49
Functioning Accumulated (FA)	−77660.39	−82936.99	−88266.36	−93649.03	−99085.52
Present Value Accumulated (NPV)	−77660.39	−82936.99	−88266.36	−93649.03	−99085.52
Present Value (NPV)	−5224.36	−5276.60	−5329.37	−5382.66	−5436.49
Energy	−16000.00	−17000.00	−18000.00	−19000.00	−20000.00
Energy Accumulated (EA)	−136000.00	−153000.00	−171000.00	−190000.00	−210000.00
Energy (NPV)	−16000.00	−17000.00	−18000.00	−19000.00	−20000.00
Accumulated Functioning (FPV)	−791699.41	−891636.41	−997902.77	−1110551.80	−1229637.32
Maintenance (*M*)	−348.29	−351.77	−355.29	−358.84	−362.43
Maintenane Accumulated (MA)	−5177.36	−5529.13	−5884.42	−6243.27	−6605.70
Present Value (MPV)	−348.29	−351.77	−355.29	−358.84	−362.43
Accumulated MPV	−5177.36	−5529.13	−5884.42	−6243.27	−6605.70
Non Production (NP)	0.00	0.00	0.00	0.00	0.00
Non Production Accumulated (NPA)	0.00	0.00	0.00	0.00	0.00
Present Value (NPPV)	0.00	0.00	0.00	0.00	0.00
Accumulated NPPV	0.00	0.00	0.00	0.00	0.00
Accumulated Total Costs (APV)	−796876.77	−897165.54	−1003787.19	−1116795.07	−1236243.02
Accumulated Total Costs	−218837.75	−241466.13	−265150.79	−289892.30	−315691.22
Benefit (*B*)	67632.85	67632.85	67632.85	67632.85	67632.85
Benefits Accumulated (BA)	1082125.60	1149758.45	1217391.30	1285024.15	1352657.00
Present Value (BPV)	67632.85	67632.85	67632.85	67632.85	67632.85
Accumulated Benefits BPV	1082125.60	1149758.45	1217391.30	1285024.15	1352657.00
BPV + FPV + MPV + NPPV + IIPV	851287.85	896292.33	940240.52	983131.86	1024965.79
MTAC (LDM)	−226524.15	−249632.93	−273797.99	−299019.90	−325299.22
MTAC (SDM)	−229184.83	−252145.79	−276126.08	−301126.27	−327146.91
MTAC (EM)	−230709.37	−253369.45	−277077.98	−301837.47	−327649.93

(Continued)

TABLE 9.3 (*Continued*)
Values Used for Simulation based on MMTAC Method Related to Energy Standards. First Simulation

Time	21	22	23	24	25
Initial Investment (II)					
Initial Investment (II)					
II (IIPV) - Return	−12000.00	−12000.00	−12000.00	−12000.00	−12000.00
Useful Life					
Residual value of equipment (RV)					
Linear Depreciation Method (Vn)	−1911.60	−1431.20	−950.80	−470.40	10.00
LDM Present Value (PV)	−1911.60	−1431.20	−950.80	−470.40	10.00
Sum of Digits Method (Vn)	−359.54	−211.72	−100.86	−26.95	10.00
SDM Present Value (PV)	−359.54	−211.72	−100.86	−26.95	10.00
Exponential Method (Vn)	−31.09	−23.42	−17.63	−13.28	−10.00
EM Present Value (Pv)	−31.09	−23.42	−17.63	−13.28	−10.00
Functioning (*F*)	−5490.86	−5545.76	−5601.22	−5657.23	−5713.81
Functioning Accumulated (FA)	−104576.37	−110122.14	−115723.36	−121380.59	−127094.40
Present Value Accumulated (NPV)	−104576.37	−110122.14	−115723.36	−121380.59	−127094.40
Present Value (NPV)	−5490.86	−5545.76	−5601.22	−5657.23	−5713.81
Energy	−21000.00	−22000.00	−23000.00	−24000.00	−25000.00
Energy Accumulated (EA)	−231000.00	−253000.00	−276000.00	−300000.00	−325000.00
Energy (NPV)	−21000.00	−22000.00	−23000.00	−24000.00	−25000.00
Accumulated Functioning (FPV)	−1355213.69	−1487335.83	−1626059.18	−1771439.78	−1923534.17
Maintenance (*M*)	−366.06	−369.72	−373.41	−377.15	−380.92
Maintenane Accumulated (MA)	−6971.76	−7341.48	−7714.89	−8092.04	−8472.96
Present Value (MPV)	−366.06	−369.72	−373.41	−377.15	−380.92
Accumulated MPV	−6971.76	−7341.48	−7714.89	−8092.04	−8472.96
Non Production (NP)	0.00	0.00	0.00	0.00	0.00
Non Production Accumulated (NPA)	0.00	0.00	0.00	0.00	0.00
Present Value (NPPV)	0.00	0.00	0.00	0.00	0.00
Accumulated NPPV	0.00	0.00	0.00	0.00	0.00
Accumulated Total Costs (APV)	−1362185.45	−1494677.30	−1633774.07	−1779531.82	−1932007.13
Accumulated Total Costs	−342548.13	−370463.61	−399438.25	−429472.63	−460567.36
Benefit (*B*)	67632.85	67632.85	67632.85	67632.85	67632.85
Benefits Accumulated (BA)	1420289.86	1487922.71	1555555.56	1623188.41	1690821.26
Present Value (BPV)	67632.85	67632.85	67632.85	67632.85	67632.85
Accumulated Benefits BPV	1420289.86	1487922.71	1555555.56	1623188.41	1690821.26
BPV + FPV + MPV + NPPV + IIPV	1065741.72	1105459.09	1144117.31	1181715.77	1218253.90
MTAC (LDM)	−352636.53	−381032.41	−410487.45	−441002.23	−472577.36
MTAC (SDM)	−354188.59	−382251.89	−411337.39	−441445.68	−472577.36
MTAC (EM)	−354517.04	−382440.20	−411420.62	−441459.35	−472557.36

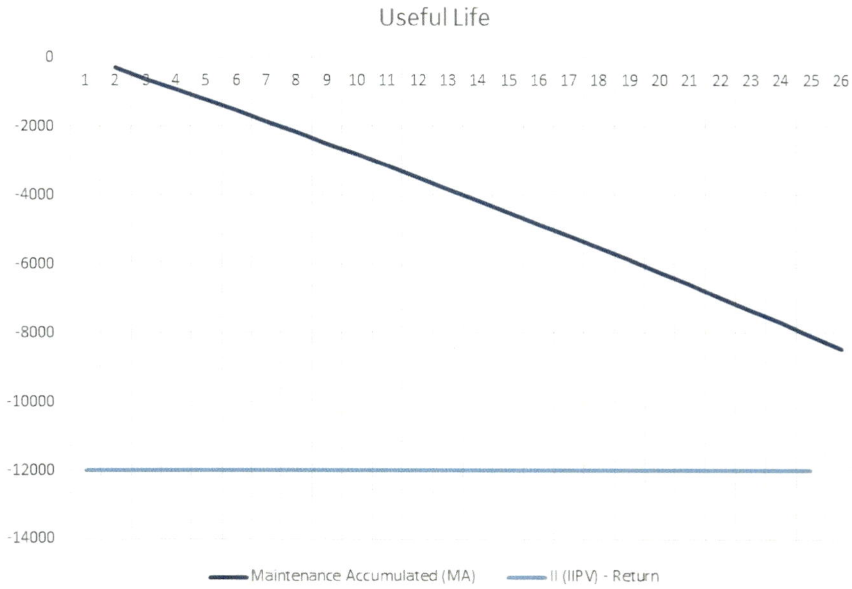

FIGURE 9.10 Physical Asset useful life.

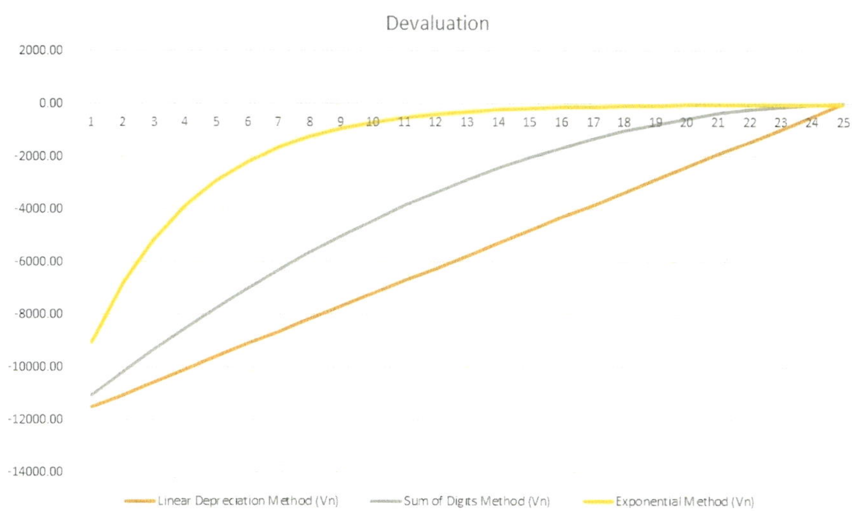

FIGURE 9.11 The three devaluation methods.

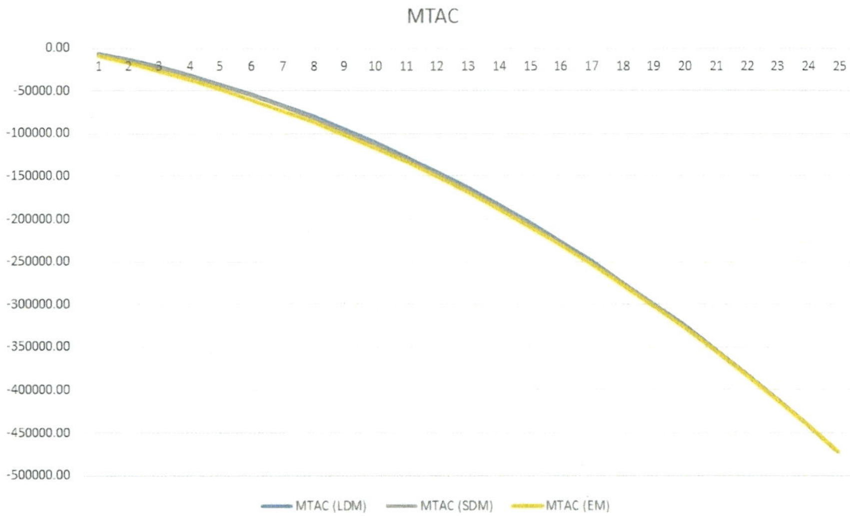

FIGURE 9.12 The MMTAC results from the three devaluation methods.

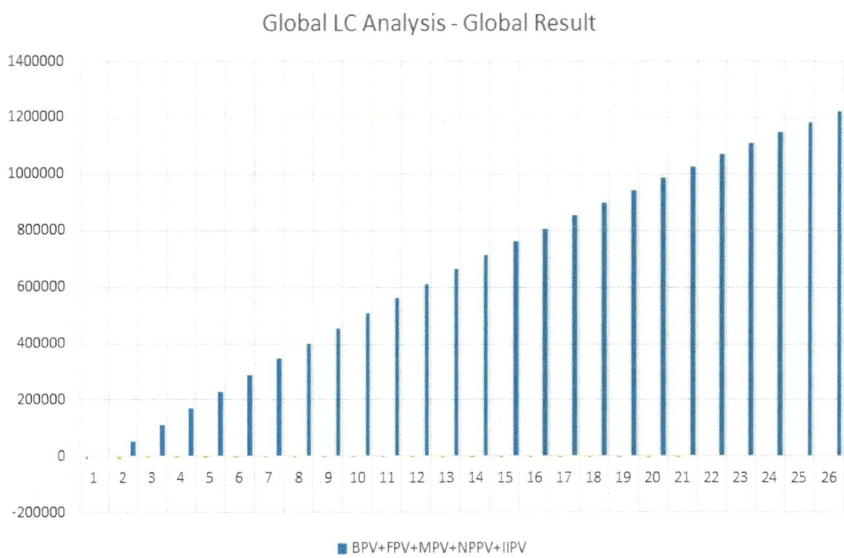

FIGURE 9.13 Life cycle analysis global result including profits.

The second simulation uses the following values (the value of Energy decreases a lot as a result of the Energy Standards):

- Initial Investment (II) –12000
- Life Time simulated 25
- Residual value of equipment (RV) 10
- Functioning (*F*) –4500
- Energy **–200**
- Maintenance (*M*) –300
- Non Production (NP) 0
- Benefit (*B*) 70000
- Internal Rate of Return (*r*) 0.0
- Risk (*R*) 0.0
- Apparent Rate 0.0
- Maintenance Increase Rate 0.01
- Availability 0.9662
- MTBF 1000
- MTTR 25
- MWT 10

The remaining values are presented in Table 9.4, and the graphical results are shown in Figures 9.14–9.17.

As can be seen, namely comparing Figures 9.13 and 9.17, with the implementation of the Energy Standards, the positive incomes increase a lot. However, these two simulations were made considering an Apparent Rate null, what does not happen in a real world. Because of this, the Reduction to Present Value (RPV) must be considered.

9.4.1.3 MMTAC Reduced to Present Value (MMTAC-RPV)

The first simulation uses the following values:

- Initial Investment (II) –12000
- Useful Life 25
- Residual value of equipment (RV) 10
- Functioning (*F*) –4500
- Energy (*E*) **–1000**
- Maintenance (*M*) –300
- Benefit (*B*) 70000
- Internal Rate of Return (*r*) 0.12
- Risk (*R*) 0.01
- Apparent Rate 0.1312
- Maintenance Increase Rate 0.01
- Availability 0.9662
- MTBF 1000
- MTTR 25
- MWT 10

TABLE 9.4
Values Used for Simulation based on MMTAC Method Related to Energy Standards. Second Simulation

Time	0	1	2	3	4	5
Initial Investment (II)	−12000					
Initial Investment (II)	−12000					
II (IIPV) - Return	−12000	−12000.00	−12000.00	−12000.00	−12000.00	−12000.00
Useful Life	25					
Residual value of equipment (RV)	10					
Linear Depreciation Method (Vn)	−12000	−11519.60	−11039.20	−10558.80	−10078.40	−9598.00
LDM Present Value (PV)		−11519.60	−11039.20	−10558.80	−10078.40	−9598.00
Sum of Digits Method (Vn)	−12000	−11076.15	−10189.26	−9339.32	−8526.34	−7750.31
SDM Present Value (PV)		−11076.15	−10189.26	−9339.32	−8526.34	−7750.31
Exponential Method (Vn)	−12000	−9036.79	−6805.29	−5124.83	−3859.33	−2906.33
EM Present Value (Pv)		−9036.79	−6805.29	−5124.83	−3859.33	−2906.33
Functioning (*F*)	−4500	−4500.00	−4545.00	−4590.45	−4636.35	−4682.72
Functioning Accumulated (FA)		−4500.00	−9045.00	−13635.45	−18271.80	−22954.52
Present Value Accumulated (NPV)		−4500.00	−9045.00	−13635.45	−18271.80	−22954.52
Present Value (NPV)		−4500.00	−4545.00	−4590.45	−4636.35	−4682.72
Energy	−200	−200.00	−400.00	−600.00	−800.00	−1000.00
Energy Accumulated (EA)		−200.00	−600.00	−1200.00	−2000.00	−3000.00
Energy (NPV)		−200.00	−400.00	−600.00	−800.00	−1000.00
Accumulated Functioning (FPV)		−16700.00	−26145.00	−40380.45	−59452.25	−83406.78
Maintenance (*M*)	−300	−300.00	−303.00	−306.03	−309.09	−312.18
Maintenane Accumulated (MA)		−300.00	−603.00	−909.03	−1218.12	−1530.30
Present Value (MPV)		−300.00	−303.00	−306.03	−309.09	−312.18
Accumulated MPV		−300.00	−603.00	−909.03	−1218.12	−1530.30
Non Production (NP)	0	0.00	0.00	0.00	0.00	0.00
Non Production Accumulated (NPA)		0.00	0.00	0.00	0.00	0.00
Present Value (NPPV)		0.00	0.00	0.00	0.00	0.00
Accumulated NPPV		0.00	0.00	0.00	0.00	0.00
Accumulated Total Costs (APV)		−17000.00	−26748.00	−41289.48	−60670.37	−84937.08
Accumulated Total Costs		−5000.00	−10248.00	−15744.48	−21489.92	−27484.82
Benefit (*B*)	70000	67632.85	67632.85	67632.85	67632.85	67632.85
Benefits Accumulated (BA)		67632.85	135265.70	202898.55	270531.40	338164.25
Present Value (BPV)		67632.85	67632.85	67632.85	67632.85	67632.85
Accumulated Benefits BPV		67632.85	135265.70	202898.55	270531.40	338164.25
BPV + FPV + MPV + NPPV + IIPV		50632.85	113017.70	175154.07	237041.48	298679.43
MTAC (LDM)		−5480.40	−11208.80	−17185.68	−23411.52	−29886.82
MTAC (SDM)		−5923.85	−12058.74	−18405.16	−24963.59	−31734.52
MTAC (EM)		−7963.21	−15442.71	−22619.65	−29630.59	−36578.49

(Continued)

TABLE 9.4 (*Continued*)

Values Used for Simulation based on MMTAC Method Related to Energy Standards. Second Simulation

Time	6	7	8	9	10
Initial Investment (II)					
Initial Investment (II)					
II (IIPV) - Return	−12000.00	−12000.00	−12000.00	−12000.00	−12000.00
Useful Life					
Residual value of equipment (RV)					
Linear Depreciation Method (Vn)	−9117.60	−8637.20	−8156.80	−7676.40	−7196.00
LDM Present Value (PV)	−9117.60	−8637.20	−8156.80	−7676.40	−7196.00
Sum of Digits Method (Vn)	−7011.23	−6309.11	−5643.94	−5015.72	−4424.46
SDM Present Value (PV)	−7011.23	−6309.11	−5643.94	−5015.72	−4424.46
Exponential Method (Vn)	−2188.66	−1648.20	−1241.20	−934.71	−703.90
EM Present Value (Pv)	−2188.66	−1648.20	−1241.20	−934.71	−703.90
Functioning (*F*)	−4729.55	−4776.84	−4824.61	−4872.86	−4921.58
Functioning Accumulated (FA)	−27684.07	−32460.91	−37285.52	−42158.37	−47079.96
Present Value Accumulated (NPV)	−27684.07	−32460.91	−37285.52	−42158.37	−47079.96
Present Value (NPV)	−4729.55	−4776.84	−4824.61	−4872.86	−4921.58
Energy	−1200.00	−1400.00	−1600.00	−1800.00	−2000.00
Energy Accumulated (EA)	−4200.00	−5600.00	−7200.00	−9000.00	−11000.00
Energy (NPV)	−1200.00	−1400.00	−1600.00	−1800.00	−2000.00
Accumulated Functioning (FPV)	−112290.84	−146151.75	−185037.27	−228995.64	−278075.60
Maintenance (*M*)	−315.30	−318.46	−321.64	−324.86	−328.11
Maintenane Accumulated (MA)	−1845.60	−2164.06	−2485.70	−2810.56	−3138.66
Present Value (MPV)	−315.30	−318.46	−321.64	−324.86	−328.11
Accumulated MPV	−1845.60	−2164.06	−2485.70	−2810.56	−3138.66
Non Production (NP)	0.00	0.00	0.00	0.00	0.00
Non Production Accumulated (NPA)	0.00	0.00	0.00	0.00	0.00
Present Value (NPPV)	0.00	0.00	0.00	0.00	0.00
Accumulated NPPV	0.00	0.00	0.00	0.00	0.00
Accumulated Total Costs (APV)	−114136.45	−148315.81	−187522.97	−231806.20	−281214.26
Accumulated Total Costs	−33729.67	−40224.97	−46971.22	−53968.93	−61218.62
Benefit (*B*)	67632.85	67632.85	67632.85	67632.85	67632.85
Benefits Accumulated (BA)	405797.10	473429.95	541062.80	608695.65	676328.50
Present Value (BPV)	67632.85	67632.85	67632.85	67632.85	67632.85
Accumulated Benefits BPV	405797.10	473429.95	541062.80	608695.65	676328.50
BPV + FPV + MPV + NPPV + IIPV	360067.43	421204.98	482091.58	542726.72	603109.88
MTAC (LDM)	−36612.07	−43587.77	−50814.42	−58292.53	−66022.62
MTAC (SDM)	−38718.44	−45915.86	−53327.28	−60953.21	−68794.16
MTAC (EM)	−43541.02	−50576.77	−57730.01	−65034.22	−72514.72

(*Continued*)

TABLE 9.4 (*Continued*)
Values Used for Simulation based on MMTAC Method Related to Energy Standards. Second Simulation

Time	11	12	13	14	15
Initial Investment (II)					
Initial Investment (II)					
II (IIPV) - Return	−12000.00	−12000.00	−12000.00	−12000.00	−12000.00
Useful Life					
Residual value of equipment (RV)					
Linear Depreciation Method (Vn)	−6715.60	−6235.20	−5754.80	−5274.40	−4794.00
LDM Present Value (PV)	−6715.60	−6235.20	−5754.80	−5274.40	−4794.00
Sum of Digits Method (Vn)	−3870.15	−3352.80	−2872.40	−2428.95	−2022.46
SDM Present Value (PV)	−3870.15	−3352.80	−2872.40	−2428.95	−2022.46
Exponential Method (Vn)	−530.08	−399.19	−300.61	−226.38	−170.48
EM Present Value (Pv)	−530.08	−399.19	−300.61	−226.38	−170.48
Functioning (*F*)	−4970.80	−5020.51	−5070.71	−5121.42	−5172.63
Functioning Accumulated (FA)	−52050.76	−57071.26	−62141.98	−67263.40	−72436.03
Present Value Accumulated (NPV)	−52050.76	−57071.26	−62141.98	−67263.40	−72436.03
Present Value (NPV)	−4970.80	−5020.51	−5070.71	−5121.42	−5172.63
Energy	−2200.00	−2400.00	−2600.00	−2800.00	−3000.00
Energy Accumulated (EA)	−13200.00	−15600.00	−18200.00	−21000.00	−24000.00
Energy (NPV)	−2200.00	−2400.00	−2600.00	−2800.00	−3000.00
Accumulated Functioning (FPV)	−332326.36	−391797.62	−456539.60	−526602.99	−602039.02
Maintenance (*M*)	−331.39	−334.70	−338.05	−341.43	−344.84
Maintenane Accumulated (MA)	−3470.05	−3804.75	−4142.80	−4484.23	−4829.07
Present Value (MPV)	−331.39	−334.70	−338.05	−341.43	−344.84
Accumulated MPV	−3470.05	−3804.75	−4142.80	−4484.23	−4829.07
Non Production (NP)	0.00	0.00	0.00	0.00	0.00
Non Production Accumulated (NPA)	0.00	0.00	0.00	0.00	0.00
Present Value (NPPV)	0.00	0.00	0.00	0.00	0.00
Accumulated NPPV	0.00	0.00	0.00	0.00	0.00
Accumulated Total Costs (APV)	−335796.41	−395602.37	−460682.39	−531087.22	−606868.09
Accumulated Total Costs	−68720.81	−76476.01	−84484.77	−92747.62	−101265.10
Benefit (*B*)	67632.85	67632.85	67632.85	67632.85	67632.85
Benefits Accumulated (BA)	743961.35	811594.20	879227.05	946859.90	1014492.75
Present Value (BPV)	67632.85	67632.85	67632.85	67632.85	67632.85
Accumulated Benefits BPV	743961.35	811594.20	879227.05	946859.90	1014492.75
BPV + FPV + MPV + NPPV + IIPV	663240.55	723118.19	782742.28	842112.28	901227.66
MTAC (LDM)	−74005.21	−82240.81	−90729.97	−99473.22	−108471.10
MTAC (SDM)	−76850.65	−85123.21	−93612.37	−102318.67	−111242.64
MTAC (EM)	−80190.73	−88076.83	−96184.16	−104521.24	−113094.62

(*Continued*)

TABLE 9.4 (*Continued*)
Values Used for Simulation based on MMTAC Method Related to Energy Standards. Second Simulation

Time	16	17	18	19	20
Initial Investment (II)					
Initial Investment (II)					
II (IIPV) - Return	−12000.00	−12000.00	−12000.00	−12000.00	−12000.00
Useful Life					
Residual value of equipment (RV)					
Linear Depreciation Method (Vn)	−4313.60	−3833.20	−3352.80	−2872.40	−2392.00
LDM Present Value (PV)	−4313.60	−3833.20	−3352.80	−2872.40	−2392.00
Sum of Digits Method (Vn)	−1652.92	−1320.34	−1024.71	−766.03	−544.31
SDM Present Value (PV)	−1652.92	−1320.34	−1024.71	−766.03	−544.31
Exponential Method (Vn)	−128.38	−96.68	−72.81	−54.83	−41.29
EM Present Value (Pv)	−128.38	−96.68	−72.81	−54.83	−41.29
Functioning (*F*)	−5224.36	−5276.60	−5329.37	−5382.66	−5436.49
Functioning Accumulated (FA)	−77660.39	−82936.99	−88266.36	−93649.03	−99085.52
Present Value Accumulated (NPV)	−77660.39	−82936.99	−88266.36	−93649.03	−99085.52
Present Value (NPV)	−5224.36	−5276.60	−5329.37	−5382.66	−5436.49
Energy	−3200.00	−3400.00	−3600.00	−3800.00	−4000.00
Energy Accumulated (EA)	−27200.00	−30600.00	−34200.00	−38000.00	−42000.00
Energy (NPV)	−3200.00	−3400.00	−3600.00	−3800.00	−4000.00
Accumulated Functioning (FPV)	−682899.41	−769236.41	−861102.77	−958551.80	−1061637.32
Maintenance (*M*)	−348.29	−351.77	−355.29	−358.84	−362.43
Maintenane Accumulated (MA)	−5177.36	−5529.13	−5884.42	−6243.27	−6605.70
Present Value (MPV)	−348.29	−351.77	−355.29	−358.84	−362.43
Accumulated MPV	−5177.36	−5529.13	−5884.42	−6243.27	−6605.70
Non Production (NP)	0.00	0.00	0.00	0.00	0.00
Non Production Accumulated (NPA)	0.00	0.00	0.00	0.00	0.00
Present Value (NPPV)	0.00	0.00	0.00	0.00	0.00
Accumulated NPPV	0.00	0.00	0.00	0.00	0.00
Accumulated Total Costs (APV)	−688076.77	−774765.54	−866987.19	−964795.07	−1068243.02
Accumulated Total Costs	−110037.75	−119066.13	−128350.79	−137892.30	−147691.22
Benefit (*B*)	67632.85	67632.85	67632.85	67632.85	67632.85
Benefits Accumulated (BA)	1082125.60	1149758.45	1217391.30	1285024.15	1352657.00
Present Value (BPV)	67632.85	67632.85	67632.85	67632.85	67632.85
Accumulated Benefits BPV	1082125.60	1149758.45	1217391.30	1285024.15	1352657.00
BPV + FPV + MPV + NPPV + IIPV	960087.85	1018692.33	1077040.52	1135131.86	1192965.79
MTAC (LDM)	−117724.15	−127232.93	−136997.99	−147019.90	−157299.22
MTAC (SDM)	−120384.83	−129745.79	−139326.08	−149126.27	−159146.91
MTAC (EM)	−121909.37	−130969.45	−140277.98	−149837.47	−159649.93

(*Continued*)

TABLE 9.4 (*Continued*)

Values Used for Simulation based on MMTAC Method Related to Energy Standards. Second Simulation

Time	21	22	23	24	25
Initial Investment (II)					
Initial Investment (II)					
II (IIPV) - Return	−12000.00	−12000.00	−12000.00	−12000.00	−12000.00
Useful Life					
Residual value of equipment (RV)					
Linear Depreciation Method (Vn)	−1911.60	−1431.20	−950.80	−470.40	10.00
LDM Present Value (PV)	−1911.60	−1431.20	−950.80	−470.40	10.00
Sum of Digits Method (Vn)	−359.54	−211.72	−100.86	−26.95	10.00
SDM Present Value (PV)	−359.54	−211.72	−100.86	−26.95	10.00
Exponential Method (Vn)	−31.09	−23.42	−17.63	−13.28	−10.00
EM Present Value (Pv)	−31.09	−23.42	−17.63	−13.28	−10.00
Functioning (*F*)	−5490.86	−5545.76	−5601.22	−5657.23	−5713.81
Functioning Accumulated (FA)	−104576.37	−110122.14	−115723.36	−121380.59	−127094.40
Present Value Accumulated (NPV)	−104576.37	−110122.14	−115723.36	−121380.59	−127094.40
Present Value (NPV)	−5490.86	−5545.76	−5601.22	−5657.23	−5713.81
Energy	−4200.00	−4400.00	−4600.00	−4800.00	−5000.00
Energy Accumulated (EA)	−46200.00	−50600.00	−55200.00	−60000.00	−65000.00
Energy (NPV)	−4200.00	−4400.00	−4600.00	−4800.00	−5000.00
Accumulated Functioning (FPV)	−1170413.69	−1284935.83	−1405259.18	−1531439.78	−1663534.17
Maintenance (*M*)	−366.06	−369.72	−373.41	−377.15	−380.92
Maintenane Accumulated (MA)	−6971.76	−7341.48	−7714.89	−8092.04	−8472.96
Present Value (MPV)	−366.06	−369.72	−373.41	−377.15	−380.92
Accumulated MPV	−6971.76	−7341.48	−7714.89	−8092.04	−8472.96
Non Production (NP)	0.00	0.00	0.00	0.00	0.00
Non Production Accumulated (NPA)	0.00	0.00	0.00	0.00	0.00
Present Value (NPPV)	0.00	0.00	0.00	0.00	0.00
Accumulated NPPV	0.00	0.00	0.00	0.00	0.00
Accumulated Total Costs (APV)	−1177385.45	−1292277.30	−1412974.07	−1539531.82	−1672007.13
Accumulated Total Costs	−157748.13	−168063.61	−178638.25	−189472.63	−200567.36
Benefit (*B*)	67632.85	67632.85	67632.85	67632.85	67632.85
Benefits Accumulated (BA)	1420289.86	1487922.71	1555555.56	1623188.41	1690821.26
Present Value (BPV)	67632.85	67632.85	67632.85	67632.85	67632.85
Accumulated Benefits BPV	1420289.86	1487922.71	1555555.56	1623188.41	1690821.26
BPV + FPV + MPV + NPPV + IIPV	1250541.72	1307859.09	1364917.31	1421715.77	1478253.90
MTAC (LDM)	−167836.53	−178632.41	−189687.45	−201002.23	−212577.36
MTAC (SDM)	−169388.59	−179851.89	−190537.39	−201445.68	−212577.36
MTAC (EM)	−169717.04	−180040.20	−190620.62	−201459.35	−212557.36

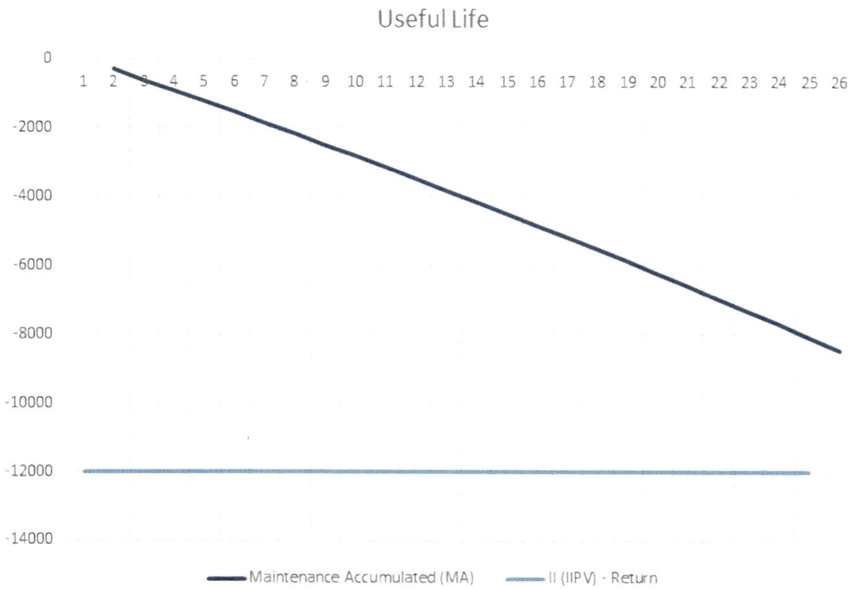

FIGURE 9.14 Physical Asset useful life.

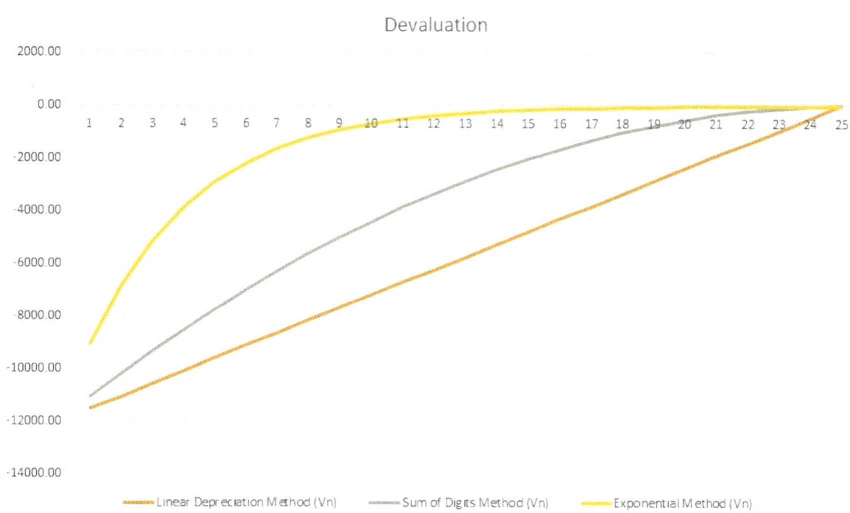

FIGURE 9.15 The three devaluation methods.

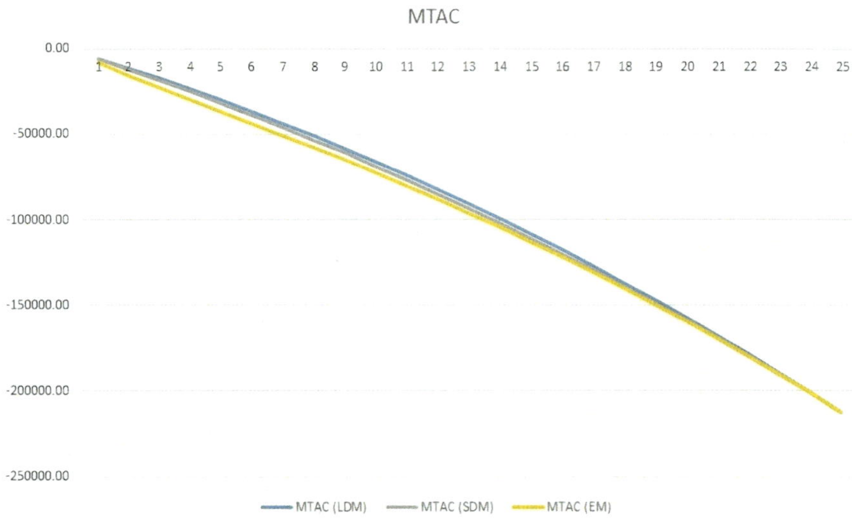

FIGURE 9.16 The MMTAC results from the three devaluation methods.

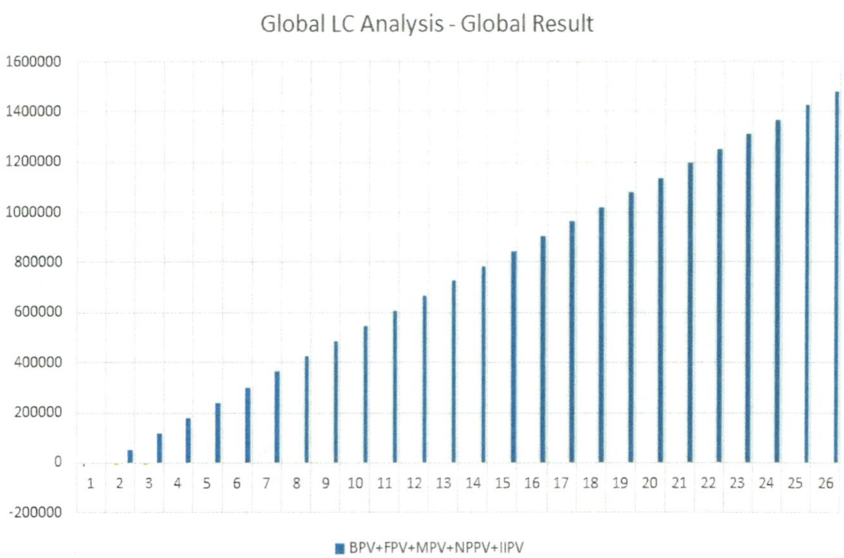

FIGURE 9.17 Life cycle analysis global result including profits.

The remaining values are presented in Table 9.5, and the graphical results are shown in Figures 9.18–9.21.

As referred in the preceding introduction of this section, the evaluation of introduction of Energy Norms, the simulation of their implementation is made through the variation of Energy expenses. Other indicators may be used, but may not be so intuitive.

The second simulation uses the following values:

• Initial Investment (II)	–12000
• Useful Life	25
• Residual value of equipment (RV)	10
• Functioning (*F*)	–4500
• Energy (*E*)	**–200**
• Maintenance (*M*)	–300
• Benefit (*B*)	70000
• Internal Rate of Return (*r*)	0.12
• Risk (*R*)	0.01
• Apparent Rate	0.1312
• Maintenance Increase Rate	0.01
• Availability	0.9662
• MTBF	1000
• MTTR	25
• MWT	10

The remaining values are presented in Table 9.6, and the graphical results are shown in Figures 9.22–9.25.

As can be seen, namely comparing Figures 9.21 and 9.25, with the decreasing of the Exergy expenses, it gets one more positive year, but the positive incomes increase a lot.

These results from the three methods are close, what reinforce a lot the robustness of the models under discussion.

9.4.2 Life Cycle Investment of Physical Assets

The way to demonstrate if the return of implementing the Standard under discussion is interesting or not, in this section it is done using the model of Life Cycle Investment, presented in Chapter 2.

The first simulation uses the following values:

• Initial Investment (II)	–12000
• Functioning (*F*)	–4500
• Energy (*E*)	**–1000**
• Maintenance (*M*)	–300
• Accumulated (AMPV)	–5000
• Non Production (NP)	0
• Benefit (*B*) including IIPV	70000

TABLE 9.5
Values Used for Simulation based on MMTAC-RPV Method Related to Energy Standards. First Simulation

Time	0	1	2	3	4	5
Initial Investment (II)	−12000					
Initial Investment (II)	−12000					
II Present Value (IIPV) - Return	−12000	−10608.20	−9377.83	−8290.16	−7328.64	−6478.65
Life Time simulated	25					
Residual value of equipment (RV)	10					
Linear Depreciation Method (Vn)	−12000	−11519.60	−11039.20	−10558.80	−10078.40	−9598.00
LDM Present Value (Vn)		−10183.52	−8626.98	−7294.51	−6155.08	−5181.84
Sum of Digits Method (Vn)	−12000	−11076.15	−10189.26	−9339.32	−8526.34	−7750.31
SDM Present Value (Vn)		−9791.51	−7962.77	−6452.04	−5207.21	−4184.29
Exponential Method (Vn)	−12000	−9036.79	−6805.29	−5124.83	−3859.33	−2906.33
EM Present Value (Pv)		−7988.67	−5318.24	−3540.47	−2356.97	−1569.09
Functioning (*F*)	−4500	−4500.00	−4545.00	−4590.45	−4636.35	−4682.72
Present Value Acummulated (NPV)		−3978.08	−7529.93	−10701.23	−13532.74	−16060.88
Present Value (NPV)		−3978.08	−3551.85	−3171.30	−2831.52	−2528.14
Energy (*E*)	−1000	−1000.00	−2000.00	−3000.00	−4000.00	−5000.00
Energy (NPV)		−884.02	−1562.97	−2072.54	−2442.88	−2699.44
Accumulated FPV		−16862.09	−25955.00	−38728.76	−54704.39	−73464.71
Maintenance (*M*)	−300	−300.00	−303.00	−306.03	−309.09	−312.18
Present Value (MPV)		−265.21	−236.79	−211.42	−188.77	−168.54
Accumulated (MPV)		−265.21	−502.00	−713.42	−902.18	−1070.73
Non Production (NP)	0	0.00	0.00	0.00	0.00	0.00
Present Value (NPPV)		0.00	0.00	0.00	0.00	0.00
Accumulated NPPV		0.00	0.00	0.00	0.00	0.00
Accumulated Total Costs (APV)		−17127.30	−26456.99	−39442.18	−55606.57	−74535.44
Accumulated Total Costs		−5800.00	−12648.00	−20544.48	−29489.92	−39484.82
Benefit (*B*)	70000	62500.00	62500.00	62500.00	62500.00	62500.00
Present Value (BPV)		55251.06	48842.88	43177.93	38170.02	33742.95
Accumulated Benefits BPV		55251.06	104093.94	147271.87	185441.89	219184.84
BPV + FPV + MPV + NPPV + IIPV		26123.76	65636.95	95829.69	117835.32	132649.40
MTAC-PV (LDM)		−39310.82	−23541.99	−19578.90	−18440.41	−18343.45
MTAC-PV (SDM)		−38918.81	−23209.88	−19298.07	−18203.45	−18143.95
MTAC-PV (EM)		−37115.97	−21887.62	−18327.55	−17490.89	−17620.91

(*Continued*)

TABLE 9.5 (*Continued*)
Values Used for Simulation based on MMTAC-RPV Method Related to Energy Standards. First Simulation

Time	6	7	8	9	10
Initial Investment (II)					
Initial Investment (II)					
II Present Value (IIPV) - Return	−5727.23	−5062.97	−4475.75	−3956.64	−3497.74
Life Time simulated					
Residual value of equipment (RV)					
Linear Depreciation Method (Vn)	−9117.60	−8637.20	−8156.80	−7676.40	−7196.00
LDM Present Value (Vn)	−4351.55	−3644.16	−3042.32	−2531.06	−2097.48
Sum of Digits Method (Vn)	−7011.23	−6309.11	−5643.94	−5015.72	−4424.46
SDM Present Value (Vn)	−3346.25	−2661.90	−2105.07	−1653.78	−1289.63
Exponential Method (Vn)	−2188.66	−1648.20	−1241.20	−934.71	−703.90
EM Present Value (Pv)	−1044.58	−695.40	−462.94	−308.19	−205.17
Functioning (*F*)	−4729.55	−4776.84	−4824.61	−4872.86	−4921.58
Present Value Acummulated (NPV)	−18318.15	−20333.57	−22133.05	−23739.73	−25174.26
Present Value (NPV)	−2257.27	−2015.42	−1799.48	−1606.68	−1434.53
Energy (*E*)	−6000.00	−7000.00	−8000.00	−9000.00	−10000.00
Energy (NPV)	−2863.62	−2953.40	−2983.84	−2967.48	−2914.78
Accumulated FPV	−94646.48	−117933.45	−143050.33	−169757.54	−197846.58
Maintenance (*M*)	−315.30	−318.46	−321.64	−324.86	−328.11
Present Value (MPV)	−150.48	−134.36	−119.97	−107.11	−95.64
Accumulated (MPV)	−1221.21	−1355.57	−1475.54	−1582.65	−1678.28
Non Production (NP)	0.00	0.00	0.00	0.00	0.00
Present Value (NPPV)	0.00	0.00	0.00	0.00	0.00
Accumulated NPPV	0.00	0.00	0.00	0.00	0.00
Accumulated Total Costs (APV)	−95867.69	−119289.02	−144525.87	−171340.19	−199524.87
Accumulated Total Costs	−50529.67	−62624.97	−75771.22	−89968.93	−105218.62
Benefit (*B*)	62500.00	62500.00	62500.00	62500.00	62500.00
Present Value (BPV)	29829.34	26369.64	23311.21	20607.51	18217.39
Accumulated Benefits BPV	249014.18	275383.82	298695.03	319302.54	337519.93
BPV + FPV + MPV + NPPV + IIPV	141146.49	144094.80	142169.17	135962.35	125995.06
MTAC-PV (LDM)	−18703.21	−19276.17	−19946.02	−20652.36	−21362.23
MTAC-PV (SDM)	−18535.66	−19135.85	−19828.87	−20554.89	−21281.45
MTAC-PV (EM)	−18152.04	−18854.92	−19623.60	−20405.38	−21173.00

(*Continued*)

TABLE 9.5 (*Continued*)
Values Used for Simulation based on MMTAC-RPV Method Related to Energy Standards. First Simulation

Time	11	12	13	14	15
Initial Investment (II)					
Initial Investment (II)					
II Present Value (IIPV) - Return	−3092.06	−2733.43	−2416.40	−2136.14	−1888.38
Life Time simulated					
Residual value of equipment (RV)					
Linear Depreciation Method (Vn)	−6715.60	−6235.20	−5754.80	−5274.40	−4794.00
LDM Present Value (Vn)	−1730.42	−1420.29	−1158.83	−938.90	−754.41
Sum of Digits Method (Vn)	−3870.15	−3352.80	−2872.40	−2428.95	−2022.46
SDM Present Value (Vn)	−997.23	−763.72	−578.41	−432.38	−318.27
Exponential Method (Vn)	−530.08	−399.19	−300.61	−226.38	−170.48
EM Present Value (Pv)	−136.59	−90.93	−60.53	−40.30	−26.83
Functioning (*F*)	−4970.80	−5020.51	−5070.71	−5121.42	−5172.63
Present Value Acummulated (NPV)	−26455.10	−27598.70	−28619.77	−29531.44	−30345.44
Present Value (NPV)	−1280.83	−1143.60	−1021.07	−911.67	−813.99
Energy (*E*)	−11000.00	−12000.00	−13000.00	−14000.00	−15000.00
Energy (NPV)	−2834.39	−2733.43	−2617.77	−2492.16	−2360.48
Accumulated FPV	−227136.06	−257468.20	−288705.73	−320729.34	−353435.26
Maintenance (*M*)	−331.39	−334.70	−338.05	−341.43	−344.84
Present Value (MPV)	−85.39	−76.24	−68.07	−60.78	−54.27
Accumulated (MPV)	−1763.67	−1839.91	−1907.98	−1968.76	−2023.03
Non Production (NP)	0.00	0.00	0.00	0.00	0.00
Present Value (NPPV)	0.00	0.00	0.00	0.00	0.00
Accumulated NPPV	0.00	0.00	0.00	0.00	0.00
Accumulated Total Costs (APV)	−228899.74	−259308.11	−290613.72	−322698.10	−355458.28
Accumulated Total Costs	−121520.81	−138876.01	−157284.77	−176747.62	−197265.10
Benefit (*B*)	62500.00	62500.00	62500.00	62500.00	62500.00
Present Value (BPV)	16104.48	14236.63	12585.42	11125.73	9835.33
Accumulated Benefits BPV	353624.40	367861.04	380446.46	391572.19	401407.52
BPV + FPV + MPV + NPPV + IIPV	112724.67	96552.93	77832.74	56874.09	33949.24
MTAC-PV (LDM)	−22057.29	−22727.37	−23367.12	−23974.07	−24547.51
MTAC-PV (SDM)	−21990.63	−22672.65	−23322.47	−23937.89	−24518.44
MTAC-PV (EM)	−21912.39	−22616.59	−23282.63	−23909.89	−24499.01

(*Continued*)

TABLE 9.5 (*Continued*)
Values Used for Simulation based on MMTAC-RPV Method Related to Energy Standards. First Simulation

Time	16	17	18	19	20
Initial Investment (II)					
Initial Investment (II)					
II Present Value (IIPV) - Return	−1669.36	−1475.75	−1304.58	−1153.27	−1019.51
Life Time simulated					
Residual value of equipment (RV)					
Linear Depreciation Method (Vn)	−4313.60	−3833.20	−3352.80	−2872.40	−2392.00
LDM Present Value (Vn)	−600.08	−471.40	−364.50	−276.06	−203.22
Sum of Digits Method (Vn)	−1652.92	−1320.34	−1024.71	−766.03	−544.31
SDM Present Value (Vn)	−229.94	−162.37	−111.40	−73.62	−46.24
Exponential Method (Vn)	−128.38	−96.68	−72.81	−54.83	−41.29
EM Present Value (Pv)	−17.86	−11.89	−7.92	−5.27	−3.51
Functioning (*F*)	−5224.36	−5276.60	−5329.37	−5382.66	−5436.49
Present Value Acummulated (NPV)	−31072.22	−31721.13	−32300.51	−32817.82	−33279.70
Present Value (NPV)	−726.78	−648.91	−579.38	−517.31	−461.88
Energy (*E*)	−16000.00	−17000.00	−18000.00	−19000.00	−20000.00
Energy (NPV)	−2225.82	−2090.64	−1956.88	−1826.02	−1699.19
Accumulated FPV	−386733.29	−420545.05	−454802.44	−489446.28	−524425.17
Maintenance (*M*)	−348.29	−351.77	−355.29	−358.84	−362.43
Present Value (MPV)	−48.45	−43.26	−38.63	−34.49	−30.79
Accumulated (MPV)	−2071.48	−2114.74	−2153.37	−2187.85	−2218.65
Non Production (NP)	0.00	0.00	0.00	0.00	0.00
Present Value (NPPV)	0.00	0.00	0.00	0.00	0.00
Accumulated NPPV	0.00	0.00	0.00	0.00	0.00
Accumulated Total Costs (APV)	−388804.77	−422659.80	−456955.81	−491634.13	−526643.81
Accumulated Total Costs	−218837.75	−241466.13	−265150.79	−289892.30	−315691.22
Benefit (*B*)	62500.00	62500.00	62500.00	62500.00	62500.00
Present Value (BPV)	8694.60	7686.18	6794.71	6006.64	5309.97
Accumulated Benefits BPV	410102.12	417788.30	424583.01	430589.65	435899.62
BPV + FPV + MPV + NPPV + IIPV	9297.35	−16871.50	−44372.80	−73044.48	−102744.20
MTAC-PV (LDM)	−25087.80	−25595.95	−26073.35	−26521.59	−26942.35
MTAC-PV (SDM)	−25064.67	−25577.77	−26059.29	−26510.93	−26934.50
MTAC-PV (EM)	−25051.41	−25568.92	−26053.54	−26507.34	−26932.37

(*Continued*)

TABLE 9.5 (*Continued*)

Values Used for Simulation based on MMTAC-RPV Method Related to Energy Standards. First Simulation

Time	21	22	23	24	25
Initial Investment (II)					
Initial Investment (II)					
II Present Value (IIPV) - Return	−901.27	−796.74	−704.33	−622.64	−550.42
Life Time simulated					
Residual value of equipment (RV)					
Linear Depreciation Method (Vn)	−1911.60	−1431.20	−950.80	−470.40	10.00
LDM Present Value (Vn)	−143.57	−95.02	−55.81	−24.41	0.46
Sum of Digits Method (Vn)	−359.54	−211.72	−100.86	−26.95	10.00
SDM Present Value (Vn)	−27.00	−14.06	−5.92	−1.40	0.46
Exponential Method (Vn)	−31.09	−23.42	−17.63	−13.28	−10.00
EM Present Value (Pv)	−2.34	−1.55	−1.03	−0.69	−0.46
Functioning (*F*)	−5490.86	−5545.76	−5601.22	−5657.23	−5713.81
Present Value Acummulated (NPV)	−33692.09	−34060.30	−34389.06	−34682.60	−34944.68
Present Value (NPV)	−412.39	−368.21	−328.76	−293.53	−262.08
Energy (*E*)	−21000.00	−22000.00	−23000.00	−24000.00	−25000.00
Energy (NPV)	−1577.22	−1460.68	−1349.96	−1245.28	−1146.71
Accumulated FPV	−559694.48	−595215.46	−630954.49	−666882.36	−702973.75
Maintenance (*M*)	−366.06	−369.72	−373.41	−377.15	−380.92
Present Value (MPV)	−27.49	−24.55	−21.92	−19.57	−17.47
Accumulated (MPV)	−2246.14	−2270.69	−2292.60	−2312.17	−2329.65
Non Production (NP)	0.00	0.00	0.00	0.00	0.00
Present Value (NPPV)	0.00	0.00	0.00	0.00	0.00
Accumulated NPPV	0.00	0.00	0.00	0.00	0.00
Accumulated Total Costs (APV)	−561940.62	−597486.15	−633247.09	−669194.53	−705303.40
Accumulated Total Costs	−342548.13	−370463.61	−399438.25	−429472.63	−460567.36
Benefit (*B*)	62500.00	62500.00	62500.00	62500.00	62500.00
Present Value (BPV)	4694.10	4149.67	3668.38	3242.91	2866.78
Accumulated Benefits BPV	440593.72	444743.39	448411.76	451654.67	454521.46
BPV + FPV + MPV + NPPV + IIPV	−133346.90	−164742.76	−196835.33	−229539.86	−262781.94
MTAC-PV (LDM)	−27337.34	−27708.24	−28056.65	−28384.12	−28692.12
MTAC-PV (SDM)	−27331.79	−27704.55	−28054.48	−28383.16	−28692.12
MTAC-PV (EM)	−27330.62	−27703.99	−28054.27	−28383.13	−28692.15

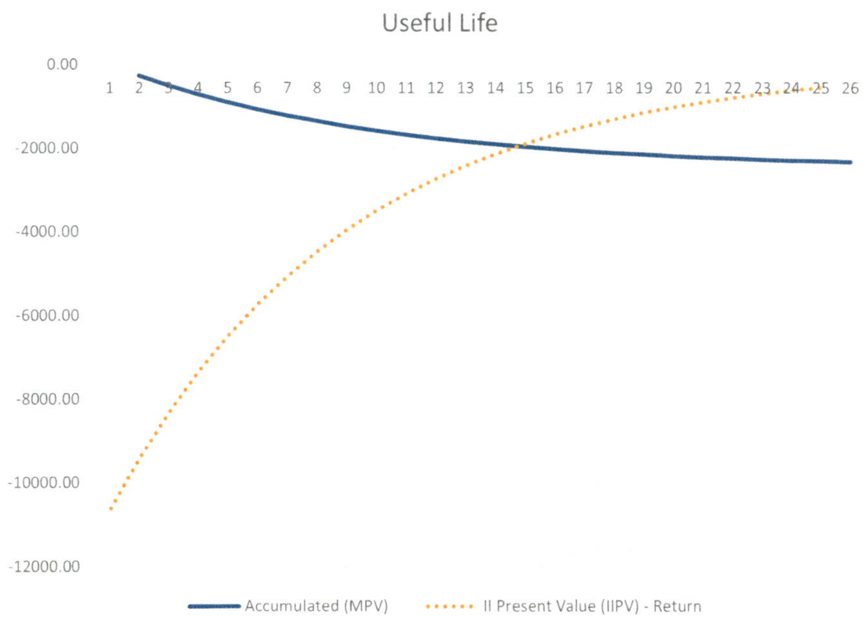

FIGURE 9.18 Physical Asset useful life.

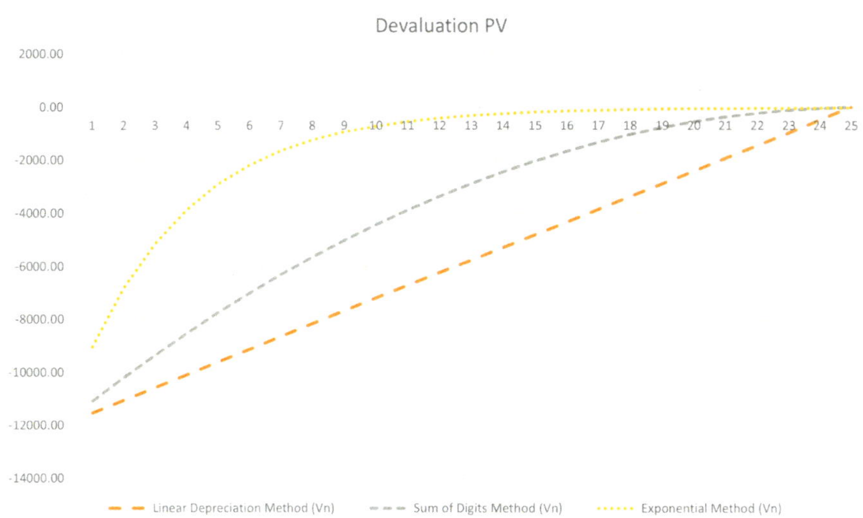

FIGURE 9.19 The three devaluation methods.

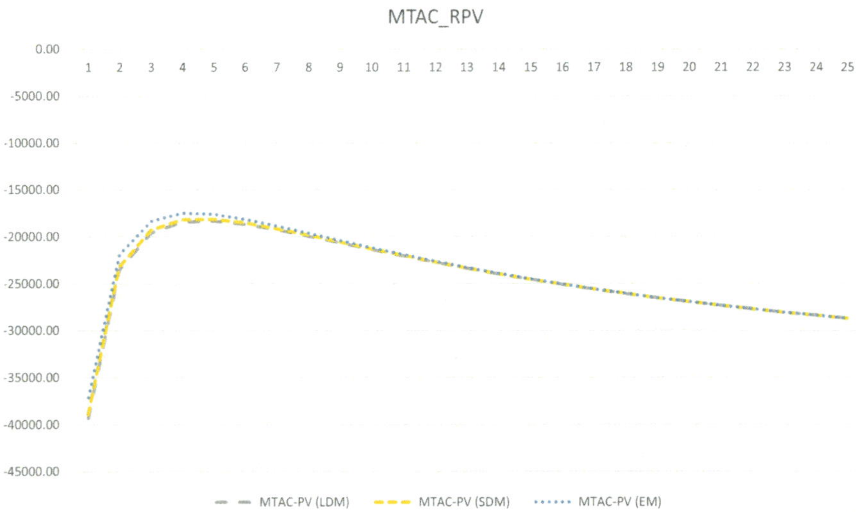

FIGURE 9.20 The MMTAC-RPV results from the three devaluation methods.

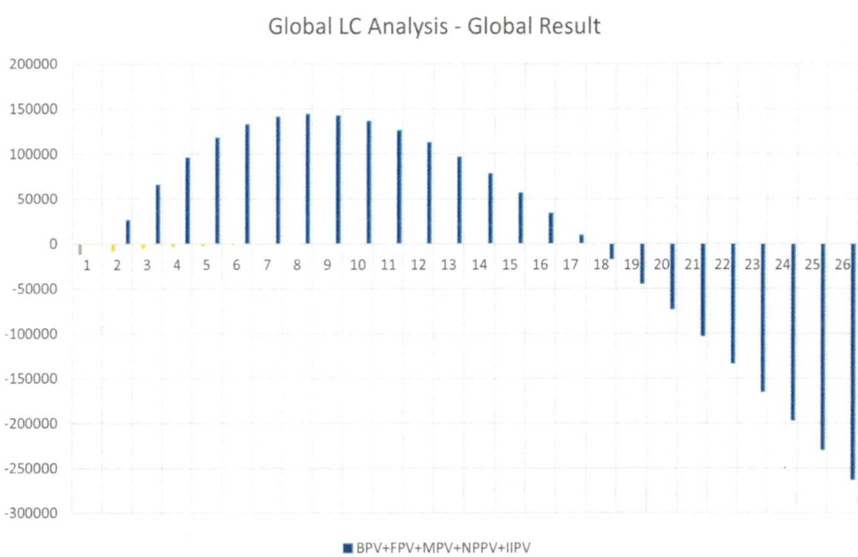

FIGURE 9.21 Life cycle analysis global result including profits.

TABLE 9.6

Values Used for Simulation based on MMTAC-RPV Method Related to Energy Standards. Second Simulation

Time	0	1	2	3	4	5
Initial Investment (II)	−12000					
Initial Investment (II)	−12000					
II Present Value (IIPV) - Return	−12000	−10608.20	−9377.83	−8290.16	−7328.64	−6478.65
Life Time simulated	25					
Residual value of equipment (RV)	10					
Linear Depreciation Method (Vn)	−12000	−11519.60	−11039.20	−10558.80	−10078.40	−9598.00
LDM Present Value (Vn)		−10183.52	−8626.98	−7294.51	−6155.08	−5181.84
Sum of Digits Method (Vn)	−12000	−11076.15	−10189.26	−9339.32	−8526.34	−7750.31
SDM Present Value (Vn)		−9791.51	−7962.77	−6452.04	−5207.21	−4184.29
Exponential Method (Vn)	−12000	−9036.79	−6805.29	−5124.83	−3859.33	−2906.33
EM Present Value (Pv)		−7988.67	−5318.24	−3540.47	−2356.97	−1569.09
Functioning (*F*)	−4500	−4500.00	−4545.00	−4590.45	−4636.35	−4682.72
Present Value Acummulated (NPV)		−3978.08	−7529.93	−10701.23	−13532.74	−16060.88
Present Value (NPV)		−3978.08	−3551.85	−3171.30	−2831.52	−2528.14
Energy (*E*)	−200	−200.00	−400.00	−600.00	−800.00	−1000.00
Energy (NPV)		−176.80	−312.59	−414.51	−488.58	−539.89
Accumulated FPV		−16154.88	−23997.40	−35113.14	−49134.46	−65735.23
Maintenance (*M*)	−300	−300.00	−303.00	−306.03	−309.09	−312.18
Present Value (MPV)		−265.21	−236.79	−211.42	−188.77	−168.54
Accumulated (MPV)		−265.21	−502.00	−713.42	−902.18	−1070.73
Non Production (NP)	0	0.00	0.00	0.00	0.00	0.00
Present Value (NPPV)		0.00	0.00	0.00	0.00	0.00
Accumulated NPPV		0.00	0.00	0.00	0.00	0.00
Accumulated Total Costs (APV)		−16420.08	−24499.40	−35826.56	−50036.64	−66805.96
Accumulated Total Costs		−5000.00	−10248.00	−15744.48	−21489.92	−27484.82
Benefit (*B*)	70000	62500.00	62500.00	62500.00	62500.00	62500.00
Present Value (BPV)		55251.06	48842.88	43177.93	38170.02	33742.95
Accumulated Benefits BPV		55251.06	104093.94	147271.87	185441.89	219184.84
BPV + FPV + MPV + NPPV + IIPV		26830.98	67594.54	99445.31	123405.25	140378.88
MTAC-PV (LDM)		−38603.61	−22563.19	−18373.69	−17047.93	−16797.56
MTAC-PV (SDM)		−38211.59	−22231.08	−18092.87	−16810.96	−16598.05
MTAC-PV (EM)		−36408.76	−20908.82	−17122.34	−16098.40	−16075.01

(Continued)

TABLE 9.6 (*Continued*)
Values Used for Simulation based on MMTAC-RPV Method Related to Energy Standards. Second Simulation

Time	6	7	8	9	10
Initial Investment (II)					
Initial Investment (II)					
II Present Value (IIPV) - Return	−5727.23	−5062.97	−4475.75	−3956.64	−3497.74
Life Time simulated					
Residual value of equipment (RV)					
Linear Depreciation Method (Vn)	−9117.60	−8637.20	−8156.80	−7676.40	−7196.00
LDM Present Value (Vn)	−4351.55	−3644.16	−3042.32	−2531.06	−2097.48
Sum of Digits Method (Vn)	−7011.23	−6309.11	−5643.94	−5015.72	−4424.46
SDM Present Value (Vn)	−3346.25	−2661.90	−2105.07	−1653.78	−1289.63
Exponential Method (Vn)	−2188.66	−1648.20	−1241.20	−934.71	−703.90
EM Present Value (Pv)	−1044.58	−695.40	−462.94	−308.19	−205.17
Functioning (*F*)	−4729.55	−4776.84	−4824.61	−4872.86	−4921.58
Present Value Acummulated (NPV)	−18318.15	−20333.57	−22133.05	−23739.73	−25174.26
Present Value (NPV)	−2257.27	−2015.42	−1799.48	−1606.68	−1434.53
Energy (*E*)	−1200.00	−1400.00	−1600.00	−1800.00	−2000.00
Energy (NPV)	−572.72	−590.68	−596.77	−593.50	−582.96
Accumulated FPV	−84626.11	−105550.36	−128280.17	−152613.40	−178370.61
Maintenance (*M*)	−315.30	−318.46	−321.64	−324.86	−328.11
Present Value (MPV)	−150.48	−134.36	−119.97	−107.11	−95.64
Accumulated (MPV)	−1221.21	−1355.57	−1475.54	−1582.65	−1678.28
Non Production (NP)	0.00	0.00	0.00	0.00	0.00
Present Value (NPPV)	0.00	0.00	0.00	0.00	0.00
Accumulated NPPV	0.00	0.00	0.00	0.00	0.00
Accumulated Total Costs (APV)	−85847.32	−106905.93	−129755.71	−154196.04	−180048.90
Accumulated Total Costs	−33729.67	−40224.97	−46971.22	−53968.93	−61218.62
Benefit (*B*)	62500.00	62500.00	62500.00	62500.00	62500.00
Present Value (BPV)	29829.34	26369.64	23311.21	20607.51	18217.39
Accumulated Benefits BPV	249014.18	275383.82	298695.03	319302.54	337519.93
BPV + FPV + MPV + NPPV + IIPV	151166.86	156477.89	156939.33	153106.50	145471.03
MTAC-PV (LDM)	−17033.14	−17507.16	−18099.75	−18747.46	−19414.64
MTAC-PV (SDM)	−16865.59	−17366.83	−17982.60	−18649.98	−19333.85
MTAC-PV (EM)	−16481.98	−17085.90	−17777.33	−18500.47	−19225.41

(Continued)

TABLE 9.6 (*Continued*)
Values Used for Simulation based on MMTAC-RPV Method Related to Energy Standards. Second Simulation

Time	11	12	13	14	15
Initial Investment (II)					
Initial Investment (II)					
II Present Value (IIPV) - Return	−3092.06	−2733.43	−2416.40	−2136.14	−1888.38
Life Time simulated					
Residual value of equipment (RV)					
Linear Depreciation Method (Vn)	−6715.60	−6235.20	−5754.80	−5274.40	−4794.00
LDM Present Value (Vn)	−1730.42	−1420.29	−1158.83	−938.90	−754.41
Sum of Digits Method (Vn)	−3870.15	−3352.80	−2872.40	−2428.95	−2022.46
SDM Present Value (Vn)	−997.23	−763.72	−578.41	−432.38	−318.27
Exponential Method (Vn)	−530.08	−399.19	−300.61	−226.38	−170.48
EM Present Value (Pv)	−136.59	−90.93	−60.53	−40.30	−26.83
Functioning (*F*)	−4970.80	−5020.51	−5070.71	−5121.42	−5172.63
Present Value Acummulated (NPV)	−26455.10	−27598.70	−28619.77	−29531.44	−30345.44
Present Value (NPV)	−1280.83	−1143.60	−1021.07	−911.67	−813.99
Energy (*E*)	−2200.00	−2400.00	−2600.00	−2800.00	−3000.00
Energy (NPV)	−566.88	−546.69	−523.55	−498.43	−472.10
Accumulated FPV	−205392.59	−233537.97	−262681.29	−292711.17	−323528.70
Maintenance (*M*)	−331.39	−334.70	−338.05	−341.43	−344.84
Present Value (MPV)	−85.39	−76.24	−68.07	−60.78	−54.27
Accumulated (MPV)	−1763.67	−1839.91	−1907.98	−1968.76	−2023.03
Non Production (NP)	0.00	0.00	0.00	0.00	0.00
Present Value (NPPV)	0.00	0.00	0.00	0.00	0.00
Accumulated NPPV	0.00	0.00	0.00	0.00	0.00
Accumulated Total Costs (APV)	−207156.26	−235377.88	−264589.28	−294679.93	−325551.73
Accumulated Total Costs	−68720.81	−76476.01	−84484.77	−92747.62	−101265.10
Benefit (*B*)	62500.00	62500.00	62500.00	62500.00	62500.00
Present Value (BPV)	16104.48	14236.63	12585.42	11125.73	9835.33
Accumulated Benefits BPV	353624.40	367861.04	380446.46	391572.19	401407.52
BPV + FPV + MPV + NPPV + IIPV	134468.15	120483.15	103857.18	84892.26	63855.79
MTAC-PV (LDM)	−20080.61	−20733.18	−21365.24	−21972.77	−22553.74
MTAC-PV (SDM)	−20013.95	−20678.47	−21320.59	−21936.59	−22524.67
MTAC-PV (EM)	−19935.71	−20622.40	−21280.75	−21908.59	−22505.24

(*Continued*)

TABLE 9.6 (*Continued*)
Values Used for Simulation based on MMTAC-RPV Method Related to Energy Standards. Second Simulation

Time	16	17	18	19	20
Initial Investment (II)					
Initial Investment (II)					
II Present Value (IIPV) - Return	−1669.36	−1475.75	−1304.58	−1153.27	−1019.51
Life Time simulated					
Residual value of equipment (RV)					
Linear Depreciation Method (Vn)	−4313.60	−3833.20	−3352.80	−2872.40	−2392.00
LDM Present Value (Vn)	−600.08	−471.40	−364.50	−276.06	−203.22
Sum of Digits Method (Vn)	−1652.92	−1320.34	−1024.71	−766.03	−544.31
SDM Present Value (Vn)	−229.94	−162.37	−111.40	−73.62	−46.24
Exponential Method (Vn)	−128.38	−96.68	−72.81	−54.83	−41.29
EM Present Value (Pv)	−17.86	−11.89	−7.92	−5.27	−3.51
Functioning (*F*)	−5224.36	−5276.60	−5329.37	−5382.66	−5436.49
Present Value Acummulated (NPV)	−31072.22	−31721.13	−32300.51	−32817.82	−33279.70
Present Value (NPV)	−726.78	−648.91	−579.38	−517.31	−461.88
Energy (*E*)	−3200.00	−3400.00	−3600.00	−3800.00	−4000.00
Energy (NPV)	−445.16	−418.13	−391.38	−365.20	−339.84
Accumulated FPV	−355046.08	−387185.33	−419877.22	−453060.24	−486679.78
Maintenance (*M*)	−348.29	−351.77	−355.29	−358.84	−362.43
Present Value (MPV)	−48.45	−43.26	−38.63	−34.49	−30.79
Accumulated (MPV)	−2071.48	−2114.74	−2153.37	−2187.85	−2218.65
Non Production (NP)	0.00	0.00	0.00	0.00	0.00
Present Value (NPPV)	0.00	0.00	0.00	0.00	0.00
Accumulated NPPV	0.00	0.00	0.00	0.00	0.00
Accumulated Total Costs (APV)	−357117.56	−389300.07	−422030.59	−455248.09	−488898.42
Accumulated Total Costs	−110037.75	−119066.13	−128350.79	−137892.30	−147691.22
Benefit (*B*)	62500.00	62500.00	62500.00	62500.00	62500.00
Present Value (BPV)	8694.60	7686.18	6794.71	6006.64	5309.97
Accumulated Benefits BPV	410102.12	417788.30	424583.01	430589.65	435899.62
BPV + FPV + MPV + NPPV + IIPV	40984.56	16488.22	−9447.58	−36658.45	−64998.81
MTAC-PV (LDM)	−23107.35	−23633.62	−24133.06	−24606.53	−25055.08
MTAC-PV (SDM)	−23084.22	−23615.44	−24119.00	−24595.88	−25047.23
MTAC-PV (EM)	−23070.96	−23606.59	−24113.25	−24592.28	−25045.10

(*Continued*)

TABLE 9.6 (*Continued*)
Values Used for Simulation based on MMTAC-RPV Method Related to Energy Standards. Second Simulation

Time	21	22	23	24	25
Initial Investment (II)					
Initial Investment (II)					
II Present Value (IIPV) - Return	−901.27	−796.74	−704.33	−622.64	−550.42
Life Time simulated					
Residual value of equipment (RV)					
Linear Depreciation Method (Vn)	−1911.60	−1431.20	−950.80	−470.40	10.00
LDM Present Value (Vn)	−143.57	−95.02	−55.81	−24.41	0.46
Sum of Digits Method (Vn)	−359.54	−211.72	−100.86	−26.95	10.00
SDM Present Value (Vn)	−27.00	−14.06	−5.92	−1.40	0.46
Exponential Method (Vn)	−31.09	−23.42	−17.63	−13.28	−10.00
EM Present Value (Pv)	−2.34	−1.55	−1.03	−0.69	−0.46
Functioning (*F*)	−5490.86	−5545.76	−5601.22	−5657.23	−5713.81
Present Value Acummulated (NPV)	−33692.09	−34060.30	−34389.06	−34682.60	−34944.68
Present Value (NPV)	−412.39	−368.21	−328.76	−293.53	−262.08
Energy (*E*)	−4200.00	−4400.00	−4600.00	−4800.00	−5000.00
Energy (NPV)	−315.44	−292.14	−269.99	−249.06	−229.34
Accumulated FPV	−520687.31	−555039.75	−589698.81	−624630.46	−659804.48
Maintenance (*M*)	−366.06	−369.72	−373.41	−377.15	−380.92
Present Value (MPV)	−27.49	−24.55	−21.92	−19.57	−17.47
Accumulated (MPV)	−2246.14	−2270.69	−2292.60	−2312.17	−2329.65
Non Production (NP)	0.00	0.00	0.00	0.00	0.00
Present Value (NPPV)	0.00	0.00	0.00	0.00	0.00
Accumulated NPPV	0.00	0.00	0.00	0.00	0.00
Accumulated Total Costs (APV)	−522933.45	−557310.44	−591991.41	−626942.63	−662134.13
Accumulated Total Costs	−157748.13	−168063.61	−178638.25	−189472.63	−200567.36
Benefit (*B*)	62500.00	62500.00	62500.00	62500.00	62500.00
Present Value (BPV)	4694.10	4149.67	3668.38	3242.91	2866.78
Accumulated Benefits BPV	440593.72	444743.39	448411.76	451654.67	454521.46
BPV + FPV + MPV + NPPV + IIPV	−94339.73	−124567.05	−155579.65	−187287.96	−219612.67
MTAC-PV (LDM)	−25479.86	−25882.07	−26262.92	−26623.63	−26965.35
MTAC-PV (SDM)	−25474.31	−25878.39	−26260.75	−26622.67	−26965.35
MTAC-PV (EM)	−25473.13	−25877.82	−26260.54	−26622.64	−26965.38

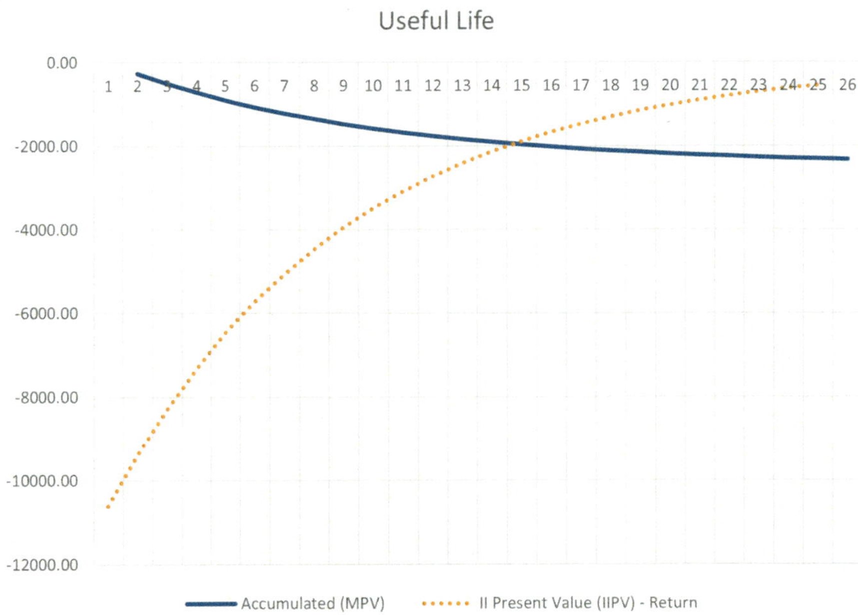

FIGURE 9.22 Physical Asset useful life.

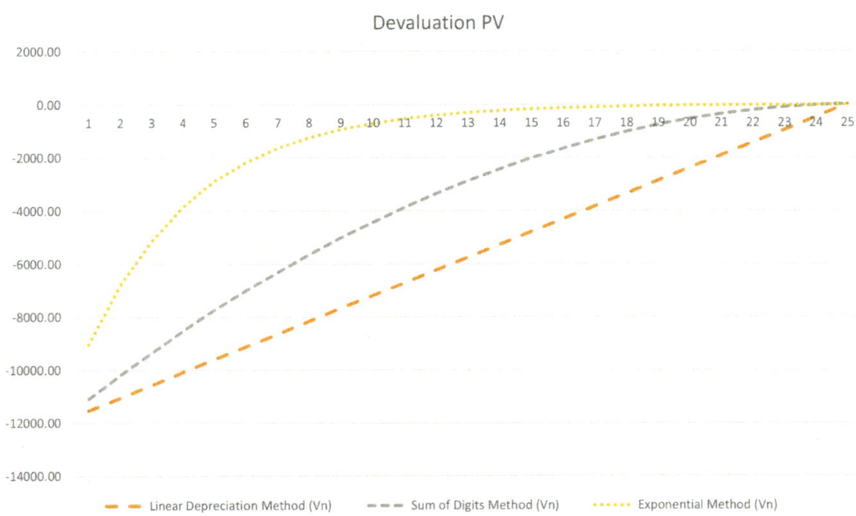

FIGURE 9.23 The three devaluation methods.

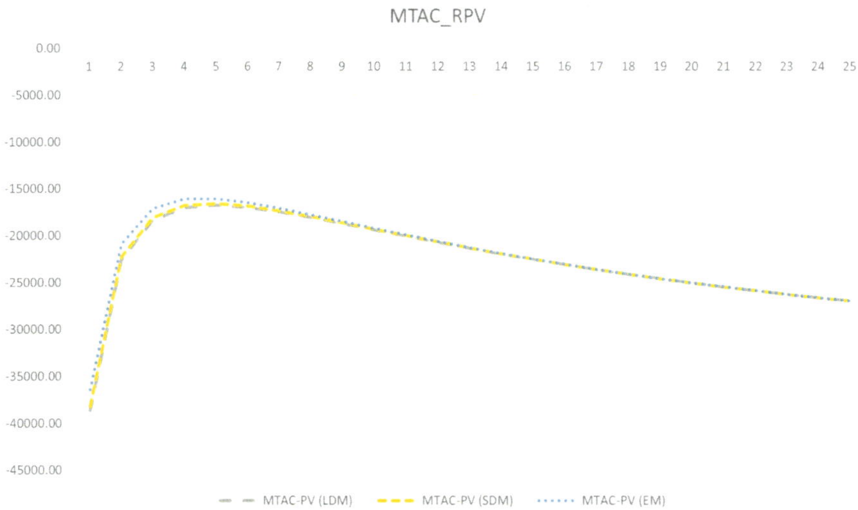

FIGURE 9.24 The MMTAC results from the three devaluation methods.

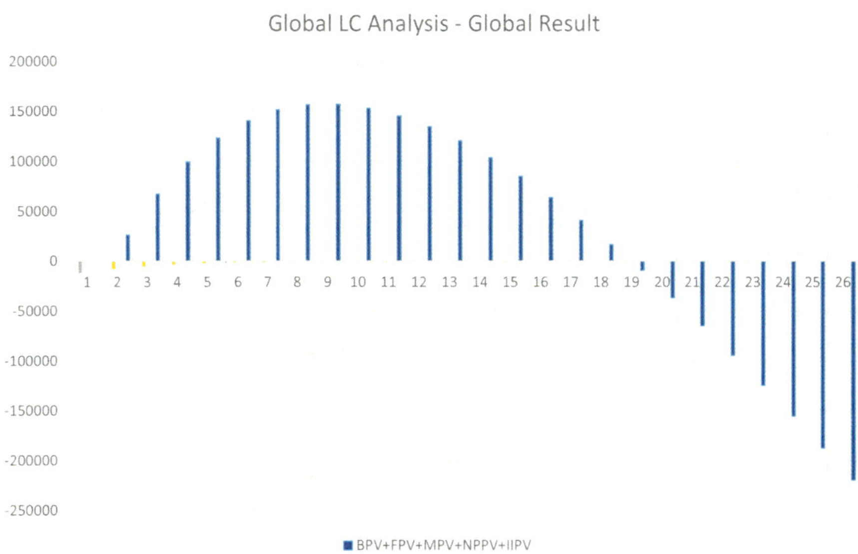

FIGURE 9.25 Life cycle analysis global result including profits.

- Internal Rate of Return (r) 0.12
- Risk (R) 0.01
- Apparent Rate 0.1312
- Maintenance Increase Rate 0.01
- Availability 0.9662
- MTBF 1000
- MTTR 25
- MWT 10

As referred in the preceding introduction of this section, the evaluation of introduction of Energy Norms, the simulation of their implementation is made through the variation of the Energy expenses. Other indicators may be used, but may not be so intuitive.

In the first scenario it is used a value of Energy of 1000, being the result showed in Table 9.7 and Figures 9.26–9.28.

The second simulation uses the following values:

- Initial Investment (II) –12000
- Functioning (F) –4500
- Energy (E) **–200**
- Maintenance (M) –300
- Accumulated (AMPV) –5000
- Non Production (NP) 0
- Benefit (B) including IIPV 70000
- Internal Rate of Return (r) 0.12
- Risk (R) 0.01
- Apparent Rate 0.1312
- Maintenance Increase Rate 0.01
- Availability 0.9662
- MTBF 1000
- MTTR 25
- MWT 10

The remaining values are presented in Table 9.8, and the graphical results are shown in Figures 9.29–9.31.

As can be seen, when comparing Figures 9.28 and 9.31, with the decreasing of the Energy expenses, the global result is more positive with the lower Energy costs, and the positive incomes increases a lot.

9.4.3 Life Cycle of Physical Assets with Recovery

The way to demonstrate if the return of implementing the Standard under discussion is interesting or not, in this section it is done using the model of Life Cycle of Physical Assets with Recovery presented in Chapter 3.

TABLE 9.7
Values Used for Simulation based on Life Cycle Investment Related to Energy Standards. First Simulation

Time	0	1	2	3	4	5	6	7	8	9	10	11	12	13
Initial Investment (II)	−12000													
II Present Value (IIPV) - Return	−12000	−10608.20	−9377.83	−8290.16	−7328.64	−6478.65	−5727.23	−5062.97	−4475.75	−3956.64	−3497.74	−3092.06	−2733.43	−2416.40
Functioning (F)	−4500	−4500.00	−4500.00	−4500.00	−4500.00	−4500.00	−4500.00	−4500.00	−4500.00	−4500.00	−4500.00	−4500.00	−4500.00	−4500.00
Present Value (PV) Accumulated		−3978.08	−7494.76	−10603.57	−13351.82	−15781.31	−17929.02	−19827.64	−21506.04	−22989.78	−24301.43	−25460.96	−26485.99	−27392.15
Energy (E)	−1000	−1000.00	−1000.00	−1000.00	−1000.00	−1000.00	−1000.00	−1000.00	−1000.00	−1000.00	−1000.00	−1000.00	−1000.00	−1000.00
Energy (PV)		−884.02	−781.49	−690.85	−610.72	−539.89	−477.27	−421.91	−372.98	−329.72	−291.48	−257.67	−227.79	−201.37
Accumulated (AFPV)		−16862.09	−8276.25	−19570.67	−33533.21	−49854.40	−68260.69	−88510.24	−110389.26	−133708.77	−158301.68	−184020.31	−210734.09	−238327.60
Maintenance (M)	−300	−300.00	−303.00	−306.03	−309.09	−312.18	−315.30	−318.46	−321.64	−324.86	−328.11	−331.39	−334.70	−338.05
Present Value (PV)		−265.21	−234.45	−207.25	−183.22	−161.97	−143.18	−126.57	−111.89	−98.92	−87.44	−77.30	−68.34	−60.41
Accumulated (AMPV)	−5800	−265.21	−499.65	−706.90	−890.12	−1052.09	−1195.27	−1321.84	−1433.74	−1532.65	−1620.10	−1697.40	−1765.73	−1826.14
Non Production (NP)	0	0.00	0.00	0.00	0.00	0.00	0.00	0.00	0.00	0.00	0.00	0.00	0.00	0.00
Present Value (PV)		0.00	0.00	0.00	0.00	0.00	0.00	0.00	0.00	0.00	0.00	0.00	0.00	0.00
Accumulated (ANPPV)		0.00	0.00	0.00	0.00	0.00	0.00	0.00	0.00	0.00	0.00	0.00	0.00	0.00
Accumulated Total Costs (ATCPV)		−17127.30	−8775.90	−20277.58	−34423.33	−50906.49	−69455.96	−89832.09	−111823.00	−135241.42	−159921.78	−185717.71	−212499.82	−240153.75
Benefit (B) including IIPV	70000	51891.80	53122.17	54209.84	55171.36	56021.35	56772.77	57437.03	58024.25	58543.36	59002.26	59407.94	59766.57	60083.60
Present Value (PV)		45873.23	41514.23	37450.70	33694.27	30245.21	27095.91	24233.50	21641.85	19302.92	17197.87	15307.74	13613.99	12098.84
Accumulated Benefits (ABPV)		45873.23	87387.46	124838.16	158532.43	188777.64	215873.55	240107.05	261748.90	281051.82	298249.69	313557.43	327171.43	339270.27
ATCPV + ABPV		28745.93	78611.56	104560.58	124109.10	137871.15	146417.58	150274.96	149925.90	145810.40	138327.91	127839.73	114671.60	99116.52

(Continued)

TABLE 9.7 (*Continued*)
Values Used for Simulation based on Life Cycle Investment Related to Energy Standards. First Simulation

Time	14	15	16	17	18	19	20	21	22	23	24	25
Initial Investment (II)												
II Present Value (IIPV) - Return	−2136.14	−1888.38	−1669.36	−1475.75	−1304.58	−1153.27	−1019.51	−901.27	−796.74	−704.33	−622.64	−550.42
Functioning (*F*)	−4500.00	−4500.00	−4500.00	−4500.00	−4500.00	−4500.00	−4500.00	−4500.00	−4500.00	−4500.00	−4500.00	−4500.00
Present Value (PV) Accumulated	−28193.20	−28901.34	−29527.35	−30080.76	−30569.98	−31002.45	−31384.77	−31722.75	−32021.52	−32285.65	−32519.14	−32725.54
Energy (*E*)	−1000.00	−1000.00	−1000.00	−1000.00	−1000.00	−1000.00	−1000.00	−1000.00	−1000.00	−1000.00	−1000.00	−1000.00
Energy (PV)	−178.01	−157.37	−139.11	−122.98	−108.72	−96.11	−84.96	−75.11	−66.39	−58.69	−51.89	−45.87
Accumulated (AFPV)	−266698.81	−295757.52	−325423.98	−355627.72	−386306.41	−417404.97	−448874.71	−480672.56	−512760.48	−545104.82	−577675.84	−610447.26
Maintenance (*M*)	−341.43	−344.84	−348.29	−351.77	−355.29	−358.84	−362.43	−366.06	−369.72	−373.41	−377.15	−380.92
Present Value (PV)	−53.40	−47.21	−41.73	−36.89	−32.61	−28.83	−25.49	−22.53	−19.92	−17.61	−15.57	−13.76
Accumulated (AMPV)	−1879.55	−1926.76	−1968.49	−2005.38	−2038.00	−2066.83	−2092.32	−2114.85	−2134.77	−2152.38	−2167.94	−2181.70
Non Production (NP)	0.00	0.00	0.00	0.00	0.00	0.00	0.00	0.00	0.00	0.00	0.00	0.00
Present Value (PV)	0.00	0.00	0.00	0.00	0.00	0.00	0.00	0.00	0.00	0.00	0.00	0.00
Accumulated (ANPPV)	0.00	0.00	0.00	0.00	0.00	0.00	0.00	0.00	0.00	0.00	0.00	0.00
Accumulated Total Costs (ATCPV)	−268578.36	−297684.27	−327392.48	−357633.11	−388344.41	−419471.80	−450967.02	−482787.41	−514895.25	−547257.20	−579843.78	−612628.96
Benefit (*B*) including IIPV	60363.86	60611.62	60830.64	61024.25	61195.42	61346.73	61480.49	61598.73	61703.26	61795.67	61877.36	61949.58
Present Value (PV)	10745.47	9538.17	8462.37	7504.69	6652.88	5895.80	5223.35	4626.41	4096.77	3627.04	3210.60	2841.54
Accumulated Benefits (ABPV)	350015.74	359553.91	368016.28	375520.97	382173.85	388069.65	393293.00	397919.41	402016.18	405643.22	408853.82	411695.36
ATCPV + ABPV	81437.38	61869.63	40623.80	17887.86	−6170.57	−31402.16	−57674.02	−84868.00	−112879.06	−141613.98	−170989.96	−200933.60

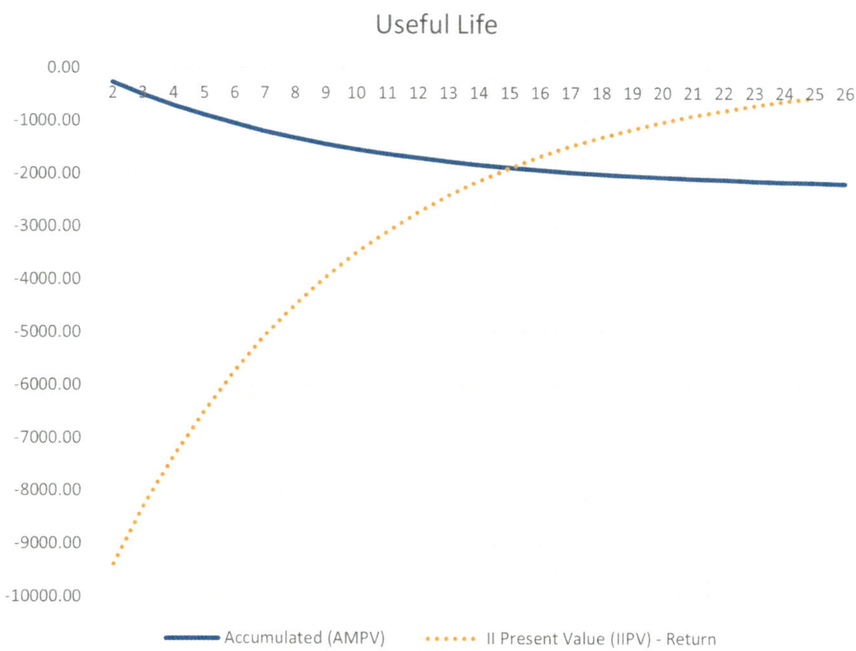

FIGURE 9.26 Physical Asset useful life.

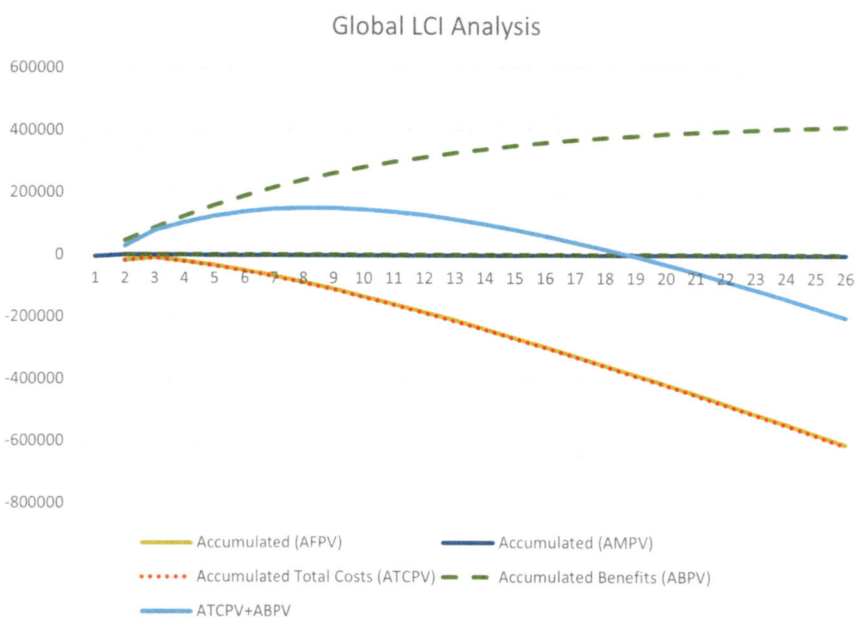

FIGURE 9.27 Global LCI analysis.

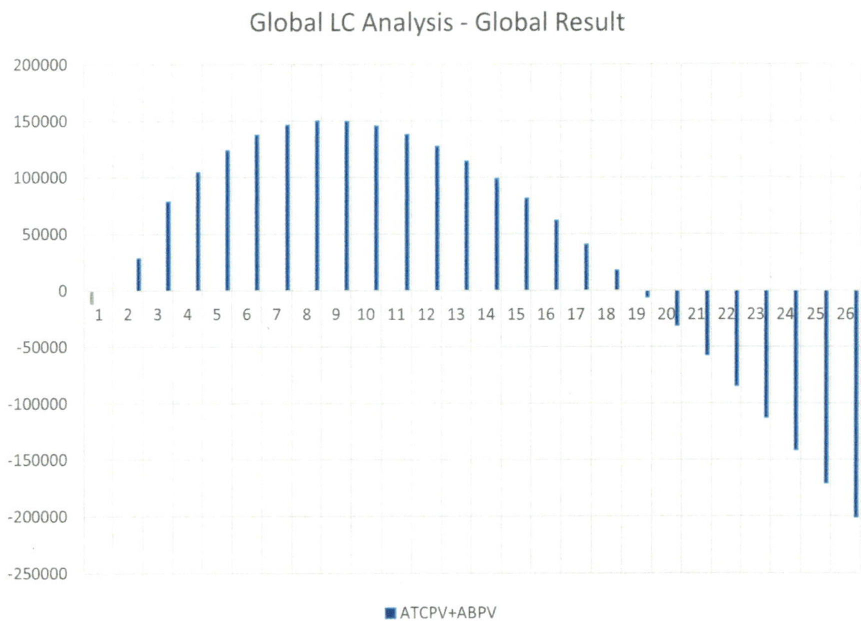

FIGURE 9.28 Life cycle analysis global result.

The first simulation uses the following values:

- Initial Investment (II) –12000
- Functioning (*F*) –4500
- Energy (*E*) **–1000**
- Maintenance (*M*) –300
- Accumulated (AMPV) –5000
- Non Production (NP) 0
- Benefit (*B*) including IIPV 70000
- Internal Rate of Return (*r*) 0.12
- Risk (R) 0.01
- Apparent Rate 0.1312
- Maintenance Increase Rate 0.01
- Availability 0.9662
- MTBF 1000
- MTTR 25
- MWT 10

As referred in the preceding introduction of this section, the evaluation of introduction of Energy Norms, the simulation of their implementation is made through the variation of the Energy expenses. Other indicators may be used, but may not be so intuitive.

TABLE 9.8
Values Used for Simulation based on Life Cycle Investment Related to Energy Standards. Second Simulation

Time	0	1	2	3	4	5	6	7	8	9	10	11	12	13
Initial Investment (II)	−12000													
II Present Value (IIPV) - Return	−12000	−10608.20	−9377.83	−8290.16	−7328.64	−6478.65	−5727.23	−5062.97	−4475.75	−3956.64	−3497.74	−3092.06	−2733.43	−2416.40
Functioning (*F*)	−4500	−4500.00	−4500.00	−4500.00	−4500.00	−4500.00	−4500.00	−4500.00	−4500.00	−4500.00	−4500.00	−4500.00	−4500.00	−4500.00
Present Value (PV) Accumulated		−3978.08	−7494.76	−10603.57	−13351.82	−15781.31	−17929.02	−19827.64	−21506.04	−22989.78	−24301.43	−25460.96	−26485.99	−27392.15
Energy (*E*)	−200	−200.00	−200.00	−200.00	−200.00	−200.00	−200.00	−200.00	−200.00	−200.00	−200.00	−200.00	−200.00	−200.00
Energy (PV)		−176.80	−156.30	−138.17	−122.14	−107.98	−95.45	−84.38	−74.60	−65.94	−58.30	−51.53	−45.56	−40.27
Accumulated (AFPV)		−16154.88	−7651.06	−18392.80	−31866.76	−47756.05	−65780.53	−85692.54	−107273.18	−130328.91	−154688.64	−180201.13	−206732.68	−234165.10
Maintenance (*M*)	−300	−300.00	−303.00	−306.03	−309.09	−312.18	−315.30	−318.46	−321.64	−324.86	−328.11	−331.39	−334.70	−338.05
Present Value (PV)		−265.21	−234.45	−207.25	−183.22	−161.97	−143.18	−126.57	−111.89	−98.92	−87.44	−77.30	−68.34	−60.41
Accumulated (AMPV)	−5000	−265.21	−499.65	−706.90	−890.12	−1052.09	−1195.27	−1321.84	−1433.74	−1532.65	−1620.10	−1697.40	−1765.73	−1826.14
Non Production (NP)	0	0.00	0.00	0.00	0.00	0.00	0.00	0.00	0.00	0.00	0.00	0.00	0.00	0.00
Present Value (PV)		0.00	0.00	0.00	0.00	0.00	0.00	0.00	0.00	0.00	0.00	0.00	0.00	0.00
Accumulated (ANPPV)		0.00	0.00	0.00	0.00	0.00	0.00	0.00	0.00	0.00	0.00	0.00	0.00	0.00
Accumulated Total Costs (ATCPV)		−16420.08	−8150.71	−19099.71	−32756.89	−48808.14	−66975.79	−87014.39	−108706.92	−131861.56	−156308.73	−181898.53	−208498.42	−235991.24
Benefit (*B*) including IIPV	70000	51891.80	53122.17	54209.84	55171.36	56021.35	56772.77	57437.03	58024.25	58543.36	59002.26	59407.94	59766.57	60083.60
Present Value (PV)		45873.23	41514.23	37450.70	33694.27	30245.21	27095.91	24233.50	21641.85	19302.92	17197.87	15307.74	13613.99	12098.84
Accumulated Benefits (ABPV)		45873.23	87387.46	124838.16	158532.43	188777.64	215873.55	240107.05	261748.90	281051.82	298249.69	313557.43	327171.43	339270.27
ATCPV + ABPV		29453.14	79236.75	105738.45	125775.54	139969.50	148897.75	153092.66	153041.98	149190.26	141940.96	131658.90	118673.01	103279.02

(Continued)

TABLE 9.8 (*Continued*)

Values Used for Simulation based on Life Cycle Investment Related to Energy Standards. Second Simulation

Time	14	15	16	17	18	19	20	21	22	23	24	25
Initial Investment (II)												
II Present Value (IIPV) - Return	−2136.14	−1888.38	−1669.36	−1475.75	−1304.58	−1153.27	−1019.51	−901.27	−796.74	−704.33	−622.64	−550.42
Functioning (*F*)	−4500.00	−4500.00	−4500.00	−4500.00	−4500.00	−4500.00	−4500.00	−4500.00	−4500.00	−4500.00	−4500.00	−4500.00
Present Value (PV) Accumulated	−28193.20	−28901.34	−29527.35	−30080.76	−30569.98	−31002.45	−31384.77	−31722.75	−32021.52	−32285.65	−32519.14	−32725.54
Energy (*E*)	−200.00	−200.00	−200.00	−200.00	−200.00	−200.00	−200.00	−200.00	−200.00	−200.00	−200.00	−200.00
Energy (PV)	−35.60	−31.47	−27.82	−24.60	−21.74	−19.22	−16.99	−15.02	−13.28	−11.74	−10.38	−9.17
Accumulated (AFPV)	−262393.90	−291326.72	−320881.89	−350987.24	−381578.96	−412600.64	−444002.40	−475740.17	−507774.98	−540072.36	−572601.88	−605336.59
Maintenance (*M*)	−341.43	−344.84	−348.29	−351.77	−355.29	−358.84	−362.43	−366.06	−369.72	−373.41	−377.15	−380.92
Present Value (PV)	−53.40	−47.21	−41.73	−36.89	−32.61	−28.83	−25.49	−22.53	−19.92	−17.61	−15.57	−13.76
Accumulated (AMPV)	−1879.55	−1926.76	−1968.49	−2005.38	−2038.00	−2066.83	−2092.32	−2114.85	−2134.77	−2152.38	−2167.94	−2181.70
Non Production (NP)	0.00	0.00	0.00	0.00	0.00	0.00	0.00	0.00	0.00	0.00	0.00	0.00
Present Value (PV)	0.00	0.00	0.00	0.00	0.00	0.00	0.00	0.00	0.00	0.00	0.00	0.00
Accumulated (ANPPV)	0.00	0.00	0.00	0.00	0.00	0.00	0.00	0.00	0.00	0.00	0.00	0.00
Accumulated Total Costs (ATCPV)	−264273.45	−293253.47	−322850.38	−352992.63	−383616.96	−414667.47	−446094.72	−477855.02	−509909.74	−542224.74	−574769.82	−607518.30
Benefit (*B*) including IIPV	60363.86	60611.62	60830.64	61024.25	61195.42	61346.73	61480.49	61598.73	61703.26	61795.67	61877.36	61949.58
Present Value (PV)	10745.47	9538.17	8462.37	7504.69	6652.88	5895.80	5223.35	4626.41	4096.77	3627.04	3210.60	2841.54
Accumulated Benefits (ABPV)	350015.74	359553.91	368016.28	375520.97	382173.85	388069.65	393293.00	397919.41	402016.18	405643.22	408853.82	411695.36
ATCPV + ABPV	85742.29	66300.43	45165.89	22528.34	−1443.12	−26597.82	−52801.72	−79935.61	−107893.56	−136581.52	−165916.00	−195822.94

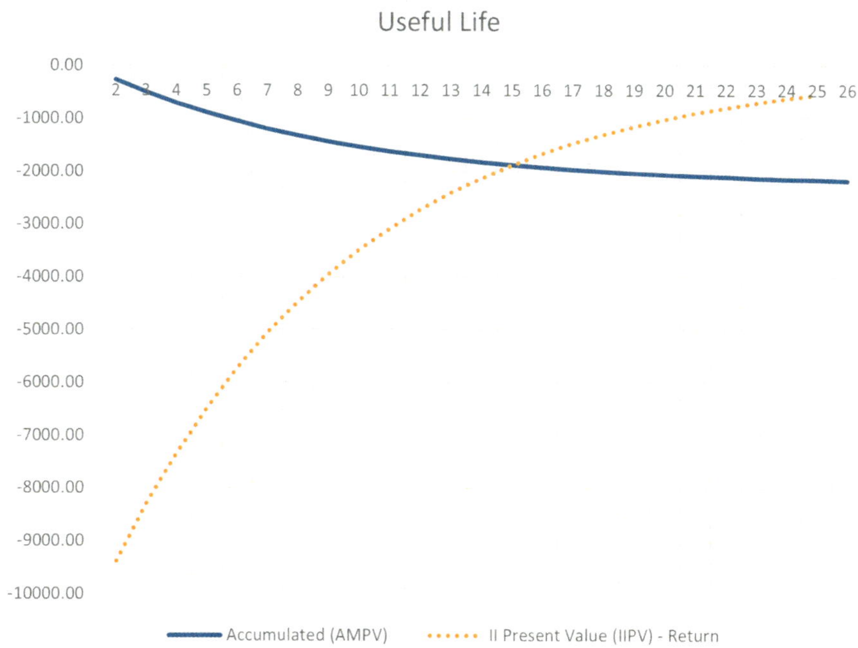

FIGURE 9.29 Physical Asset useful life.

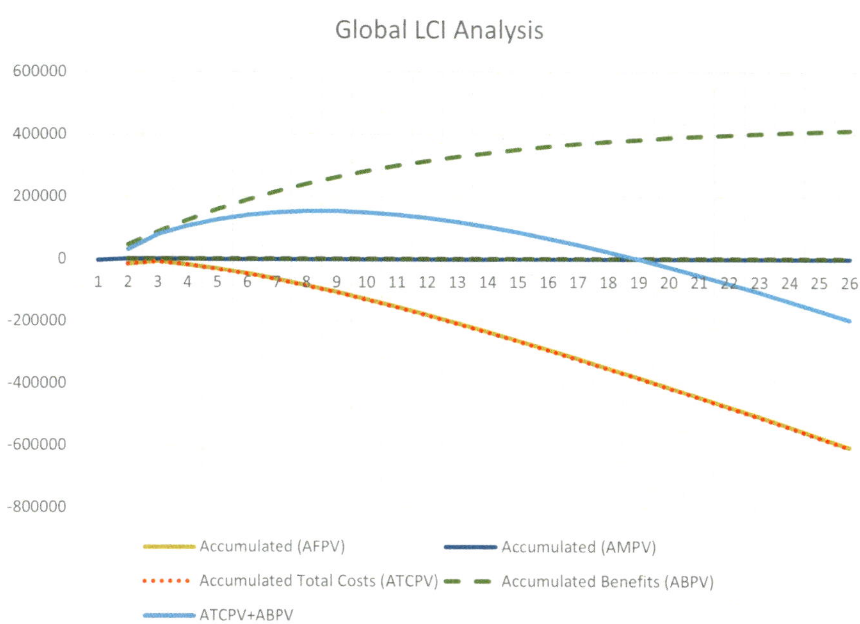

FIGURE 9.30 Global LCI analysis.

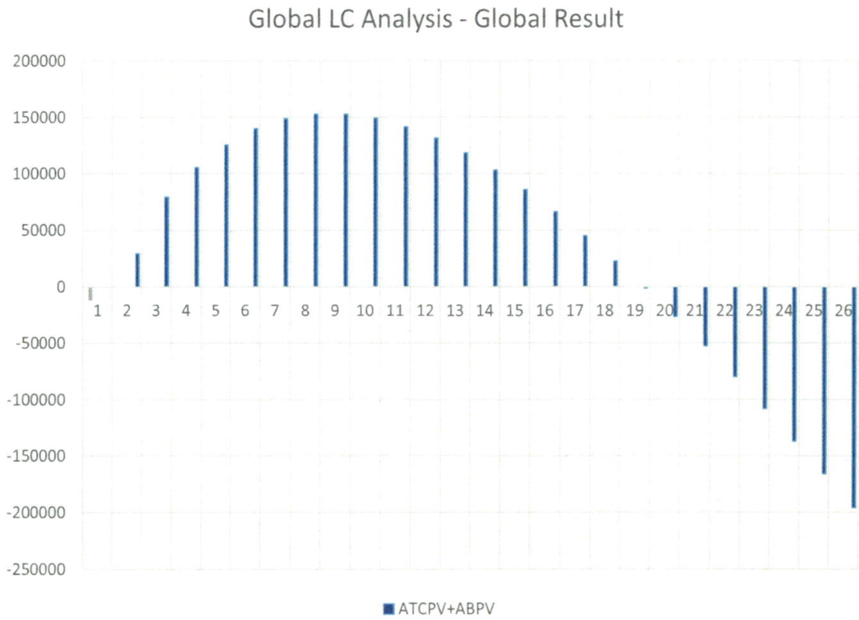

FIGURE 9.31 Life cycle analysis global result.

In the first scenario, it is used a value of Energy of 1000, being the result shown in Table 9.9 and Figures 9.32–9.34.

The second simulation uses the following values:

- Initial Investment (II) –12000
- Functioning (*F*) –4500
- Energy (*E*) **–200**
- Maintenance (*M*) –300
- Accumulated (AMPV) –5000
- Non Production (NP) 0
- Benefit (*B*) including IIPV 70000
- Internal Rate of Return (*r*) 0.12
- Risk (R) 0.01
- Apparent Rate 0.1312
- Maintenance Increase Rate 0.01
- Availability 0.9662
- MTBF 1000
- MTTR 25
- MWT 10

The remaining values are presented in Table 9.10, and the graphical results are shown in Figures 9.35–9.37.

TABLE 9.9
Values Used for Simulation based on Life Cycle with Recovery Related to Energy Standards. First Simulation

Time (t)	0	1	2	3	4	5	6	7	8	9	10	11	12	13
r – real interest rate ($1 + r$)		1.05	1.05	1.05	1.05	1.05	1.05	1.05	1.05	1.05	1.05	1.05	1.05	1.05
h – inflation rate ($1 + h$)		1.02	1.02	1.02	1.02	1.02	1.02	1.02	1.02	1.02	1.02	1.02	1.02	1.02
p – profit rate		1.05	1.05	1.05	1.05	1.05	1.05	1.05	1.05	1.05	1.05	1.05	1.05	1.05
i – money rate (nominal interest rate)		0.12	0.12	0.12	0.12	0.12	0.12	0.12	0.12	0.12	0.12	0.12	0.12	0.12
i – Apparent Rate with Risk		0.14	0.14	0.14	0.14	0.14	0.14	0.14	0.14	0.14	0.14	0.14	0.14	0.14
$1/((1+i)\wedge t))$		0.88	0.78	0.68	0.60	0.53	0.47	0.41	0.36	0.32	0.28	0.25	0.22	0.19
Initial Investment (II) (C_0)	−12000	−12000	−12000	−12000	−12000	−12000	−12000	−12000	−12000	−12000	−12000	−12000	−12000	−12000
Initial Investment (II) (C_0) with Devaluation (NPV)	−12000	−10565	−9302	−8190	−7211	−6349	−5590	−4921	−4333	−3815	−3359	−2957	−2604	−2292
Initial Investment (II) (C_0) considering NEW Asset (NPV)	−12000	−13630	−15480	−17583	−19970	−22682	−25762	−29261	−33234	−37747	−42873	−48695	−55307	−62818
Fp Financial Movement (Production)	70000	70000	70000	70000	70000	70000	70000	70000	70000	70000	70000	70000	70000	70000
Fp Financial Movement (NPV)	70000	56028	49329	43432	38239	33667	29642	26098	22978	20230	17812	15682	13807	12156
Fp Financial Movement (NPV) Accumulated	70000	56028	105357	148789	187028	220695	250336	276434	299412	319642	337454	353136	366943	379099
Functioning Costs	−4500	−4500	−4500	−4500	−4500	−4500	−4500	−4500	−4500	−4500	−4500	−4500	−4500	−4500
Functioning Costs (NPV) Accumulated	−4500	−3962	−7450	−10521	−13226	−15606	−17702	−19548	−21173	−22603	−23863	−24972	−25948	−26808
Energy	−1000	−1000	−1000	−1000	−1000	−1000	−1000	−1000	−1000	−1000	−1000	−1000	−1000	−1000
Energy Accumulated (NPV) Accumulated	−1000	−880	−1656	−2338	−2939	−3468	−3934	−4344	−4705	−5023	−5303	−5549	−5766	−5957
Maintenance Costs	−300	−300	−300	−300	−300	−300	−300	−300	−300	−300	−300	−300	−300	−300
Maintenance Costs (NPV) Accumulated	−300	−267	−502	−708	−891	−1051	−1192	−1316	−1426	−1522	−1607	−1681	−1747	−1805
Total Costs (NPV) Accumulated	−5800	−5229	−14181	−26411	−41527	−59184	−79078	−100942	−124540	−149666	−176135	−203788	−232484	−262097
FT Net Present Value (NPV) (With II Devaluated)		40234	81875	114188	138290	155162	165669	170571	170538	166162	157960	146390	131856	114711
FT Net Present Value (With II New Asset)		37170	75696	104796	125531	138829	145496	146232	141637	132230	118446	100653	79152	54185

(Continued)

TABLE 9.9 (*Continued*)
Values Used for Simulation based on Life Cycle with Recovery Related to Energy Standards. First Simulation

Time (t)	14	15	16	17	18	19	20	21	22	23	24	25
r – real interest rate $(1+r)$	1.05	1.05	1.05	1.05	1.05	1.05	1.05	1.05	1.05	1.05	1.05	1.05
h – inflation rate $(1+h)$	1.02	1.02	1.02	1.02	1.02	1.02	1.02	1.02	1.02	1.02	1.02	1.02
p – profit rate	1.05	1.05	1.05	1.05	1.05	1.05	1.05	1.05	1.05	1.05	1.05	1.05
i – money rate (nominal interest rate)	0.12	0.12	0.12	0.12	0.12	0.12	0.12	0.12	0.12	0.12	0.12	0.12
i – Apparent Rate with Risk	0.14	0.14	0.14	0.14	0.14	0.14	0.14	0.14	0.14	0.14	0.14	0.14
$1/((1+i)$ ^ $t))$	0.17	0.15	0.13	0.11	0.10	0.09	0.08	0.07	0.06	0.05	0.05	0.04
Initial Investment (II) (C_0)	−12000	−12000	−12000	−12000	−12000	−12000	−12000	−12000	−12000	−12000	−12000	−12000
Initial Investment (II) (C_0) with Devaluation (NPV)	−2018	−1777	−1565	−1377	−1213	−1068	−940	−828	−729	−642	−565	−497
Initial Investment (II) (C_0) considering NEW Asset (NPV)	−71348	−81037	−92041	−104540	−118736	−134860	−153174	−173974	−197599	−224432	−254909	−289524
Fp Financial Movement (Production)	70000	70000	70000	70000	70000	70000	70000	70000	70000	70000	70000	70000
Fp Financial Movement (NPV)	10703	9423	8297	7305	6431	5662	4985	4389	3865	3403	2996	2638
Fp Financial Movement (NPV) Accumulated	389802	399226	407522	414827	421258	426921	431906	436296	440160	443563	446559	449196
Functioning Costs	−4500	−4500	−4500	−4500	−4500	−4500	−4500	−4500	−4500	−4500	−4500	−4500
Functioning Costs (NPV) Accumulated	−27565	−28231	−28818	−29334	−29789	−30189	−30542	−30852	−31126	−31366	−31578	−31765
Energy	−1000	−1000	−1000	−1000	−1000	−1000	−1000	−1000	−1000	−1000	−1000	−1000
Energy Accumulated (NPV) Accumulated	−6125	−6274	−6404	−6519	−6620	−6709	−6787	−6856	−6917	−6970	−7017	−7059
Maintenance Costs	−300	−300	−300	−300	−300	−300	−300	−300	−300	−300	−300	−300
Maintenance Costs (NPV) Accumulated	−1856	−1901	−1940	−1975	−2006	−2033	−2056	−2077	−2096	−2112	−2126	−2139
Total Costs (NPV) Accumulated	−292517	−323649	−355407	−387716	−420511	−453733	−487332	−521262	−555483	−589961	−624666	−659569
FT Net Present Value (NPV) (With II Devaluated)	95267	73800	50551	25733	−466	−27880	−56366	−85794	−116051	−147040	−178672	−210870
FT Net Present Value (With II New Asset)	25937	−5460	−39926	−77430	−117989	−161673	−208599	−258940	−312922	−370830	−433016	−499897

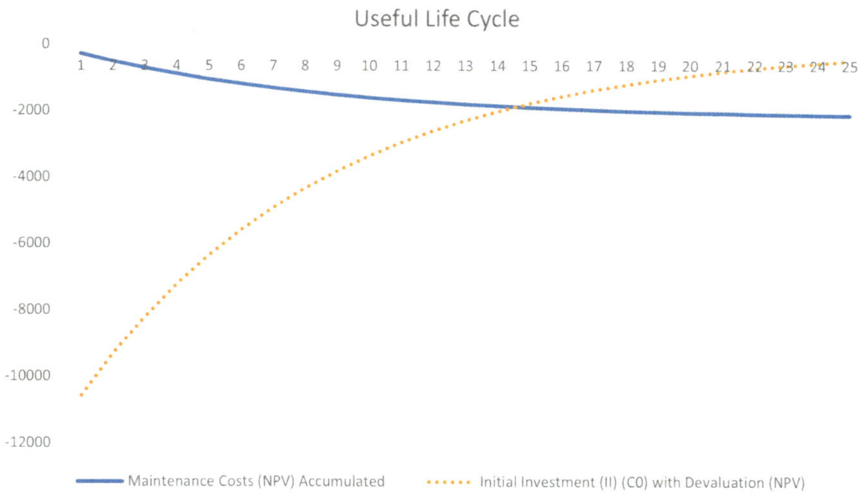

FIGURE 9.32 Physical Asset useful life.

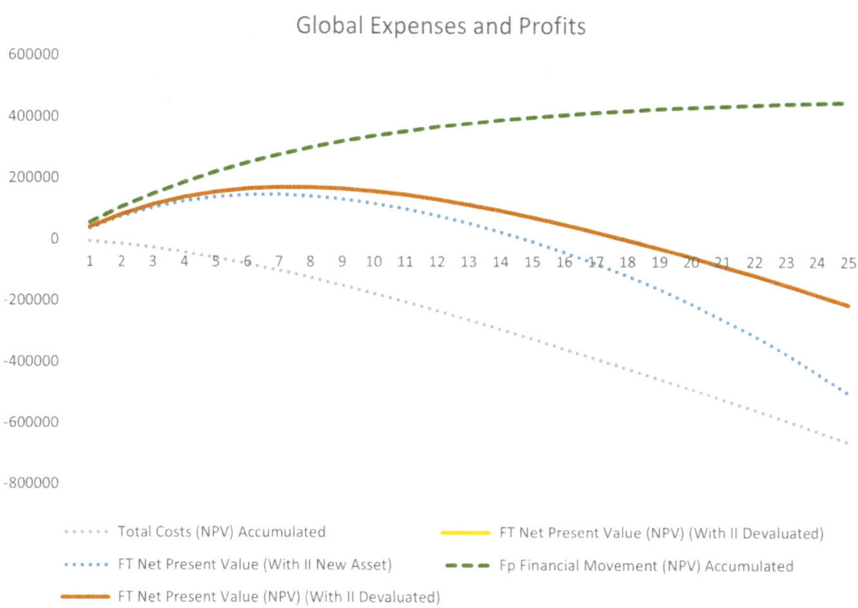

FIGURE 9.33 Global LCI analysis.

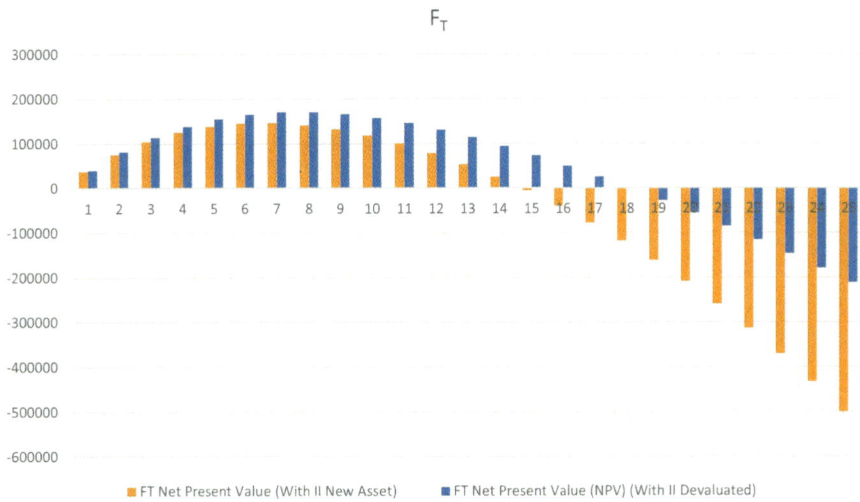

FIGURE 9.34 Life cycle analysis global result.

As can be seen, when comparing Figures 9.34 and 9.37, with the decreasing of the Energy expenses, the global result is more positive with the lower Energy costs.

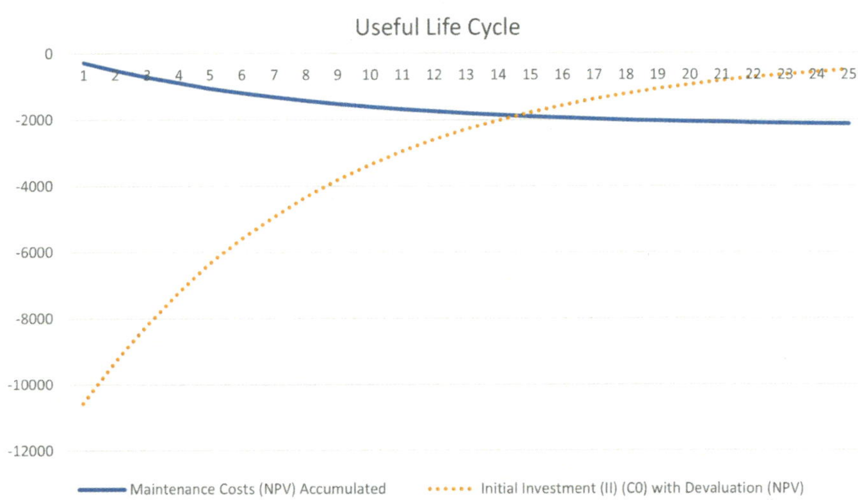

FIGURE 9.35 Physical Asset useful life.

TABLE 9.10
Values Used for Simulation based on Life Cycle with Recovery Related to Energy Standards. Second Simulation

Time (t)	0	1	2	3	4	5	6	7	8	9	10	11	12	13
r - real interest rate $(1+r)$		1.05	1.05	1.05	1.05	1.05	1.05	1.05	1.05	1.05	1.05	1.05	1.05	1.05
h - inflation rate $(1+h)$		1.02	1.02	1.02	1.02	1.02	1.02	1.02	1.02	1.02	1.02	1.02	1.02	1.02
p - profit rate		1.05	1.05	1.05	1.05	1.05	1.05	1.05	1.05	1.05	1.05	1.05	1.05	1.05
i – money rate (nominal interest rate)		0.12	0.12	0.12	0.12	0.12	0.12	0.12	0.12	0.12	0.12	0.12	0.12	0.12
i – Apparent Rate with Risk		0.14	0.14	0.14	0.14	0.14	0.14	0.14	0.14	0.14	0.14	0.14	0.14	0.14
$1/((1+i)\wedge t))$		0.88	0.78	0.68	0.60	0.53	0.47	0.41	0.36	0.32	0.28	0.25	0.22	0.19
Initial Investment (II) (C_0)	−12000	−12000	−12000	−12000	−12000	−12000	−12000	−12000	−12000	−12000	−12000	−12000	−12000	−12000
Initial Investment (II) (C_0) with Devaluation (NPV)	−12000	−10565	−9302	−8190	−7211	−6349	−5590	−4921	−4333	−3815	−3359	−2957	−2604	−2292
Initial Investment (II) (C_0) considering NEW Asset (NPV)	−12000	−13630	−15480	−17583	−19970	−22682	−25762	−29261	−33234	−37747	−42873	−48695	−55307	−62818
F_p Financial Movement (Production)	70000	70000	70000	70000	70000	70000	70000	70000	70000	70000	70000	70000	70000	70000
F_p Financial Movement (NPV)	70000	56028	49329	43432	38239	33667	29642	26098	22978	20230	17812	15682	13807	12156
F_p Financial Movement (NPV) Accumulated	70000	56028	105357	148789	187028	220695	250336	276434	299412	319642	337454	353136	366943	379099
Functioning Costs	−4500	−4500	−4500	−4500	−4500	−4500	−4500	−4500	−4500	−4500	−4500	−4500	−4500	−4500
Functioning Costs (NPV) Accumulated	−4500	−3962	−7450	−10521	−13226	−15606	−17702	−19548	−21173	−22603	−23863	−24972	−25948	−26808
Energy	−200	−200	−200	−200	−200	−200	−200	−200	−200	−200	−200	−200	−200	−200
Energy Accumulated (NPV) Accumulated	−200	−176	−331	−468	−588	−694	−787	−869	−941	−1005	−1061	−1110	−1153	−1191
Maintenance Costs	−300	−300	−300	−300	−300	−300	−300	−300	−300	−300	−300	−300	−300	−300
Maintenance Costs (NPV) Accumulated	−300	−267	−502	−708	−891	−1051	−1192	−1316	−1426	−1522	−1607	−1681	−1747	−1805
Total Costs (NPV) Accumulated	−5000	−4429	−12581	−24011	−38327	−55184	−74278	−95342	−118140	−142466	−168135	−194988	−222884	−251697
F_T Net Present Value (NPV) (With II Devaluated)		41034	83475	116588	141490	159162	170469	176171	176938	173362	165960	155190	141456	125111
F_T Net Present Value (With II New Asset)		37970	77296	107196	128731	142829	150296	151832	148037	139430	126446	109453	88752	64585

(Continued)

TABLE 9.10 (*Continued*)
Values Used for Simulation based on Life Cycle with Recovery Related to Energy Standards. Second Simulation

Time (t)	14	15	16	17	18	19	20	21	22	23	24	25
r - real interest rate $(1+r)$	1.05	1.05	1.05	1.05	1.05	1.05	1.05	1.05	1.05	1.05	1.05	1.05
h - inflation rate $(1+h)$	1.02	1.02	1.02	1.02	1.02	1.02	1.02	1.02	1.02	1.02	1.02	1.02
p - profit rate	1.05	1.05	1.05	1.05	1.05	1.05	1.05	1.05	1.05	1.05	1.05	1.05
i – money rate (nominal interest rate)	0.12	0.12	0.12	0.12	0.12	0.12	0.12	0.12	0.12	0.12	0.12	0.12
i – Apparent Rate with Risk	0.14	0.14	0.14	0.14	0.14	0.14	0.14	0.14	0.14	0.14	0.14	0.14
$1/((1+i)\wedge t))$	0.17	0.15	0.13	0.11	0.10	0.09	0.08	0.07	0.06	0.05	0.05	0.04
Initial Investment (II) (C_0)	−12000	−12000	−12000	−12000	−12000	−12000	−12000	−12000	−12000	−12000	−12000	−12000
Initial Investment (II) (C_0) with Devaluation (NPV)	−2018	−1777	−1565	−1377	−1213	−1068	−940	−828	−729	−642	−565	−497
Initial Investment (II) (C_0) considering NEW Asset (NPV)	−71348	−81037	−92041	−104540	−118736	−134860	−153174	−173974	−197599	−224432	−254909	−289524
F_p Financial Movement (Production)	70000	70000	70000	70000	70000	70000	70000	70000	70000	70000	70000	70000
F_p Financial Movement (NPV)	10703	9423	8297	7305	6431	5662	4985	4389	3865	3403	2996	2638
F_p Financial Movement (NPV) Accumulated	389802	399226	407522	414827	421258	426921	431906	436296	440160	443563	446559	449196
Functioning Costs	−4500	−4500	−4500	−4500	−4500	−4500	−4500	−4500	−4500	−4500	−4500	−4500
Functioning Costs (NPV) Accumulated	−27565	−28231	−28818	−29334	−29789	−30189	−30542	−30852	−31126	−31366	−31578	−31765
Energy	−200	−200	−200	−200	−200	−200	−200	−200	−200	−200	−200	−200
Energy Accumulated (NPV) Accumulated	−1225	−1255	−1281	−1304	−1324	−1342	−1357	−1371	−1383	−1394	−1403	−1412
Maintenance Costs	−300	−300	−300	−300	−300	−300	−300	−300	−300	−300	−300	−300
Maintenance Costs (NPV) Accumulated	−1856	−1901	−1940	−1975	−2006	−2033	−2056	−2077	−2096	−2112	−2126	−2139
Total Costs (NPV) Accumulated	−281317	−311649	−342607	−374116	−406111	−438533	−471332	−504462	−537883	−571561	−605466	−639569
F_T Net Present Value (NPV) (With II Devaluated)	106467	85800	63351	39333	13934	−12680	−40366	−68994	−98451	−128640	−159472	−190870
F_T Net Present Value (With II New Asset)	37137	6540	−27126	−63830	−103589	−146473	−192599	−242140	−295322	−352430	−413816	−479897

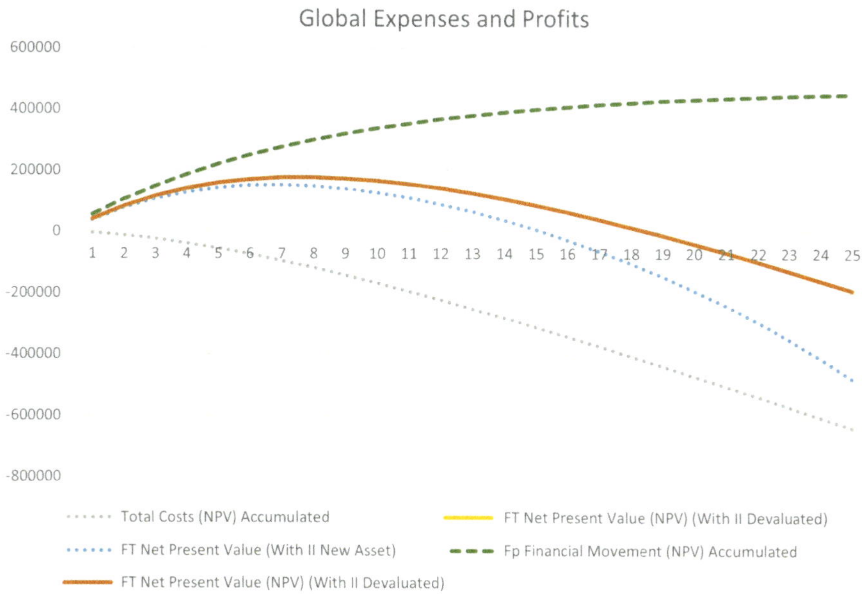

FIGURE 9.36 Global LCI analysis.

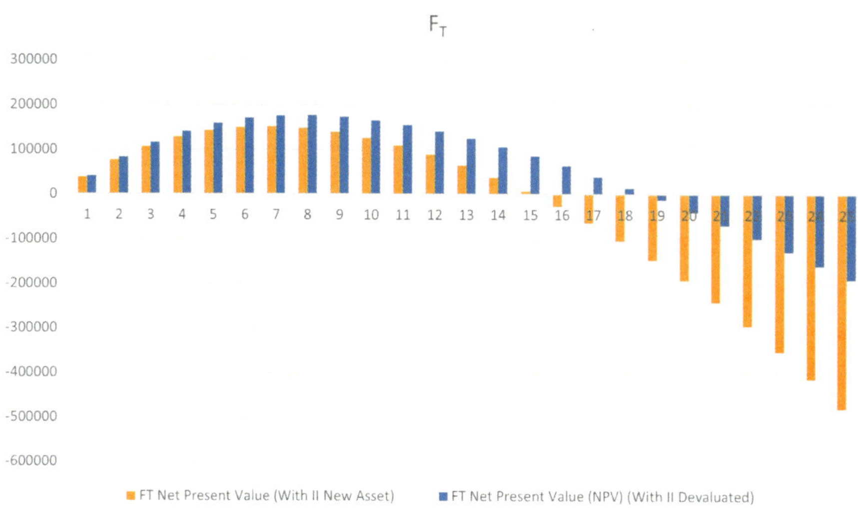

FIGURE 9.37 Life cycle analysis global result.

10 Environmental Norms

10.1 INTRODUCTION

This chapter deals with the Environmental Standards, namely the ISO 14001, according to the next main sections:

- Characterization of Environmental Norms;
- Diagnosis to Implement Environmental Norms;
- Quantitative Evaluation about the Implementation of Environmental Norms.

10.2 CHARACTERIZATION OF ENVIRONMENTAL NORMS

The ISO 14001 (Environmental management systems – Requirements with guidance for use) presents a systematic approach to environmental management aiming to provide the top management with information to build success over the long term and create options for contributing to sustainable development by:

- protecting the environment by preventing or mitigating adverse environmental impacts;
- mitigating the potential adverse effect of environmental conditions on the organization;
- assisting the organization in the fulfilment of compliance obligations;
- enhancing environmental performance;
- controlling or influencing the way the organization's products and services are designed, manufactured, distributed, consumed and disposed by using a life cycle perspective that can prevent environmental impacts from being unintentionally shifted elsewhere within the life cycle;
- achieving financial and operational benefits that can result from implementing environmentally sound alternatives that strengthen the organization's market position;
- communicating environmental information to relevant interested parties.

According to ISO 14004 (Environmental management systems – General guidelines on implementation), "sustainable development is a goal achieved by balancing the three pillars of sustainability: the environment, society and the economy".

The aim of this International Standard is to provide organizations with guidance for a common framework, in order to establish, implement, maintain and continually improve a system to support better environmental management".

The outcomes of a systematic approach to environmental management provide the top management with quantitative and qualitative data that enables informed

DOI: 10.1201/9781003395690-13

business decisions that build long-term success and create options for contributing to sustainable development. The success of the environmental management system depends on commitment from all levels and functions of the organization, led by top management. The opportunities include:

- protecting the environment, including the prevention or reduction of adverse environmental impacts;
- controlling or influencing the way products and services are designed, manufactured, distributed, used and disposed;
- using a life cycle perspective to prevent environmental impacts from being unintentionally shifted elsewhere within the cycle;
- achieving financial and operational benefits that can result from implementing environmentally sound alternatives which strengthen the organization's market position;
- communicating environmental information to relevant interested parties.

The ISO 14005 (Environmental management systems – Guidelines for a flexible approach to phased implementation) gives guidelines for a phased approach to establish, implement, maintain and improve an Environmental Management System (EMS) that organizations, including small and medium-sized enterprises (SMEs), can adopt to enhance their environmental performance.

The phased approach provides flexibility that allows organizations to develop their EMS at their own pace, over a number of phases, according to their own circumstances. Each phase consists of six consecutive stages. The system's maturity at the end of each phase can be characterized using a five-level maturity matrix (Woellner *et al.*, 2020).

10.3 DIAGNOSIS TO IMPLEMENT ENVIRONMENTAL NORMS

To implement Environmental Norms, the proposed methodology is like the other standards (see Section 6.3), that is to start with the diagnosis before giving the next step (Figure 10.1).

10.4 QUANTITATIVE EVALUATION ABOUT THE IMPLEMENTATION OF ENVIRONMENTAL NORMS

10.4.1 Life Cycle Cost of Physical Assets

To simulate the quantitative relevance about implementing this standard, the same methodology given in the preceding chapters is considered.

10.4.1.1 Uniform Annual Income Method (UAI)

The first simulation uses the following values:

- Initial Investment (II) –12000
- Life Time simulated 25
- Residual value of equipment (RV) 10

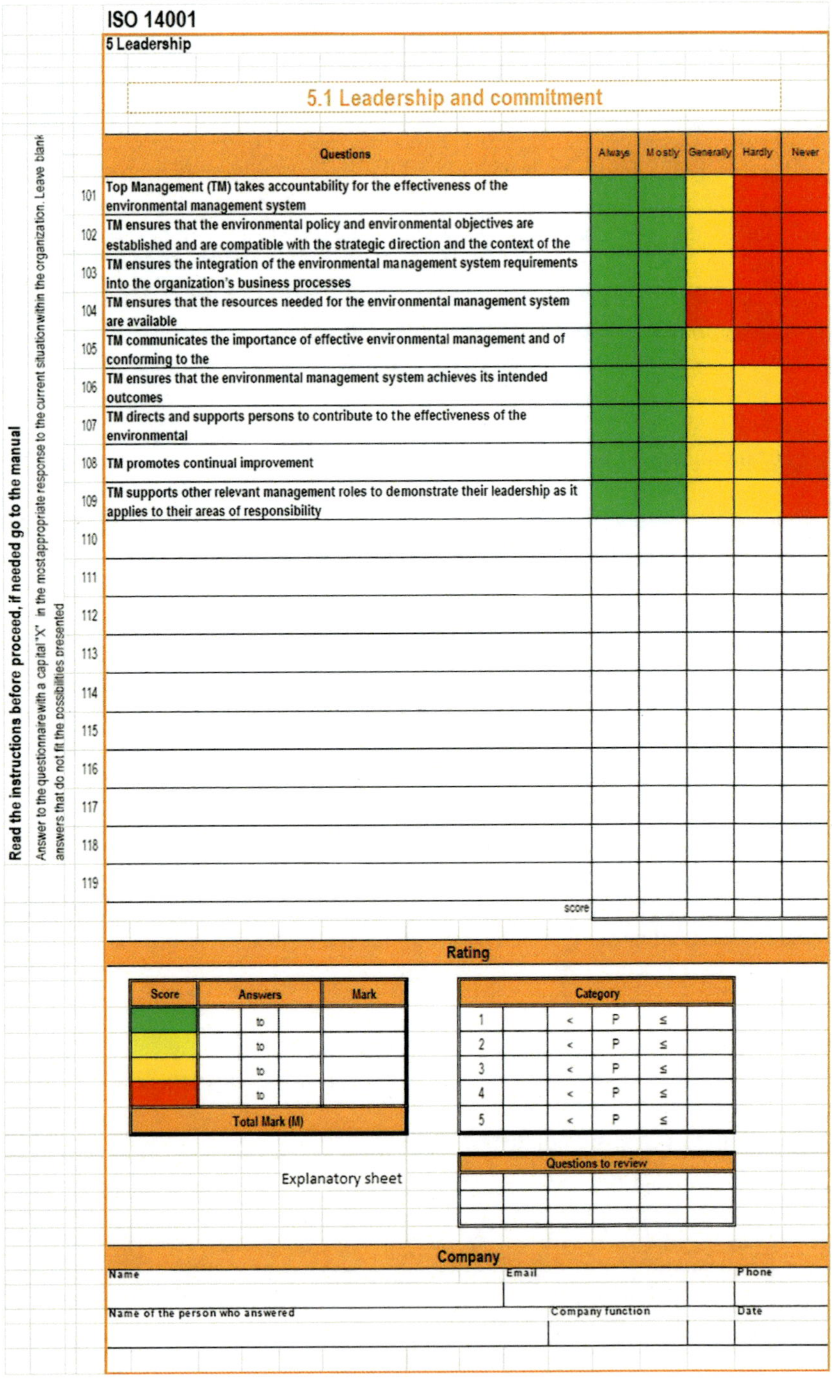

Read the instructions before proceed, if needed go to the manual

Answer to the questionnaire with a capital "X" in the most appropriate response to the current situation within the organization. Leave blank answers that do not fit the possibilities presented

ISO 14001

5 Leadership

5.1 Leadership and commitment

	Questions	Always	Mostly	Generally	Hardly	Never
101	Top Management (TM) takes accountability for the effectiveness of the environmental management system					
102	TM ensures that the environmental policy and environmental objectives are established and are compatible with the strategic direction and the context of the					
103	TM ensures the integration of the environmental management system requirements into the organization's business processes					
104	TM ensures that the resources needed for the environmental management system are available					
105	TM communicates the importance of effective environmental management and of conforming to the					
106	TM ensures that the environmental management system achieves its intended outcomes					
107	TM directs and supports persons to contribute to the effectiveness of the environmental					
108	TM promotes continual improvement					
109	TM supports other relevant management roles to demonstrate their leadership as it applies to their areas of responsibility					
110						
111						
112						
113						
114						
115						
116						
117						
118						
119						
	score					

Rating

Score	Answers			Mark
		to		
		to		
		to		
		to		
Total Mark (M)				

Category					
1		<	P	≤	
2		<	P	≤	
3		<	P	≤	
4		<	P	≤	
5		<	P	≤	

Explanatory sheet

Questions to review

Company

Name	Email	Phone

Name of the person who answered	Company function	Date

FIGURE 10.1 Diagnosis sheet.

- Functioning (*F*) –4500
- Energy –200
- Maintenance (*M*) –300
- Non Production (NP) 0
- Benefit (*B*) **50000**
- Internal Rate of Return (*r*) 0.12
- Risk (*R*) 0.01
- Apparent Rate 0.1312
- Maintenance Increase Rate 0.01
- Availability 0.8929
- MTBF 500
- MTTR 50
- MWT 10

The remaining values are presented in Table 10.1, and the graphical results are shown in Figures 10.2–10.5.

As referred in the preceding introduction to this section, the evaluation of introduction of Environmental Norms, the simulation of their implementation is made through the variation of value of benefits. Other indicators may be used, but may not be so intuitive.

Table 10.1 and Figures 10.2–10.4 present the simulation considering a benefit of 50000 monetary units.

The second simulation uses the following values (the value of energy is decreased):

- Initial Investment (II) –12000
- Life Time simulated 25
- Residual value of equipment (RV) 10
- Functioning (*F*) –4500
- Energy –200
- Maintenance (*M*) –300
- Non Production (NP) 0
- Benefit (*B*) **60000**
- Internal Rate of Return (*r*) 0.12
- Risk (*R*) 0.01
- Apparent Rate 0.1312
- Maintenance Increase Rate 0.01
- Availability 0.8929
- MTBF 500
- MTTR 50
- MWT 10

The remaining values are presented in Table 10.2, and the graphical results are shown in Figures 10.6–10.9.

As referred in the preceding introduction to this section, the evaluation of introduction of Environmental Norms, the simulation of their implementation is made

TABLE 10.1
Values Used for Simulation based on UAI Method Related to Environmental Standards. First Simulation

Time	0	1	2	3	4	5
Initial Investment (II)	−12000					
Initial Investment (II)	−12000					
II Present Value (IIPV) – Return	−12000	−10608.20	−9377.83	−8290.16	−7328.64	−6478.65
Life Time simulated	25					
Residual value of equipment (RV)	10					
Linear Depreciation Method (Vn)	−12000	−11519.60	−11039.20	−10558.80	−10078.40	−9598.00
LDM Present Value (Vn)		−10183.52	−8626.98	−7294.51	−6155.08	−5181.84
Sum of Digits Method (Vn)	−12000	−11076.15	−10189.26	−9339.32	−8526.34	−7750.31
SDM Present Value (Vn)		−9791.51	−7962.77	−6452.04	−5207.21	−4184.29
Exponential Method (Vn)	−12000	−9036.79	−6805.29	−5124.83	−3859.33	−2906.33
EM Present Value (Pv)		−7988.67	−5318.24	−3540.47	−2356.97	−1569.09
Functioning (*F*)	−4500	−4500.00	−4545.00	−4590.45	−4636.35	−4682.72
Present Value Accumulated (NPV)		−3978.08	−7529.93	−10701.23	−13532.74	−16060.88
Present Value (NPV)		−3978.08	−3551.85	−3171.30	−2831.52	−2528.14
Energy	−200	−200.00	−400.00	−600.00	−800.00	−1000.00
Energy (NPV)		−176.80	−312.59	−414.51	−488.58	−539.89
Accumulated (FPV)		−16154.88	−23997.40	−35113.14	−49134.46	−65735.23
Maintenance (*M*)	−300	−300.00	−303.00	−306.03	−309.09	−312.18
Present Value (MPV)		−265.21	−236.79	−211.42	−188.77	−168.54
Accumulated (MPV)		−265.21	−502.00	−713.42	−902.18	−1070.73
Non Production (NP)	0	0.00	0.00	0.00	0.00	0.00
Present Value (NPPV)		0.00	0.00	0.00	0.00	0.00
Accumulated NPPV		0.00	0.00	0.00	0.00	0.00
Accumulated Total Costs (APV)		−16420.08	−24499.40	−35826.56	−50036.64	−66805.96
Accumulated Total Costs		−5000.00	−10248.00	−15744.48	−21489.92	−27484.82
Benefit (*B*)	50000	44642.86	44642.86	44642.86	44642.86	44642.86
Present Value (BPV)		39465.04	34887.77	30841.38	27264.30	24102.11
Accumulated Benefits BPV		39465.04	74352.81	105194.19	132458.49	156560.60
BPV + FPV + MPV + NPPV + IIPV		11044.96	37853.41	57367.63	70421.85	77754.64
Uniform Annual Income UAIn (LDM)		−11519.60	−5179.80	−3095.68	−2074.47	−1477.59
Uniform Annual Income UAIn (SDM)		−11076.15	−4781.00	−2738.15	−1755.00	−1193.14
Uniform Annual Income UAIn (EM)		−9036.79	−3193.17	−1502.52	−794.38	−447.42

(*Continued*)

TABLE 10.1 (*Continued*)
Values Used for Simulation based on UAI Method Related to Environmental Standards. First Simulation

Time	6	7	8	9	10
Initial Investment (II)					
Initial Investment (II)					
II Present Value (IIPV) – Return	−5727.23	−5062.97	−4475.75	−3956.64	−3497.74
Life Time simulated					
Residual value of equipment (RV)					
Linear Depreciation Method (Vn)	−9117.60	−8637.20	−8156.80	−7676.40	−7196.00
LDM Present Value (Vn)	−4351.55	−3644.16	−3042.32	−2531.06	−2097.48
Sum of Digits Method (Vn)	−7011.23	−6309.11	−5643.94	−5015.72	−4424.46
SDM Present Value (Vn)	−3346.25	−2661.90	−2105.07	−1653.78	−1289.63
Exponential Method (Vn)	−2188.66	−1648.20	−1241.20	−934.71	−703.90
EM Present Value (Pv)	−1044.58	−695.40	−462.94	−308.19	−205.17
Functioning (*F*)	−4729.55	−4776.84	−4824.61	−4872.86	−4921.58
Present Value Accumulated (NPV)	−18318.15	−20333.57	−22133.05	−23739.73	−25174.26
Present Value (NPV)	−2257.27	−2015.42	−1799.48	−1606.68	−1434.53
Energy	−1200.00	−1400.00	−1600.00	−1800.00	−2000.00
Energy (NPV)	−572.72	−590.68	−596.77	−593.50	−582.96
Accumulated (FPV)	−84626.11	−105550.36	−128280.17	−152613.40	−178370.61
Maintenance (*M*)	−315.30	−318.46	−321.64	−324.86	−328.11
Present Value (MPV)	−150.48	−134.36	−119.97	−107.11	−95.64
Accumulated (MPV)	−1221.21	−1355.57	−1475.54	−1582.65	−1678.28
Non Production (NP)	0.00	0.00	0.00	0.00	0.00
Present Value (NPPV)	0.00	0.00	0.00	0.00	0.00
Accumulated NPPV	0.00	0.00	0.00	0.00	0.00
Accumulated Total Costs (APV)	−85847.32	−106905.93	−129755.71	−154196.04	−180048.90
Accumulated Total Costs	−33729.67	−40224.97	−46971.22	−53968.93	−61218.62
Benefit (*B*)	44642.86	44642.86	44642.86	44642.86	44642.86
Present Value (BPV)	21306.67	18835.46	16650.87	14719.65	13012.42
Accumulated Benefits BPV	177867.27	196702.73	213353.60	228073.24	241085.66
BPV + FPV + MPV + NPPV + IIPV	80019.95	77796.80	71597.89	61877.20	49036.77
Uniform Annual Income UAIn (LDM)	−1092.19	−827.06	−636.59	−495.43	−388.40
Uniform Annual Income UAIn (SDM)	−839.87	−604.13	−440.47	−323.71	−238.81
Uniform Annual Income UAIn (EM)	−262.18	−157.83	−96.87	−60.33	−37.99

(*Continued*)

TABLE 10.1 (*Continued*)
Values Used for Simulation based on UAI Method Related to Environmental Standards. First Simulation

Time	11	12	13	14	15
Initial Investment (II)					
Initial Investment (II)					
II Present Value (IIPV) – Return	−3092.06	−2733.43	−2416.40	−2136.14	−1888.38
Life Time simulated					
Residual value of equipment (RV)					
Linear Depreciation Method (Vn)	−6715.60	−6235.20	−5754.80	−5274.40	−4794.00
LDM Present Value (Vn)	−1730.42	−1420.29	−1158.83	−938.90	−754.41
Sum of Digits Method (Vn)	−3870.15	−3352.80	−2872.40	−2428.95	−2022.46
SDM Present Value (Vn)	−997.23	−763.72	−578.41	−432.38	−318.27
Exponential Method (Vn)	−530.08	−399.19	−300.61	−226.38	−170.48
EM Present Value (Pv)	−136.59	−90.93	−60.53	−40.30	−26.83
Functioning (*F*)	−4970.80	−5020.51	−5070.71	−5121.42	−5172.63
Present Value Accumulated (NPV)	−26455.10	−27598.70	−28619.77	−29531.44	−30345.44
Present Value (NPV)	−1280.83	−1143.60	−1021.07	−911.67	−813.99
Energy	−2200.00	−2400.00	−2600.00	−2800.00	−3000.00
Energy (NPV)	−566.88	−546.69	−523.55	−498.43	−472.10
Accumulated (FPV)	−205392.59	−233537.97	−262681.29	−292711.17	−323528.70
Maintenance (*M*)	−331.39	−334.70	−338.05	−341.43	−344.84
Present Value (MPV)	−85.39	−76.24	−68.07	−60.78	−54.27
Accumulated (MPV)	−1763.67	−1839.91	−1907.98	−1968.76	−2023.03
Non Production (NP)	0.00	0.00	0.00	0.00	0.00
Present Value (NPPV)	0.00	0.00	0.00	0.00	0.00
Accumulated NPPV	0.00	0.00	0.00	0.00	0.00
Accumulated Total Costs (APV)	−207156.26	−235377.88	−264589.28	−294679.93	−325551.73
Accumulated Total Costs	−68720.81	−76476.01	−84484.77	−92747.62	−101265.10
Benefit (*B*)	44642.86	44642.86	44642.86	44642.86	44642.86
Present Value (BPV)	11503.20	10169.02	8989.59	7946.95	7025.24
Accumulated Benefits BPV	252588.86	262757.88	271747.47	279694.42	286719.66
BPV + FPV + MPV + NPPV + IIPV	33432.60	15380.00	−4841.81	−26985.51	−50832.07
Uniform Annual Income UAIn (LDM)	−305.84	−241.31	−190.37	−149.86	−117.46
Uniform Annual Income UAIn (SDM)	−176.25	−129.76	−95.02	−69.01	−49.55
Uniform Annual Income UAIn (EM)	−24.14	−15.45	−9.94	−6.43	−4.18

(*Continued*)

TABLE 10.1 (*Continued*)
Values Used for Simulation based on UAI Method Related to Environmental Standards. First Simulation

Time	16	17	18	19	20
Initial Investment (II)					
Initial Investment (II)					
II Present Value (IIPV) – Return	−1669.36	−1475.75	−1304.58	−1153.27	−1019.51
Life Time simulated					
Residual value of equipment (RV)					
Linear Depreciation Method (Vn)	−4313.60	−3833.20	−3352.80	−2872.40	−2392.00
LDM Present Value (Vn)	−600.08	−471.40	−364.50	−276.06	−203.22
Sum of Digits Method (Vn)	−1652.92	−1320.34	−1024.71	−766.03	−544.31
SDM Present Value (Vn)	−229.94	−162.37	−111.40	−73.62	−46.24
Exponential Method (Vn)	−128.38	−96.68	−72.81	−54.83	−41.29
EM Present Value (Pv)	−17.86	−11.89	−7.92	−5.27	−3.51
Functioning (*F*)	−5224.36	−5276.60	−5329.37	−5382.66	−5436.49
Present Value Accumulated (NPV)	−31072.22	−31721.13	−32300.51	−32817.82	−33279.70
Present Value (NPV)	−726.78	−648.91	−579.38	−517.31	−461.88
Energy	−3200.00	−3400.00	−3600.00	−3800.00	−4000.00
Energy (NPV)	−445.16	−418.13	−391.38	−365.20	−339.84
Accumulated (FPV)	−355046.08	−387185.33	−419877.22	−453060.24	−486679.78
Maintenance (*M*)	−348.29	−351.77	−355.29	−358.84	−362.43
Present Value (MPV)	−48.45	−43.26	−38.63	−34.49	−30.79
Accumulated (MPV)	−2071.48	−2114.74	−2153.37	−2187.85	−2218.65
Non Production (NP)	0.00	0.00	0.00	0.00	0.00
Present Value (NPPV)	0.00	0.00	0.00	0.00	0.00
Accumulated NPPV	0.00	0.00	0.00	0.00	0.00
Accumulated Total Costs (APV)	−357117.56	−389300.07	−422030.59	−455248.09	−488898.42
Accumulated Total Costs	−110037.75	−119066.13	−128350.79	−137892.30	−147691.22
Benefit (*B*)	44642.86	44642.86	44642.86	44642.86	44642.86
Present Value (BPV)	6210.43	5490.13	4853.36	4290.46	3792.84
Accumulated Benefits BPV	292930.09	298420.21	303273.58	307564.03	311356.87
BPV + FPV + MPV + NPPV + IIPV	−76187.47	−102879.86	−130757.01	−159684.06	−189541.55
Uniform Annual Income UAIn (LDM)	−91.45	−70.52	−53.66	−40.07	−29.14
Uniform Annual Income UAIn (SDM)	−35.04	−24.29	−16.40	−10.69	−6.63
Uniform Annual Income UAIn (EM)	−2.72	−1.78	−1.17	−0.76	−0.50

(*Continued*)

TABLE 10.1 (*Continued*)
Values Used for Simulation based on UAI Method Related to Environmental Standards. First Simulation

Time	21	22	23	24	25
Initial Investment (II)					
Initial Investment (II)					
II Present Value (IIPV) – Return	−901.27	−796.74	−704.33	−622.64	−550.42
Life Time simulated					
Residual value of equipment (RV)					
Linear Depreciation Method (Vn)	−1911.60	−1431.20	−950.80	−470.40	10.00
LDM Present Value (Vn)	−143.57	−95.02	−55.81	−24.41	0.46
Sum of Digits Method (Vn)	−359.54	−211.72	−100.86	−26.95	10.00
SDM Present Value (Vn)	−27.00	−14.06	−5.92	−1.40	0.46
Exponential Method (Vn)	−31.09	−23.42	−17.63	−13.28	−10.00
EM Present Value (Pv)	−2.34	−1.55	−1.03	−0.69	−0.46
Functioning (*F*)	−5490.86	−5545.76	−5601.22	−5657.23	−5713.81
Present Value Accumulated (NPV)	−33692.09	−34060.30	−34389.06	−34682.60	−34944.68
Present Value (NPV)	−412.39	−368.21	−328.76	−293.53	−262.08
Energy	−4200.00	−4400.00	−4600.00	−4800.00	−5000.00
Energy (NPV)	−315.44	−292.14	−269.99	−249.06	−229.34
Accumulated (FPV)	−520687.31	−555039.75	−589698.81	−624630.46	−659804.48
Maintenance (*M*)	−366.06	−369.72	−373.41	−377.15	−380.92
Present Value (MPV)	−27.49	−24.55	−21.92	−19.57	−17.47
Accumulated (MPV)	−2246.14	−2270.69	−2292.60	−2312.17	−2329.65
Non Production (NP)	0.00	0.00	0.00	0.00	0.00
Present Value (NPPV)	0.00	0.00	0.00	0.00	0.00
Accumulated NPPV	0.00	0.00	0.00	0.00	0.00
Accumulated Total Costs (APV)	−522933.45	−557310.44	−591991.41	−626942.63	−662134.13
Accumulated Total Costs	−157748.13	−168063.61	−178638.25	−189472.63	−200567.36
Benefit (*B*)	44642.86	44642.86	44642.86	44642.86	44642.86
Present Value (BPV)	3352.93	2964.05	2620.27	2316.36	2047.70
Accumulated Benefits BPV	314709.80	317673.85	320294.12	322610.48	324658.18
BPV + FPV + MPV + NPPV + IIPV	−220223.65	−251636.59	−283697.29	−316332.15	−349475.94
Uniform Annual Income UAIn (LDM)	−20.37	−13.35	−7.78	−3.38	0.06
Uniform Annual Income UAIn (SDM)	−3.83	−1.98	−0.83	−0.19	0.06
Uniform Annual Income UAIn (EM)	−0.33	−0.22	−0.14	−0.10	−0.06

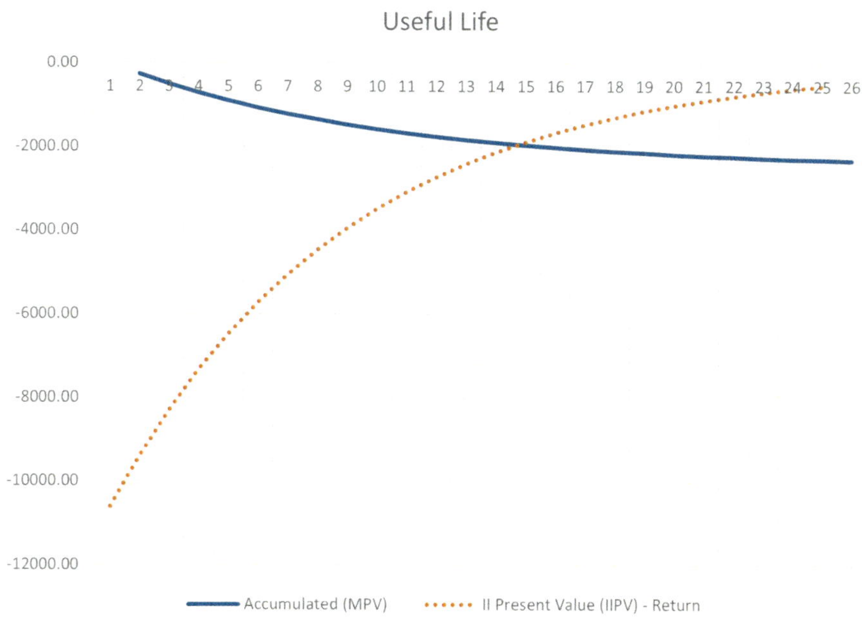

FIGURE 10.2 Physical Asset useful life.

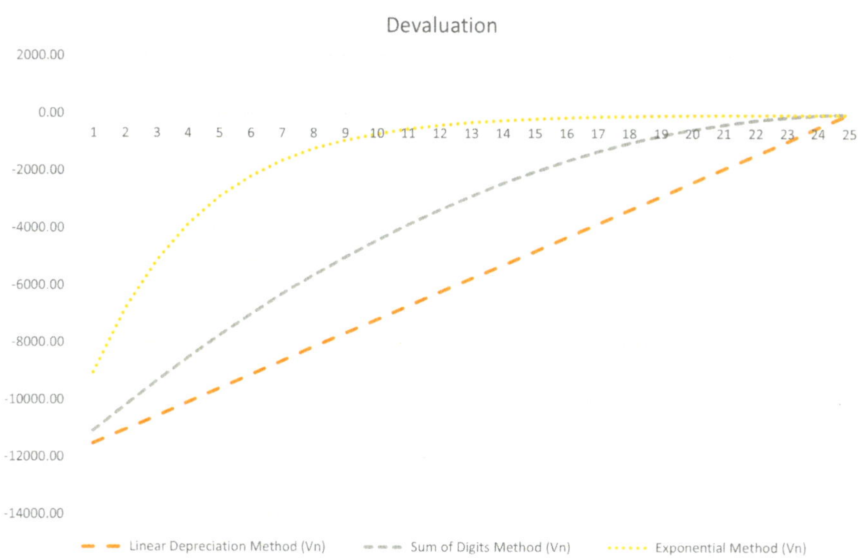

FIGURE 10.3 The three devaluation methods.

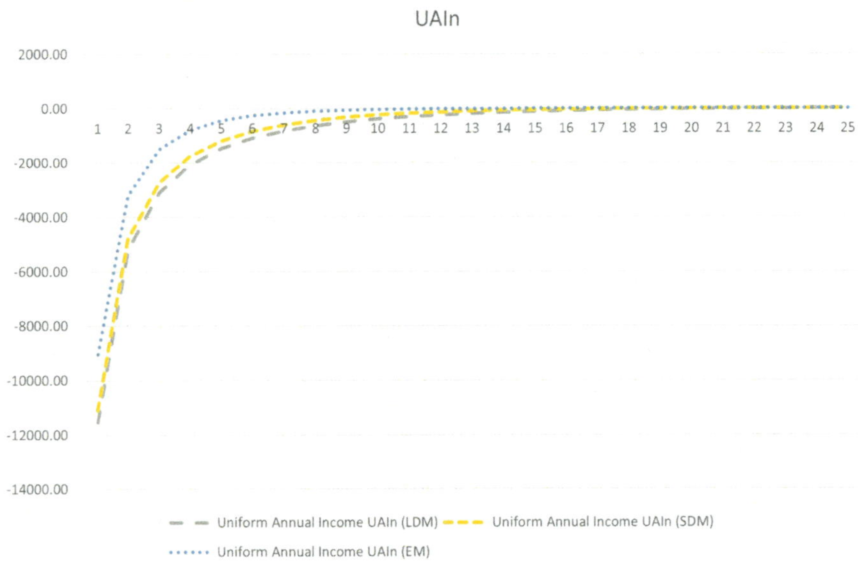

FIGURE 10.4 The UAI results from the three devaluation methods.

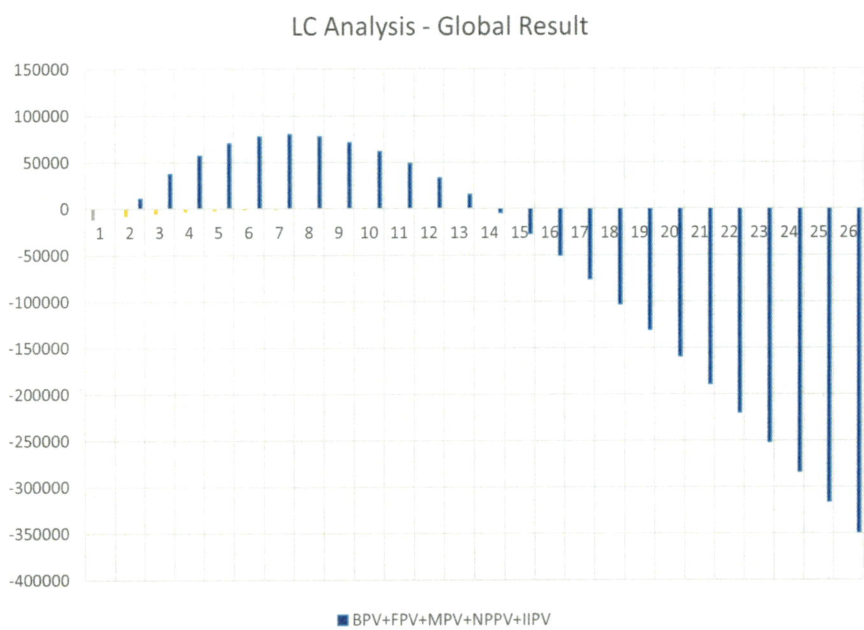

FIGURE 10.5 Life cycle analysis global result including profits.

TABLE 10.2
Values Used for Simulation based on UAI Method Related to Environmental Standards. Second Simulation

Time	0	1	2	3	4	5
Initial Investment (II)	−12000					
Initial Investment (II)	−12000					
II Present Value (IIPV) – Return	−12000	−10608.20	−9377.83	−8290.16	−7328.64	−6478.65
Life Time simulated	25					
Residual value of equipment (RV)	10					
Linear Depreciation Method (Vn)	−12000	−11519.60	−11039.20	−10558.80	−10078.40	−9598.00
LDM Present Value (Vn)		−10183.52	−8626.98	−7294.51	−6155.08	−5181.84
Sum of Digits Method (Vn)	−12000	−11076.15	−10189.26	−9339.32	−8526.34	−7750.31
SDM Present Value (Vn)		−9791.51	−7962.77	−6452.04	−5207.21	−4184.29
Exponential Method (Vn)	−12000	−9036.79	−6805.29	−5124.83	−3859.33	−2906.33
EM Present Value (Pv)		−7988.67	−5318.24	−3540.47	−2356.97	−1569.09
Functioning (*F*)	−4500	−4500.00	−4545.00	−4590.45	−4636.35	−4682.72
Present Value Accumulated (NPV)		−3978.08	−7529.93	−10701.23	−13532.74	−16060.88
Present Value (NPV)		−3978.08	−3551.85	−3171.30	−2831.52	−2528.14
Energy	−200	−200.00	−400.00	−600.00	−800.00	−1000.00
Energy (NPV)		−176.80	−312.59	−414.51	−488.58	−539.89
Accumulated (FPV)		−16154.88	−23997.40	−35113.14	−49134.46	−65735.23
Maintenance (*M*)	−300	−300.00	−303.00	−306.03	−309.09	−312.18
Present Value (MPV)		−265.21	−236.79	−211.42	−188.77	−168.54
Accumulated (MPV)		−265.21	−502.00	−713.42	−902.18	−1070.73
Non Production (NP)	0	0.00	0.00	0.00	0.00	0.00
Present Value (NPPV)		0.00	0.00	0.00	0.00	0.00
Accumulated NPPV		0.00	0.00	0.00	0.00	0.00
Accumulated Total Costs (APV)		−16420.08	−24499.40	−35826.56	−50036.64	−66805.96
Accumulated Total Costs		−5000.00	−10248.00	−15744.48	−21489.92	−27484.82
Benefit (*B*)	60000	53571.43	53571.43	53571.43	53571.43	53571.43
Present Value (BPV)		47358.05	41865.32	37009.66	32717.16	28922.53
Accumulated Benefits BPV		47358.05	89223.37	126233.03	158950.19	187872.72
BPV + FPV + MPV + NPPV + IIPV		18937.97	52723.97	78406.47	96913.55	109066.76
Uniform Annual Income UAIn (LDM)		−11519.60	−5179.80	−3095.68	−2074.47	−1477.59
Uniform Annual Income UAIn (SDM)		−11076.15	−4781.00	−2738.15	−1755.00	−1193.14
Uniform Annual Income UAIn (EM)		−9036.79	−3193.17	−1502.52	−794.38	−447.42

(Continued)

TABLE 10.2 (*Continued*)
Values Used for Simulation based on UAI Method Related to Environmental Standards. Second Simulation

Time	6	7	8	9	10
Initial Investment (II)					
Initial Investment (II)					
II Present Value (IIPV) – Return	−5727.23	−5062.97	−4475.75	−3956.64	−3497.74
Life Time simulated					
Residual value of equipment (RV)					
Linear Depreciation Method (Vn)	−9117.60	−8637.20	−8156.80	−7676.40	−7196.00
LDM Present Value (Vn)	−4351.55	−3644.16	−3042.32	−2531.06	−2097.48
Sum of Digits Method (Vn)	−7011.23	−6309.11	−5643.94	−5015.72	−4424.46
SDM Present Value (Vn)	−3346.25	−2661.90	−2105.07	−1653.78	−1289.63
Exponential Method (Vn)	−2188.66	−1648.20	−1241.20	−934.71	−703.90
EM Present Value (Pv)	−1044.58	−695.40	−462.94	−308.19	−205.17
Functioning (*F*)	−4729.55	−4776.84	−4824.61	−4872.86	−4921.58
Present Value Accumulated (NPV)	−18318.15	−20333.57	−22133.05	−23739.73	−25174.26
Present Value (NPV)	−2257.27	−2015.42	−1799.48	−1606.68	−1434.53
Energy	−1200.00	−1400.00	−1600.00	−1800.00	−2000.00
Energy (NPV)	−572.72	−590.68	−596.77	−593.50	−582.96
Accumulated (FPV)	−84626.11	−105550.36	−128280.17	−152613.40	−178370.61
Maintenance (*M*)	−315.30	−318.46	−321.64	−324.86	−328.11
Present Value (MPV)	−150.48	−134.36	−119.97	−107.11	−95.64
Accumulated (MPV)	−1221.21	−1355.57	−1475.54	−1582.65	−1678.28
Non Production (NP)	0.00	0.00	0.00	0.00	0.00
Present Value (NPPV)	0.00	0.00	0.00	0.00	0.00
Accumulated NPPV	0.00	0.00	0.00	0.00	0.00
Accumulated Total Costs (APV)	−85847.32	−106905.93	−129755.71	−154196.04	−180048.90
Accumulated Total Costs	−33729.67	−40224.97	−46971.22	−53968.93	−61218.62
Benefit (*B*)	53571.43	53571.43	53571.43	53571.43	53571.43
Present Value (BPV)	25568.01	22602.55	19981.04	17663.58	15614.90
Accumulated Benefits BPV	213440.73	236043.28	256024.31	273687.89	289302.79
BPV+FPV+MPV+NPPV+IIPV	115593.41	117137.35	114268.61	107491.85	97253.90
Uniform Annual Income UAIn (LDM)	−1092.19	−827.06	−636.59	−495.43	−388.40
Uniform Annual Income UAIn (SDM)	−839.87	−604.13	−440.47	−323.71	−238.81
Uniform Annual Income UAIn (EM)	−262.18	−157.83	−96.87	−60.33	−37.99

(*Continued*)

TABLE 10.2 (*Continued*)
Values Used for Simulation based on UAI Method Related to Environmental Standards. Second Simulation

Time	11	12	13	14	15
Initial Investment (II)					
Initial Investment (II)					
II Present Value (IIPV) – Return	−3092.06	−2733.43	−2416.40	−2136.14	−1888.38
Life Time simulated					
Residual value of equipment (RV)					
Linear Depreciation Method (Vn)	−6715.60	−6235.20	−5754.80	−5274.40	−4794.00
LDM Present Value (Vn)	−1730.42	−1420.29	−1158.83	−938.90	−754.41
Sum of Digits Method (Vn)	−3870.15	−3352.80	−2872.40	−2428.95	−2022.46
SDM Present Value (Vn)	−997.23	−763.72	−578.41	−432.38	−318.27
Exponential Method (Vn)	−530.08	−399.19	−300.61	−226.38	−170.48
EM Present Value (Pv)	−136.59	−90.93	−60.53	−40.30	−26.83
Functioning (*F*)	−4970.80	−5020.51	−5070.71	−5121.42	−5172.63
Present Value Accumulated (NPV)	−26455.10	−27598.70	−28619.77	−29531.44	−30345.44
Present Value (NPV)	−1280.83	−1143.60	−1021.07	−911.67	−813.99
Energy	−2200.00	−2400.00	−2600.00	−2800.00	−3000.00
Energy (NPV)	−566.88	−546.69	−523.55	−498.43	−472.10
Accumulated (FPV)	−205392.59	−233537.97	−262681.29	−292711.17	−323528.70
Maintenance (*M*)	−331.39	−334.70	−338.05	−341.43	−344.84
Present Value (MPV)	−85.39	−76.24	−68.07	−60.78	−54.27
Accumulated (MPV)	−1763.67	−1839.91	−1907.98	−1968.76	−2023.03
Non Production (NP)	0.00	0.00	0.00	0.00	0.00
Present Value (NPPV)	0.00	0.00	0.00	0.00	0.00
Accumulated NPPV	0.00	0.00	0.00	0.00	0.00
Accumulated Total Costs (APV)	−207156.26	−235377.88	−264589.28	−294679.93	−325551.73
Accumulated Total Costs	−68720.81	−76476.01	−84484.77	−92747.62	−101265.10
Benefit (*B*)	53571.43	53571.43	53571.43	53571.43	53571.43
Present Value (BPV)	13803.84	12202.83	10787.51	9536.34	8430.29
Accumulated Benefits BPV	303106.63	315309.46	326096.97	335633.31	344063.59
BPV + FPV + MPV + NPPV + IIPV	83950.37	67931.58	49507.69	28953.37	6511.86
Uniform Annual Income UAIn (LDM)	−305.84	−241.31	−190.37	−149.86	−117.46
Uniform Annual Income UAIn (SDM)	−176.25	−129.76	−95.02	−69.01	−49.55
Uniform Annual Income UAIn (EM)	−24.14	−15.45	−9.94	−6.43	−4.18

(*Continued*)

TABLE 10.2 (*Continued*)
Values Used for Simulation based on UAI Method Related to Environmental Standards. Second Simulation

Time	16	17	18	19	20
Initial Investment (II)					
Initial Investment (II)					
II Present Value (IIPV) – Return	−1669.36	−1475.75	−1304.58	−1153.27	−1019.51
Life Time simulated					
Residual value of equipment (RV)					
Linear Depreciation Method (Vn)	−4313.60	−3833.20	−3352.80	−2872.40	−2392.00
LDM Present Value (Vn)	−600.08	−471.40	−364.50	−276.06	−203.22
Sum of Digits Method (Vn)	−1652.92	−1320.34	−1024.71	−766.03	−544.31
SDM Present Value (Vn)	−229.94	−162.37	−111.40	−73.62	−46.24
Exponential Method (Vn)	−128.38	−96.68	−72.81	−54.83	−41.29
EM Present Value (Pv)	−17.86	−11.89	−7.92	−5.27	−3.51
Functioning (*F*)	−5224.36	−5276.60	−5329.37	−5382.66	−5436.49
Present Value Accumulated (NPV)	−31072.22	−31721.13	−32300.51	−32817.82	−33279.70
Present Value (NPV)	−726.78	−648.91	−579.38	−517.31	−461.88
Energy	−3200.00	−3400.00	−3600.00	−3800.00	−4000.00
Energy (NPV)	−445.16	−418.13	−391.38	−365.20	−339.84
Accumulated (FPV)	−355046.08	−387185.33	−419877.22	−453060.24	−486679.78
Maintenance (*M*)	−348.29	−351.77	−355.29	−358.84	−362.43
Present Value (MPV)	−48.45	−43.26	−38.63	−34.49	−30.79
Accumulated (MPV)	−2071.48	−2114.74	−2153.37	−2187.85	−2218.65
Non Production (NP)	0.00	0.00	0.00	0.00	0.00
Present Value (NPPV)	0.00	0.00	0.00	0.00	0.00
Accumulated NPPV	0.00	0.00	0.00	0.00	0.00
Accumulated Total Costs (APV)	−357117.56	−389300.07	−422030.59	−455248.09	−488898.42
Accumulated Total Costs	−110037.75	−119066.13	−128350.79	−137892.30	−147691.22
Benefit (*B*)	53571.43	53571.43	53571.43	53571.43	53571.43
Present Value (BPV)	7452.52	6588.15	5824.04	5148.55	4551.40
Accumulated Benefits BPV	351516.11	358104.26	363928.29	369076.84	373628.24
BPV + FPV + MPV + NPPV + IIPV	−17601.45	−43195.82	−70102.29	−98171.25	−127270.18
Uniform Annual Income UAIn (LDM)	−91.45	−70.52	−53.66	−40.07	−29.14
Uniform Annual Income UAIn (SDM)	−35.04	−24.29	−16.40	−10.69	−6.63
Uniform Annual Income UAIn (EM)	−2.72	−1.78	−1.17	−0.76	−0.50

(*Continued*)

TABLE 10.2 (*Continued*)

Values Used for Simulation based on UAI Method Related to Environmental Standards. Second Simulation

Time	21	22	23	24	25
Initial Investment (II)					
Initial Investment (II)					
II Present Value (IIPV) – Return	−901.27	−796.74	−704.33	−622.64	−550.42
Life Time simulated					
Residual value of equipment (RV)					
Linear Depreciation Method (Vn)	−1911.60	−1431.20	−950.80	−470.40	10.00
LDM Present Value (Vn)	−143.57	−95.02	−55.81	−24.41	0.46
Sum of Digits Method (Vn)	−359.54	−211.72	−100.86	−26.95	10.00
SDM Present Value (Vn)	−27.00	−14.06	−5.92	−1.40	0.46
Exponential Method (Vn)	−31.09	−23.42	−17.63	−13.28	−10.00
EM Present Value (Pv)	−2.34	−1.55	−1.03	−0.69	−0.46
Functioning (*F*)	−5490.86	−5545.76	−5601.22	−5657.23	−5713.81
Present Value Accumulated (NPV)	−33692.09	−34060.30	−34389.06	−34682.60	−34944.68
Present Value (NPV)	−412.39	−368.21	−328.76	−293.53	−262.08
Energy	−4200.00	−4400.00	−4600.00	−4800.00	−5000.00
Energy (NPV)	−315.44	−292.14	−269.99	−249.06	−229.34
Accumulated (FPV)	−520687.31	−555039.75	−589698.81	−624630.46	−659804.48
Maintenance (*M*)	−366.06	−369.72	−373.41	−377.15	−380.92
Present Value (MPV)	−27.49	−24.55	−21.92	−19.57	−17.47
Accumulated (MPV)	−2246.14	−2270.69	−2292.60	−2312.17	−2329.65
Non Production (NP)	0.00	0.00	0.00	0.00	0.00
Present Value (NPPV)	0.00	0.00	0.00	0.00	0.00
Accumulated NPPV	0.00	0.00	0.00	0.00	0.00
Accumulated Total Costs (APV)	−522933.45	−557310.44	−591991.41	−626942.63	−662134.13
Accumulated Total Costs	−157748.13	−168063.61	−178638.25	−189472.63	−200567.36
Benefit (*B*)	53571.43	53571.43	53571.43	53571.43	53571.43
Present Value (BPV)	4023.52	3556.86	3144.32	2779.63	2457.24
Accumulated Benefits BPV	377651.76	381208.62	384352.94	387132.58	389589.82
BPV + FPV + MPV + NPPV + IIPV	−157281.69	−188101.82	−219638.47	−251810.06	−284544.31
Uniform Annual Income UAIn (LDM)	−20.37	−13.35	−7.78	−3.38	0.06
Uniform Annual Income UAIn (SDM)	−3.83	−1.98	−0.83	−0.19	0.06
Uniform Annual Income UAIn (EM)	−0.33	−0.22	−0.14	−0.10	−0.06

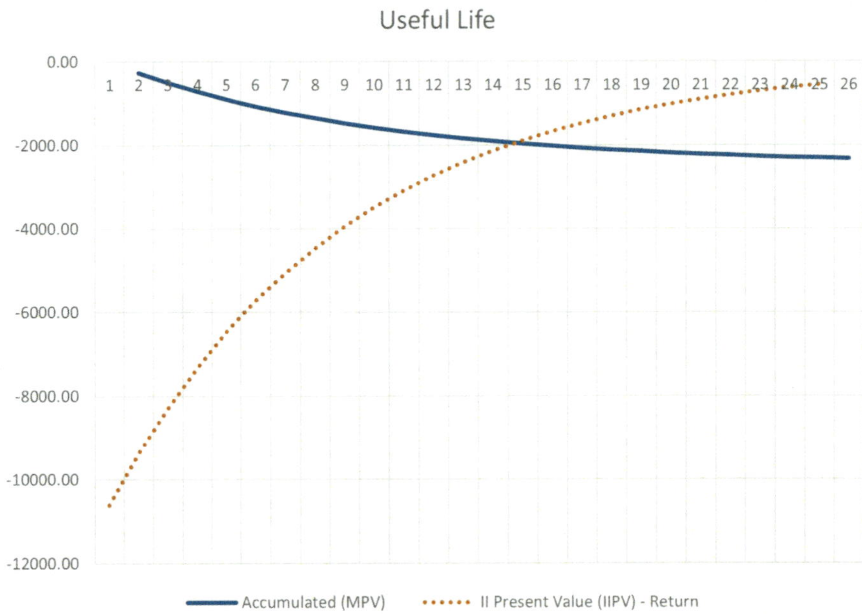

FIGURE 10.6 Physical Asset useful life.

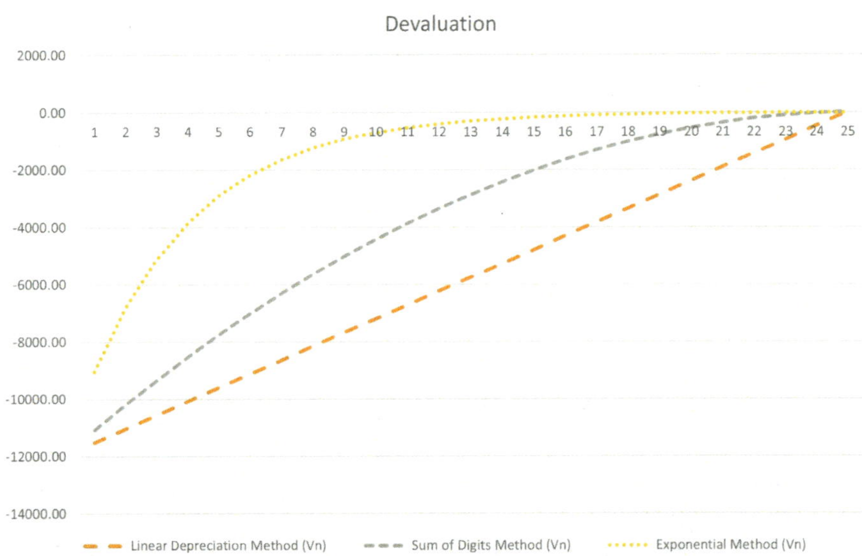

FIGURE 10.7 The three devaluation methods.

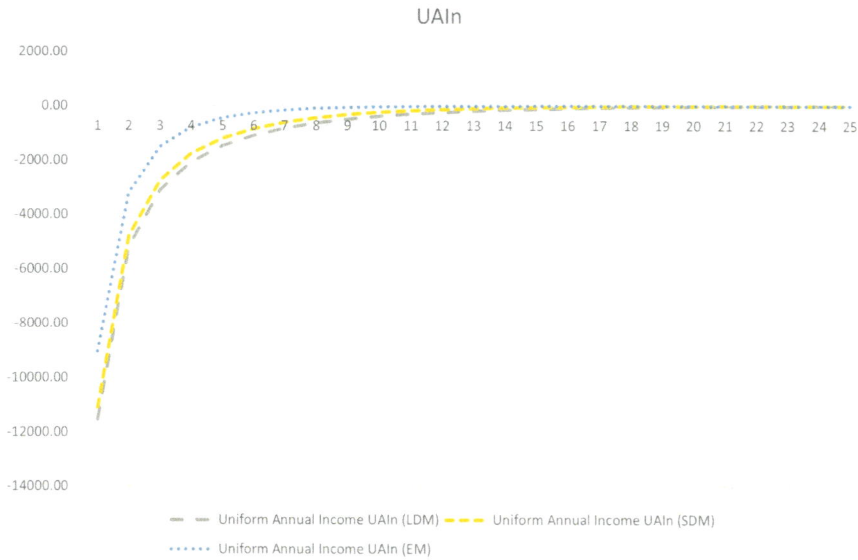

FIGURE 10.8 The UAI results from the three devaluation methods.

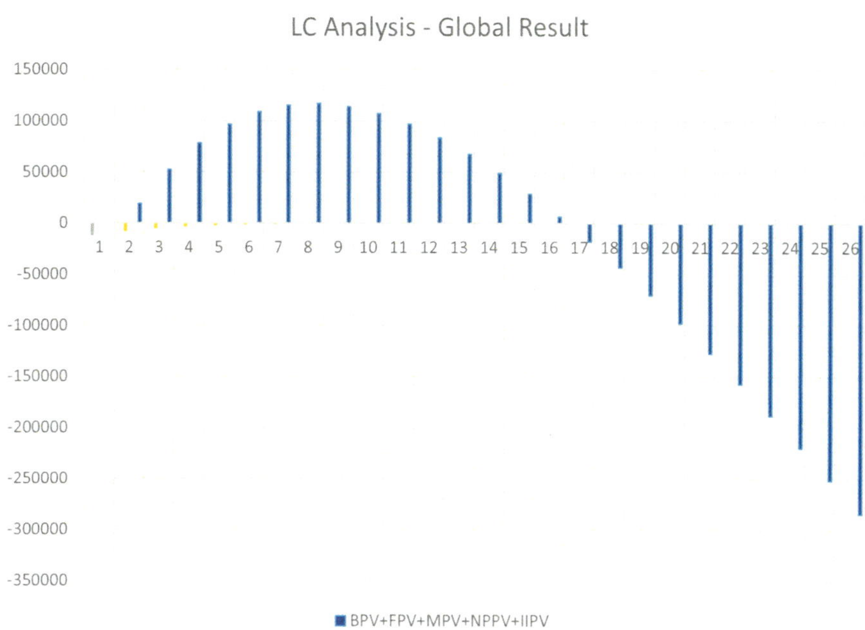

FIGURE 10.9 Life cycle analysis global result including profits.

through the variation of value of Benefits. Other indicators may be used, but may not be so intuitive.

As can be seen, namely comparing Figures 10.5 and 10.9, with the implementation of Energy Standards, it is gotten three more positive years, and the positive incomes increase a lot.

10.4.1.2 Minimization of Total Average Cost Method (MMTAC)

The first simulation uses the following values:

- Initial Investment (II) –12000
- Life Time simulated 25
- Residual value of equipment (RV) 10
- Functioning (*F*) –4500
- Energy –200
- Maintenance (*M*) –300
- Non Production (NP) 0
- Benefit (*B*) **50000**
- Internal Rate of Return (*r*) 0.12
- Risk (*R*) 0.01
- Apparent Rate 0.1312
- Maintenance Increase Rate 0.01
- Availability 0.9662
- MTBF 1000
- MTTR 25
- MWT 10

The remaining values are presented in Table 10.3 and the graphical results are shown in Figures 10.10–10.13.

As referred in the preceding introduction of this section, the evaluation of introduction of Environmental Norms, the simulation of their implementation is made through the increase of Benefits. Other indicators may be used, but may not be so intuitive.

The second simulation uses the following values (the value of Energy decreases a lot as a result of the Energy Standards):

- Initial Investment (II) –12000
- Life Time simulated 25
- Residual value of equipment (RV) 10
- Functioning (*F*) –4500
- Energy –200
- Maintenance (*M*) –300
- Non Production (NP) 0
- Benefit (*B*) **60000**
- Internal Rate of Return (*r*) 0.0
- Risk (*R*) 0.0
- Apparent Rate 0.0
- Maintenance Increase Rate 0.01
- Availability 0.9662

TABLE 10.3
Values Used for Simulation based on MMTAC Method Related to Environmental Standards. First Simulation

Time	0	1	2	3	4	5
Initial Investment (II)	−12000					
Initial Investment (II)	−12000					
II (IIPV) – Return	−12000	−12000.00	−12000.00	−12000.00	−12000.00	−12000.00
Useful Life	25					
Residual value of equipment (RV)	10					
Linear Depreciation Method (Vn)	−12000	−11519.60	−11039.20	−10558.80	−10078.40	−9598.00
LDM Present Value (PV)		−11519.60	−11039.20	−10558.80	−10078.40	−9598.00
Sum of Digits Method (Vn)	−12000	−11076.15	−10189.26	−9339.32	−8526.34	−7750.31
SDM Present Value (PV)		−11076.15	−10189.26	−9339.32	−8526.34	−7750.31
Exponential Method (Vn)	−12000	−9036.79	−6805.29	−5124.83	−3859.33	−2906.33
EM Present Value (Pv)		−9036.79	−6805.29	−5124.83	−3859.33	−2906.33
Functioning (*F*)	−4500	−4500.00	−4545.00	−4590.45	−4636.35	−4682.72
Functioning Accumulated (FA)		−4500.00	−9045.00	−13635.45	−18271.80	−22954.52
Present Value Accumulated (NPV)		−4500.00	−9045.00	−13635.45	−18271.80	−22954.52
Present Value (NPV)		−4500.00	−4545.00	−4590.45	−4636.35	−4682.72
Energy	−200	−200.00	−400.00	−600.00	−800.00	−1000.00
Energy Accumulated (EA)		−200.00	−600.00	−1200.00	−2000.00	−3000.00
Energy (NPV)		−200.00	−400.00	−600.00	−800.00	−1000.00
Accumulated Functioning (FPV)		−16700.00	−26145.00	−40380.45	−59452.25	−83406.78
Maintenance (*M*)	−300	−300.00	−303.00	−306.03	−309.09	−312.18
Maintenane Accumulated (MA)		−300.00	−603.00	−909.03	−1218.12	−1530.30
Present Value (MPV)		−300.00	−303.00	−306.03	−309.09	−312.18
Accumulated MPV		−300.00	−603.00	−909.03	−1218.12	−1530.30
Non Production (NP)	0	0.00	0.00	0.00	0.00	0.00
Non Production Accumulated (NPA)		0.00	0.00	0.00	0.00	0.00
Present Value (NPPV)		0.00	0.00	0.00	0.00	0.00
Accumulated NPPV		0.00	0.00	0.00	0.00	0.00
Accumulated Total Costs (APV)		−17000.00	−26748.00	−41289.48	−60670.37	−84937.08
Accumulated Total Costs		−5000.00	−10248.00	−15744.48	−21489.92	−27484.82
Benefit (*B*)	50000	48309.18	48309.18	48309.18	48309.18	48309.18
Benefits Accumulated (BA)		48309.18	96618.36	144927.54	193236.71	241545.89
Present Value (BPV)		48309.18	48309.18	48309.18	48309.18	48309.18
Accumulated Benefits BPV		48309.18	96618.36	144927.54	193236.71	241545.89
BPV+FPV+MPV+NPPV+IIPV		31309.18	74370.36	117183.06	159746.79	202061.07
MTAC (LDM)		−5480.40	−11208.80	−17185.68	−23411.52	−29886.82
MTAC (SDM)		−5923.85	−12058.74	−18405.16	−24963.59	−31734.52
MTAC (EM)		−7963.21	−15442.71	−22619.65	−29630.59	−36578.49

(*Continued*)

TABLE 10.3 (*Continued*)
Values Used for Simulation based on MMTAC Method Related to Environmental Standards. First Simulation

Time	6	7	8	9	10
Initial Investment (II)					
Initial Investment (II)					
II (IIPV) – Return	−12000.00	−12000.00	−12000.00	−12000.00	−12000.00
Useful Life					
Residual value of equipment (RV)					
Linear Depreciation Method (Vn)	−9117.60	−8637.20	−8156.80	−7676.40	−7196.00
LDM Present Value (PV)	−9117.60	−8637.20	−8156.80	−7676.40	−7196.00
Sum of Digits Method (Vn)	−7011.23	−6309.11	−5643.94	−5015.72	−4424.46
SDM Present Value (PV)	−7011.23	−6309.11	−5643.94	−5015.72	−4424.46
Exponential Method (Vn)	−2188.66	−1648.20	−1241.20	−934.71	−703.90
EM Present Value (Pv)	−2188.66	−1648.20	−1241.20	−934.71	−703.90
Functioning (*F*)	−4729.55	−4776.84	−4824.61	−4872.86	−4921.58
Functioning Accumulated (FA)	−27684.07	−32460.91	−37285.52	−42158.37	−47079.96
Present Value Accumulated (NPV)	−27684.07	−32460.91	−37285.52	−42158.37	−47079.96
Present Value (NPV)	−4729.55	−4776.84	−4824.61	−4872.86	−4921.58
Energy	−1200.00	−1400.00	−1600.00	−1800.00	−2000.00
Energy Accumulated (EA)	−4200.00	−5600.00	−7200.00	−9000.00	−11000.00
Energy (NPV)	−1200.00	−1400.00	−1600.00	−1800.00	−2000.00
Accumulated Functioning (FPV)	−112290.84	−146151.75	−185037.27	−228995.64	−278075.60
Maintenance (*M*)	−315.30	−318.46	−321.64	−324.86	−328.11
Maintenane Accumulated (MA)	−1845.60	−2164.06	−2485.70	−2810.56	−3138.66
Present Value (MPV)	−315.30	−318.46	−321.64	−324.86	−328.11
Accumulated MPV	−1845.60	−2164.06	−2485.70	−2810.56	−3138.66
Non Production (NP)	0.00	0.00	0.00	0.00	0.00
Non Production Accumulated (NPA)	0.00	0.00	0.00	0.00	0.00
Present Value (NPPV)	0.00	0.00	0.00	0.00	0.00
Accumulated NPPV	0.00	0.00	0.00	0.00	0.00
Accumulated Total Costs (APV)	−114136.45	−148315.81	−187522.97	−231806.20	−281214.26
Accumulated Total Costs	−33729.67	−40224.97	−46971.22	−53968.93	−61218.62
Benefit (*B*)	48309.18	48309.18	48309.18	48309.18	48309.18
Benefits Accumulated (BA)	289855.07	338164.25	386473.43	434782.61	483091.79
Present Value (BPV)	48309.18	48309.18	48309.18	48309.18	48309.18
Accumulated Benefits BPV	289855.07	338164.25	386473.43	434782.61	483091.79
BPV + FPV + MPV + NPPV + IIPV	244125.40	285939.28	327502.21	368813.68	409873.17
MTAC (LDM)	−36612.07	−43587.77	−50814.42	−58292.53	−66022.62
MTAC (SDM)	−38718.44	−45915.86	−53327.28	−60953.21	−68794.16
MTAC (EM)	−43541.02	−50576.77	−57730.01	−65034.22	−72514.72

(Continued)

TABLE 10.3 (*Continued*)
Values Used for Simulation based on MMTAC Method Related to Environmental Standards. First Simulation

Time	11	12	13	14	15
Initial Investment (II)					
Initial Investment (II)					
II (IIPV) – Return	−12000.00	−12000.00	−12000.00	−12000.00	−12000.00
Useful Life					
Residual value of equipment (RV)					
Linear Depreciation Method (Vn)	−6715.60	−6235.20	−5754.80	−5274.40	−4794.00
LDM Present Value (PV)	−6715.60	−6235.20	−5754.80	−5274.40	−4794.00
Sum of Digits Method (Vn)	−3870.15	−3352.80	−2872.40	−2428.95	−2022.46
SDM Present Value (PV)	−3870.15	−3352.80	−2872.40	−2428.95	−2022.46
Exponential Method (Vn)	−530.08	−399.19	−300.61	−226.38	−170.48
EM Present Value (Pv)	−530.08	−399.19	−300.61	−226.38	−170.48
Functioning (*F*)	−4970.80	−5020.51	−5070.71	−5121.42	−5172.63
Functioning Accumulated (FA)	−52050.76	−57071.26	−62141.98	−67263.40	−72436.03
Present Value Accumulated (NPV)	−52050.76	−57071.26	−62141.98	−67263.40	−72436.03
Present Value (NPV)	−4970.80	−5020.51	−5070.71	−5121.42	−5172.63
Energy	−2200.00	−2400.00	−2600.00	−2800.00	−3000.00
Energy Accumulated (EA)	−13200.00	−15600.00	−18200.00	−21000.00	−24000.00
Energy (NPV)	−2200.00	−2400.00	−2600.00	−2800.00	−3000.00
Accumulated Functioning (FPV)	−332326.36	−391797.62	−456539.60	−526602.99	−602039.02
Maintenance (*M*)	−331.39	−334.70	−338.05	−341.43	−344.84
Maintenane Accumulated (MA)	−3470.05	−3804.75	−4142.80	−4484.23	−4829.07
Present Value (MPV)	−331.39	−334.70	−338.05	−341.43	−344.84
Accumulated MPV	−3470.05	−3804.75	−4142.80	−4484.23	−4829.07
Non Production (NP)	0.00	0.00	0.00	0.00	0.00
Non Production Accumulated (NPA)	0.00	0.00	0.00	0.00	0.00
Present Value (NPPV)	0.00	0.00	0.00	0.00	0.00
Accumulated NPPV	0.00	0.00	0.00	0.00	0.00
Accumulated Total Costs (APV)	−335796.41	−395602.37	−460682.39	−531087.22	−606868.09
Accumulated Total Costs	−68720.81	−76476.01	−84484.77	−92747.62	−101265.10
Benefit (*B*)	48309.18	48309.18	48309.18	48309.18	48309.18
Benefits Accumulated (BA)	531400.97	579710.14	628019.32	676328.50	724637.68
Present Value (BPV)	48309.18	48309.18	48309.18	48309.18	48309.18
Accumulated Benefits BPV	531400.97	579710.14	628019.32	676328.50	724637.68
BPV + FPV + MPV + NPPV + IIPV	450680.16	491234.13	531534.55	571580.88	611372.58
MTAC (LDM)	−74005.21	−82240.81	−90729.97	−99473.22	−108471.10
MTAC (SDM)	−76850.65	−85123.21	−93612.37	−102318.67	−111242.64
MTAC (EM)	−80190.73	−88076.83	−96184.16	−104521.24	−113094.62

(Continued)

TABLE 10.3 (*Continued*)
Values Used for Simulation based on MMTAC Method Related to Environmental Standards. First Simulation

Time	16	17	18	19	20
Initial Investment (II)					
Initial Investment (II)					
II (IIPV) – Return	−12000.00	−12000.00	−12000.00	−12000.00	−12000.00
Useful Life					
Residual value of equipment (RV)					
Linear Depreciation Method (Vn)	−4313.60	−3833.20	−3352.80	−2872.40	−2392.00
LDM Present Value (PV)	−4313.60	−3833.20	−3352.80	−2872.40	−2392.00
Sum of Digits Method (Vn)	−1652.92	−1320.34	−1024.71	−766.03	−544.31
SDM Present Value (PV)	−1652.92	−1320.34	−1024.71	−766.03	−544.31
Exponential Method (Vn)	−128.38	−96.68	−72.81	−54.83	−41.29
EM Present Value (Pv)	−128.38	−96.68	−72.81	−54.83	−41.29
Functioning (*F*)	−5224.36	−5276.60	−5329.37	−5382.66	−5436.49
Functioning Accumulated (FA)	−77660.39	−82936.99	−88266.36	−93649.03	−99085.52
Present Value Accumulated (NPV)	−77660.39	−82936.99	−88266.36	−93649.03	−99085.52
Present Value (NPV)	−5224.36	−5276.60	−5329.37	−5382.66	−5436.49
Energy	−3200.00	−3400.00	−3600.00	−3800.00	−4000.00
Energy Accumulated (EA)	−27200.00	−30600.00	−34200.00	−38000.00	−42000.00
Energy (NPV)	−3200.00	−3400.00	−3600.00	−3800.00	−4000.00
Accumulated Functioning (FPV)	−682899.41	−769236.41	−861102.77	−958551.80	−1061637.32
Maintenance (*M*)	−348.29	−351.77	−355.29	−358.84	−362.43
Maintenane Accumulated (MA)	−5177.36	−5529.13	−5884.42	−6243.27	−6605.70
Present Value (MPV)	−348.29	−351.77	−355.29	−358.84	−362.43
Accumulated MPV	−5177.36	−5529.13	−5884.42	−6243.27	−6605.70
Non Production (NP)	0.00	0.00	0.00	0.00	0.00
Non Production Accumulated (NPA)	0.00	0.00	0.00	0.00	0.00
Present Value (NPPV)	0.00	0.00	0.00	0.00	0.00
Accumulated NPPV	0.00	0.00	0.00	0.00	0.00
Accumulated Total Costs (APV)	−688076.77	−774765.54	−866987.19	−964795.07	−1068243.02
Accumulated Total Costs	−110037.75	−119066.13	−128350.79	−137892.30	−147691.22
Benefit (*B*)	48309.18	48309.18	48309.18	48309.18	48309.18
Benefits Accumulated (BA)	772946.86	821256.04	869565.22	917874.40	966183.57
Present Value (BPV)	48309.18	48309.18	48309.18	48309.18	48309.18
Accumulated Benefits BPV	772946.86	821256.04	869565.22	917874.40	966183.57
BPV + FPV + MPV + NPPV + IIPV	650909.11	690189.91	729214.43	767982.10	806492.36
MTAC (LDM)	−117724.15	−127232.93	−136997.99	−147019.90	−157299.22
MTAC (SDM)	−120384.83	−129745.79	−139326.08	−149126.27	−159146.91
MTAC (EM)	−121909.37	−130969.45	−140277.98	−149837.47	−159649.93

(*Continued*)

TABLE 10.3 (*Continued*)
Values Used for Simulation based on MMTAC Method Related to Environmental Standards. First Simulation

Time	21	22	23	24	25
Initial Investment (II)					
Initial Investment (II)					
II (IIPV) – Return	−12000.00	−12000.00	−12000.00	−12000.00	−12000.00
Useful Life					
Residual value of equipment (RV)					
Linear Depreciation Method (Vn)	−1911.60	−1431.20	−950.80	−470.40	10.00
LDM Present Value (PV)	−1911.60	−1431.20	−950.80	−470.40	10.00
Sum of Digits Method (Vn)	−359.54	−211.72	−100.86	−26.95	10.00
SDM Present Value (PV)	−359.54	−211.72	−100.86	−26.95	10.00
Exponential Method (Vn)	−31.09	−23.42	−17.63	−13.28	−10.00
EM Present Value (Pv)	−31.09	−23.42	−17.63	−13.28	−10.00
Functioning (*F*)	−5490.86	−5545.76	−5601.22	−5657.23	−5713.81
Functioning Accumulated (FA)	−104576.37	−110122.14	−115723.36	−121380.59	−127094.40
Present Value Accumulated (NPV)	−104576.37	−110122.14	−115723.36	−121380.59	−127094.40
Present Value (NPV)	−5490.86	−5545.76	−5601.22	−5657.23	−5713.81
Energy	−4200.00	−4400.00	−4600.00	−4800.00	−5000.00
Energy Accumulated (EA)	−46200.00	−50600.00	−55200.00	−60000.00	−65000.00
Energy (NPV)	−4200.00	−4400.00	−4600.00	−4800.00	−5000.00
Accumulated Functioning (FPV)	−1170413.69	−1284935.83	−1405259.18	−1531439.78	−1663534.17
Maintenance (*M*)	−366.06	−369.72	−373.41	−377.15	−380.92
Maintenane Accumulated (MA)	−6971.76	−7341.48	−7714.89	−8092.04	−8472.96
Present Value (MPV)	−366.06	−369.72	−373.41	−377.15	−380.92
Accumulated MPV	−6971.76	−7341.48	−7714.89	−8092.04	−8472.96
Non Production (NP)	0.00	0.00	0.00	0.00	0.00
Non Production Accumulated (NPA)	0.00	0.00	0.00	0.00	0.00
Present Value (NP.PV)	0.00	0.00	0.00	0.00	0.00
Accumulated NPPV	0.00	0.00	0.00	0.00	0.00
Accumulated Total Costs (APV)	−1177385.45	−1292277.30	−1412974.07	−1539531.82	−1672007.13
Accumulated Total Costs	−157748.13	−168063.61	−178638.25	−189472.63	−200567.36
Benefit (*B*)	48309.18	48309.18	48309.18	48309.18	48309.18
Benefits Accumulated (BA)	1014492.75	1062801.93	1111111.11	1159420.29	1207729.47
Present Value (BPV)	48309.18	48309.18	48309.18	48309.18	48309.18
Accumulated Benefits BPV	1014492.75	1062801.93	1111111.11	1159420.29	1207729.47
BPV + FPV + MPV + NPPV + IIPV	844744.62	882738.32	920472.86	957947.66	995162.11
MTAC (LDM)	−167836.53	−178632.41	−189687.45	−201002.23	−212577.36
MTAC (SDM)	−169388.59	−179851.89	−190537.39	−201445.68	−212577.36
MTAC (EM)	−169717.04	−180040.20	−190620.62	−201459.35	−212557.36

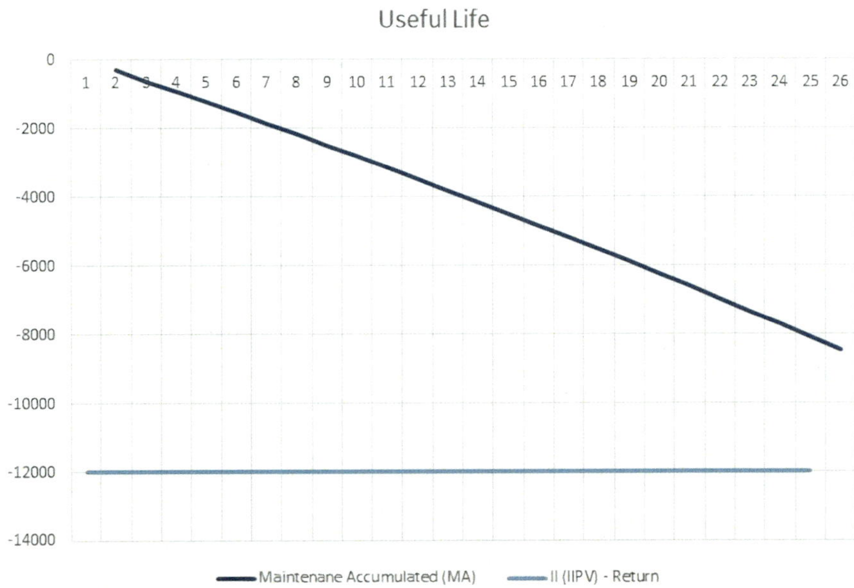

FIGURE 10.10 Physical Asset useful life.

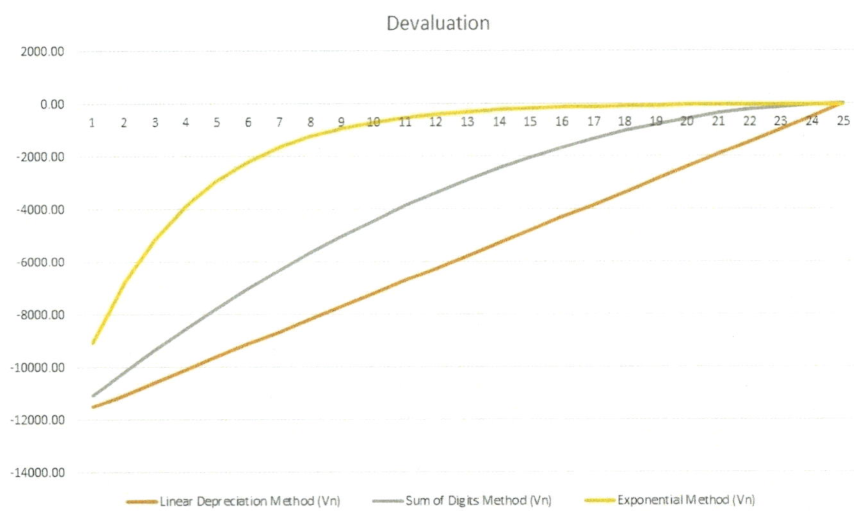

FIGURE 10.11 The three devaluation methods.

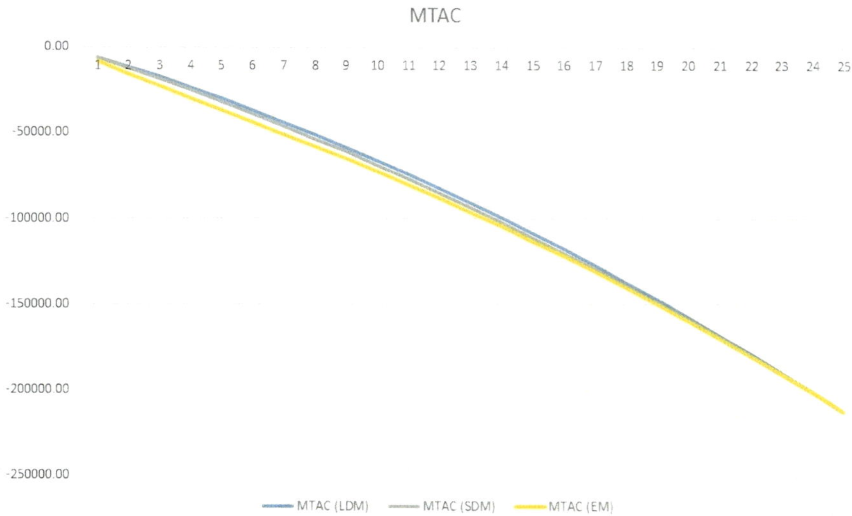

FIGURE 10.12 The MMTAC results from the three devaluation methods.

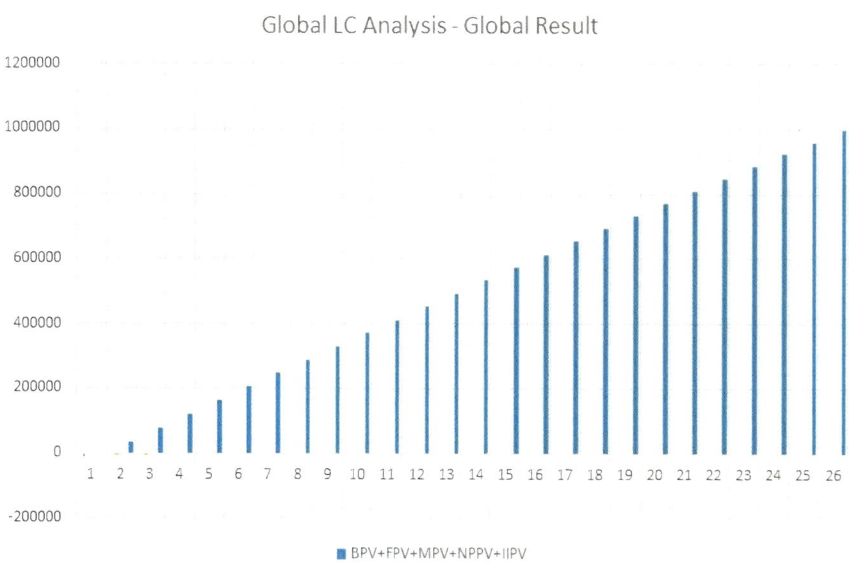

FIGURE 10.13 Life cycle analysis global result including profits.

- MTBF 1000
- MTTR 25
- MWT 10

The remaining values are presented in Table 10.4, and the graphical results are shown in Figures 10.14–10.17.

As can be seen, namely comparing Figures 10.13 and 10.17, with the implementation of the Environmental Standards, it is gotten positive incomes increase a lot. However, in this simulation, it is considered an Apparent rate null, what does not happen in a real world. Because of this, the Reduction to Present Value and the other models must be considered.

10.4.1.3 MMTAC Reduced to Present Value (MMTAC-RPV)

The first simulation uses the following values:

- Initial Investment (II) –12000
- Life Time simulated 25
- Residual value of equipment (RV) 10
- Functioning (*F*) –4500
- Energy (*E*) –200
- Maintenance (*M*) –300
- Non Production (NP) 0
- Benefit (*B*) **50000**
- Internal Rate of Return (*r*) 0.12
- Risk (*R*) 0.01
- Apparent Rate 0.1312
- Maintenance Increase Rate 0.01
- Availability 0.9662
- MTBF 1000
- MTTR 25
- MWT 10

The remaining values are presented in Table 10.5, and the graphical results are shown in Figures 10.18–10.21.

As referred in the preceding introduction on this section, the evaluation of introduction of Environmental Norms, the simulation of their implementation is made through the variation of Benefits. Other indicators may be used, but may not be so intuitive.

The second simulation uses the following values:

- Initial Investment (II) –12000
- Life Time simulated 25
- Residual value of equipment (RV) 10
- Functioning (*F*) –4500
- Energy (*E*) –200
- Maintenance (*M*) –300

TABLE 10.4
Values Used for Simulation based on MMTAC Method Related to Environmental Standards. Second Simulation

Time	0	1	2	3	4	5
Initial Investment (II)	−12000					
Initial Investment (II)	−12000					
II (IIPV) − Return	−12000	−12000.00	−12000.00	−12000.00	−12000.00	−12000.00
Useful Life	25					
Residual value of equipment (RV)	10					
Linear Depreciation Method (Vn)	−12000	−11519.60	−11039.20	−10558.80	−10078.40	−9598.00
LDM Present Value (PV)		−11519.60	−11039.20	−10558.80	−10078.40	−9598.00
Sum of Digits Method (Vn)	−12000	−11076.15	−10189.26	−9339.32	−8526.34	−7750.31
SDM Present Value (PV)		−11076.15	−10189.26	−9339.32	−8526.34	−7750.31
Exponential Method (Vn)	−12000	−9036.79	−6805.29	−5124.83	−3859.33	−2906.33
EM Present Value (Pv)		−9036.79	−6805.29	−5124.83	−3859.33	−2906.33
Functioning (*F*)	−4500	−4500.00	−4545.00	−4590.45	−4636.35	−4682.72
Functioning Accumulated (FA)		−4500.00	−9045.00	−13635.45	−18271.80	−22954.52
Present Value Accumulated (NPV)		−4500.00	−9045.00	−13635.45	−18271.80	−22954.52
Present Value (NPV)		−4500.00	−4545.00	−4590.45	−4636.35	−4682.72
Energy	−200	−200.00	−400.00	−600.00	−800.00	−1000.00
Energy Accumulated (EA)		−200.00	−600.00	−1200.00	−2000.00	−3000.00
Energy (NPV)		−200.00	−400.00	−600.00	−800.00	−1000.00
Accumulated Functioning (FPV)		−16700.00	−26145.00	−40380.45	−59452.25	−83406.78
Maintenance (*M*)	−300	−300.00	−303.00	−306.03	−309.09	−312.18
Maintenane Accumulated (MA)		−300.00	−603.00	−909.03	−1218.12	−1530.30
Present Value (MPV)		−300.00	−303.00	−306.03	−309.09	−312.18
Accumulated MPV		−300.00	−603.00	−909.03	−1218.12	−1530.30
Non Production (NP)	0	0.00	0.00	0.00	0.00	0.00
Non Production Accumulated (NPA)		0.00	0.00	0.00	0.00	0.00
Present Value (NPPV)		0.00	0.00	0.00	0.00	0.00
Accumulated NPPV		0.00	0.00	0.00	0.00	0.00
Accumulated Total Costs (APV)		−17000.00	−26748.00	−41289.48	−60670.37	−84937.08
Accumulated Total Costs		−5000.00	−10248.00	−15744.48	−21489.92	−27484.82
Benefit (*B*)	60000	57971.01	57971.01	57971.01	57971.01	57971.01
Benefits Accumulated (BA)		57971.01	115942.03	173913.04	231884.06	289855.07
Present Value (BPV)		57971.01	57971.01	57971.01	57971.01	57971.01
Accumulated Benefits BPV		57971.01	115942.03	173913.04	231884.06	289855.07
BPV + FPV + MPV + NPPV + IIPV		40971.01	93694.03	146168.56	198394.13	250370.25
MTAC (LDM)		−5480.40	−11208.80	−17185.68	−23411.52	−29886.82
MTAC (SDM)		−5923.85	−12058.74	−18405.16	−24963.59	−31734.52
MTAC (EM)		−7963.21	−15442.71	−22619.65	−29630.59	−36578.49

(Continued)

TABLE 10.4 (*Continued*)
Values Used for Simulation based on MMTAC Method Related to Environmental Standards. Second Simulation

Time	6	7	8	9	10
Initial Investment (II)					
Initial Investment (II)					
II (IIPV) – Return	−12000.00	−12000.00	−12000.00	−12000.00	−12000.00
Useful Life					
Residual value of equipment (RV)					
Linear Depreciation Method (Vn)	−9117.60	−8637.20	−8156.80	−7676.40	−7196.00
LDM Present Value (PV)	−9117.60	−8637.20	−8156.80	−7676.40	−7196.00
Sum of Digits Method (Vn)	−7011.23	−6309.11	−5643.94	−5015.72	−4424.46
SDM Present Value (PV)	−7011.23	−6309.11	−5643.94	−5015.72	−4424.46
Exponential Method (Vn)	−2188.66	−1648.20	−1241.20	−934.71	−703.90
EM Present Value (Pv)	−2188.66	−1648.20	−1241.20	−934.71	−703.90
Functioning (*F*)	−4729.55	−4776.84	−4824.61	−4872.86	−4921.58
Functioning Accumulated (FA)	−27684.07	−32460.91	−37285.52	−42158.37	−47079.96
Present Value Accumulated (NPV)	−27684.07	−32460.91	−37285.52	−42158.37	−47079.96
Present Value (NPV)	−4729.55	−4776.84	−4824.61	−4872.86	−4921.58
Energy	−1200.00	−1400.00	−1600.00	−1800.00	−2000.00
Energy Accumulated (EA)	−4200.00	−5600.00	−7200.00	−9000.00	−11000.00
Energy (NPV)	−1200.00	−1400.00	−1600.00	−1800.00	−2000.00
Accumulated Functioning (FPV)	−112290.84	−146151.75	−185037.27	−228995.64	−278075.60
Maintenance (*M*)	−315.30	−318.46	−321.64	−324.86	−328.11
Maintenane Accumulated (MA)	−1845.60	−2164.06	−2485.70	−2810.56	−3138.66
Present Value (MPV)	−315.30	−318.46	−321.64	−324.86	−328.11
Accumulated MPV	−1845.60	−2164.06	−2485.70	−2810.56	−3138.66
Non Production (NP)	0.00	0.00	0.00	0.00	0.00
Non Production Accumulated (NPA)	0.00	0.00	0.00	0.00	0.00
Present Value (NPPV)	0.00	0.00	0.00	0.00	0.00
Accumulated NPPV	0.00	0.00	0.00	0.00	0.00
Accumulated Total Costs (APV)	−114136.45	−148315.81	−187522.97	−231806.20	−281214.26
Accumulated Total Costs	−33729.67	−40224.97	−46971.22	−53968.93	−61218.62
Benefit (*B*)	57971.01	57971.01	57971.01	57971.01	57971.01
Benefits Accumulated (BA)	347826.09	405797.10	463768.12	521739.13	579710.14
Present Value (BPV)	57971.01	57971.01	57971.01	57971.01	57971.01
Accumulated Benefits BPV	347826.09	405797.10	463768.12	521739.13	579710.14
BPV + FPV + MPV + NPPV + IIPV	302096.41	353572.13	404796.90	455770.20	506491.52
MTAC (LDM)	−36612.07	−43587.77	−50814.42	−58292.53	−66022.62
MTAC (SDM)	−38718.44	−45915.86	−53327.28	−60953.21	−68794.16
MTAC (EM)	−43541.02	−50576.77	−57730.01	−65034.22	−72514.72

(*Continued*)

TABLE 10.4 (*Continued*)
Values Used for Simulation based on MMTAC Method Related to Environmental Standards. Second Simulation

Time	11	12	13	14	15
Initial Investment (II)					
Initial Investment (II)					
II (IIPV) – Return	−12000.00	−12000.00	−12000.00	−12000.00	−12000.00
Useful Life					
Residual value of equipment (RV)					
Linear Depreciation Method (Vn)	−6715.60	−6235.20	−5754.80	−5274.40	−4794.00
LDM Present Value (PV)	−6715.60	−6235.20	−5754.80	−5274.40	−4794.00
Sum of Digits Method (Vn)	−3870.15	−3352.80	−2872.40	−2428.95	−2022.46
SDM Present Value (PV)	−3870.15	−3352.80	−2872.40	−2428.95	−2022.46
Exponential Method (Vn)	−530.08	−399.19	−300.61	−226.38	−170.48
EM Present Value (Pv)	−530.08	−399.19	−300.61	−226.38	−170.48
Functioning (*F*)	−4970.80	−5020.51	−5070.71	−5121.42	−5172.63
Functioning Accumulated (FA)	−52050.76	−57071.26	−62141.98	−67263.40	−72436.03
Present Value Accumulated (NPV)	−52050.76	−57071.26	−62141.98	−67263.40	−72436.03
Present Value (NPV)	−4970.80	−5020.51	−5070.71	−5121.42	−5172.63
Energy	−2200.00	−2400.00	−2600.00	−2800.00	−3000.00
Energy Accumulated (EA)	−13200.00	−15600.00	−18200.00	−21000.00	−24000.00
Energy (NPV)	−2200.00	−2400.00	−2600.00	−2800.00	−3000.00
Accumulated Functioning (FPV)	−332326.36	−391797.62	−456539.60	−526602.99	−602039.02
Maintenance (*M*)	−331.39	−334.70	−338.05	−341.43	−344.84
Maintenane Accumulated (MA)	−3470.05	−3804.75	−4142.80	−4484.23	−4829.07
Present Value (MPV)	−331.39	−334.70	−338.05	−341.43	−344.84
Accumulated MPV	−3470.05	−3804.75	−4142.80	−4484.23	−4829.07
Non Production (NP)	0.00	0.00	0.00	0.00	0.00
Non Production Accumulated (NPA)	0.00	0.00	0.00	0.00	0.00
Present Value (NPPV)	0.00	0.00	0.00	0.00	0.00
Accumulated NPPV	0.00	0.00	0.00	0.00	0.00
Accumulated Total Costs (APV)	−335796.41	−395602.37	−460682.39	−531087.22	−606868.09
Accumulated Total Costs	−68720.81	−76476.01	−84484.77	−92747.62	−101265.10
Benefit (*B*)	57971.01	57971.01	57971.01	57971.01	57971.01
Benefits Accumulated (BA)	637681.16	695652.17	753623.19	811594.20	869565.22
Present Value (BPV)	57971.01	57971.01	57971.01	57971.01	57971.01
Accumulated Benefits BPV	637681.16	695652.17	753623.19	811594.20	869565.22
BPV + FPV + MPV + NPPV + IIPV	556960.35	607176.16	657138.41	706846.58	756300.12
MTAC (LDM)	−74005.21	−82240.81	−90729.97	−99473.22	−108471.10
MTAC (SDM)	−76850.65	−85123.21	−93612.37	−102318.67	−111242.64
MTAC (EM)	−80190.73	−88076.83	−96184.16	−104521.24	−113094.62

(*Continued*)

TABLE 10.4 (*Continued*)
Values Used for Simulation based on MMTAC Method Related to Environmental Standards. Second Simulation

Time	16	17	18	19	20
Initial Investment (II)					
Initial Investment (II)					
II (IIPV) – Return	−12000.00	−12000.00	−12000.00	−12000.00	−12000.00
Useful Life					
Residual value of equipment (RV)					
Linear Depreciation Method (Vn)	−4313.60	−3833.20	−3352.80	−2872.40	−2392.00
LDM Present Value (PV)	−4313.60	−3833.20	−3352.80	−2872.40	−2392.00
Sum of Digits Method (Vn)	−1652.92	−1320.34	−1024.71	−766.03	−544.31
SDM Present Value (PV)	−1652.92	−1320.34	−1024.71	−766.03	−544.31
Exponential Method (Vn)	−128.38	−96.68	−72.81	−54.83	−41.29
EM Present Value (Pv)	−128.38	−96.68	−72.81	−54.83	−41.29
Functioning (*F*)	−5224.36	−5276.60	−5329.37	−5382.66	−5436.49
Functioning Accumulated (FA)	−77660.39	−82936.99	−88266.36	−93649.03	−99085.52
Present Value Accumulated (NPV)	−77660.39	−82936.99	−88266.36	−93649.03	−99085.52
Present Value (NPV)	−5224.36	−5276.60	−5329.37	−5382.66	−5436.49
Energy	−3200.00	−3400.00	−3600.00	−3800.00	−4000.00
Energy Accumulated (EA)	−27200.00	−30600.00	−34200.00	−38000.00	−42000.00
Energy (NPV)	−3200.00	−3400.00	−3600.00	−3800.00	−4000.00
Accumulated Functioning (FPV)	−682899.41	−769236.41	−861102.77	−958551.80	−1061637.32
Maintenance (*M*)	−348.29	−351.77	−355.29	−358.84	−362.43
Maintenane Accumulated (MA)	−5177.36	−5529.13	−5884.42	−6243.27	−6605.70
Present Value (MPV)	−348.29	−351.77	−355.29	−358.84	−362.43
Accumulated MPV	−5177.36	−5529.13	−5884.42	−6243.27	−6605.70
Non Production (NP)	0.00	0.00	0.00	0.00	0.00
Non Production Accumulated (NPA)	0.00	0.00	0.00	0.00	0.00
Present Value (NPPV)	0.00	0.00	0.00	0.00	0.00
Accumulated NPPV	0.00	0.00	0.00	0.00	0.00
Accumulated Total Costs (APV)	−688076.77	−774765.54	−866987.19	−964795.07	−1068243.02
Accumulated Total Costs	−110037.75	−119066.13	−128350.79	−137892.30	−147691.22
Benefit (*B*)	57971.01	57971.01	57971.01	57971.01	57971.01
Benefits Accumulated (BA)	927536.23	985507.25	1043478.26	1101449.28	1159420.29
Present Value (BPV)	57971.01	57971.01	57971.01	57971.01	57971.01
Accumulated Benefits BPV	927536.23	985507.25	1043478.26	1101449.28	1159420.29
BPV + FPV + MPV + NPPV + IIPV	805498.48	854441.12	903127.47	951556.98	999729.07
MTAC (LDM)	−117724.15	−127232.93	−136997.99	−147019.90	−157299.22
MTAC (SDM)	−120384.83	−129745.79	−139326.08	−149126.27	−159146.91
MTAC (EM)	−121909.37	−130969.45	−140277.98	−149837.47	−159649.93

(*Continued*)

TABLE 10.4 (*Continued*)
Values Used for Simulation based on MMTAC Method Related to Environmental Standards. Second Simulation

Time	21	22	23	24	25
Initial Investment (II)					
Initial Investment (II)					
II (IIPV) – Return	−12000.00	−12000.00	−12000.00	−12000.00	−12000.00
Useful Life					
Residual value of equipment (RV)					
Linear Depreciation Method (Vn)	−1911.60	−1431.20	−950.80	−470.40	10.00
LDM Present Value (PV)	−1911.60	−1431.20	−950.80	−470.40	10.00
Sum of Digits Method (Vn)	−359.54	−211.72	−100.86	−26.95	10.00
SDM Present Value (PV)	−359.54	−211.72	−100.86	−26.95	10.00
Exponential Method (Vn)	−31.09	−23.42	−17.63	−13.28	−10.00
EM Present Value (Pv)	−31.09	−23.42	−17.63	−13.28	−10.00
Functioning (*F*)	−5490.86	−5545.76	−5601.22	−5657.23	−5713.81
Functioning Accumulated (FA)	−104576.37	−110122.14	−115723.36	−121380.59	−127094.40
Present Value Accumulated (NPV)	−104576.37	−110122.14	−115723.36	−121380.59	−127094.40
Present Value (NPV)	−5490.86	−5545.76	−5601.22	−5657.23	−5713.81
Energy	−4200.00	−4400.00	−4600.00	−4800.00	−5000.00
Energy Accumulated (EA)	−46200.00	−50600.00	−55200.00	−60000.00	−65000.00
Energy (NPV)	−4200.00	−4400.00	−4600.00	−4800.00	−5000.00
Accumulated Functioning (FPV)	−1170413.69	−1284935.83	−1405259.18	−1531439.78	−1663534.17
Maintenance (*M*)	−366.06	−369.72	−373.41	−377.15	−380.92
Maintenane Accumulated (MA)	−6971.76	−7341.48	−7714.89	−8092.04	−8472.96
Present Value (MPV)	−366.06	−369.72	−373.41	−377.15	−380.92
Accumulated MPV	−6971.76	−7341.48	−7714.89	−8092.04	−8472.96
Non Production (NP)	0.00	0.00	0.00	0.00	0.00
Non Production Accumulated (NPA)	0.00	0.00	0.00	0.00	0.00
Present Value (NPPV)	0.00	0.00	0.00	0.00	0.00
Accumulated NPPV	0.00	0.00	0.00	0.00	0.00
Accumulated Total Costs (APV)	−1177385.45	−1292277.30	−1412974.07	−1539531.82	−1672007.13
Accumulated Total Costs	−157748.13	−168063.61	−178638.25	−189472.63	−200567.36
Benefit (*B*)	57971.01	57971.01	57971.01	57971.01	57971.01
Benefits Accumulated (BA)	1217391.30	1275362.32	1333333.33	1391304.35	1449275.36
Present Value (BPV)	57971.01	57971.01	57971.01	57971.01	57971.01
Accumulated Benefits BPV	1217391.30	1275362.32	1333333.33	1391304.35	1449275.36
BPV + FPV + MPV + NPPV + IIPV	1047643.17	1095298.71	1142695.08	1189831.72	1236708.00
MTAC (LDM)	−167836.53	−178632.41	−189687.45	−201002.23	−212577.36
MTAC (SDM)	−169388.59	−179851.89	−190537.39	−201445.68	−212577.36
MTAC (EM)	−169717.04	−180040.20	−190620.62	−201459.35	−212557.36

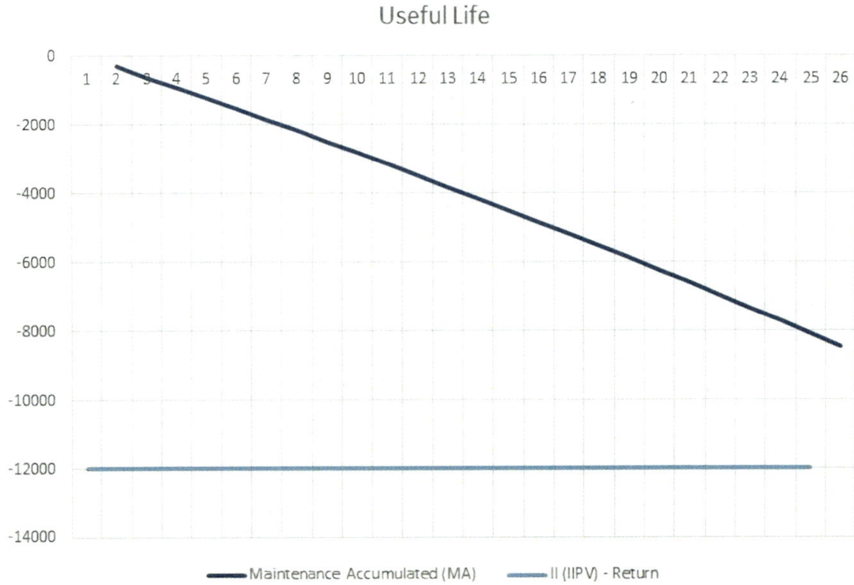

FIGURE 10.14 Physical Asset useful life.

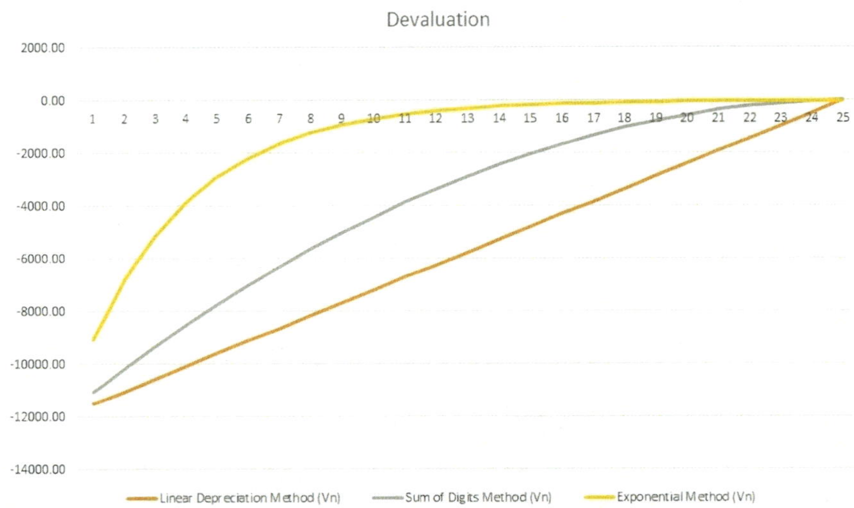

FIGURE 10.15 The three devaluation methods.

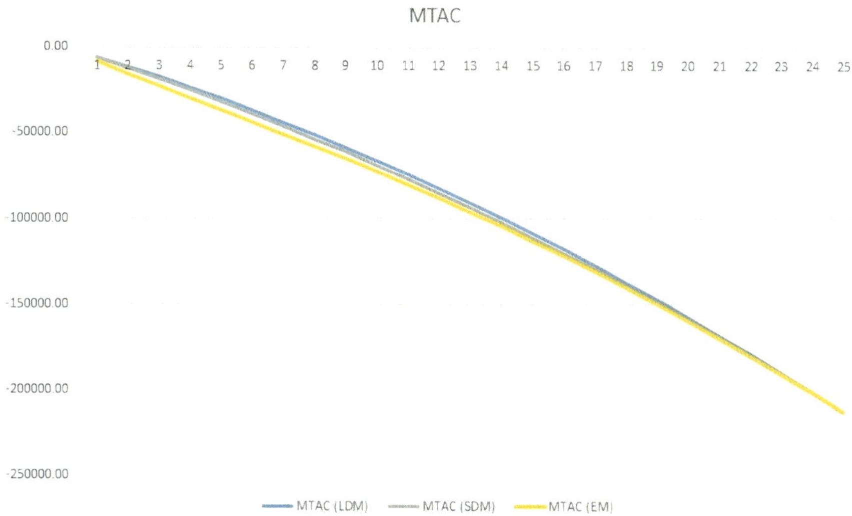

FIGURE 10.16 The MMTAC results from the three devaluation methods.

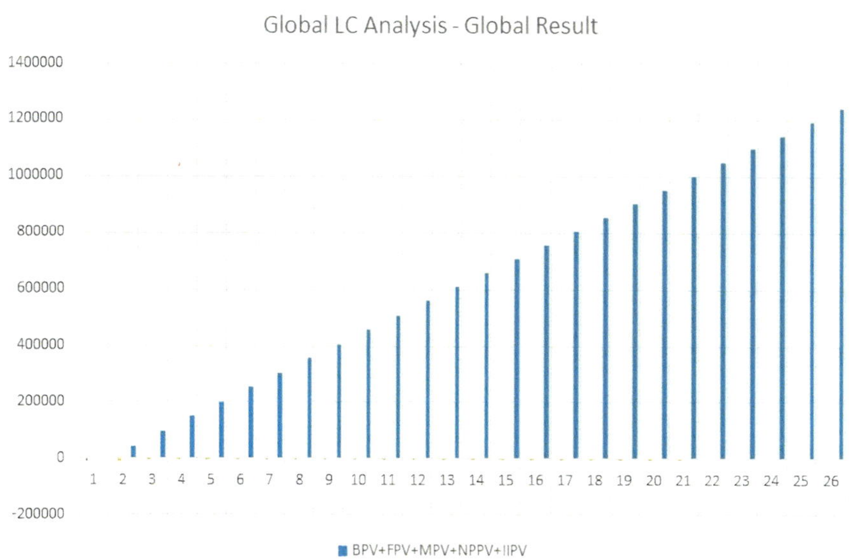

FIGURE 10.17 Life cycle analysis global result including profits.

TABLE 10.5
Values Used for Simulation based on MMTAC-RPV Method Related to Environmental Standards. First Simulation

Time	0	1	2	3	4	5
Initial Investment (II)	−12000					
Initial Investment (II)	−12000					
II Present Value (IIPV) – Return	−12000	−10608.20	−9377.83	−8290.16	−7328.64	−6478.65
Life Time simulated	25					
Residual value of equipment (RV)	10					
Linear Depreciation Method (Vn)	−12000	−11519.60	−11039.20	−10558.80	−10078.40	−9598.00
LDM Present Value (Vn)		−10183.52	−8626.98	−7294.51	−6155.08	−5181.84
Sum of Digits Method (Vn)	−12000	−11076.15	−10189.26	−9339.32	−8526.34	−7750.31
SDM Present Value (Vn)		−9791.51	−7962.77	−6452.04	−5207.21	−4184.29
Exponential Method (Vn)	−12000	−9036.79	−6805.29	−5124.83	−3859.33	−2906.33
EM Present Value (Pv)		−7988.67	−5318.24	−3540.47	−2356.97	−1569.09
Functioning (*F*)	−4500	−4500.00	−4545.00	−4590.45	−4636.35	−4682.72
Present Value Acummulated (NPV)		−3978.08	−7529.93	−10701.23	−13532.74	−16060.88
Present Value (NPV)		−3978.08	−3551.85	−3171.30	−2831.52	−2528.14
Energy (*E*)	−200	−200.00	−400.00	−600.00	−800.00	−1000.00
Energy (NPV)		−176.80	−312.59	−414.51	−488.58	−539.89
Accumulated FPV		−16154.88	−23997.40	−35113.14	−49134.46	−65735.23
Maintenance (*M*)	−300	−300.00	−303.00	−306.03	−309.09	−312.18
Present Value (MPV)		−265.21	−236.79	−211.42	−188.77	−168.54
Accumulated (MPV)		−265.21	−502.00	−713.42	−902.18	−1070.73
Non Production (NP)	0	0.00	0.00	0.00	0.00	0.00
Present Value (NPPV)		0.00	0.00	0.00	0.00	0.00
Accumulated NPPV		0.00	0.00	0.00	0.00	0.00
Accumulated Total Costs (APV)		−16420.08	−24499.40	−35826.56	−50036.64	−66805.96
Accumulated Total Costs		−5000.00	−10248.00	−15744.48	−21489.92	−27484.82
Benefit (*B*)	50000	44642.86	44642.86	44642.86	44642.86	44642.86
Present Value (BPV)		39465.04	34887.77	30841.38	27264.30	24102.11
Accumulated Benefits BPV		39465.04	74352.81	105194.19	132458.49	156560.60
BPV + FPV + MPV + NPPV + IIPV		11044.96	37853.41	57367.63	70421.85	77754.64
MTAC-PV (LDM)		−38603.61	−22563.19	−18373.69	−17047.93	−16797.56
MTAC-PV (SDM)		−38211.59	−22231.08	−18092.87	−16810.96	−16598.05
MTAC-PV (EM)		−36408.76	−20908.82	−17122.34	−16098.40	−16075.01

(Continued)

TABLE 10.5 (*Continued*)
Values Used for Simulation based on MMTAC-RPV Method Related to Environmental Standards. First Simulation

Time	6	7	8	9	10
Initial Investment (II)					
Initial Investment (II)					
II Present Value (IIPV) – Return	−5727.23	−5062.97	−4475.75	−3956.64	−3497.74
Life Time simulated					
Residual value of equipment (RV)					
Linear Depreciation Method (Vn)	−9117.60	−8637.20	−8156.80	−7676.40	−7196.00
LDM Present Value (Vn)	−4351.55	−3644.16	−3042.32	−2531.06	−2097.48
Sum of Digits Method (Vn)	−7011.23	−6309.11	−5643.94	−5015.72	−4424.46
SDM Present Value (Vn)	−3346.25	−2661.90	−2105.07	−1653.78	−1289.63
Exponential Method (Vn)	−2188.66	−1648.20	−1241.20	−934.71	−703.90
EM Present Value (Pv)	−1044.58	−695.40	−462.94	−308.19	−205.17
Functioning (*F*)	−4729.55	−4776.84	−4824.61	−4872.86	−4921.58
Present Value Acummulated (NPV)	−18318.15	−20333.57	−22133.05	−23739.73	−25174.26
Present Value (NPV)	−2257.27	−2015.42	−1799.48	−1606.68	−1434.53
Energy (*E*)	−1200.00	−1400.00	−1600.00	−1800.00	−2000.00
Energy (NPV)	−572.72	−590.68	−596.77	−593.50	−582.96
Accumulated FPV	−84626.11	−105550.36	−128280.17	−152613.40	−178370.61
Maintenance (*M*)	−315.30	−318.46	−321.64	−324.86	−328.11
Present Value (MPV)	−150.48	−134.36	−119.97	−107.11	−95.64
Accumulated (MPV)	−1221.21	−1355.57	−1475.54	−1582.65	−1678.28
Non Production (NP)	0.00	0.00	0.00	0.00	0.00
Present Value (NPPV)	0.00	0.00	0.00	0.00	0.00
Accumulated NPPV	0.00	0.00	0.00	0.00	0.00
Accumulated Total Costs (APV)	−85847.32	−106905.93	−129755.71	−154196.04	−180048.90
Accumulated Total Costs	−33729.67	−40224.97	−46971.22	−53968.93	−61218.62
Benefit (*B*)	44642.86	44642.86	44642.86	44642.86	44642.86
Present Value (BPV)	21306.67	18835.46	16650.87	14719.65	13012.42
Accumulated Benefits BPV	177867.27	196702.73	213353.60	228073.24	241085.66
BPV + FPV + MPV + NPPV + IIPV	80019.95	77796.80	71597.89	61877.20	49036.77
MTAC-PV (LDM)	−17033.14	−17507.16	−18099.75	−18747.46	−19414.64
MTAC-PV (SDM)	−16865.59	−17366.83	−17982.60	−18649.98	−19333.85
MTAC-PV (EM)	−16481.98	−17085.90	−17777.33	−18500.47	−19225.41

(*Continued*)

TABLE 10.5 (*Continued*)
Values Used for Simulation based on MMTAC-RPV Method Related to Environmental Standards. First Simulation

Time	11	12	13	14	15
Initial Investment (II)					
Initial Investment (II)					
II Present Value (IIPV) – Return	−3092.06	−2733.43	−2416.40	−2136.14	−1888.38
Life Time simulated					
Residual value of equipment (RV)					
Linear Depreciation Method (Vn)	−6715.60	−6235.20	−5754.80	−5274.40	−4794.00
LDM Present Value (Vn)	−1730.42	−1420.29	−1158.83	−938.90	−754.41
Sum of Digits Method (Vn)	−3870.15	−3352.80	−2872.40	−2428.95	−2022.46
SDM Present Value (Vn)	−997.23	−763.72	−578.41	−432.38	−318.27
Exponential Method (Vn)	−530.08	−399.19	−300.61	−226.38	−170.48
EM Present Value (Pv)	−136.59	−90.93	−60.53	−40.30	−26.83
Functioning (*F*)	−4970.80	−5020.51	−5070.71	−5121.42	−5172.63
Present Value Acummulated (NPV)	−26455.10	−27598.70	−28619.77	−29531.44	−30345.44
Present Value (NPV)	−1280.83	−1143.60	−1021.07	−911.67	−813.99
Energy (*E*)	−2200.00	−2400.00	−2600.00	−2800.00	−3000.00
Energy (NPV)	−566.88	−546.69	−523.55	−498.43	−472.10
Accumulated FPV	−205392.59	−233537.97	−262681.29	−292711.17	−323528.70
Maintenance (*M*)	−331.39	−334.70	−338.05	−341.43	−344.84
Present Value (MPV)	−85.39	−76.24	−68.07	−60.78	−54.27
Accumulated (MPV)	−1763.67	−1839.91	−1907.98	−1968.76	−2023.03
Non Production (NP)	0.00	0.00	0.00	0.00	0.00
Present Value (NPPV)	0.00	0.00	0.00	0.00	0.00
Accumulated NPPV	0.00	0.00	0.00	0.00	0.00
Accumulated Total Costs (APV)	−207156.26	−235377.88	−264589.28	−294679.93	−325551.73
Accumulated Total Costs	−68720.81	−76476.01	−84484.77	−92747.62	−101265.10
Benefit (*B*)	44642.86	44642.86	44642.86	44642.86	44642.86
Present Value (BPV)	11503.20	10169.02	8989.59	7946.95	7025.24
Accumulated Benefits BPV	252588.86	262757.88	271747.47	279694.42	286719.66
BPV + FPV + MPV + NPPV + IIPV	33432.60	15380.00	−4841.81	−26985.51	−50832.07
MTAC-PV (LDM)	−20080.61	−20733.18	−21365.24	−21972.77	−22553.74
MTAC-PV (SDM)	−20013.95	−20678.47	−21320.59	−21936.59	−22524.67
MTAC-PV (EM)	−19935.71	−20622.40	−21280.75	−21908.59	−22505.24

(*Continued*)

TABLE 10.5 (*Continued*)
Values Used for Simulation based on MMTAC-RPV Method Related to Environmental Standards. First Simulation

Time	16	17	18	19	20
Initial Investment (II)					
Initial Investment (II)					
II Present Value (IIPV) – Return	−1669.36	−1475.75	−1304.58	−1153.27	−1019.51
Life Time simulated					
Residual value of equipment (RV)					
Linear Depreciation Method (Vn)	−4313.60	−3833.20	−3352.80	−2872.40	−2392.00
LDM Present Value (Vn)	−600.08	−471.40	−364.50	−276.06	−203.22
Sum of Digits Method (Vn)	−1652.92	−1320.34	−1024.71	−766.03	−544.31
SDM Present Value (Vn)	−229.94	−162.37	−111.40	−73.62	−46.24
Exponential Method (Vn)	−128.38	−96.68	−72.81	−54.83	−41.29
EM Present Value (Pv)	−17.86	−11.89	−7.92	−5.27	−3.51
Functioning (*F*)	−5224.36	−5276.60	−5329.37	−5382.66	−5436.49
Present Value Acummulated (NPV)	−31072.22	−31721.13	−32300.51	−32817.82	−33279.70
Present Value (NPV)	−726.78	−648.91	−579.38	−517.31	−461.88
Energy (*E*)	−3200.00	−3400.00	−3600.00	−3800.00	−4000.00
Energy (NPV)	−445.16	−418.13	−391.38	−365.20	−339.84
Accumulated FPV	−355046.08	−387185.33	−419877.22	−453060.24	−486679.78
Maintenance (*M*)	−348.29	−351.77	−355.29	−358.84	−362.43
Present Value (MPV)	−48.45	−43.26	−38.63	−34.49	−30.79
Accumulated (MPV)	−2071.48	−2114.74	−2153.37	−2187.85	−2218.65
Non Production (NP)	0.00	0.00	0.00	0.00	0.00
Present Value (NPPV)	0.00	0.00	0.00	0.00	0.00
Accumulated NPPV	0.00	0.00	0.00	0.00	0.00
Accumulated Total Costs (APV)	−357117.56	−389300.07	−422030.59	−455248.09	−488898.42
Accumulated Total Costs	−110037.75	−119066.13	−128350.79	−137892.30	−147691.22
Benefit (*B*)	44642.86	44642.86	44642.86	44642.86	44642.86
Present Value (BPV)	6210.43	5490.13	4853.36	4290.46	3792.84
Accumulated Benefits BPV	292930.09	298420.21	303273.58	307564.03	311356.87
BPV + FPV + MPV + NPPV + IIPV	−76187.47	−102879.86	−130757.01	−159684.06	−189541.55
MTAC-PV (LDM)	−23107.35	−23633.62	−24133.06	−24606.53	−25055.08
MTAC-PV (SDM)	−23084.22	−23615.44	−24119.00	−24595.88	−25047.23
MTAC-PV (EM)	−23070.96	−23606.59	−24113.25	−24592.28	−25045.10

(*Continued*)

TABLE 10.5 (*Continued*)
Values Used for Simulation based on MMTAC-RPV Method Related to Environmental Standards. First Simulation

Time	21	22	23	24	25
Initial Investment (II)					
Initial Investment (II)					
II Present Value (IIPV) – Return	−901.27	−796.74	−704.33	−622.64	−550.42
Life Time simulated					
Residual value of equipment (RV)					
Linear Depreciation Method (Vn)	−1911.60	−1431.20	−950.80	−470.40	10.00
LDM Present Value (Vn)	−143.57	−95.02	−55.81	−24.41	0.46
Sum of Digits Method (Vn)	−359.54	−211.72	−100.86	−26.95	10.00
SDM Present Value (Vn)	−27.00	−14.06	−5.92	−1.40	0.46
Exponential Method (Vn)	−31.09	−23.42	−17.63	−13.28	−10.00
EM Present Value (Pv)	−2.34	−1.55	−1.03	−0.69	−0.46
Functioning (*F*)	−5490.86	−5545.76	−5601.22	−5657.23	−5713.81
Present Value Acummulated (NPV)	−33692.09	−34060.30	−34389.06	−34682.60	−34944.68
Present Value (NPV)	−412.39	−368.21	−328.76	−293.53	−262.08
Energy (*E*)	−4200.00	−4400.00	−4600.00	−4800.00	−5000.00
Energy (NPV)	−315.44	−292.14	−269.99	−249.06	−229.34
Accumulated FPV	−520687.31	−555039.75	−589698.81	−624630.46	−659804.48
Maintenance (*M*)	−366.06	−369.72	−373.41	−377.15	−380.92
Present Value (MPV)	−27.49	−24.55	−21.92	−19.57	−17.47
Accumulated (MPV)	−2246.14	−2270.69	−2292.60	−2312.17	−2329.65
Non Production (NP)	0.00	0.00	0.00	0.00	0.00
Present Value (NPPV)	0.00	0.00	0.00	0.00	0.00
Accumulated NPPV	0.00	0.00	0.00	0.00	0.00
Accumulated Total Costs (APV)	−522933.45	−557310.44	−591991.41	−626942.63	−662134.13
Accumulated Total Costs	−157748.13	−168063.61	−178638.25	−189472.63	−200567.36
Benefit (*B*)	44642.86	44642.86	44642.86	44642.86	44642.86
Present Value (BPV)	3352.93	2964.05	2620.27	2316.36	2047.70
Accumulated Benefits BPV	314709.80	317673.85	320294.12	322610.48	324658.18
BPV + FPV + MPV + NPPV + IIPV	−220223.65	−251636.59	−283697.29	−316332.15	−349475.94
MTAC-PV (LDM)	−25479.86	−25882.07	−26262.92	−26623.63	−26965.35
MTAC-PV (SDM)	−25474.31	−25878.39	−26260.75	−26622.67	−26965.35
MTAC-PV (EM)	−25473.13	−25877.82	−26260.54	−26622.64	−26965.38

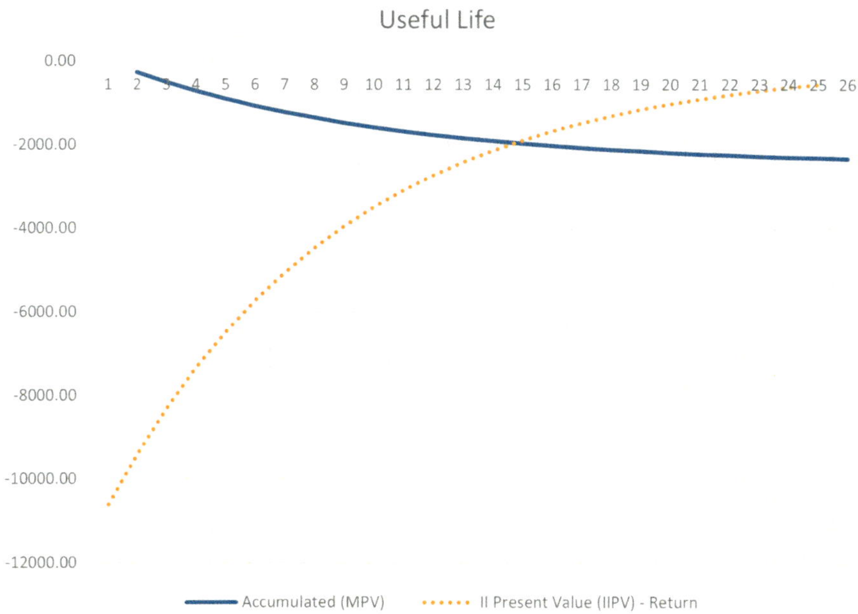

FIGURE 10.18 Physical Asset useful life.

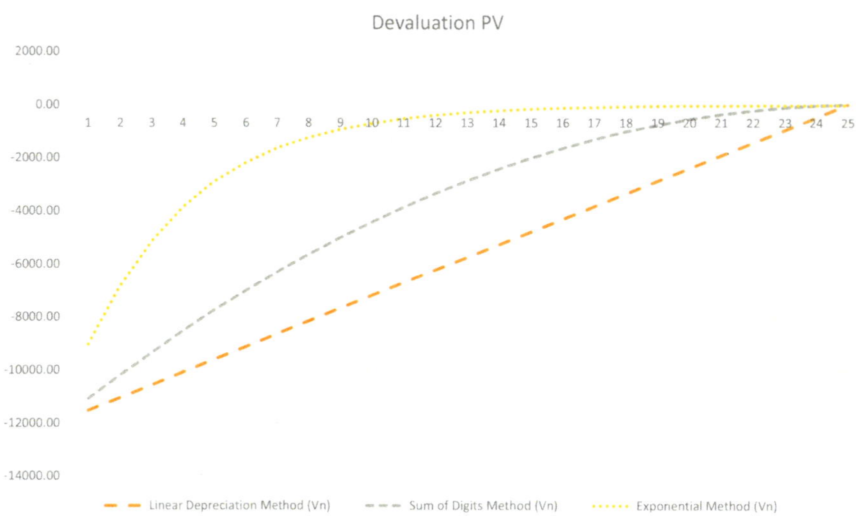

FIGURE 10.19 The three devaluation methods.

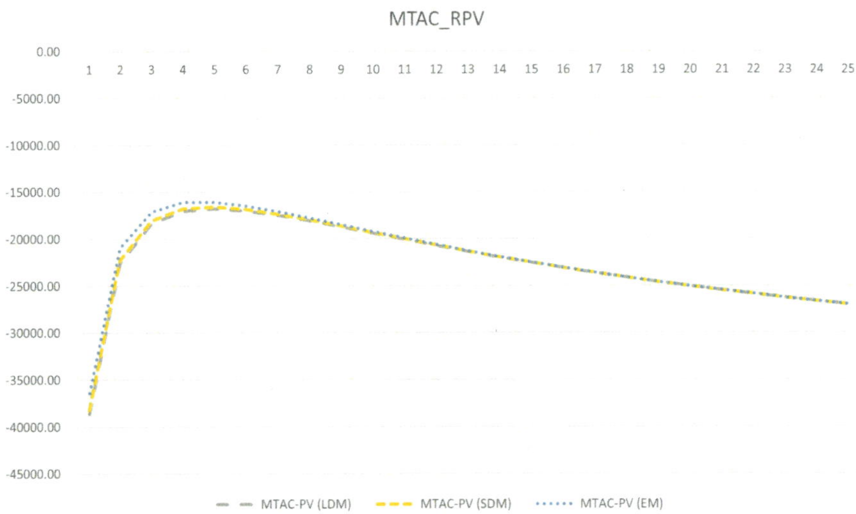

FIGURE 10.20 The MMTAC-RPV results from the three devaluation methods.

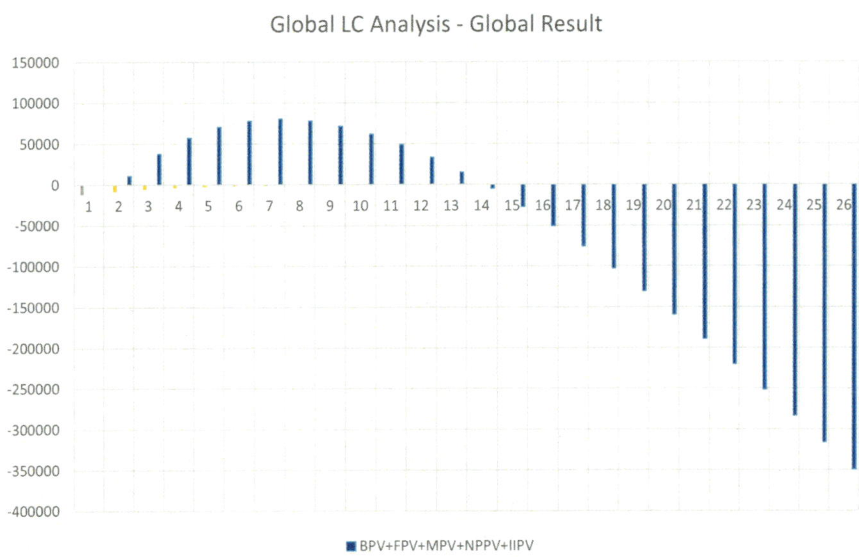

FIGURE 10.21 Life cycle analysis global result including profits.

- Non Production (NP) 0
- Benefit (B) **60000**
- Internal Rate of Return (r) 0.12
- Risk (R) 0.01
- Apparent Rate 0.1312
- Maintenance Increase Rate 0.01
- Availability 0.9662
- MTBF 1000
- MTTR 25
- MWT 10

The remaining values are presented in Table 10.6, and the graphical results are shown in Figures 10.22–10.25.

As can be seen, namely comparing Figures 10.21 and 10.25, with the implementing of Environmental standards, it is get three more positive years, but the positive incomes increase a lot.

These results from the three methods are close, which reinforce a lot the robustness of the models under discussion.

10.4.2 Life Cycle Investment of Physical Assets

The way to demonstrate if the return of implementing the Standard under discussion is interesting or not, in this section, is done using the model of Life Cycle Investment, presented in Chapter 2.

The first simulation uses the following values:

- Initial Investment (II) –12000
- Functioning (F) –4500
- Energy (E) –1000
- Maintenance (M) –300
- Accumulated (AMPV) –5800
- Non Production (NP) 0
- Benefit (B) including IIPV **50000**
- Internal Rate of Return (r) 0.12
- Risk (R) 0.01
- Apparent Rate 0.1312
- Maintenance Increase Rate 0.01
- Availability 0.9662
- MTBF 1000
- MTTR 25
- MWT 10

As referred in the preceding introduction to this section, the evaluation of introduction of Environmental Norms, the simulation of their implementation is made through the variation of the Benefits. Other indicators may be used, but may not be so intuitive.

TABLE 10.6
Values Used for Simulation based on MMTAC-RPV Method Related to Environmental Standards. Second Simulation

Time	0	1	2	3	4	5
Initial Investment (II)	−12000					
Initial Investment (II)	−12000					
II Present Value (IIPV) – Return	−12000	−10608.20	−9377.83	−8290.16	−7328.64	−6478.65
Life Time simulated	25					
Residual value of equipment (RV)	10					
Linear Depreciation Method (Vn)	−12000	−11519.60	−11039.20	−10558.80	−10078.40	−9598.00
LDM Present Value (Vn)		−10183.52	−8626.98	−7294.51	−6155.08	−5181.84
Sum of Digits Method (Vn)	−12000	−11076.15	−10189.26	−9339.32	−8526.34	−7750.31
SDM Present Value (Vn)		−9791.51	−7962.77	−6452.04	−5207.21	−4184.29
Exponential Method (Vn)	−12000	−9036.79	−6805.29	−5124.83	−3859.33	−2906.33
EM Present Value (Pv)		−7988.67	−5318.24	−3540.47	−2356.97	−1569.09
Functioning (*F*)	−4500	−4500.00	−4545.00	−4590.45	−4636.35	−4682.72
Present Value Acummulated (NPV)		−3978.08	−7529.93	−10701.23	−13532.74	−16060.88
Present Value (NPV)		−3978.08	−3551.85	−3171.30	−2831.52	−2528.14
Energy (*E*)	−200	−200.00	−400.00	−600.00	−800.00	−1000.00
Energy (NPV)		−176.80	−312.59	−414.51	−488.58	−539.89
Accumulated FPV		−16154.88	−23997.40	−35113.14	−49134.46	−65735.23
Maintenance (*M*)	−300	−300.00	−303.00	−306.03	−309.09	−312.18
Present Value (MPV)		−265.21	−236.79	−211.42	−188.77	−168.54
Accumulated (MPV)		−265.21	−502.00	−713.42	−902.18	−1070.73
Non Production (NP)	0	0.00	0.00	0.00	0.00	0.00
Present Value (NPPV)		0.00	0.00	0.00	0.00	0.00
Accumulated NPPV		0.00	0.00	0.00	0.00	0.00
Accumulated Total Costs (APV)		−16420.08	−24499.40	−35826.56	−50036.64	−66805.96
Accumulated Total Costs		−5000.00	−10248.00	−15744.48	−21489.92	−27484.82
Benefit (*B*)	60000	53571.43	53571.43	53571.43	53571.43	53571.43
Present Value (BPV)		47358.05	41865.32	37009.66	32717.16	28922.53
Accumulated Benefits BPV		47358.05	89223.37	126233.03	158950.19	187872.72
BPV + FPV + MPV + NPPV + IIPV		18937.97	52723.97	78406.47	96913.55	109066.76
MTAC−PV (LDM)		−38603.61	−22563.19	−18373.69	−17047.93	−16797.56
MTAC−PV (SDM)		−38211.59	−22231.08	−18092.87	−16810.96	−16598.05
MTAC−PV (EM)		−36408.76	−20908.82	−17122.34	−16098.40	−16075.01

(Continued)

TABLE 10.6 (*Continued*)
Values Used for Simulation based on MMTAC-RPV Method Related to Environmental Standards. Second Simulation

Time	6	7	8	9	10
Initial Investment (II)					
Initial Investment (II)					
II Present Value (IIPV) – Return	−5727.23	−5062.97	−4475.75	−3956.64	−3497.74
Life Time simulated					
Residual value of equipment (RV)					
Linear Depreciation Method (Vn)	−9117.60	−8637.20	−8156.80	−7676.40	−7196.00
LDM Present Value (Vn)	−4351.55	−3644.16	−3042.32	−2531.06	−2097.48
Sum of Digits Method (Vn)	−7011.23	−6309.11	−5643.94	−5015.72	−4424.46
SDM Present Value (Vn)	−3346.25	−2661.90	−2105.07	−1653.78	−1289.63
Exponential Method (Vn)	−2188.66	−1648.20	−1241.20	−934.71	−703.90
EM Present Value (Pv)	−1044.58	−695.40	−462.94	−308.19	−205.17
Functioning (*F*)	−4729.55	−4776.84	−4824.61	−4872.86	−4921.58
Present Value Acummulated (NPV)	−18318.15	−20333.57	−22133.05	−23739.73	−25174.26
Present Value (NPV)	−2257.27	−2015.42	−1799.48	−1606.68	−1434.53
Energy (*E*)	−1200.00	−1400.00	−1600.00	−1800.00	−2000.00
Energy (NPV)	−572.72	−590.68	−596.77	−593.50	−582.96
Accumulated FPV	−84626.11	−105550.36	−128280.17	−152613.40	−178370.61
Maintenance (*M*)	−315.30	−318.46	−321.64	−324.86	−328.11
Present Value (MPV)	−150.48	−134.36	−119.97	−107.11	−95.64
Accumulated (MPV)	−1221.21	−1355.57	−1475.54	−1582.65	−1678.28
Non Production (NP)	0.00	0.00	0.00	0.00	0.00
Present Value (NPPV)	0.00	0.00	0.00	0.00	0.00
Accumulated NPPV	0.00	0.00	0.00	0.00	0.00
Accumulated Total Costs (APV)	−85847.32	−106905.93	−129755.71	−154196.04	−180048.90
Accumulated Total Costs	−33729.67	−40224.97	−46971.22	−53968.93	−61218.62
Benefit (*B*)	53571.43	53571.43	53571.43	53571.43	53571.43
Present Value (BPV)	25568.01	22602.55	19981.04	17663.58	15614.90
Accumulated Benefits BPV	213440.73	236043.28	256024.31	273687.89	289302.79
BPV + FPV + MPV + NPPV + IIPV	115593.41	117137.35	114268.61	107491.85	97253.90
MTAC−PV (LDM)	−17033.14	−17507.16	−18099.75	−18747.46	−19414.64
MTAC−PV (SDM)	−16865.59	−17366.83	−17982.60	−18649.98	−19333.85
MTAC−PV (EM)	−16481.98	−17085.90	−17777.33	−18500.47	−19225.41

(*Continued*)

TABLE 10.6 (*Continued*)
Values Used for Simulation based on MMTAC-RPV Method Related to Environmental Standards. Second Simulation

Time	11	12	13	14	15
Initial Investment (II)					
Initial Investment (II)					
II Present Value (IIPV) – Return	−3092.06	−2733.43	−2416.40	−2136.14	−1888.38
Life Time simulated					
Residual value of equipment (RV)					
Linear Depreciation Method (Vn)	−6715.60	−6235.20	−5754.80	−5274.40	−4794.00
LDM Present Value (Vn)	−1730.42	−1420.29	−1158.83	−938.90	−754.41
Sum of Digits Method (Vn)	−3870.15	−3352.80	−2872.40	−2428.95	−2022.46
SDM Present Value (Vn)	−997.23	−763.72	−578.41	−432.38	−318.27
Exponential Method (Vn)	−530.08	−399.19	−300.61	−226.38	−170.48
EM Present Value (Pv)	−136.59	−90.93	−60.53	−40.30	−26.83
Functioning (*F*)	−4970.80	−5020.51	−5070.71	−5121.42	−5172.63
Present Value Acummulated (NPV)	−26455.10	−27598.70	−28619.77	−29531.44	−30345.44
Present Value (NPV)	−1280.83	−1143.60	−1021.07	−911.67	−813.99
Energy (*E*)	−2200.00	−2400.00	−2600.00	−2800.00	−3000.00
Energy (NPV)	−566.88	−546.69	−523.55	−498.43	−472.10
Accumulated FPV	−205392.59	−233537.97	−262681.29	−292711.17	−323528.70
Maintenance (*M*)	−331.39	−334.70	−338.05	−341.43	−344.84
Present Value (MPV)	−85.39	−76.24	−68.07	−60.78	−54.27
Accumulated (MPV)	−1763.67	−1839.91	−1907.98	−1968.76	−2023.03
Non Production (NP)	0.00	0.00	0.00	0.00	0.00
Present Value (NPPV)	0.00	0.00	0.00	0.00	0.00
Accumulated NPPV	0.00	0.00	0.00	0.00	0.00
Accumulated Total Costs (APV)	−207156.26	−235377.88	−264589.28	−294679.93	−325551.73
Accumulated Total Costs	−68720.81	−76476.01	−84484.77	−92747.62	−101265.10
Benefit (*B*)	53571.43	53571.43	53571.43	53571.43	53571.43
Present Value (BPV)	13803.84	12202.83	10787.51	9536.34	8430.29
Accumulated Benefits BPV	303106.63	315309.46	326096.97	335633.31	344063.59
BPV + FPV + MPV + NPPV + IIPV	83950.37	67931.58	49507.69	28953.37	6511.86
MTAC−PV (LDM)	−20080.61	−20733.18	−21365.24	−21972.77	−22553.74
MTAC−PV (SDM)	−20013.95	−20678.47	−21320.59	−21936.59	−22524.67
MTAC−PV (EM)	−19935.71	−20622.40	−21280.75	−21908.59	−22505.24

(*Continued*)

TABLE 10.6 (*Continued*)
Values Used for Simulation based on MMTAC-RPV Method Related to Environmental Standards. Second Simulation

Time	16	17	18	19	20
Initial Investment (II)					
Initial Investment (II)					
II Present Value (IIPV) – Return	−1669.36	−1475.75	−1304.58	−1153.27	−1019.51
Life Time simulated					
Residual value of equipment (RV)					
Linear Depreciation Method (Vn)	−4313.60	−3833.20	−3352.80	−2872.40	−2392.00
LDM Present Value (Vn)	−600.08	−471.40	−364.50	−276.06	−203.22
Sum of Digits Method (Vn)	−1652.92	−1320.34	−1024.71	−766.03	−544.31
SDM Present Value (Vn)	−229.94	−162.37	−111.40	−73.62	−46.24
Exponential Method (Vn)	−128.38	−96.68	−72.81	−54.83	−41.29
EM Present Value (Pv)	−17.86	−11.89	−7.92	−5.27	−3.51
Functioning (*F*)	−5224.36	−5276.60	−5329.37	−5382.66	−5436.49
Present Value Acummulated (NPV)	−31072.22	−31721.13	−32300.51	−32817.82	−33279.70
Present Value (NPV)	−726.78	−648.91	−579.38	−517.31	−461.88
Energy (*E*)	−3200.00	−3400.00	−3600.00	−3800.00	−4000.00
Energy (NPV)	−445.16	−418.13	−391.38	−365.20	−339.84
Accumulated FPV	−355046.08	−387185.33	−419877.22	−453060.24	−486679.78
Maintenance (*M*)	−348.29	−351.77	−355.29	−358.84	−362.43
Present Value (MPV)	−48.45	−43.26	−38.63	−34.49	−30.79
Accumulated (MPV)	−2071.48	−2114.74	−2153.37	−2187.85	−2218.65
Non Production (NP)	0.00	0.00	0.00	0.00	0.00
Present Value (NPPV)	0.00	0.00	0.00	0.00	0.00
Accumulated NPPV	0.00	0.00	0.00	0.00	0.00
Accumulated Total Costs (APV)	−357117.56	−389300.07	−422030.59	−455248.09	−488898.42
Accumulated Total Costs	−110037.75	−119066.13	−128350.79	−137892.30	−147691.22
Benefit (*B*)	53571.43	53571.43	53571.43	53571.43	53571.43
Present Value (BPV)	7452.52	6588.15	5824.04	5148.55	4551.40
Accumulated Benefits BPV	351516.11	358104.26	363928.29	369076.84	373628.24
BPV + FPV + MPV + NPPV + IIPV	−17601.45	−43195.82	−70102.29	−98171.25	−127270.18
MTAC–PV (LDM)	−23107.35	−23633.62	−24133.06	−24606.53	−25055.08
MTAC–PV (SDM)	−23084.22	−23615.44	−24119.00	−24595.88	−25047.23
MTAC–PV (EM)	−23070.96	−23606.59	−24113.25	−24592.28	−25045.10

(*Continued*)

TABLE 10.6 (*Continued*)

Values Used for Simulation based on MMTAC-RPV Method Related to Environmental Standards. Second Simulation

Time	21	22	23	24	25
Initial Investment (II)					
Initial Investment (II)					
II Present Value (IIPV) – Return	−901.27	−796.74	−704.33	−622.64	−550.42
Life Time simulated					
Residual value of equipment (RV)					
Linear Depreciation Method (Vn)	−1911.60	−1431.20	−950.80	−470.40	10.00
LDM Present Value (Vn)	−143.57	−95.02	−55.81	−24.41	0.46
Sum of Digits Method (Vn)	−359.54	−211.72	−100.86	−26.95	10.00
SDM Present Value (Vn)	−27.00	−14.06	−5.92	−1.40	0.46
Exponential Method (Vn)	−31.09	−23.42	−17.63	−13.28	−10.00
EM Present Value (Pv)	−2.34	−1.55	−1.03	−0.69	−0.46
Functioning (*F*)	−5490.86	−5545.76	−5601.22	−5657.23	−5713.81
Present Value Acummulated (NPV)	−33692.09	−34060.30	−34389.06	−34682.60	−34944.68
Present Value (NPV)	−412.39	−368.21	−328.76	−293.53	−262.08
Energy (*E*)	−4200.00	−4400.00	−4600.00	−4800.00	−5000.00
Energy (NPV)	−315.44	−292.14	−269.99	−249.06	−229.34
Accumulated FPV	−520687.31	−555039.75	−589698.81	−624630.46	−659804.48
Maintenance (*M*)	−366.06	−369.72	−373.41	−377.15	−380.92
Present Value (MPV)	−27.49	−24.55	−21.92	−19.57	−17.47
Accumulated (MPV)	−2246.14	−2270.69	−2292.60	−2312.17	−2329.65
Non Production (NP)	0.00	0.00	0.00	0.00	0.00
Present Value (NPPV)	0.00	0.00	0.00	0.00	0.00
Accumulated NPPV	0.00	0.00	0.00	0.00	0.00
Accumulated Total Costs (APV)	−522933.45	−557310.44	−591991.41	−626942.63	−662134.13
Accumulated Total Costs	−157748.13	−168063.61	−178638.25	−189472.63	−200567.36
Benefit (*B*)	53571.43	53571.43	53571.43	53571.43	53571.43
Present Value (BPV)	4023.52	3556.86	3144.32	2779.63	2457.24
Accumulated Benefits BPV	377651.76	381208.62	384352.94	387132.58	389589.82
BPV + FPV + MPV + NPPV + IIPV	−157281.69	−188101.82	−219638.47	−251810.06	−284544.31
MTAC−PV (LDM)	−25479.86	−25882.07	−26262.92	−26623.63	−26965.35
MTAC−PV (SDM)	−25474.31	−25878.39	−26260.75	−26622.67	−26965.35
MTAC−PV (EM)	−25473.13	−25877.82	−26260.54	−26622.64	−26965.38

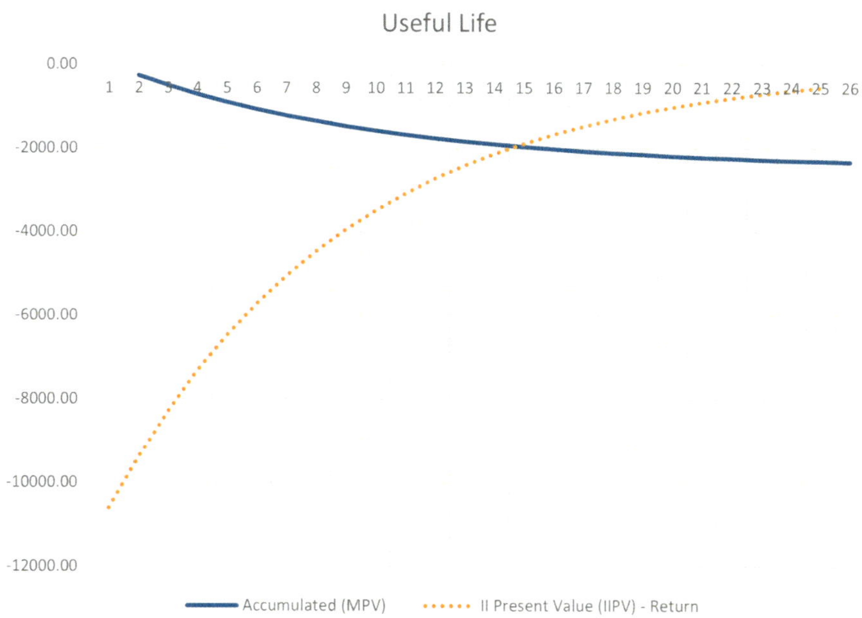

FIGURE 10.22 Physical Asset useful life.

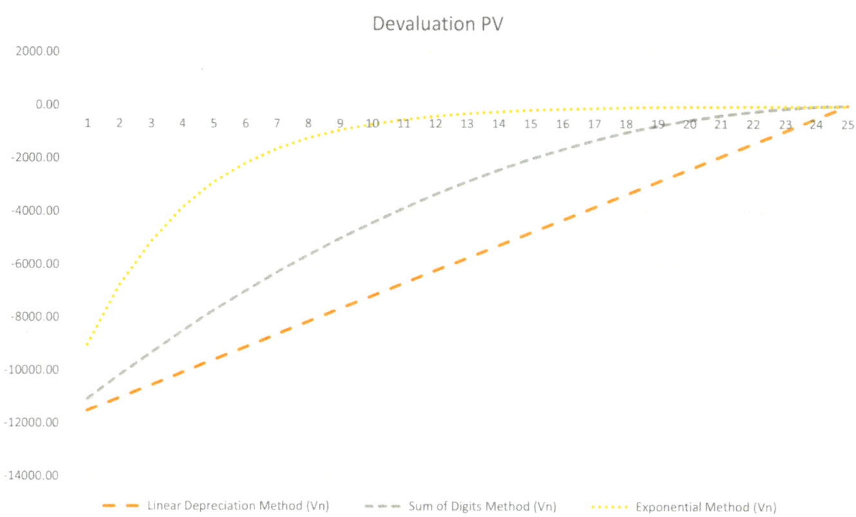

FIGURE 10.23 The three devaluation methods.

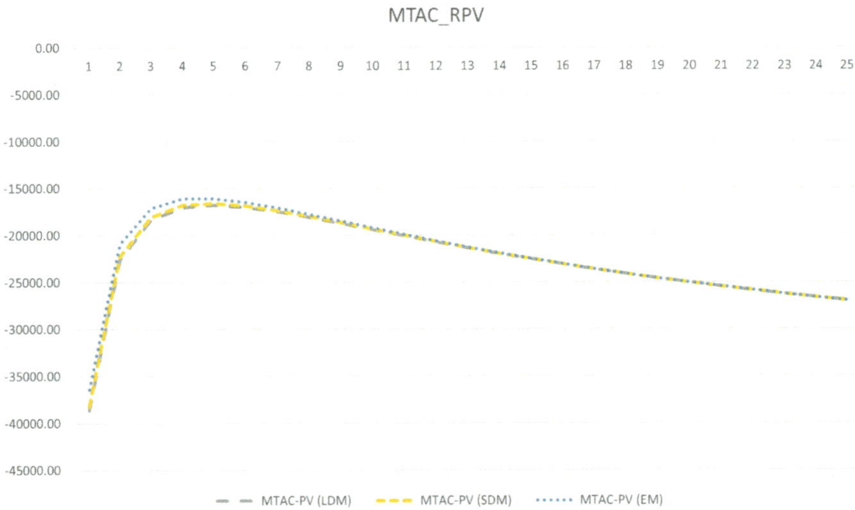

FIGURE 10.24 The MMTAC results from the three devaluation methods.

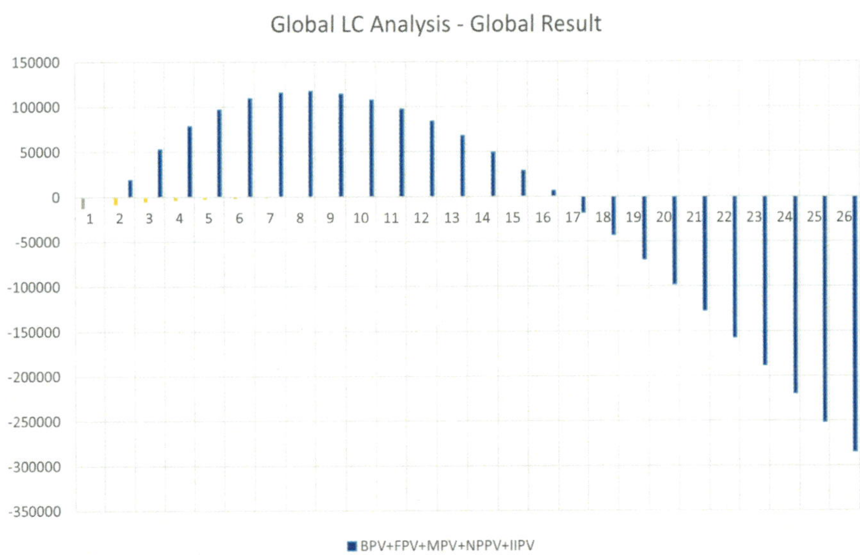

FIGURE 10.25 Life cycle analysis global result including profits.

In the first scenario, a value of Benefits of 50000 is used, being the result shown in Table 10.7 and Figures 10.26–10.28.

The second simulation uses the following values:

- Initial Investment (II) –12000
- Functioning (F) –4500
- Energy (E) –1000
- Maintenance (M) –300
- Accumulated (AMPV) –5800
- Non Production (NP) 0
- Benefit (B) including IIPV **60000**
- Internal Rate of Return (r) 0.12
- Risk (R) 0.01
- Apparent Rate 0.1312
- Maintenance Increase Rate 0.01
- Availability 0.9662
- MTBF 1000
- MTTR 25
- MWT 10

The remaining values are presented in Table 10.8, and the graphical results are shown in Figures 10.29–10.31.

As can be seen, when comparing Figures 10.28 and 10.31, with the implementation of Environmental standards, the positive cycle increases in three years, and the positive incomes increase a lot.

10.4.3 Life Cycle of Physical Assets with Recovery

The way to demonstrate if the return of implementing the Standard under discussion is interesting or not, in this section, is done using the model of Life Cycle of Physical Assets with Recovery, presented in Chapter 3.

The first simulation uses the following values:

- Interest Rate (r) 0.05
- Inflation Rate (h) 0.02
- Profit Rate (p) 0.05
- Maintenance Increase Rate 0.01
- MTBF 1000.0
- MTTR 30.0
- Risk (R) 0.01
- Time (t) 0
- Initial Investment (II) (C_0) –12000
- F_p Financial Movement (Production) **50000**
- Functioning Costs –4500
- Energy –1000
- Maintenance Costs –300

TABLE 10.7
Values Used for Simulation based on Life Cycle Investment Related to Environmental Standards. First Simulation

Time	0	1	2	3	4	5	6	7	8	9	10	11	12	13
Initial Investment (II)	−12000													
II Present Value (IIPV) – Return	−12000	−10608.20	−9377.83	−8290.16	−7328.64	−6478.65	−5727.23	−5062.97	−4475.75	−3956.64	−3497.74	−3092.06	−2733.43	−2416.40
Functioning (*F*)	−4500	−4500.00	−4500.00	−4500.00	−4500.00	−4500.00	−4500.00	−4500.00	−4500.00	−4500.00	−4500.00	−4500.00	−4500.00	−4500.00
Present Value (PV) Accumulated		−3978.08	−7494.76	−10603.57	−13351.82	−15781.31	−17929.02	−19827.64	−21506.04	−22989.78	−24301.43	−25460.96	−26485.99	−27392.15
Energy (*E*)	−200	−200.00	−200.00	−200.00	−200.00	−200.00	−200.00	−200.00	−200.00	−200.00	−200.00	−200.00	−200.00	−200.00
Energy (PV)		−176.80	−156.30	−138.17	−122.14	−107.98	−95.45	−84.38	−74.60	−65.94	−58.30	−51.53	−45.56	−40.27
Accumulated (AFPV)		−16154.88	−7651.06	−18392.80	−31866.76	−47756.05	−65780.53	−85692.54	−107273.18	−130328.91	−154688.64	−180201.13	−206732.68	−234165.10
Maintenance (*M*)	−300	−300.00	−303.00	−306.03	−309.09	−312.18	−315.30	−318.46	−321.64	−324.86	−328.11	−331.39	−334.70	−338.05
Present Value (PV)		−265.21	−234.45	−207.25	−183.22	−161.97	−143.18	−126.57	−111.89	−98.92	−87.44	−77.30	−68.34	−60.41
Accumulated (AMPV)	−5000	−265.21	−499.65	−706.90	−890.12	−1052.09	−1195.27	−1321.84	−1433.74	−1532.65	−1620.10	−1697.40	−1765.73	−1826.14
Non Production (NP)	0	0.00	0.00	0.00	0.00	0.00	0.00	0.00	0.00	0.00	0.00	0.00	0.00	0.00
Present Value (PV)		0.00	0.00	0.00	0.00	0.00	0.00	0.00	0.00	0.00	0.00	0.00	0.00	0.00
Accumulated (ANPPV)		0.00	0.00	0.00	0.00	0.00	0.00	0.00	0.00	0.00	0.00	0.00	0.00	0.00
Accumulated Total Costs (ATCPV)		−16420.08	−8150.71	−19099.71	−32756.89	−48808.14	−66975.79	−87014.39	−108706.92	−131861.56	−156308.73	−181898.53	−208498.42	−235991.24
Benefit (*B*) including IIPV	50000	34034.65	35265.03	36352.69	37314.21	38164.21	38915.62	39579.89	40167.10	40686.22	41145.12	41550.80	41909.42	42226.46
Present Value (PV)		30087.21	27559.12	25114.15	22788.55	20604.37	18573.24	16699.32	14981.50	13415.06	11992.90	10706.46	9546.38	8503.01
Accumulated Benefits (ABPV)		30087.21	57646.34	82760.48	105549.03	126153.40	144726.64	161425.96	176407.46	189822.52	201815.43	212521.89	222068.27	230571.28
ATCPV+ABPV		13667.13	49495.62	63660.77	72792.15	77345.26	77750.84	74411.57	67700.54	57960.96	45506.69	30623.36	13569.86	−5419.97

(Continued)

TABLE 10.7 (*Continued*)
Values Used for Simulation based on Life Cycle Investment Related to Environmental Standards. First Simulation

Time	14	15	16	17	18	19	20	21	22	23	24	25
Initial Investment (II)												
II Present Value (IIPV) – Return	−2136.14	−1888.38	−1669.36	−1475.75	−1304.58	−1153.27	−1019.51	−901.27	−796.74	−704.33	−622.64	−550.42
Functioning (*F*)	−4500.00	−4500.00	−4500.00	−4500.00	−4500.00	−4500.00	−4500.00	−4500.00	−4500.00	−4500.00	−4500.00	−4500.00
Present Value (PV) Accumulated	−28193.20	−28901.34	−29527.35	−30080.76	−30569.98	−31002.45	−31384.77	−31722.75	−32021.52	−32285.65	−32519.14	−32725.54
Energy (*E*)	−200.00	−200.00	−200.00	−200.00	−200.00	−200.00	−200.00	−200.00	−200.00	−200.00	−200.00	−200.00
Energy (PV)	−35.60	−31.47	−27.82	−24.60	−21.74	−19.22	−16.99	−15.02	−13.28	−11.74	−10.38	−9.17
Accumulated (AFPV)	−262393.90	−291326.72	−320881.89	−350987.24	−381578.96	−412600.64	−444002.40	−475740.17	−507774.98	−540072.36	−572601.88	−605336.59
Maintenance (*M*)	−341.43	−344.84	−348.29	−351.77	−355.29	−358.84	−362.43	−366.06	−369.72	−373.41	−377.15	−380.92
Present Value (PV)	−53.40	−47.21	−41.73	−36.89	−32.61	−28.83	−25.49	−22.53	−19.92	−17.61	−15.57	−13.76
Accumulated (AMPV)	−1879.55	−1926.76	−1968.49	−2005.38	−2038.00	−2066.83	−2092.32	−2114.85	−2134.77	−2152.38	−2167.94	−2181.70
Non Production (NP)	0.00	0.00	0.00	0.00	0.00	0.00	0.00	0.00	0.00	0.00	0.00	0.00
Present Value (PV)	0.00	0.00	0.00	0.00	0.00	0.00	0.00	0.00	0.00	0.00	0.00	0.00
Accumulated (ANPPV)	0.00	0.00	0.00	0.00	0.00	0.00	0.00	0.00	0.00	0.00	0.00	0.00
Accumulated Total Costs (ATCPV)	−264273.45	−293253.47	−322850.38	−352992.63	−383616.96	−414667.47	−446094.72	−477855.02	−509909.74	−542224.74	−574769.82	−607518.30
Benefit (*B*) including IIPV	42506.72	42754.47	42973.49	43167.11	43338.27	43489.58	43623.34	43741.59	43846.12	43938.53	44020.22	44092.43
Present Value (PV)	7566.69	6728.07	5978.20	5308.64	4711.54	4179.62	3706.22	3285.24	2911.15	2578.93	2284.06	2022.46
Accumulated Benefits (ABPV)	238137.97	244866.04	250844.24	256152.88	260864.42	265044.03	268750.25	272035.49	274946.64	277525.57	279809.63	281832.08
ATCPV+ABPV	−26135.48	−48387.43	−72006.14	−96839.75	−122752.55	−149623.44	−177344.47	−205819.53	−234963.10	−264699.17	−294960.19	−325686.21

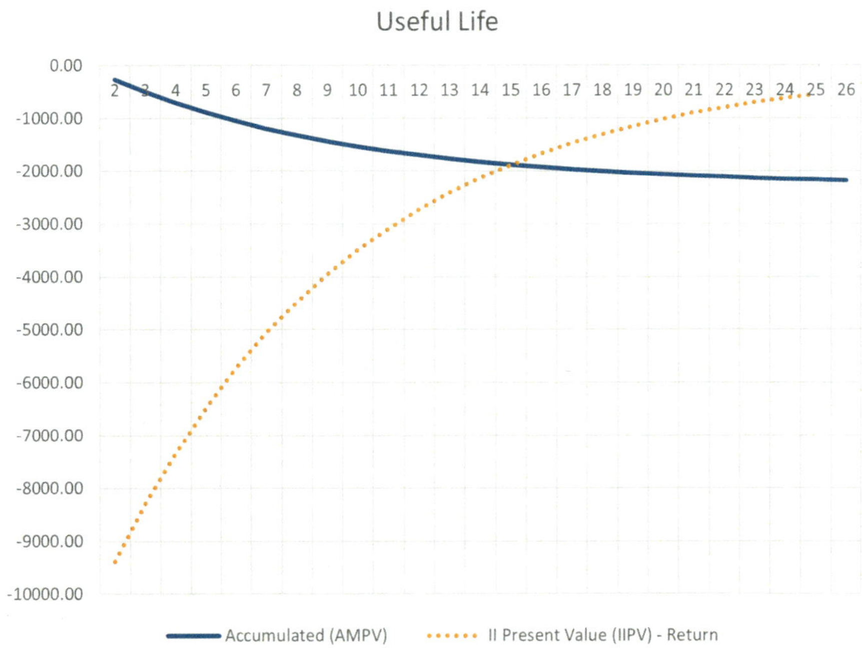

FIGURE 10.26 Physical Asset useful life.

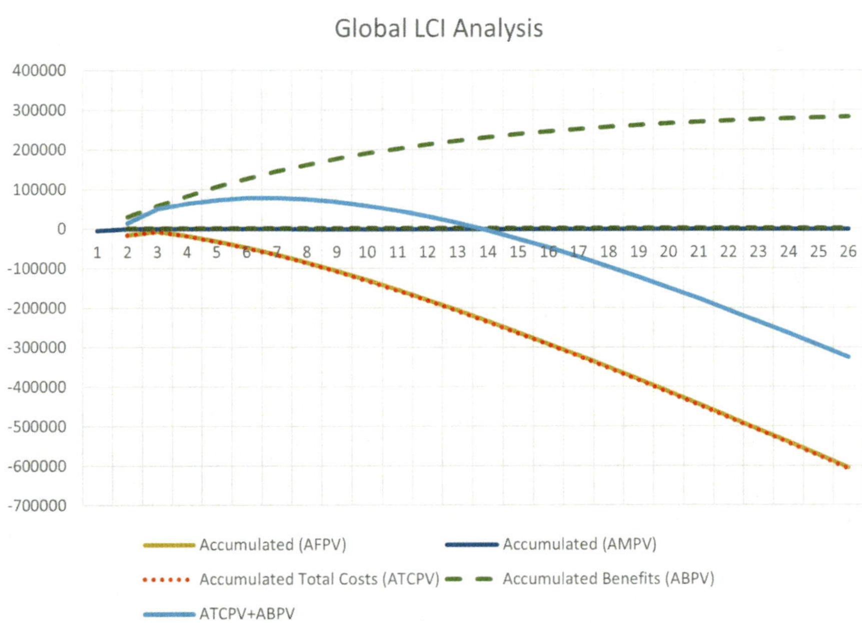

FIGURE 10.27 Global LCI analysis.

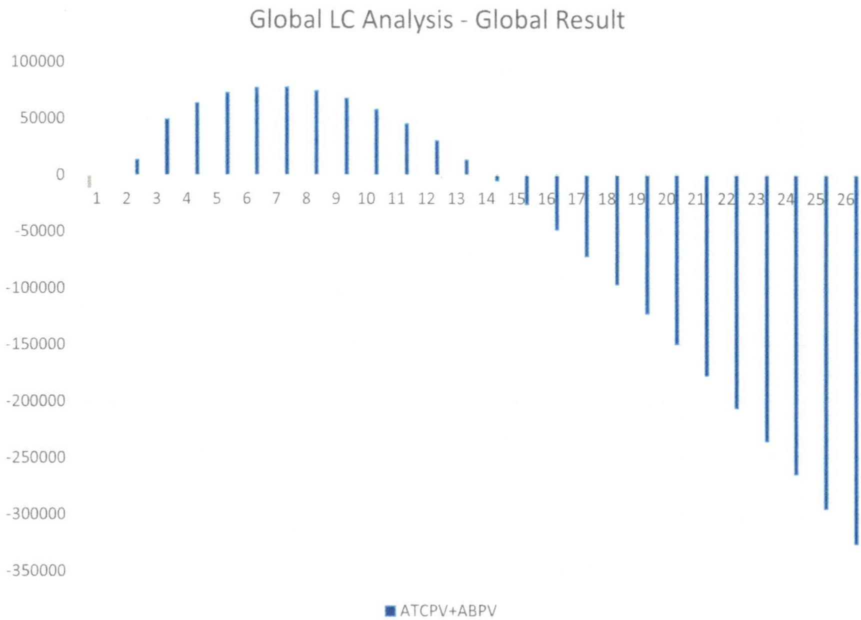

FIGURE 10.28 Life cycle analysis global result.

As referred in the preceding introduction to this section, the evaluation of introduction of Environmental Norms, the simulation of their implementation is made through the variation of the value of benefits. Other indicators may be used, but may not be so intuitive.

In the first scenario, a value of Environmental Benefit of 50000 it is used, being the result shown in Table 10.9 and Figures 10.32–10.34.

The second simulation uses the following values:

- Interest Rate (r) 0.05
- Inflation Rate (h) 0.02
- Profit Rate (p) 0.05
- Maintenance Increase Rate 0.01
- MTBF 1000.0
- MTTR 30.0
- Risk (R) 0.01
- Time (t) 0
- Initial Investment (II) (C_0) –12000
- F_p Financial Movement (Production) **60000**
- Functioning Costs –4500
- Energy –1000
- Maintenance Costs –300

The remaining values are presented in Table 10.10 and the graphical results are shown in Figures 10.35–10.37.

TABLE 10.8
Values Used for Simulation based on Life Cycle Investment Related to Environmental Standards. Second Simulation

Time	0	1	2	3	4	5	6	7	8	9	10	11	12	13
Initial Investment (II)	−12000													
II Present Value (IIPV) – Return	−12000	−10608.20	−9377.83	−8290.16	−7328.64	−6478.65	−5727.23	−5062.97	−4475.75	−3956.64	−3497.74	−3092.06	−2733.43	−2416.40
Functioning (*F*)	−4500	−4500.00	−4500.00	−4500.00	−4500.00	−4500.00	−4500.00	−4500.00	−4500.00	−4500.00	−4500.00	−4500.00	−4500.00	−4500.00
Present Value (PV) Accumulated		−3978.08	−7494.76	−10603.57	−13351.82	−15781.31	−17929.02	−19827.64	−21506.04	−22989.78	−24301.43	−25460.96	−26485.99	−27392.15
Energy (*E*)	−200	−200.00	−200.00	−200.00	−200.00	−200.00	−200.00	−200.00	−200.00	−200.00	−200.00	−200.00	−200.00	−200.00
Energy (PV)		−176.80	−156.30	−138.17	−122.14	−107.98	−95.45	−84.38	−74.60	−65.94	−58.30	−51.53	−45.56	−40.27
Accumulated (AFPV)		−16154.88	−7651.06	−18392.80	−31866.76	−47756.05	−65780.53	−85692.54	−107273.18	−130328.91	−154688.64	−180201.13	−206732.68	−234165.10
Maintenance (*M*)	−300	−300.00	−303.00	−306.03	−309.09	−312.18	−315.30	−318.46	−321.64	−324.86	−328.11	−331.39	−334.70	−338.05
Present Value (PV)		−265.21	−234.45	−207.25	−183.22	−161.97	−143.18	−126.57	−111.89	−98.92	−87.44	−77.30	−68.34	−60.41
Accumulated (AMPV)	−5000	−265.21	−499.65	−706.90	−890.12	−1052.09	−1195.27	−1321.84	−1433.74	−1532.65	−1620.10	−1697.40	−1765.73	−1826.14
Non Production (NP)	0	0.00	0.00	0.00	0.00	0.00	0.00	0.00	0.00	0.00	0.00	0.00	0.00	0.00
Present Value (PV)		0.00	0.00	0.00	0.00	0.00	0.00	0.00	0.00	0.00	0.00	0.00	0.00	0.00
Accumulated (ANPPV)		0.00	0.00	0.00	0.00	0.00	0.00	0.00	0.00	0.00	0.00	0.00	0.00	0.00
Accumulated Total Costs (ATCPV)		−16420.08	−8150.71	−19099.71	−32756.89	−48808.14	−66975.79	−87014.39	−108706.92	−131861.56	−156308.73	−181898.53	−208498.42	−235991.24
Benefit (*B*) including IIPV	60000	42963.22	44193.60	45281.27	46242.78	47092.78	47844.20	48508.46	49095.68	49614.79	50073.69	50479.37	50838.00	51155.03
Present Value (PV)		37980.22	34536.68	31282.42	28241.41	25424.79	22834.57	20466.41	18311.68	16358.99	14595.39	13007.10	11580.19	10300.92
Accumulated Benefits (ABPV)		37980.22	72516.90	103799.32	132040.73	157465.52	180300.09	200766.50	219078.18	235437.17	250032.56	263039.66	274619.85	284920.77
ATCPV+ABPV		21560.14	64366.19	84699.61	99283.84	108657.38	113324.30	113752.12	110371.26	103575.61	93723.82	81141.13	66121.43	48929.53

(Continued)

TABLE 10.8 (*Continued*)

Values Used for Simulation based on Life Cycle Investment Related to Environmental Standards. Second Simulation

Time	14	15	16	17	18	19	20	21	22	23	24	25
Initial Investment (II)												
II Present Value (IIPV) – Return	−2136.14	−1888.38	−1669.36	−1475.75	−1304.58	−1153.27	−1019.51	−901.27	−796.74	−704.33	−622.64	−550.42
Functioning (*F*)	−4500.00	−4500.00	−4500.00	−4500.00	−4500.00	−4500.00	−4500.00	−4500.00	−4500.00	−4500.00	−4500.00	−4500.00
Present Value (PV) Accumulated	−28193.20	−28901.34	−29527.35	−30080.76	−30569.98	−31002.45	−31384.77	−31722.75	−32021.52	−32285.65	−32519.14	−32725.54
Energy (*E*)	−200.00	−200.00	−200.00	−200.00	−200.00	−200.00	−200.00	−200.00	−200.00	−200.00	−200.00	−200.00
Energy (PV)	−35.60	−31.47	−27.82	−24.60	−21.74	−19.22	−16.99	−15.02	−13.28	−11.74	−10.38	−9.17
Accumulated (AFPV)	−262393.90	−291326.72	−320881.89	−350987.24	−381578.96	−412600.64	−444002.40	−475740.17	−507774.98	−540072.36	−572601.88	−605336.59
Maintenance (*M*)	−341.43	−344.84	−348.29	−351.77	−355.29	−358.84	−362.43	−366.06	−369.72	−373.41	−377.15	−380.92
Present Value (PV)	−53.40	−47.21	−41.73	−36.89	−32.61	−28.83	−25.49	−22.53	−19.92	−17.61	−15.57	−13.76
Accumulated (AMPV)	−1879.55	−1926.76	−1968.49	−2005.38	−2038.00	−2066.83	−2092.32	−2114.85	−2134.77	−2152.38	−2167.94	−2181.70
Non Production (NP)	0.00	0.00	0.00	0.00	0.00	0.00	0.00	0.00	0.00	0.00	0.00	0.00
Present Value (PV)	0.00	0.00	0.00	0.00	0.00	0.00	0.00	0.00	0.00	0.00	0.00	0.00
Accumulated (ANPPV)	0.00	0.00	0.00	0.00	0.00	0.00	0.00	0.00	0.00	0.00	0.00	0.00
Accumulated Total Costs (ATCPV)	−264273.45	−293253.47	−322850.38	−352992.63	−383616.96	−414667.47	−446094.72	−477855.02	−509909.74	−542224.74	−574769.82	−607518.30
Benefit (*B*) including IIPV	51435.29	51683.04	51902.07	52095.68	52266.84	52418.15	52551.91	52670.16	52774.69	52867.10	52948.79	53021.01
Present Value (PV)	9156.08	8133.12	7220.28	6406.66	5682.21	5037.71	4464.79	3955.83	3503.96	3102.98	2747.33	2432.00
Accumulated Benefits (ABPV)	294076.85	302209.97	309430.26	315836.92	321519.13	326556.84	331021.63	334977.45	338481.41	341584.40	344331.72	346763.72
ATCPV+ABPV	29803.41	8956.50	−13420.12	−37155.71	−62097.83	−88110.63	−115073.10	−142877.57	−171428.33	−200640.34	−230438.09	−260754.58

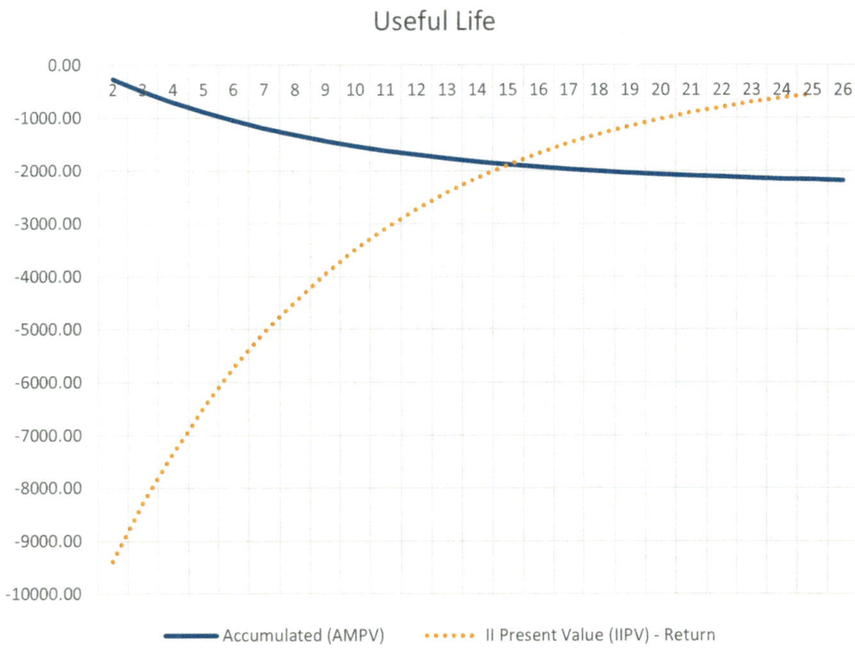

FIGURE 10.29 Physical Asset useful life.

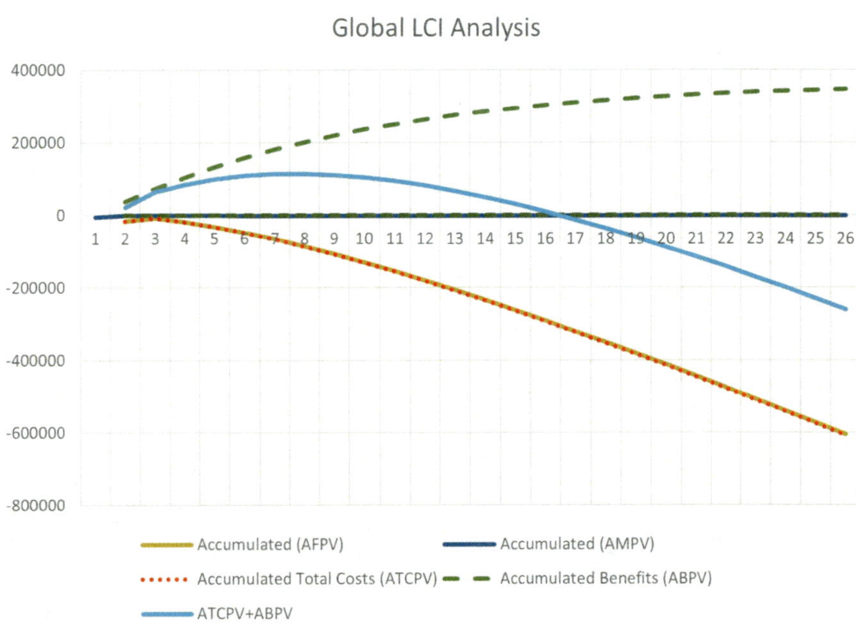

FIGURE 10.30 Global LCI analysis.

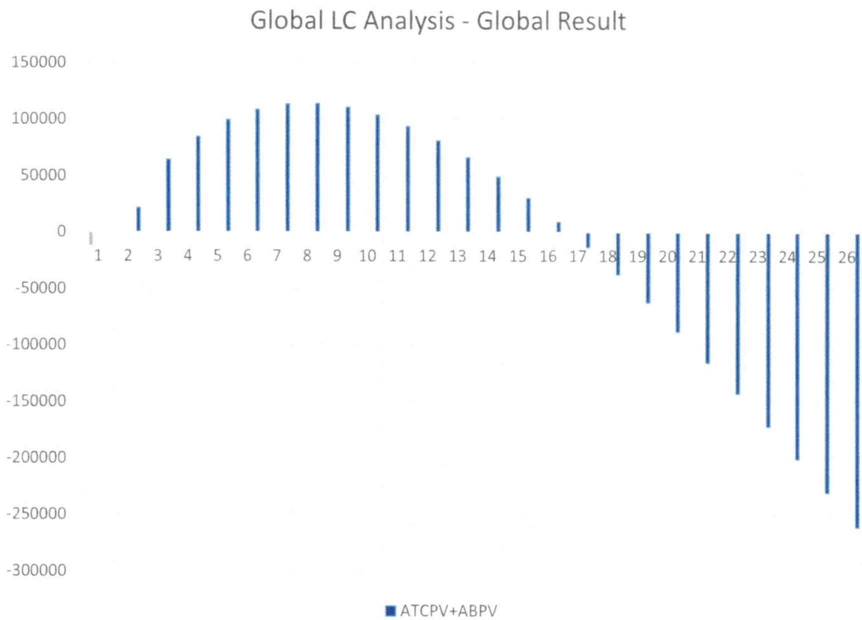

FIGURE 10.31 Life cycle analysis global result.

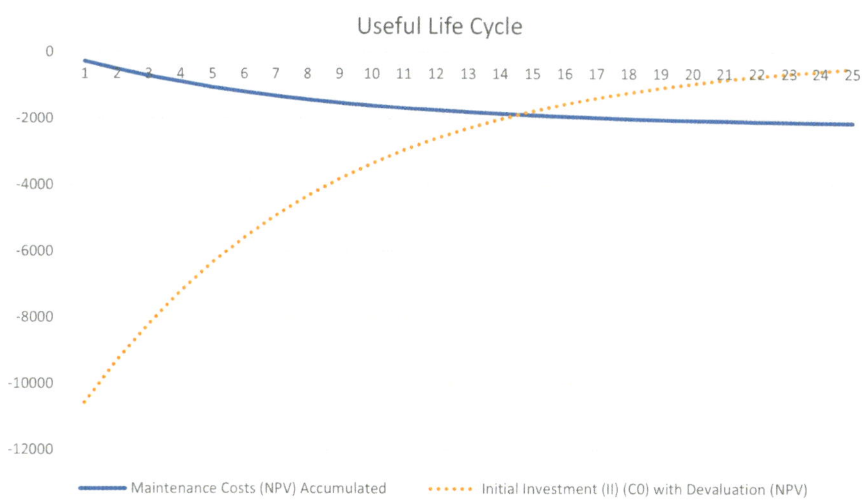

FIGURE 10.32 Physical Asset useful life.

TABLE 10.9

Values Used for Simulation based on Life Cycle with Recovery Related to Environmental Standards. First Simulation

Time (t)	0	1	2	3	4	5	6	7	8	9	10	11	12	13
r - real interest rate $(1+r)$		1.05	1.05	1.05	1.05	1.05	1.05	1.05	1.05	1.05	1.05	1.05	1.05	1.05
h - inflation rate $(1+h)$		1.02	1.02	1.02	1.02	1.02	1.02	1.02	1.02	1.02	1.02	1.02	1.02	1.02
p - profit rate		1.05	1.05	1.05	1.05	1.05	1.05	1.05	1.05	1.05	1.05	1.05	1.05	1.05
i - money rate (nominal interest rate)		0.12	0.12	0.12	0.12	0.12	0.12	0.12	0.12	0.12	0.12	0.12	0.12	0.12
i – Apparent Rate with Risk		0.14	0.14	0.14	0.14	0.14	0.14	0.14	0.14	0.14	0.14	0.14	0.14	0.14
$1/((1+i)$ ^ $t))$		0.88	0.78	0.68	0.60	0.53	0.47	0.41	0.36	0.32	0.28	0.25	0.22	0.19
Initial Investment (II) (C_0)	−12000	−12000	−12000	−12000	−12000	−12000	−12000	−12000	−12000	−12000	−12000	−12000	−12000	−12000
Initial Investment (II) (C_0) with Devaluation (NPV)	−12000	−10565	−9302	−8190	−7211	−6349	−5590	−4921	−4333	−3815	−3359	−2957	−2604	−2292
Initial Investment (II) (C_0) considering NEW Asset (NPV)	−12000	−13630	−15480	−17583	−19970	−22682	−25762	−29261	−33234	−37747	−42873	−48695	−55307	−62818
F_p Financial Movement (Production)	50000	50000	50000	50000	50000	50000	50000	50000	50000	50000	50000	50000	50000	50000
F_p Financial Movement (NPV)	50000	40020	35235	31023	27313	24048	21173	18641	16413	14450	12723	11201	9862	8683
F_p Financial Movement (NPV) Accumulated	50000	40020	75255	106278	133591	157639	178812	197453	213866	228316	241039	252240	262102	270785
Functioning Costs	−4500	−4500	−4500	−4500	−4500	−4500	−4500	−4500	−4500	−4500	−4500	−4500	−4500	−4500
Functioning Costs (NPV) Accumulated	−4500	−3962	−7450	−10521	−13226	−15606	−17702	−19548	−21173	−22603	−23863	−24972	−25948	−26808
Energy	−1000	−1000	−1000	−1000	−1000	−1000	−1000	−1000	−1000	−1000	−1000	−1000	−1000	−1000
Energy Accumulated (NPV) Accumulated	−1000	−880	−1656	−2338	−2939	−3468	−3934	−4344	−4705	−5023	−5303	−5549	−5766	−5957
Maintenance Costs	−300	−300	−300	−300	−300	−300	−300	−300	−300	−300	−300	−300	−300	−300
Maintenance Costs (NPV) Accumulated	−300	−267	−502	−708	−891	−1051	−1192	−1316	−1426	−1522	−1607	−1681	−1747	−1805
Total Costs (NPV) Accumulated	−5800	−5229	−14181	−26411	−41527	−59184	−79078	−100942	−124540	−149666	−176135	−203788	−232484	−262097
F_T Net Present Value (NPV) (With II Devaluated)		24226	51772	71677	84854	92107	94144	91590	84992	74835	61544	45494	27015	6396
F_T Net Present Value (With II New Asset)		21162	45594	62285	72094	75773	73972	67250	56091	40903	22030	−243	−25689	−54129

(Continued)

TABLE 10.9 (*Continued*)
Values Used for Simulation based on Life Cycle with Recovery Related to Environmental Standards. First Simulation

Time (t)	14	15	16	17	18	19	20	21	22	23	24	25
r - real interest rate $(1+r)$	1.05	1.05	1.05	1.05	1.05	1.05	1.05	1.05	1.05	1.05	1.05	1.05
h - inflation rate $(1+h)$	1.02	1.02	1.02	1.02	1.02	1.02	1.02	1.02	1.02	1.02	1.02	1.02
p - profit rate	1.05	1.05	1.05	1.05	1.05	1.05	1.05	1.05	1.05	1.05	1.05	1.05
i - money rate (nominal interest rate)	0.12	0.12	0.12	0.12	0.12	0.12	0.12	0.12	0.12	0.12	0.12	0.12
i – Apparent Rate with Risk	0.14	0.14	0.14	0.14	0.14	0.14	0.14	0.14	0.14	0.14	0.14	0.14
$1/((1+i)\wedge t))$	0.17	0.15	0.13	0.11	0.10	0.09	0.08	0.07	0.06	0.05	0.05	0.04
Initial Investment (II) (C_0)	−12000	−12000	−12000	−12000	−12000	−12000	−12000	−12000	−12000	−12000	−12000	−12000
Initial Investment (II) (C_0) with Devaluation (NPV)	−2018	−1777	−1565	−1377	−1213	−1068	−940	−828	−729	−642	−565	−497
Initial Investment (II) (C_0) considering NEW Asset (NPV)	−71348	−81037	−92041	−104540	−118736	−134860	−153174	−173974	−197599	−224432	−254909	−289524
F_p Financial Movement (Production)	50000	50000	50000	50000	50000	50000	50000	50000	50000	50000	50000	50000
F_p Financial Movement (NPV)	7645	6731	5926	5218	4594	4045	3561	3135	2760	2430	2140	1884
F_p Financial Movement (NPV) Accumulated	278430	285161	291087	296305	300899	304943	308505	311640	314400	316831	318970	320854
Functioning Costs	−4500	−4500	−4500	−4500	−4500	−4500	−4500	−4500	−4500	−4500	−4500	−4500
Functioning Costs (NPV) Accumulated	−27565	−28231	−28818	−29334	−29789	−30189	−30542	−30852	−31126	−31366	−31578	−31765
Energy	−1000	−1000	−1000	−1000	−1000	−1000	−1000	−1000	−1000	−1000	−1000	−1000
Energy Accumulated (NPV) Accumulated	−6125	−6274	−6404	−6519	−6620	−6709	−6787	−6856	−6917	−6970	−7017	−7059
Maintenance Costs	−300	−300	−300	−300	−300	−300	−300	−300	−300	−300	−300	−300
Maintenance Costs (NPV) Accumulated	−1856	−1901	−1940	−1975	−2006	−2033	−2056	−2077	−2096	−2112	−2126	−2139
Total Costs (NPV) Accumulated	−292517	−323649	−355407	−387716	−420511	−453733	−487332	−521262	−555483	−589961	−624666	−659569
F_T Net Present Value (NPV) (With II Devaluated)	−16105	−40265	−65884	−92789	−120825	−149858	−179767	−210450	−241812	−273772	−306260	−339212
F_T Net Present Value (With II New Asset)	−85435	−119525	−156361	−195952	−238349	−283650	−332001	−383596	−438682	−497563	−560604	−628239

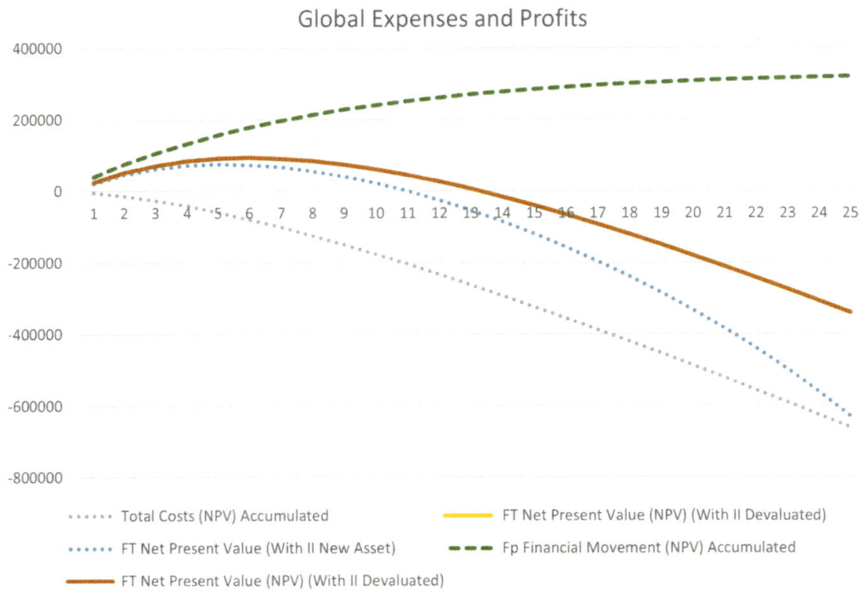

FIGURE 10.33 Global LCI analysis.

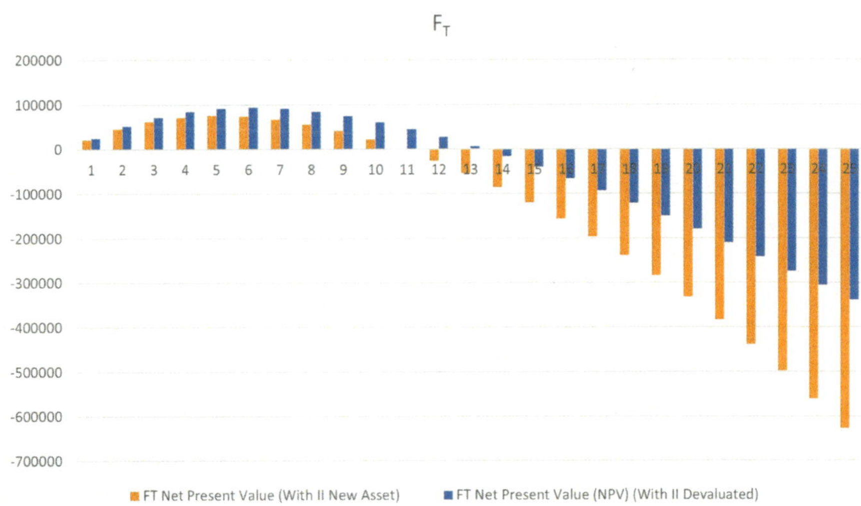

FIGURE 10.34 Life cycle analysis global result.

TABLE 10.10
Values Used for Simulation based on Life Cycle with Recovery Related to Environmental Standards. Second Simulation

Time (t)	0	1	2	3	4	5	6	7	8	9	10	11	12	13
r - real interest rate $(1+r)$		1.05	1.05	1.05	1.05	1.05	1.05	1.05	1.05	1.05	1.05	1.05	1.05	1.05
h - inflation rate $(1+h)$		1.02	1.02	1.02	1.02	1.02	1.02	1.02	1.02	1.02	1.02	1.02	1.02	1.02
p - profit rate		1.05	1.05	1.05	1.05	1.05	1.05	1.05	1.05	1.05	1.05	1.05	1.05	1.05
i – money rate (nominal interest rate)		0.12	0.12	0.12	0.12	0.12	0.12	0.12	0.12	0.12	0.12	0.12	0.12	0.12
i – Apparent Rate with Risk		0.14	0.14	0.14	0.14	0.14	0.14	0.14	0.14	0.14	0.14	0.14	0.14	0.14
$1/((1+i)$ ^ $t))$		0.88	0.78	0.68	0.60	0.53	0.47	0.41	0.36	0.32	0.28	0.25	0.22	0.19
Initial Investment (II) (C_0)	−12000	−12000	−12000	−12000	−12000	−12000	−12000	−12000	−12000	−12000	−12000	−12000	−12000	−12000
Initial Investment (II) (C_0) with Devaluation (NPV)	−12000	−10565	−9302	−8190	−7211	−6349	−5590	−4921	−4333	−3815	−3359	−2957	−2604	−2292
Initial Investment (II) (C_0) considering NEW Asset (NPV)	−12000	−13630	−15480	−17583	−19970	−22682	−25762	−29261	−33234	−37747	−42873	−48695	−55307	−62818
F_p Financial Movement (Production)	60000	60000	60000	60000	60000	60000	60000	60000	60000	60000	60000	60000	60000	60000
F_p Financial Movement (NPV)	60000	48024	42282	37227	32776	28857	25407	22370	19695	17340	15267	13442	11835	10420
F_p Financial Movement (NPV) Accumulated	60000	48024	90306	127533	160309	189167	214574	236944	256639	273979	289246	302688	314523	324942
Functioning Costs	−4500	−4500	−4500	−4500	−4500	−4500	−4500	−4500	−4500	−4500	−4500	−4500	−4500	−4500
Functioning Costs (NPV) Accumulated	−4500	−3962	−7450	−10521	−13226	−15606	−17702	−19548	−21173	−22603	−23863	−24972	−25948	−26808
Energy	−1000	−1000	−1000	−1000	−1000	−1000	−1000	−1000	−1000	−1000	−1000	−1000	−1000	−1000
Energy Accumulated (NPV) Accumulated	−1000	−880	−1656	−2338	−2939	−3468	−3934	−4344	−4705	−5023	−5303	−5549	−5766	−5957
Maintenance Costs	−300	−300	−300	−300	−300	−300	−300	−300	−300	−300	−300	−300	−300	−300
Maintenance Costs (NPV) Accumulated	−300	−267	−502	−708	−891	−1051	−1192	−1316	−1426	−1522	−1607	−1681	−1747	−1805
Total Costs (NPV) Accumulated	−5800	−5229	−14181	−26411	−41527	−59184	−79078	−100942	−124540	−149666	−176135	−203788	−232484	−262097
F_T Net Present Value (NPV) (With II Devaluated)		32230	66824	92933	111572	123634	129906	131080	127765	120499	109752	95942	79435	60554
F_T Net Present Value (With II New Asset)		29166	60645	83540	98813	107301	109734	106741	98864	86566	70238	50205	26732	28

(*Continued*)

TABLE 10.10 (*Continued*)
Values Used for Simulation based on Life Cycle with Recovery Related to Environmental Standards. Second Simulation

Time (t)	14	15	16	17	18	19	20	21	22	23	24	25
r - real interest rate $(1+r)$	1.05	1.05	1.05	1.05	1.05	1.05	1.05	1.05	1.05	1.05	1.05	1.05
h - inflation rate $(1+h)$	1.02	1.02	1.02	1.02	1.02	1.02	1.02	1.02	1.02	1.02	1.02	1.02
p - profit rate	1.05	1.05	1.05	1.05	1.05	1.05	1.05	1.05	1.05	1.05	1.05	1.05
i – money rate (nominal interest rate)	0.12	0.12	0.12	0.12	0.12	0.12	0.12	0.12	0.12	0.12	0.12	0.12
i – Apparent Rate with Risk	0.14	0.14	0.14	0.14	0.14	0.14	0.14	0.14	0.14	0.14	0.14	0.14
$1/((1+i)\wedge t))$	0.17	0.15	0.13	0.11	0.10	0.09	0.08	0.07	0.06	0.05	0.05	0.04
Initial Investment (II) (C_0)	−12000	−12000	−12000	−12000	−12000	−12000	−12000	−12000	−12000	−12000	−12000	−12000
Initial Investment (II) (C_0) with Devaluation (NPV)	−2018	−1777	−1565	−1377	−1213	−1068	−940	−828	−729	−642	−565	−497
Initial Investment (II) (C_0) considering NEW Asset (NPV)	−71348	−81037	−92041	−104540	−118736	−134860	−153174	−173974	−197599	−224432	−254909	−289524
F_p Financial Movement (Production)	60000	60000	60000	60000	60000	60000	60000	60000	60000	60000	60000	60000
F_p Financial Movement (NPV)	9174	8077	7111	6261	5513	4854	4273	3762	3312	2916	2568	2261
F_p Financial Movement (NPV) Accumulated	334116	342193	349305	355566	361079	365932	370205	373968	377280	380197	382764	385025
Functioning Costs	−4500	−4500	−4500	−4500	−4500	−4500	−4500	−4500	−4500	−4500	−4500	−4500
Functioning Costs (NPV) Accumulated	−27565	−28231	−28818	−29334	−29789	−30189	−30542	−30852	−31126	−31366	−31578	−31765
Energy	−1000	−1000	−1000	−1000	−1000	−1000	−1000	−1000	−1000	−1000	−1000	−1000
Energy Accumulated (NPV) Accumulated	−6125	−6274	−6404	−6519	−6620	−6709	−6787	−6856	−6917	−6970	−7017	−7059
Maintenance Costs	−300	−300	−300	−300	−300	−300	−300	−300	−300	−300	−300	−300
Maintenance Costs (NPV) Accumulated	−1856	−1901	−1940	−1975	−2006	−2033	−2056	−2077	−2096	−2112	−2126	−2139
Total Costs (NPV) Accumulated	−292517	−323649	−355407	−387716	−420511	−453733	−487332	−521262	−555483	−589961	−624666	−659569
F_T Net Present Value (NPV) (With II Devaluated)	39581	16767	−7667	−33528	−60645	−88869	−118067	−148122	−178931	−210406	−242466	−275041
F_T Net Present Value (With II New Asset)	−29749	−62493	−98144	−136691	−178169	−222661	−270300	−321268	−375802	−434196	−496810	−564068

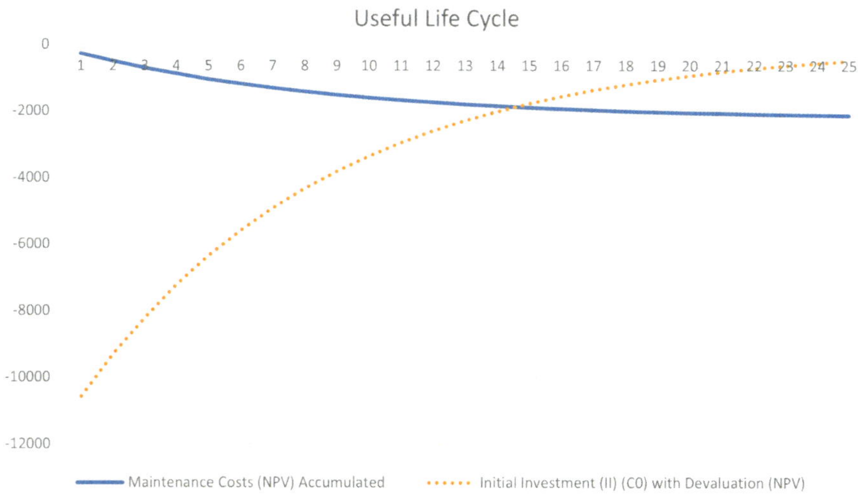

FIGURE 10.35 Physical Asset useful life.

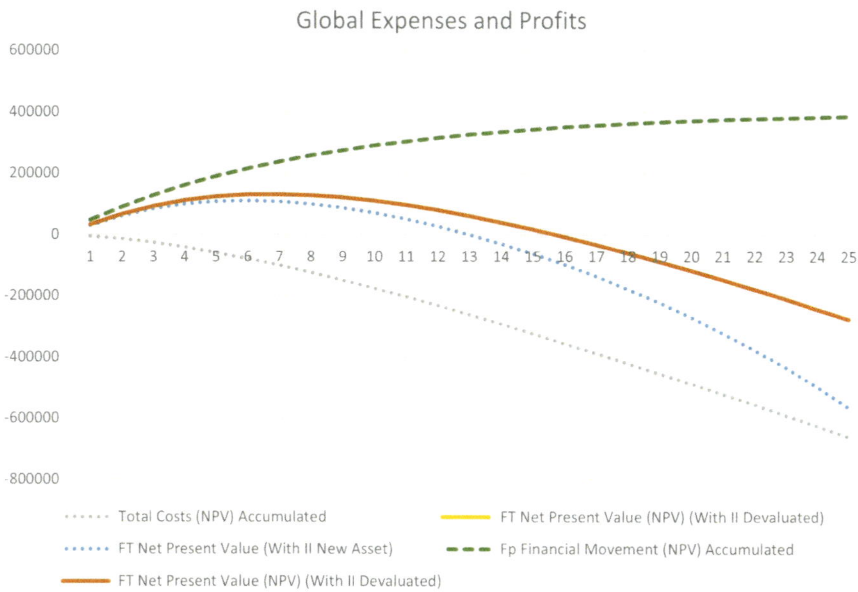

FIGURE 10.36 Global LCI analysis.

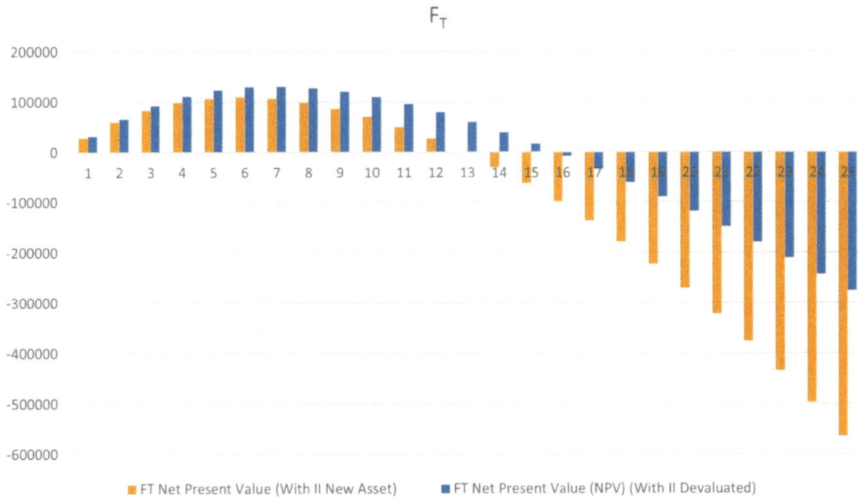

FIGURE 10.37 Life cycle analysis global result.

As can be seen, when comparing Figures 10.34 and 10.37, with the increasing of the benefits, the global result is positive in two and three more years, considering devaluation and a new Physical Asset.

REFERENCE

Woellner, R. A., Voorhees, J., & Bell, C. L. (2020) *International Environmental Risk Management* (2nd Edition). London: CRC Press.

Index

apparent rate 6–8, 10–13, 18–19, 27, 44, 47, 60–63, 65–66, 69–70, 123, 138, 145, 153, 168, 177–178, 181–182, 187, 194, 209, 224, 232, 243–245, 248–249, 254, 256, 264, 279, 287, 302, 313–315, 318–320, 323, 332, 347, 355, 370, 374, 378–380, 383–384, 389, 404, 412, 427, 435, 444–445, 447–448
artificial intelligence 90–91, 96, 102–103
availability 9, 17, 19, 47, 49, 50–53, 58, 62–63, 81, 123, 138, 145, 153, 166, 168, 187, 194, 209–210, 224, 232, 256, 264, 279, 287, 302, 323, 332, 347, 355, 370, 374, 378, 389, 404, 412, 427, 435

benefits 11–12, 18, 20–24, 28–32, 36–40, 44–46, 49, 51–52, 55–57, 124–128, 131–135, 139–143, 146–150, 154–158, 161–165, 169–170, 173–174, 187–191, 194–199, 202–206, 209–214, 217–221, 224–229, 232–235, 238–241, 257–261, 265–269, 272–276, 280–284, 288–292, 295–299, 303–306, 308–310, 325–329, 333–337, 340–344, 348–352, 356–360, 363–367, 371–372, 375–376, 386–387, 389–394, 397–401, 404–410, 412–417, 420–424, 427–432, 435–437, 439–441, 450

computerized maintenance management system (CMMS) 87–88
condition based maintenance (CBM) 76, 78, 82–83, 90, 160
contactless sensors 82, 84

diagnosis sheets 111–112, 186, 255, 388

EAM 88
economic life cycle 6, 10–11, 14, 44
expert system 103

failures 77–78, 97
fault 10, 49, 76, 80, 87–88, 96, 104, 253
fault tree analysis 253

holistic diagnosis model (HDM) 109

inflation rate 6–8, 59–60, 64–67, 69–70, 172, 176–178, 181–182, 237, 242–245, 248–249, 254, 307, 312–315, 318–320, 379–380, 383–384, 435, 439, 444–445, 447–448
information system 87

key performance indicators (KPI) 8, 49, 88–89, 107–108, 121

life cycle cost (LCC) 3, 18, 44, 49, 58–59, 121, 187, 254, 323
life cycle investment (LCI) 44, 58–59, 160, 167, 169–160, 173–174, 184, 224, 233–235, 238–240, 287, 303–305, 308–310, 355, 371–372, 375–376, 427, 436–437, 440–441

mean time between failures (MTBF) 9, 17, 19, 47, 49–50, 53, 62, 64, 68, 123, 138, 153, 166, 168, 172, 176, 187, 194, 209–210, 224, 232, 237, 242, 256, 264, 279, 302, 307, 312, 323, 332, 347, 355, 370, 374, 378, 389, 404, 412, 427, 435, 439
MTTR 9, 14–17, 19, 47, 49–50, 53, 62, 64, 68, 123, 138, 153, 166, 168, 172, 176, 187, 194, 209–210, 224, 232, 237, 242, 256, 264, 279, 302, 307, 312, 323, 332, 347, 355, 370, 374, 378, 389, 404, 412, 427, 435, 439
MWT 9–10, 19, 47, 49–50, 53, 123, 138, 153, 168, 187, 194, 209–210, 224, 232, 256, 264, 279, 302, 323, 332, 347, 355, 370, 374, 378, 389, 404, 412, 427, 435

net present value (NPV) 11, 60–66, 69–70, 177–178, 181–182, 243–245, 248–249, 313–315, 318–320, 379–380, 383–384, 444–445, 447–448
neural networks 89, 96–100, 102–104

oil analysis 77

questionnaires 109–112, 115–116, 119, 121

radar map 119, 120
reliability centred maintenance (RCM) 253

sensors 78–89
SMIT 88

time between failures (TBF) 9, 49
time to repair (TTR) 9, 15–18, 49

vibration analysis 77, 86

wired sensors 82–84
wireless sensors 82–83

Made in the USA
Las Vegas, NV
17 September 2026

2df8fadc-54cf-473f-848a-4987decadb6bR04